Horizons of Fractal Geometry and Complex Dimensions

CONTEMPORARY MATHEMATICS

731

Horizons of Fractal Geometry and Complex Dimensions

2016 Summer School
Fractal Geometry and Complex Dimensions,
In celebration of the 60th birthday of Michel Lapidus
June 21–29, 2016
California Polytechnic State University,
San Luis Obispo, California

Robert G. Niemeyer
Erin P. J. Pearse
John A. Rock
Tony Samuel
Editors

AMS
AMERICAN MATHEMATICAL SOCIETY

2010 *Mathematics Subject Classification.* Primary 11M26, 26A30, 28D05, 30D10, 31A10, 35P20, 37B10, 52A39, 52C23, 60G50.

Library of Congress Cataloging-in-Publication Data

Names: Summer School on Fractal Geometry and Complex Dimensions (2016 : San Luis Obispo, Calif.)| Niemeyer, Robert G., 1983- editor.

Title: Horizons of fractal geometry and complex dimensions : 2016 Summer School on Fractal Geometry and Complex Dimensions, June 21–29, 2016, California Polytechnic State University, San Luis Obispo, California/Robert G. Niemeyer [and three others], editors.

Description: Providence, Rhode Island : American Mathematical Society, [2019] | Series: Contemporary mathematics ; volume 731 | Includes bibliographical references.

Identifiers: LCCN 2018047353 | ISBN 9781470435813 (alk. paper)

Subjects: LCSH: Fractals – Congresses.| AMS: Number theory – Zeta and L-functions: analytic theory – Nonreal zeros of $\zeta(s)$ and $L(s,\chi)$; Riemann and other hypotheses. msc | Real functions – Functions of one variable – Singular functions, Cantor functions, functions with other special properties. msc | Measure and integration – Measure-theoretic ergodic theory – Measure-preserving transformations. msc | Functions of a complex variable – Entire and meromorphic functions, and related topics – Representations of entire functions by series and integrals. msc | Potential theory – Two-dimensional theory – Integral representations, integral operators, integral equations methods. msc | Partial differential equations – Spectral theory and eigenvalue problems – Asymptotic distribution of eigenvalues and eigenfunctions. msc | Dynamical systems and ergodic theory – Topological dynamics – Symbolic dynamics. msc | Convex and discrete geometry – General convexity – Mixed volumes and related topics. msc | Convex and discrete geometry – Discrete geometry – Quasicrystals, aperiodic tilings. msc | Probability theory and stochastic processes – Stochastic processes – Sums of independent random variables; random walks. msc

Classification: LCC QA614.86.S86 2016 | DDC 514/.742–dc23

LC record available at https://lccn.loc.gov/2018047353

Contemporary Mathematics ISSN: 0271-4132 (print); ISSN: 1098-3627 (online)

DOI: https://doi.org/10.1090/conm/731

∞ The paper used in this book is acid-free and falls within the guidelines
established to ensure permanence and durability.
Visit the AMS home page at `https://www.ams.org/`

10 9 8 7 6 5 4 3 2 1   24 23 22 21 20 19

Contents

Preface: Horizons of Fractal Geometry and Complex Dimensions

Here we showcase various recent results from the fertile and active nexus of research topics: dynamical systems, fractal geometry, number theory and quasicrystals. The contributions presented reflect different aspects of these topics and form new connections; they have been carefully selected, peer reviewed and are aimed at both the specialist and those that are new to the fields.

This volume complements and records the outcomes of the 2016 Summer School on Fractal Geometry and Complex Dimensions, an event which was funded by the National Science Foundation and hosted by the Mathematics Department of California Polytechnic State University, and which, in part, celebrated the 60th birthday of Prof. M. L. Lapidus. The event successfully brought together experts from 14 countries and was attended by over 100 academics. During this event new collaborative projects were formed — results of which are featured within. We hope to have captured the exciting atmosphere of the event in this issue through the original research, survey and expository articles.

We are indebted and grateful to a number of anonymous referees for their invaluable help and suggestions in preparing this issue. We express our gratitude to the National Science Foundation for sponsoring the summer school; without this support, this issue and the event could not have come to fruition.

This issue is divided into four parts, each centered around a different theme: dimension gaps and the mass transfer principle, fractal strings and complex dimensions, Laplacians on fractal domains and SDEs with fractal noise, and aperiodic order (Delone sets and tilings).

We begin with the subject of dimension gaps and the mass transfer principle. A goal often encountered in mathematics is to develop relationships between independently defined invariants of classes of mathematical objects. In conformal dynamics, there is a large body of work devoted to showing that the critical exponent of the Poincaré series associated to a conformal dynamical system, for instance Kleinian groups acting on hyperbolic space or rational maps acting on the Riemann sphere, coincides with the Hausdorff dimension of the corresponding limit set or Julia set, respectively. The article by K. Falk gives a survey of this topic, concentrating on an important notion in geometric group theory, namely amenability, as a tool to determine if the aforementioned invariants coincide or not. This work is complemented by a comprehensive survey article by D. Allen and S. Troscheit on the celebrated mass transfer principle of V. Beresnevich and S. Velani, and its applications to metric number theory. This result was originally motivated by a conjecture of Duffin and Schaeffer, and allows one to transfer a Lebesgue measure statement for a different lim sup set of balls to a f-dimensional Hausdorff measure statement for a lim sup set of balls which are obtained by 'shrinking' the original balls in a certain manner according to the dimension function f. This is a remarkable result given that the Lebesgue measure can be considered to be 'coarser' than the Hausdorff measure.

Our second theme concerns the theory of fractal strings and complex dimensions. The theory of complex dimensions, as introduced by M. L. Lapidus and M. van Frankenhuijsen, is a $\mathbb{C}$-valued extension of non-integer notions of dimension such as the Hausdorff dimension and the Minkowski dimension. These non-integer notions of dimension assign a single number to a given set $U \subseteq \mathbb{R}^d$. By contrast, the set of complex dimensions provides a richly structured geometric invariant of U, typically an infinite but discrete subset of $\mathbb{C}$ of which the Minkowski dimension is a distinguished member. The complex dimensions describe not just the order of magnitude of the scaling properties of U, but also the oscillatory aspects. The research expository article by M. L. Lapidus gives a detailed overview bringing the reader to the frontiers of the subject. The author includes a historical account of the subject, an introduction to a higher dimensional theory as well as a presentation of two new concepts: quantized number theory and fractal cohomology. Moreover, many conjectures, open problems and future research directions are stated.

In our third theme we turn to the closely related topic of Laplacians on fractal domains and SDEs with fractal noise. We begin with the article of P. Pollack and C. Pomerance concerning the asymptotic behaviour of $\mathcal{N}$ – the eigenvlaue counting function of a Laplacian on a domain with a fractal boundary. The first asymptotic term of $\mathcal{N}$ is given by the dimension of the open domain and depends on its volume. The Weyl-Berry-Lapidus conjecture states that if the boundary of the open domain is Minkowski measurable in dimension d, then the second asymptotic term is of order d and that it depends on the Minkowski content in dimension d. This conjecture has been shown to hold in $\mathbb{R}$. The present article shows that in $\mathbb{R}^2$ it is false. Indeed, here the authors present two fractal sprays which possess the same volume and whose boundaries have the same Minkowski content, but with different spectral counting functions. Indeed, they show that the two zeta functions never coincide on the interval $(1, 2)$. This leads us to the article of T. Eichinger and S. Winter where they investigate the second asymptotic term of $\mathcal{N}$ for Laplacians on domains with fractal boundaries which are not Minkowski measurable. Returning to the one-dimensional setting, in M. van Frankenhuijsen's contribution to this issue, a description of how fractal strings can be used to understand the Riemann hypothesis through the relationship between geometric and spectral oscillations is given. This exposition illuminates a number of open questions, some of which are explicitly stated, while others are only suggested.

Complementing these work are the articles by M. Kesseböhmer, T. Samuel and H. Weyer, and M. Zähle and E. Schneider. In the first of these articles, measure geometric Laplacians for discrete distributions are investigated. In the article of M. Zähle and E. Schneider, uniform local contraction properties of the fractional integral operators are established and a higher-dimensional variant of the Doss-Sussman approach to path-wise global solutions is presented.

Our final theme concerns aspects of quasicrystals, specifically tilings and Delone sets. Here we have two contributions, one focusing on tilings associated with iterated function systems, and the other on Delone sets. In the latter of these two articles, a connection with the Duffin-Schaeffer conjecture is also made.

Iterated function systems were pioneered in the 1980s by M. F. Barnsley and J. Hutchinson as a way to generate fractals. This is still one of the most widely used method and has brought about a rich body of work. The article by M. F. Barnsley and A. Vince examines tilings associated with iterated function systems consisting of similitudes and gives conditions under which the resulting tiling is periodic, quasi-periodic and non-periodic.

Delone sets are ubiquitous in the study of aperiodic order (mathemtical quasicrystals) and have applications in coding theory as well as in algorithms that find approximate solutions to NP-hard optimisation problems. M. Baake and A. Haynes provide elegant and interesting generalisations of some of the most classical results in metric number theory, in particular Khinchin's theorem and the Duffin-Schaeffer conjecture, to the case of Diophantine approximations where the distance to the nearest integer is replaced by the distance to a Delone set.

We hope that you enjoy reading this volume as much as we have enjoyed compiling it.

R. G. Niemeyer
E. P. J. Pearse
J. A. Rock
T. Samuel

List of Participants

L. Barnsley
Australian National University
Australia

M. Barnsley
Australian National University
Australia

Y. Du
Universidade de São Paulo
Brazil

C. Ma
Macau University of Science and Technology
China

L. Fang
South China University of Technology
China

B. Li
South China University of Technology
China

Y. Li
Nanjing Audit University
China

J. Luo
Chongqing University
China

J. J. Miao
East China Normal University
China

H. Qiu
Nanjing University
China

F. Yang
Ocean University of China
China

G. Radunovic
University of Zagreb
Croatia

M. Resman
University of Zagreb
Croatia

D. Vlah
University of Zagreb
Croatia

C. Bandt
Universität Greifswald
Germany

K. Falk
Universität Bremen
Germany

U. Freiberg
Universität Stuttgart
Germany

E. Hauser
Universität Stuttgart
Germany

M. Keßeböhmer
Universität Bremen
Germany

S. Kohl
Universität Stuttgart
Germany

S. Kombrink
Universität zu Lübeck, Germany
Germany

B. Munz
Universität Stuttgart
Germany

H. Peña
Universität Greifswald
Germany

M. Rauch
Universität Jena
Germany

K. Sender
Universität Bremen
Germany

H. Weyer
Universität Bremen
Germany

S. Winter
Karlsruhe Institute of Technology
Germany

M. Zähle
Universität Jena
Germany

A. Priyadarshi
Indian Institute of Technology Delhi
India

D. Carfí
University of Messina
Italy

R. Peirone
Universitá di Roma Tor Vergata
Italy

J. Kigami
Kyoto University
Japan

T. Otsuka
Tokyo Metropolitan University
Japan

N. Langeveld
Universiteit Leiden
The Netherlands

M. Khan
Institute of Business Management
Pakistan

A. Soos
Babes Bolyai University
Romania

K. Tsougkas
Uppsala University
Sweden

K. Falconer
University of St Andrews
UK

B. Hambly
University of Oxford
UK

A. Haynes
University of York
UK

S. Troscheit
University of St Andrews
UK

J. Walton
University of York
UK

B. Afeyan
Polymath research Inc.
USA

E. Alhajjar
George Mason University
USA

A. Arauza
UC Riverside
USA

E. Bullard
California State Polytechnic University
USA

A. Carmichael
California Polytechnic State University
USA

A. Carta
University of New Hampshire
USA

A. Chau
California State Polytechnic University
USA

J. Chen
University of Connecticut
USA

T. Cobler
Fullerton College
USA

A. Daw
California Polytechnic State University
USA

S. Dent
University of Southern Mississippi
USA

A. Diaz-Cabrera
California State Polytechnic University
USA

M. van Frankenhuijsen
Utah Valley University
USA

R. Han
UC Irvine
USA

A. Henderson
UC Riverside
USA

H. Herichi
Santa Monica College
USA

J. Humphrey
SLAC National Accelerator Laboratory
USA

C. Hurley
California Polytechnic State University
USA

S. Kandasamy
California State Polytechnic University
USA

D. Kelleher
Purdue University
USA

M. Kitofee
University of Guam
USA

F. Kloster
UC Riverside
USA

M. Kobayashi
California State Polytechnic University
USA

J. Lee
California Polytechnic State University
USA

N. Lal
Occidental College
USA

M. Landeros
California State Polytechnic University
USA

M. L. Lapidus
UC Riverside
USA

T. Lazarus
UC Davis
USA

G. Lewis
California Polytechnic State University
USA

Z. Li
Stony Brook University
USA

J. Lindgren
California Polytechnic State University
USA

W. Liu
UC Irvine
USA

F. G. Lopez
UC Riverside
USA

M. Maroun
UC Riverside
USA

D. Nguyen
California State Polytechnic University
USA

R. G. Niemeyer
University of Maine
USA

A. Oudich
California State Polytechnic University
USA

E. P. J. Pearse
California Polytechnic State University
USA

C. Pomerance
Dartmouth College
USA

R. Reyna
California State Polytechnic University
USA

A. Richard
University of Cincinnati
USA

S. Roby
UC Riverside
USA

J. A. Rock
California State Polytechnic University
USA

L. Rogers
University of Connecticut
USA

M. Roychowdhury
University of Texas Rio Grande Valley
USA

T. Samuel
California Polytechnic State University
USA

J. Silva
California State Polytechnic University
USA

J. Spalding
UC Riverside
USA

J. Swift
Northern Arizona University
USA

Y. Takahashi
UC Irvine
USA

A. Teplyaev
University of Connecticut
USA

T. Todorov
California Polytechnic State University
USA

J. Tramontano
MIT
USA

A. Vince
University of Florida
USA

E. Voskanian
UC Riverside
USA

S. Warnert
California Polytechnic State University
USA

S. Watson
UC Riverside
USA

Q. Xia
UC Davis
USA

Contemporary Mathematics
Volume **731**, 2019
https://doi.org/10.1090/conm/731/14670

The Mass Transference Principle: Ten years on

Demi Allen and Sascha Troscheit

ABSTRACT. In this article we discuss the Mass Transference Principle due to Beresnevich and Velani and survey several generalisations and variants, both deterministic and random. Using a Hausdorff measure analogue of the inhomogeneous Khintchine–Groshev Theorem, proved recently via an extension of the Mass Transference Principle to systems of linear forms, we give an alternative proof of (most cases of) a general inhomogeneous Jarník–Besicovitch Theorem which was originally proved by Levesley. We additionally show that without monotonicity Levesley's theorem no longer holds in general. Thereafter, we discuss recent advances by Wang, Wu and Xu towards mass transference principles where one transitions from lim sup sets defined by balls to lim sup sets defined by rectangles (rather than from "balls to balls" as is the case in the original Mass Transference Principle). Furthermore, we consider mass transference principles for transitioning from rectangles to rectangles and extend known results using a slicing technique. We end this article with a brief survey of random analogues of the Mass Transference Principle.

1. Introduction

Since its discovery by Beresnevich and Velani in 2006, the Mass Transference Principle has become an important tool in metric number theory. Originally motivated by the desire for a Hausdorff measure version of the Duffin–Schaeffer conjecture, the Mass Transference Principle allows us to transfer a Lebesgue measure statement for a lim sup set defined by a sequence of balls in $\mathbb{R}^k$ to a Hausdorff measure statement for a related lim sup set. Over the past few years a number of generalisations have been proved and more general settings have been considered.[1] In this article we survey several of these recent developments and consider some of their applications, mostly in the field of metric number theory.

1.1. Notation and Basic Definitions. Throughout, by a *dimension function* we mean a continuous, non-decreasing function $f : \mathbb{R}^+ \to \mathbb{R}^+$ such that $f(r) \to 0$ as $r \to 0$. Recall that $\mathbb{R}^+ = [0, \infty)$. If there exists a constant $\lambda > 1$ such that for $x > 0$ we have $f(2x) \leq \lambda f(x)$ then we say that f is *doubling*.

2010 *Mathematics Subject Classification.* Primary 11J83, 28A78; Secondary 11K60, 11J13, 60D05.

Key words and phrases. Mass Transference Principle, Diophantine approximation, linear forms, random covering sets.

[1]Since the initial writing of this paper, further generalisations of the Mass Transference Principle have been established in [**1**] and [**48**].

Definition 1.1. *Let $F \subseteq \mathbb{R}^k$ and let $\delta > 0$. The* δ-Hausdorff pre-measure *of F with respect to the dimension function f, denoted $\mathcal{H}^f_\delta(F)$, is given by*

$$\mathcal{H}^f_\delta(F) = \inf\left\{\sum_{i=1}^{\infty} f(\operatorname{diam}(U_i)) \; : \; F \subseteq \bigcup_{i=1}^{\infty} U_i \text{ and } \operatorname{diam}(U_i) \leq \delta \text{ for all } i\right\},$$

where the infimum is taken over all countable collections $\{U_i\}$ of open sets. The Hausdorff content *$\mathcal{H}^f_\infty$ with respect to f is*

$$\mathcal{H}^f_\infty(F) = \inf_{\delta>0} \mathcal{H}^f_\delta(F).$$

The Hausdorff measure *$\mathcal{H}^f$ with respect to f is defined by*

$$\mathcal{H}^f(F) = \lim_{\delta\to 0} \mathcal{H}^f_\delta(F).$$

A simple consequence of the definition of $\mathcal{H}^f$ is the following useful fact (see, for example, [**24**]).

Lemma 1.2. *If f and g are two dimension functions such that the ratio $f(r)/g(r) \to 0$ as $r \to 0$, then $\mathcal{H}^f(F) = 0$ whenever $\mathcal{H}^g(F) < \infty$.*

We note that for all dimension functions f, and all bounded subsets $F \subset \mathbb{R}^k$, the Hausdorff content satisfies $\mathcal{H}^f_\infty(F) \leq f(\operatorname{diam}(F))$ and for all $\delta > 0$

$$\mathcal{H}^f_\infty(F) \leq \mathcal{H}^f_\delta(F) < \infty.$$

We also observe that, for a given f, the δ-Hausdorff pre-measure $\mathcal{H}^f_\delta(F)$ is non-decreasing as $\delta \to 0$. So, using monotone convergence, the limit $\lim_{\delta\to 0} \mathcal{H}^f_\delta(F)$ exists but may be infinite.

Often we are interested in Hausdorff dimension and the classical Hausdorff s-measure. The Hausdorff s-measure, which we will usually denote by $\mathcal{H}^s$, can be obtained by letting $f(r) = r^s$. The Hausdorff dimension of a set F is then defined as follows.

Definition 1.3. *Let $F \subseteq \mathbb{R}^k$. The* Hausdorff dimension *of F is*

$$\dim_{\mathrm{H}} F = \inf\left\{s > 0 \; : \; \mathcal{H}^s(F) = 0\right\}.$$

One interesting property of the Hausdorff measure is that for subsets of $\mathbb{R}^k$, $\mathcal{H}^k$ is comparable to the k-dimensional Lebesgue measure. For a set $X \subset \mathbb{R}^k$ we denote the k-dimensional Lebesgue measure by $|X|$. Lebesgue null sets, *i.e.* sets X with $|X| = 0$, can still have intricate geometric structure and in many cases we are able to appeal to Hausdorff dimension to discriminate between their respective 'sizes'. For further information regarding Hausdorff measures and dimension we refer the reader to [**24, 51, 57**]. Finally, we recall the notion of a lim sup set.

Definition 1.4. *Let $(A_i)_{i\in\mathbb{N}}$ be a collection of subsets of a set Y. Then*

$$\limsup_i A_i = \bigcap_{k=1}^{\infty} \bigcup_{i=k}^{\infty} A_i.$$

Equivalently,

$$\limsup_i A_i = \{x \in Y \; : \; x \in A_i \text{ for infinitely many } i \in \mathbb{N}\}.$$

2. The Mass Transference Principle

The main object of study in this article is the *Mass Transference Principle* and its generalisations and variants. The original Mass Transference Principle was developed by Beresnevich and Velani in [**8**] and was motivated by a conjecture of Duffin and Schaeffer.

Given an approximating function $\psi : \mathbb{N} \to \mathbb{R}^+$, for $k \in \mathbb{N}$, let

$$\mathcal{A}_k(\psi) := \left\{ \mathbf{x} \in \mathbb{I}^k : \max_{1 \le i \le k} \left| x_i - \frac{p_i}{q} \right| < \frac{\psi(q)}{q} \text{ for infinitely many } (\mathbf{p}, q) \in \mathbb{Z}^k \times \mathbb{N} \right\},$$

where $\mathbf{p} = (p_1, p_2, \ldots, p_k)$, be the *simultaneously ψ-approximable points* in the unit cube, $\mathbb{I}^k = [0,1]^k$, and consider the following classical theorem by Khintchine [**46**].

THEOREM 2.1 (Khintchine [**46**]). *Let $\psi : \mathbb{N} \to \mathbb{R}^+$ be an approximating function. Then*

$$|\mathcal{A}_1(\psi)| = \begin{cases} 0 & \text{if } \sum_{q=1}^{\infty} \psi(q) < \infty, \\ \\ 1 & \text{if } \sum_{q=1}^{\infty} \psi(q) = \infty \text{ and } \psi \text{ is monotonic.} \end{cases}$$

Khintchine also extended this result to the simultaneously ψ-approximable points in higher dimensions.

THEOREM 2.2 (Khintchine [**47**]). *Let $\psi : \mathbb{N} \to \mathbb{R}^+$ be an approximating function. Then*

$$|\mathcal{A}_k(\psi)| = \begin{cases} 0 & \text{if } \sum_{q=1}^{\infty} \psi(q)^k < \infty, \\ \\ 1 & \text{if } \sum_{q=1}^{\infty} \psi(q)^k = \infty \text{ and } \psi \text{ is monotonic.} \end{cases}$$

In the one-dimensional case Duffin and Schaeffer [**18**] constructed a counter-example showing that the full Lebesgue measure statement can fail for non-monotonic ψ. They also posed a conjecture on what should be true when considering general (not necessarily monotonic) approximating functions.

Given an approximating function $\psi : \mathbb{N} \to \mathbb{R}^+$ and an integer $k \ge 1$, let us denote by $\mathcal{A}'_k(\psi)$ the set of points $\mathbf{x} \in \mathbb{I}^k$ such that

$$\max_{1 \le i \le k} \left| x_i - \frac{\mathbf{p}_i}{q} \right| < \frac{\psi(q)}{q} \tag{2.1}$$

for infinitely many $(\mathbf{p}, q) \in \mathbb{Z}^k \times \mathbb{N}$ with $\gcd(\mathbf{p}, q) := \gcd(p_1, \ldots, p_k, q) = 1$.

CONJECTURE 2.3 (Duffin–Schaeffer Conjecture [**18**]). *Let $\psi : \mathbb{N} \to \mathbb{R}^+$ be any approximating function and denote by $\phi(q)$ the Euler function. If*

$$\sum_{q=1}^{\infty} \phi(q) \frac{\psi(q)}{q} = \infty \quad \textit{then} \quad |\mathcal{A}'_1(\psi)| = 1.$$

For $k \ge 2$ the analogous conjecture was formulated by Sprindžuk [**61**, Chapter 1, Section 8]. The conjecture depends again on slightly different coprimality conditions. Therefore, for any approximating function $\psi : \mathbb{N} \to \mathbb{R}^+$, let us denote by $\mathcal{A}''_k(\psi)$ the set of points $\mathbf{x} \in \mathbb{I}^k$ for which the inequality (2.1) is satisfied for infinitely many $(\mathbf{p}, q) \in \mathbb{Z}^k \times \mathbb{N}$ which also have $\gcd(p_i, q) = 1$ for all $1 \le i \le k$.

CONJECTURE 2.4 (Higher-Dimensional Duffin–Schaeffer Conjecture [**61**]). *Let $\psi : \mathbb{N} \to \mathbb{R}^+$ be any approximating function and denote by $\phi(q)$ the Euler function. If*

$$\sum_{q=1}^{\infty} \phi(q)^k \frac{\psi(q)^k}{q^k} = \infty \quad \text{then} \quad |\mathcal{A}_k''(\psi)| = 1.$$

For $k > 1$ Sprindžuk's conjecture (Conjecture 2.4) was proved in the affirmative by Pollington and Vaughan [**56**].

Finding a general Hausdorff measure analogue of the Duffin–Schaeffer conjecture inspired the Mass Transference Principle that we will now state. Let f be a dimension function and let $\mathcal{H}^f(\,\cdot\,)$ denote Hausdorff f-measure. Given a ball $B = B(x,r)$ in $\mathbb{R}^k$ of radius r centred at x let $B^f = B(x, f(r)^{\frac{1}{k}})$. We write B^s instead of B^f if $f(x) = x^s$ for some $s > 0$. In particular, we have $B^k = B$.

THEOREM 2.5 (**Mass Transference Principle**, Beresnevich – Velani [**8**]). *Let $\{B_i = B(x_i, r_i)\}_{i\in\mathbb{N}}$ be a sequence of balls in $\mathbb{R}^k$ with $r_i \to 0$ as $i \to \infty$. Let f be a dimension function such that $x^{-k}f(x)$ is monotonic and let Ω be a ball in $\mathbb{R}^k$. Suppose that, for any ball B in Ω,*

$$\mathcal{H}^k\big(B \cap \limsup_{i\to\infty} B_i^f\,\big) = \mathcal{H}^k(B)\ .$$

Then, for any ball B in Ω,

$$\mathcal{H}^f\big(B \cap \limsup_{i\to\infty} B_i^k\,\big) = \mathcal{H}^f(B)\ .$$

REMARK. *Strictly speaking, the statement of the Mass Transference Principle given initially by Beresnevich and Velani,* [**8**, Theorem 2], *corresponds to the case where Ω is taken to be $\mathbb{R}^k$ in Theorem* 2.5. *The statement we have opted to give above is a consequence of* [**8**, Theorem 2].

The Mass Transference Principle allows us therefore to transfer a Lebesgue measure statement for a lim sup set of balls to a Hausdorff measure statement for a lim sup set of balls which are obtained by "shrinking" the original balls in a certain manner according to f. This is a remarkable result given that Lebesgue measure can be considered to be much 'coarser' than Hausdorff measure.

Before we continue our discussion of the Mass Transference Principle, we take this opportunity to note that some related results, which are more specialised but still similar in spirit to the Mass Transference Principle, were proved earlier by Jaffard in [**36, 37**]. Jaffard's results allow for the transference of full Lebesgue measure statements for a certain type of lim sup set in $[0,1]^k$ to Hausdorff f-measure statements for related lim sup sets where the dimension function f takes a specific form. A one-dimensional statement is given in [**36**] and is extended to higher dimensions in [**37**]. More precisely, suppose we are given a sequence of points $(\mathbf{x}_n)_{n\in N}$ in $[0,1]^k$ and a sequence of positive real numbers $(r_n)_{n\in\mathbb{N}}$. Given $a > 0$, define

$$E_a := \limsup_{n\to\infty} B(\mathbf{x}_n, r_n^a).$$

THEOREM 2.6 (Jaffard, [**37**]). *For $c > 0$, let $f_c : \mathbb{R}^+ \to \mathbb{R}^+$ be defined by $f_c(r) = (\log r)^2 r^c$. If $|E_a| = 1$, then for every $b \in \mathbb{R}$ such that $b > a$ we have*

$$\mathcal{H}^{f_{ka/b}}(E_b) > 0.$$

In particular, for $b > a$, we have $\dim_{\mathrm{H}}(E_b) \geq \frac{ka}{b}$.

Returning to our earlier discussion, the Mass Transference Principle (Theorem 2.5) was used to show that the Duffin–Schaeffer conjecture for Lebesgue measure gives rise to an analogous statement for Hausdorff measures.

CONJECTURE 2.7 (Hausdorff Measure Duffin–Schaeffer Conjecture [**8**]). *Let $\psi : \mathbb{N} \to \mathbb{R}^+$ be any approximating function and let f be a dimension function such that $r^{-k}f(r)$ is monotonic. If*

$$\sum_{q=1}^{\infty} \phi(q)^k f\left(\frac{\psi(q)}{q}\right) = \infty \quad \textit{then} \quad \mathcal{H}^f(\mathcal{A}_k''(\psi)) = \mathcal{H}^f(\mathbb{I}^k).$$

Setting $f(r) = r^k$ in the above we see that we immediately recover Conjecture 2.4. What is much more surprising is that, using the Mass Transference Principle (Theorem 2.5), Beresnevich and Velani proved that Conjecture 2.4 implies Conjecture 2.7 and hence that they are equivalent. In particular, Conjecture 2.7 holds for $k > 1$ for general approximating functions ψ and for $k = 1$ if ψ is furthermore monotonic, see [**8**].

Two further easy yet surprising consequences of the Mass Transference Principle, which are also mentioned in [**8**], are that Khintchine's Theorem implies Jarník's Theorem and also that Dirichlet's Theorem implies the Jarník–Besicovitch Theorem. We shall elaborate briefly on these examples here, however for further details and proofs we refer the reader to [**7**, **8**].

Let us consider what Khintchine's Theorem (Theorem 2.1) can tell us when our approximating function $\psi : \mathbb{N} \to \mathbb{R}^+$ is given by $\psi(q) = q^{-\tau}$ for some $\tau > 1$. In this case we will write $\mathcal{A}(\tau)$ in place of $\mathcal{A}_1(\psi)$ and we will refer to the points in $\mathcal{A}(\tau)$ as *τ-approximable points*. For any $\tau > 1$ it can be seen that the sum of interest in Khintchine's Theorem converges and so all we can infer is that $|\mathcal{A}(\tau)| = 0$ for all values of $\tau > 1$. However, in this case, we are still able to distinguish the "sizes" of these sets thanks to the Jarník–Besicovitch Theorem. Jarník and Besicovitch both independently proved the following result regarding the Hausdorff dimension of the τ-approximable points.

THEOREM 2.8 (Jarník [**38**], Besicovitch [**12**]). *Let $\tau > 1$. Then*

$$\dim_{\mathrm{H}}(\mathcal{A}(\tau)) = \frac{2}{\tau + 1}.$$

In fact it turns out that, using the Mass Transference Principle, the Jarník–Besicovitch Theorem can be extracted from Dirichlet's theorem. Jarník later proved a much stronger statement, regarding the Hausdorff measures of more general sets of ψ-approximable points, which can be viewed as the Hausdorff measure analogue of Khintchine's Theorem (Theorem 2.2). We state below a modern version of Jarník's Theorem, see [**5**, Theorem 11] for a greater discussion of the derivation of this statement.

THEOREM 2.9 (Jarník [**39**]). *Let $\psi : \mathbb{N} \to \mathbb{R}^+$ be an approximating function and let f be a dimension function such that $r^{-k}f(r)$ is monotonic. Then*

$$\mathcal{H}^f(\mathcal{A}_k(\psi)) = \begin{cases} 0 & \textit{if } \sum_{q=1}^{\infty} q^k f\left(\frac{\psi(q)}{q}\right) < \infty, \\ \\ \mathcal{H}^f(\mathbb{I}^k) & \textit{if } \sum_{q=1}^{\infty} q^k f\left(\frac{\psi(q)}{q}\right) = \infty \textit{ and } \psi \textit{ is monotonic.} \end{cases}$$

Setting $\psi(q) = q^{-\tau}$ and $f(r) - r^s$ in Jarník's Theorem we recover the Jarník–Besicovitch Theorem and additionally gain knowledge of the Hausdorff measure at the critical value $s_0 = 2/(\tau+1)$, *i.e.* $\mathcal{H}^{s_0}(\mathcal{A}(\tau)) = \infty$. Although it may at first be surprising, (the original statement of) Jarník's Theorem follows directly from (the original statement of) Khintchine's Theorem using the Mass Transference Principle. For a proof see, for example, [**7**,**8**]. We remark here that in the original versions of Theorems 2.1, 2.2 and 2.9 various stronger monotonicity conditions were required and note that this is of some relevance when using the Mass Transference Principle to deduce Jarník's Theorem from Khintchine's Theorem. It is possible to deduce Theorem 2.9 from Theorem 2.2 via the Mass Transference Principle but an additional constraint is required on the monotonicity of ψ in this case.

Apart from these important applications in number theory, the Mass Transference Principle can be used to determine Hausdorff dimension and Hausdorff measure statements for many other constructions.

We end this section by stating the most general variant of the Mass Transference Principle in the original article of Beresnevich and Velani [**8**] and mentioning one of its applications. Let (X,d) be a locally compact metric space. Let g be a doubling dimension function and suppose that X is g-Ahlfors regular, *i.e.* there exist constants $0 < c_1 < 1 < c_2 < \infty$ and $r_0 > 0$ such that

$$c_1 g(r) \leq \mathcal{H}^g(B(x,r)) \leq c_2 g(r)$$

for any ball $B = B(x,r)$ with centre $x \in X$ and radius $r \leq r_0$. In this case, given a ball $B = B(x,r)$ and any dimension function f we define $B^{f,g} = B(x, g^{-1}f(r))$. Note that $B^{g,g} = B$.

THEOREM 2.10 (Beresnevich – Velani [**8**]). *Let (X,d) be a locally compact metric space and let g be a doubling dimension function as above. Let $\{B_i = B(x_i,r_i)\}_{i\in\mathbb{N}}$ be a sequence of balls in X with $r_i \to 0$ as $i \to \infty$ and let f be a dimension function such that $f(x)/g(x)$ is monotonic. Suppose that, for any ball B in X,*

$$\mathcal{H}^g(B \cap \limsup_{i\to\infty} B_i^{f,g}) = \mathcal{H}^g(B).$$

Then, for any ball B in X, we have

$$\mathcal{H}^f(B \cap \limsup_{i\to\infty} B_i^{g,g}) = \mathcal{H}^f(B).$$

As an example, Theorem 2.10 is applicable when X is, say, the standard middle-third Cantor set which we denote by K (*i.e.* K is the set of $x \in [0,1]$ which have a ternary expansion containing only 0s and 2s). In fact, in this case, Levesley, Salp, and Velani [**50**] have used Theorem 2.10 as a tool for proving an assertion of Mahler on the existence of very well approximable numbers in the middle-third Cantor set. It is well known that

$$|K| = 0 \quad \text{and} \quad \dim_{\mathrm{H}} K = \frac{\log 2}{\log 3}.$$

As a result of Dirichlet's Theorem, we know that for any $x \in \mathbb{R}$ there exist infinitely many pairs $(p,q) \in \mathbb{Z} \times \mathbb{N}$ for which

$$\left| x - \frac{p}{q} \right| < \frac{1}{q^2}.$$

If the exponent in the denominator of the right-hand side of the above can be improved (*i.e.* increased) for some $x \in \mathbb{R}$ then x is said to be *very well approximable*; that is, a real number x is said to be very well approximable if there exists some $\varepsilon > 0$ such that

$$\left| x - \frac{p}{q} \right| < \frac{1}{q^{2+\varepsilon}} \tag{2.2}$$

for infinitely many pairs $(p, q) \in \mathbb{Z} \times \mathbb{N}$. We will denote the set of very well approximable numbers by $\mathcal{W}$. If, further, (2.2) is satisfied for every $\varepsilon > 0$ for some $x \in \mathbb{R}$ then x is called a *Liouville number*. We will denote by $\mathcal{L}$ the set of all Liouville numbers.

It is known that

$$|\mathcal{W}| = 0, \ \dim_{\mathrm{H}}(\mathcal{W}) = 1,$$
$$|\mathcal{L}| = 0, \ \text{and} \ \dim_{\mathrm{H}}(\mathcal{L}) = 0.$$

Regarding the intersection of $\mathcal{W}$ with the middle-third Cantor set, Mahler is attributed with having made the following claim.

MAHLER'S ASSERTION. *There exist very well approximable numbers, other than Liouville numbers, in the middle-third Cantor set; i.e.*

$$(\mathcal{W} \setminus \mathcal{L}) \cap K \neq \emptyset.$$

REMARK. *We refer the reader to* [**50**] *for discussion of the precise origin of this claim and also for some discussion regarding why it is natural/necessary to exclude Liouville numbers from Mahler's assertion.*

Now, let $\mathcal{B} = \{3^n : n = 0, 1, 2, \dots\}$ and, given an approximating function $\psi : \mathbb{N} \to \mathbb{R}^+$, consider the set

$$\mathcal{A}_{\mathcal{B}}(\psi) := \left\{ x \in [0,1] : \left| x - \frac{p}{q} \right| < \psi(q) \text{ for infinitely many } (p,q) \in \mathbb{Z} \times \mathcal{B} \right\}.$$

Levesley, Salp and Velani have used the general Mass Transference Principle (Theorem 2.10) as a tool for establishing the following statement regarding Hausdorff measures of the set $\mathcal{A}_{\mathcal{B}}(\psi) \cap K$ in [**50**].

THEOREM 2.11 (Levesley – Salp – Velani [**50**]). *Let f be a dimension function such that $r^{-\frac{\log 2}{\log 3}} f(r)$ is monotonic. Then,*

$$\mathcal{H}^f(\mathcal{A}_{\mathcal{B}}(\psi) \cap K) = \begin{cases} 0 & \text{if } \sum_{n=1}^{\infty} f(\psi(3^n)) \times (3^n)^{\frac{\log 2}{\log 3}} < \infty, \\ \mathcal{H}^f(K) & \text{if } \sum_{n=1}^{\infty} f(\psi(3^n)) \times (3^n)^{\frac{\log 2}{\log 3}} = \infty. \end{cases}$$

As a consequence of Theorem 2.11 the following corollary may be deduced, for details of how we refer the reader to [**50**].

COROLLARY 2.12 (Levesley – Salp – Velani [**50**]). *We have*

$$\dim_{\mathrm{H}}((\mathcal{W} \setminus \mathcal{L}) \cap K) \geq \frac{\log 2}{2 \log 3}.$$

The truth of Mahler's assertion follows immediately from this corollary.

We conclude this section by noting that both the original Mass Transference Principle (Theorem 2.5) and its generalisation given by Theorem 2.10 concern lim sup sets arising from sequences of balls. In subsequent sections we will explore what happens when this condition is relaxed. More precisely, we will consider linear forms (Section 3) and rectangles (Section 4) in the deterministic setting and arbitrary Lebesgue measurable sets in the random setting (Section 5).

Inevitably, there are various aspects of the Mass Transference Principle that are not covered in this survey. For example, we have not touched upon the fundamental connections between the Mass Transference Principle set up and the ubiquitous systems framework as developed in [**6**] — in short, the so-called KGB-Lemma [**8**, Lemma 5] is very much at the heart of both. Although a ubiquitous framework was developed in [**6**], we remark that the idea of a ubiquitous system was introduced earlier in [**17**] and was further developed in [**11**]. For an overview of ubiquity and some of its applications also see [**6, 20**] and references within. Another omission from this survey is any mention of mass transference principles in the multifractal setting — see, for example, [**4, 28**]. In the interest of brevity we have ultimately opted against the inclusion of such topics and chosen here to only focus on the aspects mentioned above.

3. Extension to systems of linear forms

In this section we will consider an extension of the Mass Transference Principle to systems of linear forms and mention some of the associated consequences in the theory of Diophantine approximation.

3.1. A mass transference principle for systems of linear forms. Let $k, m \geq 1$ and $l \geq 0$ be integers such that $k = m + l$. Let $\mathcal{R} = (R_n)_{n\in\mathbb{N}}$ be a family of planes in $\mathbb{R}^k$ of common dimension l. For every $n \in \mathbb{N}$ and $\delta \geq 0$, define

$$\Delta(R_n, \delta) := \{\mathbf{x} \in \mathbb{R}^k : \operatorname{dist}(\mathbf{x}, R_n) < \delta\},$$

where $\operatorname{dist}(\mathbf{x}, R_n) = \inf\{\|\mathbf{x} - \mathbf{y}\| : \mathbf{y} \in R_n\}$ and $\|\cdot\|$ is a norm on $\mathbb{R}^k$.

Let $\Upsilon : \mathbb{N} \to \mathbb{R}$ be a non-negative real-valued function $n \mapsto \Upsilon_n$ on $\mathbb{N}$ such that $\Upsilon_n \to 0$ as $n \to \infty$. Consider the lim sup set

$$\Lambda(\Upsilon) := \{\mathbf{x} \in \mathbb{R}^k : \mathbf{x} \in \Delta(R_n, \Upsilon_n) \text{ for infinitely many } n \in \mathbb{N}\}.$$

In 2006, Beresnevich and Velani also established the following extension of the Mass Transference Principle to systems of linear forms [**9**].

THEOREM 3.1 (Beresnevich – Velani [**9**]). *Let $\mathcal{R}$ and Υ be as given above. Let V be a linear subspace of $\mathbb{R}^k$ such that* $\dim V = m = \operatorname{codim} \mathcal{R}$,

(i) *$V \cap R_n \neq \emptyset$ for all $n \in \mathbb{N}$, and*
(ii) $\sup_{n\in\mathbb{N}} \operatorname{diam}(V \cap \Delta(R_n, 1)) < \infty$.

Let f and $g : r \mapsto g(r) := r^{-l} f(r)$ be dimension functions such that $r^{-k} f(r)$ is monotonic and let Ω be a ball in $\mathbb{R}^k$. Suppose for any ball B in Ω that

$$\mathcal{H}^k\left(B \cap \Lambda\left(g(\Upsilon)^{\frac{1}{m}}\right)\right) = \mathcal{H}^k(B).$$

Then, for any ball B in Ω,

$$\mathcal{H}^f(B \cap \Lambda(\Upsilon)) = \mathcal{H}^f(B).$$

REMARK. *Note that when $l = 0$ in Theorem 3.1 we recover the Mass Transference Principle (Theorem 2.5).*

Conditions (i) and (ii) essentially say that V should intersect every plane and that the angle of intersection between V and each plane should be bounded away from 0. In other words, every plane R_n ought not to be parallel to V and should intersect V in precisely one place. These conditions are technical and come about as a consequence of the "slicing" technique used by Beresnevich and Velani to prove Theorem 3.1 in [**9**] (for a simple demonstration of the idea of "slicing" see the proofs of Propositions 4.7 and 4.8 in Section 4). It was conjectured by Beresnevich et al. [**5**, Conjecture E] that Theorem 3.1 should also be true without conditions (i) and (ii). Recently, this conjecture has been settled by Allen and Beresnevich in [**2**] by using a proof closer in strategy to that used by Beresnevich and Velani to prove the original Mass Transference Principle in [**8**], rather than "slicing".

THEOREM 3.2 (Allen – Beresnevich [**2**]). *Let $\mathcal{R}$ and Υ be as given above. Let f and $g : r \mapsto g(r) := r^{-l}f(r)$ be dimension functions such that $r^{-k}f(r)$ is monotonic and let Ω be a ball in $\mathbb{R}^k$. Suppose for any ball B in Ω that*

$$\mathcal{H}^k\left(B \cap \Lambda\left(g(\Upsilon)^{\frac{1}{m}}\right)\right) = \mathcal{H}^k(B).$$

Then, for any ball B in Ω,

$$\mathcal{H}^f(B \cap \Lambda(\Upsilon)) = \mathcal{H}^f(B).$$

Although Theorem 3.1 itself has some interesting consequences, see [**5, 9**], it seems likely that there will be a number of applications of Theorem 3.2 which are unachievable using Theorem 3.1. In particular, in Section 3.4 we record some very general statements obtained in [**2**] which essentially rephrase Theorem 3.2 as statements for transferring Lebesgue measure statements to Hausdorff measure statements for sets of ψ-approximable (and Ψ-approximable) points.

Before that we state some more concrete applications of Theorems 3.1 and 3.2; namely, we mention Hausdorff measure analogues of the homogeneous and inhomogeneous Khintchine–Groshev Theorems obtained in [**2, 5**]. We also use the Hausdorff measure analogue of the inhomogeneous Khintchine–Groshev Theorem to make some remarks on a theorem of Levesley [**49**].

3.2. Hausdorff measure Khintchine–Groshev statements. Throughout this section, let $n \geq 1$ and $m \geq 1$ be integers and denote by $\mathbb{I}^{nm}$ the unit cube $[0,1]^{nm} \subset \mathbb{R}^{nm}$.

Given a function $\psi : \mathbb{N} \to \mathbb{R}^+$, let $\mathcal{A}_{n,m}(\psi)$ denote the set of $\mathbf{x} \in \mathbb{I}^{nm}$ such that

$$|\mathbf{qx} + \mathbf{p}| < \psi(|\mathbf{q}|)$$

for infinitely many $(\mathbf{p}, \mathbf{q}) \in \mathbb{Z}^m \times \mathbb{Z}^n \setminus \{\mathbf{0}\}$.

Here, $|\cdot|$ denotes the supremum norm and we think of $\mathbf{x} = (x_{ij})$ as an $n \times m$ matrix and of $\mathbf{p}$ and $\mathbf{q}$ as row vectors. So $\mathbf{qx}$ represents a point in $\mathbb{R}^m$ given by the system

$$q_1x_{1j} + \cdots + q_nx_{nj} \quad (1 \leq j \leq m)$$

of m real linear forms in n variables. We say that the points in $\mathcal{A}_{n,m}(\psi)$ are *ψ-approximable.* As with many sets of interest in Diophantine approximation, $\mathcal{A}_{n,m}(\psi)$ satisfies an elegant zero-one law with respect to Lebesgue measure.

THEOREM 3.3 (Khintchine–Groshev Theorem [**10**]). *Let $\psi : \mathbb{N} \to \mathbb{R}^+$ and let $nm > 1$. Then*

$$|\mathcal{A}_{n,m}(\psi)| = \begin{cases} 0 & \text{if } \sum_{q=1}^{\infty} q^{n-1}\psi(q)^m < \infty, \\ \\ 1 & \text{if } \sum_{q=1}^{\infty} q^{n-1}\psi(q)^m = \infty. \end{cases}$$

The earliest versions of this theorem are attributed to Khintchine and Groshev [**33, 46**]. These were subject to extra assumptions including monotonicity of ψ. Due to the famous counter-example of Duffin and Schaeffer [**18**] we know that if we have $m = n = 1$ then monotonicity cannot be removed. However, when we insist that $nm > 1$ monotonicity of ψ is unnecessary. In the case that $n = 1$ or $n \geq 3$ this follows, respectively, from results due to Gallagher [**31**] and Schmidt [**59**]. In the case that $n \geq 3$ this also follows from a result of Sprindžuk [**61**, Chapter 1, Section 5], stated as Theorem 3.5 below. For further information we refer the reader to the detailed survey [**5**]. It was conjectured by Beresnevich et al. in [**5**, Conjecture A] that monotonicity should also be unnecessary when $n = 2$. This conjecture was finally settled by Beresnevich and Velani in [**10**] leaving the above modern statement of the Khintchine–Groshev Theorem, which is the best possible.

Regarding the Hausdorff measure theory, combining Theorem 3.2 with Theorem 3.3 yields the following.

THEOREM 3.4 (Hausdorff measure Khintchine–Groshev Theorem [**2, 5**]). *Let $\psi : \mathbb{N} \to \mathbb{R}^+$ be any approximating function and let $nm > 1$. Let f and $g : r \mapsto g(r) = r^{-m(n-1)}f(r)$ be dimension functions such that $r^{-nm}f(r)$ is monotonic. Then,*

$$\mathcal{H}^f(\mathcal{A}_{n,m}(\psi)) = \begin{cases} 0 & \text{if } \sum_{q=1}^{\infty} q^{n+m-1} g\left(\frac{\psi(q)}{q}\right) < \infty, \\ \\ \mathcal{H}^f(\mathbb{I}^{nm}) & \text{if } \sum_{q=1}^{\infty} q^{n+m-1} g\left(\frac{\psi(q)}{q}\right) = \infty. \end{cases}$$

For completeness, we remark here that before the above statement appeared in [**5**], the Hausdorff measures and dimension of the sets $\mathcal{A}_{n,m}(\psi)$ had already been studied by a number of people. Indeed, earlier similar results, albeit subject to further constraints, had already been established. In particular, the first Hausdorff measure result in this direction was obtained by Dickinson and Velani [**15**] and, even before that, the first Hausdorff dimension results had already been established by Bovey and Dodson [**13**].

Returning to Theorem 3.4 we note that the statement in [**5**] additionally required ψ to be monotonic when $n = 2$. At that time it was still unproven that the Khintchine–Groshev Theorem was true without monotonicity in the case that $n = 2$. However, it was conjectured in [**5**] that, subject to the validity of the Khintchine–Groshev Theorem without assuming monotonicity when $n = 2$ (*i.e.* Theorem 3.3), it should be possible to use Theorem 3.1 to remove this final monotonicity condition also from the Hausdorff measure version of the Khintchine–Groshev theorem, giving Theorem 3.4. This conjecture has been verified in [**2**] where, in fact, two proofs of Theorem 3.4 are given. The first uses a combination of Theorem 3.1 and "slicing", thus verifying the conjecture, and the second uses Theorem 3.2.

In [**2**], a Hausdorff measure version of the inhomogeneous analogue of the Khintchine–Groshev Theorem is also established. If we are given an approximating function $\psi : \mathbb{N} \to \mathbb{R}^+$ and a fixed $\mathbf{y} \in \mathbb{I}^m$ then we will denote by $\mathcal{A}^{\mathbf{y}}_{n,m}(\psi)$ the set of $\mathbf{x} \in \mathbb{I}^{nm}$ such that

$$|\mathbf{q}\mathbf{x} + \mathbf{p} - \mathbf{y}| < \psi(|\mathbf{q}|)$$

for infinitely many $(\mathbf{p}, \mathbf{q}) \in \mathbb{Z}^m \times \mathbb{Z}^n \setminus \{\mathbf{0}\}$.

Supposing we are given an approximating function $\Psi : \mathbb{Z}^n \setminus \{\mathbf{0}\} \to \mathbb{R}^+$ which can depend on $\mathbf{q}$ rather than just $|\mathbf{q}|$, we will let $\mathcal{A}_{n,m}(\Psi)$ be the set of $\mathbf{x} \in \mathbb{I}^{nm}$ such that

$$|\mathbf{q}\mathbf{x} + \mathbf{p}| < \Psi(\mathbf{q})$$

for infinitely many $(\mathbf{p}, \mathbf{q}) \in \mathbb{Z}^m \times \mathbb{Z}^n \setminus \{\mathbf{0}\}$, we will call these Ψ*-approximable points.* If we are also given a fixed $\mathbf{y} \in \mathbb{I}^m$ we will denote by $\mathcal{A}^{\mathbf{y}}_{n,m}(\Psi)$ the set of $\mathbf{x} \in \mathbb{I}^{nm}$ such that

$$|\mathbf{q}\mathbf{x} + \mathbf{p} - \mathbf{y}| < \Psi(\mathbf{q})$$

for infinitely many $(\mathbf{p}, \mathbf{q}) \in \mathbb{Z}^m \times \mathbb{Z}^n \setminus \{\mathbf{0}\}$.

For $n \geq 2$ we will represent the set of primitive vectors in $\mathbb{Z}^n$ by $P(\mathbb{Z}^n)$; that is, non-zero integer vectors $(v_1, v_2, \ldots, v_n) \in \mathbb{Z}^n \setminus \{\mathbf{0}\}$ with $\gcd(v_1, v_2, \ldots, v_n) = 1$. Regarding the Lebesgue measure of sets of inhomogeneously Ψ-approximable points, we have the following result due to Sprindžuk [**61**, Chapter 1, Section 5].

THEOREM 3.5 (Sprindžuk [**61**]). *Let $m \geq 1$ and $n \geq 2$ be integers. Let $\Psi : \mathbb{Z}^n \to \mathbb{R}^+$ be an approximating function such that $\Psi(\mathbf{q}) = 0$ whenever $\mathbf{q} \notin P(\mathbb{Z}^n)$ and let $\mathbf{y} \in \mathbb{I}^m$ be fixed. Then,*

$$|\mathcal{A}^{\mathbf{y}}_{n,m}(\Psi)| = \begin{cases} 0 & \text{if } \sum_{\mathbf{q} \in \mathbb{Z}^n} \Psi^m(\mathbf{q}) < \infty, \\ \\ 1 & \text{if } \sum_{\mathbf{q} \in \mathbb{Z}^n} \Psi^m(\mathbf{q}) = \infty. \end{cases}$$

Restricting the approximating function Ψ so that it only depends on $|\mathbf{q}|$ yields the following inhomogeneous analogue of the Khintchine–Groshev Theorem (Theorem 3.3).

THEOREM 3.6 (Inhomogeneous Khintchine–Groshev Theorem). *Let $m, n \geq 1$ be integers and let $\mathbf{y} \in \mathbb{I}^m$. If $\psi : \mathbb{N} \to \mathbb{R}^+$ is an approximating function which is assumed to be monotonic if $n = 1$ or $n = 2$, then*

$$|\mathcal{A}^{\mathbf{y}}_{n,m}(\psi)| = \begin{cases} 0 & \text{if } \sum_{q=1}^{\infty} q^{n-1}\psi(q)^m < \infty, \\ \\ 1 & \text{if } \sum_{q=1}^{\infty} q^{n-1}\psi(q)^m = \infty. \end{cases}$$

When $n \geq 3$ the above theorem is a consequence of Theorem 3.5. In the other cases, where monotonicity of ψ is imposed, the above statement follows from results of Beresnevich, Dickinson and Velani [**6**, Section 12]. For more detailed discussion we refer the reader to, for example, [**5**].

By combining Theorem 3.2 with Theorem 3.6 the following Hausdorff measure analogue of Theorem 3.6 may be obtained.

THEOREM 3.7 (Allen–Beresnevich [**2**]). *Let $m, n \geq 1$ be integers, let $\mathbf{y} \in \mathbb{I}^m$, and let $\psi : \mathbb{N} \to \mathbb{R}^+$ be an approximating function. Let f and $g : r \mapsto g(r) := r^{-m(n-1)} f(r)$*

be dimension functions such that $r^{-nm}f(r)$ is monotonic. In the case that $n=1$ or $n=2$ suppose also that $rg\left(\frac{\psi(r)}{r}\right)^{\frac{1}{m}}$ is monotonic. Then,

$$\mathcal{H}^f(\mathcal{A}^{\mathbf{y}}_{n,m}(\psi)) = \begin{cases} 0 & \text{if } \sum_{q=1}^{\infty} q^{n+m-1} g\left(\frac{\psi(q)}{q}\right) < \infty, \\ \\ \mathcal{H}^f(\mathbb{I}^{nm}) & \text{if } \sum_{q=1}^{\infty} q^{n+m-1} g\left(\frac{\psi(q)}{q}\right) = \infty. \end{cases}$$

REMARK 3.8. *In the case when $n=2$, it is possible to prove Theorem 3.7 subject to the more aesthetically pleasing monotonicity constraint that ψ is monotonically decreasing. This can be achieved by combining Theorem 3.2 with Theorem 3.5, rather than Theorem 3.6.*

REMARK 3.9. *We note here that in both Theorems 3.6 and 3.7 the monotonicity condition on ψ when $n=1$ or $n=2$ is only required for the divergence cases. For both of these theorems the proofs of the convergence parts follow from standard covering arguments for which no monotonicity conditions need to be imposed.*

In the next section we show how we can use Theorem 3.7 to provide an alternative proof of most cases of a general inhomogeneous Jarník–Besicovitch Theorem proved by Levesley [**49**]. Furthermore, we are able to comment on the necessity of the monotonicity condition imposed in this theorem of Levesley.

3.3. A Theorem of Levesley. The Hausdorff dimension of $\mathcal{A}^{\mathbf{y}}_{n,m}(\psi)$, in the general inhomogeneous setting, was determined by Levesley in [**49**]. Given a function $f:\mathbb{N}\to\mathbb{R}^+$, the *lower order at infinity of f*, usually denoted by λ, is

$$\lambda(f) = \liminf_{q\to\infty} \frac{\log(f(q))}{\log(q)}.$$

THEOREM 3.10 (Levesley, [**49**]). *Let $m,n\in\mathbb{N}$ and let $\psi:\mathbb{N}\to\mathbb{R}^+$ be a monotonically decreasing function. Let λ be the lower order at infinity of $1/\psi$. Then, for any $\mathbf{y}\in\mathbb{I}^m$,*

$$\dim_{\mathrm{H}}(\mathcal{A}^{\mathbf{y}}_{n,m}(\psi)) = \begin{cases} m(n-1)+\frac{m+n}{\lambda+1} & \text{when } \lambda \geq \frac{n}{m}, \\ \\ nm & \text{when } \lambda < \frac{n}{m}. \end{cases}$$

REMARK. *In the homogeneous case, when $\mathbf{y}=\mathbf{0}$, this result was previously established by Dodson* [**16**].

Levesley proved the above theorem by considering the cases of $n=1$ and $n\geq 2$ separately. In both cases his argument uses ideas from ubiquitous systems. These are combined with ideas from uniform distribution in the former case and with a more statistical ("mean-variance") argument in the latter case.

Using Theorem 3.7, we can give an alternative proof of this theorem in the case that $n\geq 2$. That is, we will prove:

THEOREM 3.11. *Let $m\geq 1$ and $n\geq 2$ be integers. Let $\psi:\mathbb{N}\to\mathbb{R}^+$ be a monotonically decreasing function and let λ be the lower order at infinity of $1/\psi$. Then, for any $\mathbf{y}\in\mathbb{I}^m$,*

$$\dim_{\mathrm{H}}(\mathcal{A}^{\mathbf{y}}_{n,m}(\psi)) = \begin{cases} m(n-1)+\frac{m+n}{\lambda+1} & \text{when } \lambda \geq \frac{n}{m}, \\ \\ nm & \text{when } \lambda < \frac{n}{m}. \end{cases}$$

Recall that in Remark 3.8 we noted that it was sufficient in Theorem 3.7 to assume that ψ is monotonically decreasing in the case that $n = 2$. Throughout this section, we shall assume any mention of Theorem 3.7 refers to a statement including this nicer monotonicity condition for the $n = 2$ case.

To prove Theorem 3.11 using Theorem 3.7 we first establish a useful lemma.

LEMMA 3.12. *Let $\psi : \mathbb{N} \to \mathbb{R}^+$ be monotonic and bounded. Then,*

$$\liminf_{q\to\infty} \frac{-\log(\psi(q))}{\log q} = \liminf_{t\to\infty} \frac{-\log\left(\psi(2^t)\right)}{\log 2^t}.$$

PROOF. Assume first that ψ is non-increasing. Note that $(2^t)_{t=1}^\infty$ is a subsequence of $(q)_{q=1}^\infty$ and so

$$\liminf_{q\to\infty} \frac{-\log(\psi(q))}{\log q} \le \liminf_{t\to\infty} \frac{-\log\left(\psi(2^t)\right)}{\log 2^t}.$$

It remains to prove the reverse inequality. Suppose for now that $\psi(q) \ge 1$ for all $q \in \mathbb{N}$. In this case, since $\psi(q) \to c$ for some $c \ge 1$ by monotone convergence,

$$\liminf_{q\to\infty} \frac{-\log(\psi(q))}{\log q} = 0 = \liminf_{t\to\infty} \frac{-\log(\psi(2^t))}{\log 2^t}.$$

Thus, we may assume that $\psi(q) < 1$ for all sufficiently large q. Given $q \in \mathbb{N}$, set t_q to be the unique integer satisfying $2^{t_q} \le q < 2^{t_q+1}$. Then $\psi(2^{t_q}) \ge \psi(q)$ and $\log(\psi(2^{t_q})) \ge \log(\psi(q))$. Since further $q < 2^{t_q+1}$ and so $\log q < \log 2^{t_q+1}$, we obtain

$$\begin{aligned}
\liminf_{q\to\infty} \frac{-\log(\psi(q))}{\log q} &\ge \liminf_{q\to\infty} \frac{-\log\left(\psi(2^{t_q})\right)}{\log 2^{t_q+1}} \\
&= \liminf_{q\to\infty} \frac{-\log\left(\psi(2^{t_q})\right)}{\log 2^{t_q} + \log 2} \\
&= \liminf_{t\to\infty} \frac{-\log\left(\psi(2^t)\right)}{\log 2^t}
\end{aligned}$$

as required.

For non-decreasing ψ the proof is similar. By the same argument as above, it is again sufficient to show that

$$\liminf_{q\to\infty} \frac{-\log(\psi(q))}{\log q} \ge \liminf_{t\to\infty} \frac{-\log\left(\psi(2^t)\right)}{\log 2^t}.$$

We note that if $\psi(q) \ge 1$ for all sufficiently large q then, since ψ is bounded,

$$\liminf_{q\to\infty} \frac{-\log(\psi(q))}{\log q} = 0 = \liminf_{t\to\infty} \frac{-\log\left(\psi(2^t)\right)}{\log 2^t}.$$

Therefore, we may assume that $\psi(q) < 1$ for all $q \in \mathbb{N}$. Now, along the same lines as the argument given above, given $q \in \mathbb{N}$ let t'_q be the unigue integer for which $2^{t'_q-1} \le q < 2^{t'_q}$. Thus, we have

$$\log 2^{t'_q} > \log q \quad \text{and} \quad \log \psi(2^{t'_q}) \ge \log\left(\psi(q)\right).$$

Hence, it follows that

$$\liminf_{q\to\infty}\frac{-\log(\psi(q))}{\log q}\geq\liminf_{q\to\infty}\frac{-\log(\psi(2^{t'_q}))}{\log 2^{t'_q}}$$
$$=\liminf_{t\to\infty}\frac{-\log(\psi(2^{t}))}{\log 2^{t}},$$

and the proof is thus complete. □

Proof of Theorem 3.11 using Theorem 3.7. To avoid confusion throughout the proof, for approximating functions $\psi:\mathbb{N}\to\mathbb{R}^+$ we will write λ_ψ to denote the lower order at infinity of $1/\psi$. However, when there is no ambiguity we will just write λ and omit the additional subscript.

We observe that, since ψ is assumed to be monotonically decreasing, we must have $\lambda_\psi\geq 0$. To see this, suppose that $\lambda_\psi<0$. Then, by the definition of the lower order at infinity, it follows that for any $\varepsilon>0$ we must have $\psi(q)\geq q^{-(\lambda_\psi+\varepsilon)}$ for infinitely many values of q. In particular, this is true for every $0<\varepsilon<|\lambda_\psi|$ and so we conclude that ψ cannot be monotonically decreasing if $\lambda_\psi<0$.

We will now show that if the result stated in Theorem 3.11 is true for approximating functions with $\lambda=\frac{n}{m}$, then this implies the validity of the result for approximating functions with $0\leq\lambda<\frac{n}{m}$. We will then establish the result for approximating functions with $\lambda\geq\frac{n}{m}$.

For the time being, assume that the conclusion in Theorem 3.11 holds for any monotonically decreasing approximating function with $\lambda=\frac{n}{m}$ and let $\psi:\mathbb{N}\to\mathbb{R}^+$ be a monotonically decreasing approximating function such that $\lambda_\psi<\frac{n}{m}$. Consider the function $\Psi:\mathbb{N}\to\mathbb{R}^+$ defined by $\Psi(q)=\min\{\psi(q),q^{-\frac{n}{m}}\}$. Note that Ψ is a monotonically decreasing function (since it is the minimum of two monotonically decreasing functions) and that $\Psi(q)\leq\psi(q)$ for all $q\in\mathbb{N}$. In particular, we have $\dim_{\mathrm{H}}(\mathcal{A}^{\mathbf{y}}_{n,m}(\Psi))\leq\dim_{\mathrm{H}}(\mathcal{A}^{\mathbf{y}}_{n,m}(\psi))$. Next, note that it follows from the fact that $\Psi(q)\leq q^{-\frac{n}{m}}$ for all $q\in\mathbb{N}$ that $\lambda_\Psi\geq\frac{n}{m}$. On the other hand, since $\lambda_\psi<\frac{n}{m}$ we know that $\psi(q)\geq q^{-\frac{n}{m}}$ for infinitely many values of q. In particular, this implies that we must have $\Psi(q)=q^{-\frac{n}{m}}$ infinitely often and, consequently, that $\lambda_\Psi\leq\frac{n}{m}$. Hence, $\lambda_\Psi=\frac{n}{m}$ and so, by our assumption, we see that

$$\dim_{\mathrm{H}}(\mathcal{A}^{\mathbf{y}}_{n,m}(\psi))\geq\dim_{\mathrm{H}}(\mathcal{A}^{\mathbf{y}}_{n,m}(\Psi))=m(n-1)+\frac{n+m}{\lambda_\Psi+1}=nm.$$

Combining this with the trivial upper bound we see that $\dim_{\mathrm{H}}(\mathcal{A}^{\mathbf{y}}_{n,m}(\psi))=nm$, as required.

It remains to be shown that $\dim_{\mathrm{H}}(\mathcal{A}^{\mathbf{y}}_{n,m}(\psi))=m(n-1)+\frac{n+m}{\lambda+1}$ for monotonically decreasing approximating functions $\psi:\mathbb{N}\to\mathbb{R}^+$ with $\lambda_\psi=\lambda\geq\frac{n}{m}$. To this end, suppose ψ is such an approximating function.

Let $s_0=m(n-1)+\frac{m+n}{\lambda+1}$ and consider $f_\delta(r)=r^{s_0+\delta}$ where $-\frac{n+m}{\lambda+1}<\delta<\frac{n+m}{\lambda+1}$. We aim to show that

$$\mathcal{H}^{s_0+\delta}(\mathcal{A}^{\mathbf{y}}_{n,m}(\psi))=\begin{cases}0 & \text{if}\quad\delta>0\,,\\[2ex] \mathcal{H}^{s_0+\delta}(\mathbb{I}^{nm}) & \text{if}\quad\delta<0,\end{cases}$$

from which the result would follow.

Note that $f_\delta(r)$ is a dimension function and $r^{-nm}f_\delta(r)$ is monotonic. Let $g_\delta(r)=r^{-m(n-1)}f_\delta(r)=r^{-m(n-1)+s_0+\delta}$. Since $\delta>-\frac{n+m}{\lambda+1}$, and consequently

$-m(n-1)+s_0+\delta>0$, the function $g_\delta(r)$ is a dimension function. Thus, f_δ and g_δ satisfy the hypotheses of Theorem 3.7.

It follows from the definition of the lower order at infinity that, for any $\varepsilon>0$,

$$\psi(q)\le q^{-(\lambda-\varepsilon)} \text{ for all large enough } q \text{ and}$$

$$\psi(q)\ge q^{-(\lambda+\varepsilon)} \text{ for infinitely many } q\in\mathbb{N}. \tag{3.1}$$

Combining this with Lemma 3.12, we have

$$\psi(2^t)\le 2^{-t(\lambda-\varepsilon)} \text{ for large enough } t, \text{ and}$$

$$\psi(2^t)\ge 2^{-t(\lambda+\varepsilon)} \text{ for infinitely many } t. \tag{3.2}$$

By Theorem 3.7 it follows that to determine $\mathcal{H}^{f_\delta}(\mathcal{A}^{\mathbf{y}}_{n,m}(\psi))$ we are interested in the behaviour of the sum

$$\sum_{q=1}^{\infty} q^{n+m-1} g_\delta\left(\frac{\psi(q)}{q}\right)=\sum_{q=1}^{\infty} q^{n+m-1}\left(\frac{\psi(q)}{q}\right)^{-m(n-1)+s_0+\delta}. \tag{3.3}$$

Observe that, by the conditions imposed on δ, $-m(n-1)+s_0+\delta>0$ and also that, by (3.1), we have $\psi(q)\le q^{-(\lambda-\varepsilon)}$ for sufficiently large q. Thus, (3.3) will converge if

$$\sum_{q=1}^{\infty} q^{n+m-1}\left(q^{-(\lambda-\varepsilon)-1}\right)^{-m(n-1)+s_0+\delta}=\sum_{q=1}^{\infty} q^{n+m-1+(\lambda+1-\varepsilon)(m(n-1)-s_0-\delta)}<\infty. \tag{3.4}$$

This will be the case if

$$n+m-1+(\lambda+1-\varepsilon)(m(n-1)-s_0-\delta)<-1,$$

which is true if and only if

$$\frac{n+m}{\lambda+1-\varepsilon}+m(n-1)<s_0+\delta.$$

If $\delta>0$ we can force the above to be true by taking ε to be sufficiently small. Thus we conclude that, for $\delta>0$, (3.3) converges and consequently $\mathcal{H}^{s_0+\delta}(\mathcal{A}^{\mathbf{y}}_{n,m}(\psi))=0$.

Next we establish that (3.3) diverges when $-\frac{n+m}{\lambda+1}<\delta<0$. First we note, since ψ is monotonically decreasing, that

$$\begin{aligned}\sum_{q=1}^{\infty} q^{n+m-1}\left(\frac{\psi(q)}{q}\right)^{-m(n-1)+s_0+\delta} &=\sum_{t=1}^{\infty}\sum_{2^{t-1}\le q<2^t} q^{n+m-1}\left(\frac{\psi(q)}{q}\right)^{-m(n-1)+s_0+\delta}\\ &\ge\sum_{t=1}^{\infty}\sum_{2^{t-1}\le q<2^t}(2^{t-1})^{n+m-1}\left(\frac{\psi(2^t)}{2^t}\right)^{-m(n-1)+s_0+\delta}\\ &=\sum_{t=1}^{\infty}2^{t-1}(2^{t-1})^{n+m-1}\left(\frac{\psi(2^t)}{2^t}\right)^{-m(n-1)+s_0+\delta}\\ &=\frac{1}{2^{m+n}}\sum_{t=1}^{\infty}2^{t(n+m)}\left(\frac{\psi(2^t)}{2^t}\right)^{-m(n-1)+s_0+\delta}. \qquad (3.5)\end{aligned}$$

We proceed by showing that, when $\delta < 0$, we have for infinitely many t that

$$2^{t(m+n)} \left(\frac{\psi(2^t)}{2^t}\right)^{-m(n-1)+s_0+\delta} \geq 1. \tag{3.6}$$

For any $\delta < 0$ we can choose $\varepsilon > 0$ small enough such that

$$\frac{m+n}{\lambda+1+\varepsilon} + m(n-1) \geq s_0 + \delta.$$

Note that such an ε exists since we are assuming that δ is negative. Rearranging,

$$m+n-(\lambda+\varepsilon+1)(-m(n-1)+s_0+\delta) \geq 0$$

and then, exponentiating,

$$2^{t(m+n)} \left(\frac{2^{-t(\lambda+\varepsilon)}}{2^t}\right)^{-m(n-1)+s_0+\delta} \geq 1.$$

Now, by (3.2) we have $\psi(2^t) \geq 2^{-t(\lambda+\varepsilon)}$ infinitely often and so (3.6) holds, thus proving the divergence of (3.5) and hence also the divergence of (3.3).

Hence, we have shown that

$$\mathcal{H}^{s_0+\delta}(\mathcal{A}^{\mathbf{y}}_{n,m}(\psi)) = \begin{cases} 0 & \text{if} \quad \delta > 0\,, \\ \\ \mathcal{H}^{s_0+\delta}(\mathbb{I}^{nm}) & \text{if} \quad \delta < 0. \end{cases}$$

If $s_0 \leq nm$ then $\mathcal{H}^{s_0+\delta}(\mathbb{I}^{nm}) = \infty$ whenever $\delta < 0$. From this it would follow that $\dim_{\mathrm{H}}(\mathcal{A}^{\mathbf{y}}_{n,m}(\psi)) = s_0$. The proof is completed upon noting that $s_0 \leq nm$ is equivalent to $\lambda \geq \frac{n}{m}$. □

In Theorems 3.10 and 3.11 the approximating function ψ is assumed to be monotonic. However, the main tool in our proof of Theorem 3.11 is Theorem 3.7 which requires no monotonicity assumptions on ψ for $n \geq 3$. This leads immediately to the natural question of whether this monotonicity assumption is indeed necessary in Levesley's Theorem (Theorem 3.10).

In an attempt to address this question, let us consider general (not necessarily monotonic) approximating functions $\psi : \mathbb{N} \to \mathbb{R}^+$ with λ, the lower order at infinity of $1/\psi$, satisfying $\lambda > n/m$. Assuming no monotonicity conditions on ψ, and applying similar arguments to those which we have employed here to prove Theorem 3.11, we obtain the following bounds on the Hausdorff dimension of $\mathcal{A}^{\mathbf{y}}_{n,m}(\psi)$. Although, in the interest of brevity, we omit proof.

Proposition 3.13. *Let $m \geq 1$ and $n \geq 3$ be integers. If $\psi : \mathbb{N} \to \mathbb{R}^+$ is any function and λ is the lower order at infinity of $1/\psi$ then, for any $\mathbf{y} \in \mathbb{I}^m$, if $\lambda > n/m$ we have*

$$m(n-1) + \frac{m+n-1}{\lambda+1} \leq \dim_{\mathrm{H}}(\mathcal{A}^{\mathbf{y}}_{n,m}(\psi)) \leq m(n-1) + \frac{m+n}{\lambda+1}.$$

We see that the upper and lower bounds for $\dim_{\mathrm{H}}(\mathcal{A}^{\mathbf{y}}_{n,m}(\psi))$ in Proposition 3.13 do not coincide. Interestingly, it turns out that these bounds are the best possible if one does not assume monotonicity of ψ — as we will now show. To the best of our knowledge the following result has not been considered before, even in the homogeneous setting.

THEOREM 3.14. *Let $m, n \geq 1$ be integers. Let $\alpha > n/m$ be arbitrary and let s_0 be such that*

$$m(n-1) + \frac{m+n-1}{\alpha+1} < s_0 < m(n-1) + \frac{m+n}{\alpha+1}.$$

There exists an approximating function $\psi : \mathbb{N} \to \mathbb{R}^+$ such that for every $\mathbf{y} \in \mathbb{I}^m$ we have $\dim_{\mathrm{H}}(\mathcal{A}^{\mathbf{y}}_{n,m}(\psi)) = s_0$ and $\lambda_\psi = \alpha$ (where λ_ψ is the lower order at infinity of $1/\psi$).

PROOF. Fix s_0 satisfying the inequality in the statement of the theorem and let $\mathbf{y} \in \mathbb{I}^m$ be arbitrary. Then, let $J := \{a_k : k \in \mathbb{N}\}$, where $a_k = \lceil k^{-\gamma} \rceil$,

$$\gamma := \frac{2}{n+m-1-(\alpha+1)\left(\frac{n+m}{\beta+1}\right)} \quad \text{and} \quad \beta := \frac{n+m}{s_0 - m(n-1)} - 1.$$

Note that $\gamma < 0$. Define $\psi : \mathbb{N} \to \mathbb{R}^+$ by

$$\psi(q) = \begin{cases} q^{-\alpha} & \text{if} \quad q \in J\,, \\ q^{-\beta} & \text{if} \quad q \notin J. \end{cases}$$

We show that ψ is an approximating function which satisfies the desired properties of the theorem. First, note that

$$m(n-1) + \frac{n+m}{\alpha+1} > s_0,$$

which implies that

$$\frac{n+m}{s_0 - m(n-1)} - 1 > \alpha.$$

In turn, this implies that $\beta > \alpha$ and so $\liminf_{q\to\infty} -\log(\psi(q))/\log(q) = \alpha$, giving $\lambda_\psi = \alpha$, as required.

Recall that if $\lambda_\psi = \alpha$ then for any $\varepsilon > 0$ there exists some $N \in \mathbb{N}$ such that $\psi(q) \leq q^{-(\alpha-\varepsilon)}$ for all $q \geq N$, and $\psi(q) \geq q^{-(\alpha+\varepsilon)}$ for infinitely many $q \in \mathbb{N}$.

To establish that the Hausdorff dimension is s_0 we note that $\dim_{\mathrm{H}}(\mathcal{A}^{\mathbf{y}}_{n,m}(\psi)) \geq \dim_{\mathrm{H}}(\mathcal{A}^{\mathbf{y}}_{n,m}(q \mapsto q^{-\beta}))$ since $\psi(q) \geq q^{-\beta}$ for all q. Furthermore, since $q \mapsto q^{-\beta}$ is a monotonic function with $\lambda_{(q \mapsto q^{-\beta})} = \beta$, by Theorem 3.10 we have

$$\dim_{\mathrm{H}}(\mathcal{A}^{\mathbf{y}}_{n,m}(q \mapsto q^{-\beta})) = m(n-1) + \frac{m+n}{\beta+1} = s_0.$$

Therefore, $\dim_{\mathrm{H}}(\mathcal{A}^{\mathbf{y}}_{n,m}(\psi)) \geq s_0$ and it remains to show that $\dim_{\mathrm{H}}(\mathcal{A}^{\mathbf{y}}_{n,m}(\psi)) \leq s_0$.

As a consequence of Theorem 3.7 (and Remark 3.9), we only need to verify that for all $\delta > 0$ we have

$$\sum_{q=1}^{\infty} q^{n+m-1} \left(\frac{\psi(q)}{q}\right)^{-m(n-1)+s_0+\delta} < \infty$$

since this would imply that $\mathcal{H}^{s_0+\delta}(\mathcal{A}^{\mathbf{y}}_{n,m}(\psi)) = 0$ and $\dim_{\mathrm{H}}(\mathcal{A}^{\mathbf{y}}_{n,m}(\psi)) \leq s_0 + \delta$.

We note that

$$\begin{aligned}
&\sum_{q=1}^{\infty} q^{n+m-1}\left(\frac{\psi(q)}{q}\right)^{-m(n-1)+s_0+\delta} \\
&= \sum_{q\in J} q^{n+m-1}(q^{-\alpha-1})^{-m(n-1)+s_0+\delta} + \sum_{q\notin J} q^{n+m-1}(q^{-\beta-1})^{-m(n-1)+s_0+\delta} \\
&= \sum_{q\in J} q^{n+m-1-(\alpha+1)(s_0+\delta-m(n-1))} + \sum_{q\notin J} q^{n+m-1-(\beta+1)(s_0+\delta-m(n-1))}. \qquad (3.7)
\end{aligned}$$

We consider each of the terms on the right-hand side of (3.7) separately and show that each of them converges. We first consider the second sum on the right-hand side of (3.7). Since $\delta > 0$ we have $s_0 - m(n-1) < s_0 + \delta - m(n-1)$ and hence

$$n+m < \left(\frac{n+m}{s_0 - m(n-1)}\right)(s_0+\delta-m(n-1)).$$

Recalling that

$$\beta = \frac{n+m}{s_0-m(n-1)} - 1$$

it follows that

$$n+m-1-(\beta+1)(s_0+\delta-m(n-1)) < -1,$$

which is sufficient for the second sum on the right-hand side of (3.7) to converge.

For the first sum on the right-hand side of (3.7) we make the following observations. First of all, notice that

$$n+m-1-(\alpha+1)\left(\frac{n+m}{\beta+1}\right) = n+m-1-(\alpha+1)(s_0-m(n-1)).$$

Also note that

$$\frac{n+m-1}{\alpha+1} + m(n-1) < s_0 \quad \text{gives} \quad n+m-1-(\alpha+1)(s_0-m(n-1)) < 0.$$

Thus, provided that δ is sufficiently small,

$$\begin{aligned}
\sum_{q\in J} q^{n+m-1-(\alpha+1)(s_0+\delta-m(n-1))} &= \sum_{k=1}^{\infty} a_k^{n+m-1-(\alpha+1)(s_0+\delta-m(n-1))} \\
&= \sum_{k=1}^{\infty} \lceil k^{-\gamma}\rceil^{n+m-1-(\alpha+1)(s_0+\delta-m(n-1))} \\
&\le \sum_{k=1}^{\infty} (k^{-\gamma})^{n+m-1-(\alpha+1)(s_0+\delta-m(n-1))} \qquad (3.8)
\end{aligned}$$

as $n+m-1-(\alpha+1)(s_0+\delta-m(n-1)) < 0$ and $\gamma < 0$.

Now, for $\delta > 0$,

$$\begin{aligned}
\frac{2}{\gamma} &= n+m-1-(\alpha+1)(s_0-m(n-1)) \\
&> n+m-1-(\alpha+1)(s_0+\delta-m(n-1)).
\end{aligned}$$

Hence,

$$2 < \gamma(n+m-1-(\alpha+1)(s_0+\delta-m(n-1)))$$

since $\gamma < 0$. Therefore $(k^{-\gamma})^{n+m-1-(\alpha+1)(s_0+\delta-m(n-1))} < k^{-2}$ and so (3.8) converges. Consequently, since both the component sums converge, it follows that (3.7) converges, *i.e.*

$$\sum_{q=1}^{\infty} q^{n+m-1}\left(\frac{\psi(q)}{q}\right)^{-m(n-1)+s_0+\delta} < \infty,$$

and we conclude that $\dim_{\mathrm{H}}(\mathcal{A}_{n,m}^{\mathbf{y}}(\psi)) \leq s_0 + \delta$. The desired result follows upon noticing that $\delta > 0$ can be taken to be arbitrarily small. □

3.4. General statements. We conclude this section on the extension of the Mass Transference Principle to systems of linear forms by recording a couple of very general statements established in [**2**]. Recall that given an approximating function $\Psi : \mathbb{Z}^n \setminus \{\mathbf{0}\} \to \mathbb{R}^+$, which can depend on $\mathbf{q}$ rather than just $|\mathbf{q}|$, and a fixed inhomogeneous parameter $\mathbf{y} \in \mathbb{I}^m$, we define $\mathcal{A}_{n,m}^{\mathbf{y}}(\Psi)$ to be the set of $\mathbf{x} \in \mathbb{I}^{nm}$ such that

$$|\mathbf{q}\mathbf{x} + \mathbf{p} - \mathbf{y}| < \Psi(\mathbf{q})$$

for infinitely many $(\mathbf{p}, \mathbf{q}) \in \mathbb{Z}^m \times \mathbb{Z}^n \setminus \{\mathbf{0}\}$.

Considering the Ψ-approximable points we have the following statement.

THEOREM 3.15 (Allen – Beresnevich [**2**]). *Let $\Psi : \mathbb{Z}^n \setminus \{\mathbf{0}\} \to \mathbb{R}^+$ be an approximating function and let $\mathbf{y} \in \mathbb{I}^m$ be fixed. Let f and $g : r \mapsto g(r) = r^{-m(n-1)}f(r)$ be dimension functions such that $r^{-nm}f(r)$ is monotonic. Let*

$$\Theta : \mathbb{Z}^n \setminus \{\mathbf{0}\} \to \mathbb{R}^+ \qquad \textit{be defined by} \qquad \Theta(\mathbf{q}) := |\mathbf{q}|\, g\left(\frac{\Psi(\mathbf{q})}{|\mathbf{q}|}\right)^{\frac{1}{m}}$$

Then

$$|\mathcal{A}_{n,m}^{\mathbf{y}}(\Theta)| = 1 \qquad \textit{implies} \qquad \mathcal{H}^f(\mathcal{A}_{n,m}^{\mathbf{y}}(\Psi)) = \mathcal{H}^f(\mathbb{I}^{nm}).$$

Supposing we are interested in the case where we have approximating functions $\psi : \mathbb{N} \to \mathbb{R}^+$ which depend only on $|\mathbf{q}|$ (*i.e.* $\Psi(\mathbf{q}) = \psi(|\mathbf{q}|)$) we can extract the following statement as a corollary to Theorem 3.15.

THEOREM 3.16 (Allen – Beresnevich [**2**]). *Let $\psi : \mathbb{N} \to \mathbb{R}^+$ be an approximating function, let $\mathbf{y} \in \mathbb{I}^m$ be fixed, and let f and $g : r \mapsto g(r) = r^{-m(n-1)}f(r)$ be dimension functions such that $r^{-nm}f(r)$ is monotonic. Let*

$$\theta : \mathbb{N} \to \mathbb{R}^+ \qquad \textit{be defined by} \qquad \theta(r) := r\, g\left(\frac{\psi(r)}{r}\right)^{\frac{1}{m}}.$$

Then

$$|\mathcal{A}_{n,m}^{\mathbf{y}}(\theta)| = 1 \qquad \textit{implies} \qquad \mathcal{H}^f(\mathcal{A}_{n,m}^{\mathbf{y}}(\psi)) = \mathcal{H}^f(\mathbb{I}^{nm}).$$

It is observed in [**2**] that Theorems 3.4 and 3.7 follow as corollaries from Theorem 3.16. In fact, in some sense, Theorems 3.15 and 3.16 are fairly natural reformulations of Theorem 3.2 in terms of, respectively, Ψ and ψ-approximable points. In essentially the same way that Theorem 3.2 may be used to prove Theorem 3.15 a more general statement can also be obtained. Namely, suppose we are now given a function $\Psi : \mathbb{Z}^m \times \mathbb{Z}^n \setminus \{\mathbf{0}\} \to \mathbb{R}^+$ which can depend upon both $\mathbf{p}$ and $\mathbf{q}$. Furthermore, suppose we are also given fixed $\Phi \in \mathbb{I}^{mm}$ and $\mathbf{y} \in \mathbb{I}^m$. We denote by $\mathcal{M}_{n,m}^{\mathbf{y},\Phi}(\Psi)$ the set of $\mathbf{x} \in \mathbb{I}^{nm}$ for which

$$|\mathbf{q}\mathbf{x} + \mathbf{p}\Phi - \mathbf{y}| < \Psi(\mathbf{p}, \mathbf{q})$$

holds for $(\mathbf{p},\mathbf{q}) \in \mathbb{Z}^m \times \mathbb{Z}^n \setminus \{\mathbf{0}\}$ with $|\mathbf{q}|$ arbitrarily large.

The following statement, which actually includes Theorems 3.15 and 3.16, can be made.

THEOREM 3.17 (Allen – Beresnevich [**2**]). *Let $\Psi : \mathbb{Z}^m \times \mathbb{Z}^n \setminus \{\mathbf{0}\} \to \mathbb{R}^+$ be such that*

$$\lim_{|\mathbf{q}|\to\infty} \sup_{\mathbf{p}\in\mathbb{Z}^m} \frac{\Psi(\mathbf{p},\mathbf{q})}{|\mathbf{q}|} = 0\,,$$

and let $\mathbf{y} \in \mathbb{I}^m$ and $\Phi \in \mathbb{I}^{mm}$ be fixed. Let f and $g : r \mapsto g(r) = r^{-m(n-1)}f(r)$ be dimension functions such that $r^{-nm}f(r)$ is monotonic. Let

$$\Theta : \mathbb{Z}^m \times \mathbb{Z}^n \setminus \{\mathbf{0}\} \to \mathbb{R}^+ \qquad \textit{be defined by} \qquad \Theta(\mathbf{p},\mathbf{q}) = |\mathbf{q}|\, g\left(\frac{\Psi(\mathbf{p},\mathbf{q})}{|\mathbf{q}|}\right)^{\frac{1}{m}}.$$

Then

$$|\mathcal{M}^{\mathbf{y},\Phi}_{n,m}(\Theta)| = 1 \qquad \textit{implies} \qquad \mathcal{H}^f(\mathcal{M}^{\mathbf{y},\Phi}_{n,m}(\Psi)) = \mathcal{H}^f(\mathbb{I}^{nm})\,.$$

The above theorem not only allows us to consider the usual homogeneous and inhomogeneous settings of Diophantine approximation for systems of linear forms (see [**5**]) but also allows us to consider Hausdorff measure statements where we may have some restrictions on our "approximating points" $(\mathbf{p},\mathbf{q})$. As an example, recently Dani, Laurent and Nogueira have established Lebesgue measure "Khintchine–Groshev" type statements for sets of ψ-approximable points where they have imposed certain primitivity conditions on their "approximating points" [**14**]. In [**2**], Theorem 3.17 has been used to establish Hausdorff measure versions of these results.

4. Extension to rectangles

Another very natural situation, not covered by the setting of systems of linear forms, for which we might hope for some kind of mass transference principle is when our lim sup sets of interest are defined by sequences of rectangles. Recently some progress has been made in this direction by Wang, Wu and Xu [**64**]. For example, this is of interest when we consider weighted simultaneous approximation.

4.1. A mass transference principle from balls to rectangles. Throughout this section let $k \in \mathbb{N}$ and, as usual, denote by $\mathbb{I}^k$ the unit cube $[0,1]^k$ in $\mathbb{R}^k$. Given a ball $B = B(\mathbf{x}, r)$ in $\mathbb{R}^k$ of radius r centred at $\mathbf{x}$ and a k-dimensional real vector $\mathbf{a} = (a_1, a_2, \ldots, a_k)$ we will denote by $B^{\mathbf{a}}$ the rectangle with centre $\mathbf{x}$ and side-lengths $(r^{a_1}, r^{a_2}, \ldots, r^{a_k})$. Given a sequence $(\mathbf{x}_n)_{n\in\mathbb{N}}$ of points in $\mathbb{I}^k$ and a sequence $(r_n)_{n\in\mathbb{N}}$ of positive real numbers such that $r_n \to 0$ as $n \to \infty$ we define

$$W_0 = \{\mathbf{x} \in \mathbb{I}^k : \mathbf{x} \in B_n = B(\mathbf{x}_n, r_n) \text{ for infinitely many } n \in \mathbb{N}\}.$$

For any $\mathbf{a} \in \mathbb{R}^k$ we will also write

$$W_{\mathbf{a}} = \{\mathbf{x} \in \mathbb{I}^k : \mathbf{x} \in B_n^{\mathbf{a}} \text{ for infinitely many } n \in \mathbb{N}\}.$$

In [**64**], Wang, Wu and Xu established the following mass transference principle.

THEOREM 4.1 (Wang – Wu – Xu [**64**]). *Let $(\mathbf{x}_n)_{n\in\mathbb{N}}$ be a sequence of points in $\mathbb{I}^k$ and let $(r_n)_{n\in\mathbb{N}}$ be a sequence of positive real numbers such that $r_n \to 0$ as*

$n \to \infty$. *Let* $\mathbf{a} = (a_1, a_2, \ldots, a_k) \in \mathbb{R}^k$ *be such that* $1 \leq a_1 \leq a_2 \leq \cdots \leq a_k$. *Suppose that* $|W_0| = 1$. *Then,*

$$\dim_{\mathrm{H}} W_{\mathbf{a}} \geq \min_{1 \leq j \leq k} \left\{ \frac{k + ja_j - \sum_{i=1}^{j} a_i}{a_j} \right\}.$$

Furthermore, if we have the additional constraint $a_d > 1$, Wang, Wu and Xu are also able to say something about the Hausdorff measure of $W_{\mathbf{a}}$ at the critical value

$$s := \min_{1 \leq j \leq k} \left\{ \frac{k + ja_j - \sum_{i=1}^{j} a_i}{a_j} \right\}. \tag{4.1}$$

THEOREM 4.2 (Wang – Wu – Xu [**64**]). *Assume the same conditions as in Theorem* 4.1. *If the additional constraint that* $a_d > 1$ *holds, then*

$$\mathcal{H}^s(W_{\mathbf{a}}) = \infty.$$

Essentially, the results of Wang, Wu and Xu allow us to pass from a full Lebesgue measure statement for a lim sup set defined by a sequence of balls to a Hausdorff measure statement for a lim sup set defined by an associated sequence of rectangles. As an application, Wang, Wu and Xu demonstrate how Theorem 4.1 may be applied to obtain the Hausdorff dimension of the following set of weighted simultaneously well-approximable points. Let $\boldsymbol{\tau} = (\tau_1, \tau_2, \ldots, \tau_k) \in \mathbb{R}^k$ be such that $\tau_i > 0$ for $1 \leq i \leq k$ and denote by $W_k(\boldsymbol{\tau})$ the set of points $\mathbf{x} = (x_1, x_2, \ldots, x_k) \in \mathbb{I}^k$ such that

$$|qx_i + p_i| < q^{-\tau_i}, \quad 1 \leq i \leq k, \tag{4.2}$$

for infinitely many $(\mathbf{p}, q) \in \mathbb{Z}^k \times \mathbb{N}$. The following is derived in [**64**] as a corollary to Theorem 4.1.

COROLLARY 4.3 (Wang – Wu – Xu [**64**]). *Let* $\boldsymbol{\tau} = (\tau_1, \tau_2, \ldots, \tau_k) \in \mathbb{R}^k$ *be such that* $\frac{1}{k} \leq \tau_1 \leq \tau_2 \leq \cdots \leq \tau_k$, *then*

$$\dim_{\mathrm{H}}(W_k(\boldsymbol{\tau})) = \min_{1 \leq j \leq k} \left\{ \frac{k + 1 + j\tau_j - \sum_{i=1}^{j} \tau_i}{1 + \tau_j} \right\}.$$

While the proof of Corollary 4.3 given in [**64**] is novel, and is a neat application of Theorem 4.1, the result itself was already previously known. In fact, Corollary 4.3 is a special case of an earlier more general theorem due to Rynne [**58**] which we now state.

Suppose Q is an arbitrary infinite set of natural numbers and, given $\boldsymbol{\tau} \in \mathbb{R}^k$, let $W_k^Q(\boldsymbol{\tau})$ denote the set of points $\mathbf{x} \in \mathbb{I}^k$ for which the inequalities in (4.2) hold for infinitely many pairs $(\mathbf{p}, q) \in \mathbb{Z}^k \times Q$, hence $W_k^{\mathbb{N}}(\boldsymbol{\tau}) = W_k(\boldsymbol{\tau})$. Define

$$\nu(Q) = \inf \left\{ \nu \in \mathbb{R} : \sum_{q \in Q} q^{-\nu} < \infty \right\}$$

and let $\sigma(\boldsymbol{\tau}) = \sum_{i=1}^{k} \tau_i$.

THEOREM 4.4 (Rynne [**58**]). *Let $\boldsymbol{\tau} = (\tau_1, \tau_2, \dots, \tau_k) \in \mathbb{R}^k$ be such that $0 < \tau_1 \leq \tau_2 \leq \cdots \leq \tau_k$. Let Q be an arbitrary infinite subset of $\mathbb{N}$ and suppose that $\sigma(\boldsymbol{\tau}) \geq \nu(Q)$. Then,*

$$\dim_{\mathrm{H}} W_k^Q(\boldsymbol{\tau}) = \min_{1 \leq j \leq k} \left\{ \frac{k + \nu(Q) + j\tau_j - \sum_{i=1}^{j} \tau_i}{1 + \tau_j} \right\}.$$

We may easily recover Corollary 4.3 by taking $Q = \mathbb{N}$ in Theorem 4.4 and noting that $\nu(\mathbb{N}) = 1$. Since the hypotheses of Corollary 4.3 demand that $\tau_i \geq \frac{1}{k}$ for all $1 \leq i \leq k$, we see that the condition $\sigma(\boldsymbol{\tau}) \geq \nu(Q)$ in Theorem 4.4 is also satisfied.

Sets such as $W_k(\boldsymbol{\tau})$ and variations on $W_k^Q(\boldsymbol{\tau})$ have been studied in some depth, with particular attention paid to the question of determining their Hausdorff dimension, even before the work of Rynne [**58**]. For example, consider $\tau \in \mathbb{R}$ for some $\tau > 1$. Then the set $W_1^{\mathbb{N}}(\tau) = W_1(\tau)$ coincides precisely with the set $\mathcal{A}(\tau)$ considered in the Jarník–Besicovitch Theorem (Theorem 2.8). For an overview of some other earlier work in this direction we direct the reader to the discussion given in [**58**] and references therein.

4.2. Rectangles to rectangles. The original Mass Transference Principle (Theorem 2.5) allows us to transition from Lebesgue to Hausdorff measure statements when our original and "transformed" lim sup sets are defined by sequences of balls, *i.e.* it allows us to go from "balls to balls". Theorem 4.1 allows us to go from "balls to rectangles". Another goal which we might like to achieve, which is not covered by any of the frameworks mentioned so far, would be to prove a similar mass transference principle where we both start and finish with lim sup sets arising from sequences of rectangles, *i.e.* from "rectangles to rectangles".

PROBLEM 4.5. *Does there exist a mass transference principle, similar to Theorem 2.5 or Theorem 4.1, where both the original and transformed* lim sup *sets are defined by sequences of rectangles?*

Although in the most general settings this problem remains open, we survey what can be said in a few special cases.

In [**9**] Beresnevich and Velani employ a "slicing" technique, which uses a combination of a slicing lemma and the original Mass Transference Principle, to prove Theorem 3.1. We show how an appropriate combination of these two results can also be applied to considering the problem of proving a mass transference principle for rectangles. We proceed by stating the "Slicing Lemma" as given by Beresnevich and Velani in [**9**].

LEMMA 4.6 (Slicing Lemma [**9**]). *Let $l, k \in \mathbb{N}$ be such that $l \leq k$ and let f and $g : r \mapsto r^{-l} f(r)$ be dimension functions. Let $A \subset \mathbb{R}^k$ be a Borel set and let V be a $(k-l)$-dimensional linear subspace of $\mathbb{R}^k$. If for a subset S of $V^{\perp}$ of positive $\mathcal{H}^l$-measure*

$$\mathcal{H}^g(A \cap (V + b)) = \infty \quad \text{for all } b \in S,$$

then $\mathcal{H}^f(A) = \infty$.

Suppose that $(\mathbf{x}_n)_n = (x_{n,1}, x_{n,2}, \dots, x_{n,k})_n$ is a sequence of points in $[0,1]^k$. Let $(r_n^1)_n, (r_n^2)_n, \dots, (r_n^k)_n$ be sequences of positive real numbers and suppose that

$r^1_n \to 0$ as $n \to \infty$. Let

$$H_n = \prod_{i=1}^{k} B(x_{n,i}, r^i_n)$$

be a sequence of rectangles in $[0,1]^k$, where $\prod_{i=1}^k A_i = A_1 \times A_2 \times \cdots \times A_k$ is the Cartesian product of subsets A_i of $\mathbb{R}^k$. Let $\alpha > 1$ be a real number and define another sequence of rectangles by

$$h_n = B(x_{n,1}, (r^1_n)^\alpha) \times \prod_{i=2}^{k} B(x_{n,i}, r^i_n).$$

So, h_n is essentially a "shrunk" rectangle corresponding to H_n from the original sequence. Note that in this case we only allow shrinking of the original rectangle in one direction. Then, we are able to establish the following.

PROPOSITION 4.7. *Let the sequences H_n and h_n be as given above and further suppose that* $|\limsup_{n\to\infty} H_n| = 1$. *Then,*

$$\dim_{\mathrm{H}}\left(\limsup_{n\to\infty} h_n\right) \geq \frac{1}{\alpha} + k - 1.$$

PROOF. Let $V = \{\mathbf{x} = (x_1, \dots, x_k) \in [0,1]^k : x_i = 0 \quad \text{for all } i \neq 1\}$. Since $|\limsup_{n\to\infty} H_n| = 1$, for Lebesgue almost every

$$\mathbf{b} \in \{\mathbf{x} = (x_1, \dots, x_k) \in [0,1]^k : x_1 = 0\}$$

we have

$$|(V + \mathbf{b}) \cap \limsup_{n\to\infty} H_n| = 1.$$

Let us fix a $\mathbf{b}$ for which this holds and let $W = V + \mathbf{b}$. Now, $\limsup_{n\to\infty} H_n \cap W$ can be written as the lim sup set of a sequence of balls $B_j = B(x_{n_j,1}, r^1_{n_j})$ with radii $r^1_{n_j}$. Note that $|\limsup_{j\to\infty} B_j \cap W| = 1$. For each j also let $b_j = B(x_{n_j,1}, (r^1_{n_j})^\alpha)$ and note that

$$\limsup_{j\to\infty} b_j \cap W = \limsup_{n\to\infty} h_n \cap W.$$

In accordance with our earlier notation, $b^s_j = B(x_{n_j,1}, (r^1_{n_j})^{\alpha s})$. Therefore, if $s \leq \frac{1}{\alpha}$ then $(r^1_{n_j})^{\alpha s} \geq r^1_{n_j}$ for sufficiently large j and so

$$b^s_j \supseteq B_j \quad \text{and} \quad |\limsup_{j\to\infty} b^s_j \cap W| = 1.$$

Thus, for any $s \leq \frac{1}{\alpha}$ we may use the Mass Transference Principle to conclude that for any ball $B \subseteq W$ we have

$$\mathcal{H}^s(\limsup_{j\to\infty} b_j \cap B) = \mathcal{H}^s(B).$$

In particular, since $s \leq \frac{1}{\alpha} < 1$, this means

$$\mathcal{H}^s(\limsup_{n\to\infty} h_n \cap W) = \mathcal{H}^s(W) = \infty.$$

Since this is the case for Lebesgue almost every $\mathbf{b} \in \{\mathbf{x} = (x_1, \dots, x_k) : x_1 = 0\}$, we can use the Slicing Lemma (Lemma 4.6) to conclude that

$$\mathcal{H}^{s'}(\limsup_{n\to\infty} h_n) = \infty$$

for all $s' \leq \frac{1}{\alpha} + k - 1$. Therefore, it follows that

$$\dim_{\mathrm{H}}\left(\limsup_{n\to\infty} h_n\right) \geq \frac{1}{\alpha} + k - 1.$$

□

Using Theorem 4.2 in place of Theorem 2.5, we are actually able to extend this argument a little further. Again, let $(\mathbf{x}_n)_n = (x_{n,1}, x_{n,2}, \ldots, x_{n,k})_n$ be a sequence of points in $[0,1]^k$ and let $(r_n^1)_n, (r_n^2)_n, \ldots, (r_n^k)_n$ be sequences of positive real numbers. Suppose that for some $1 \leq k_0 \leq k$ we have $r_n^1 = r_n^2 = \cdots = r_n^{k_0}$ for all $n \in \mathbb{N}$ and also that $r_n^1 \to 0$ as $n \to \infty$. Let

$$H_n = \prod_{i=1}^{k} B(x_{n,i}, r_n^i)$$

be a sequence of rectangles in $[0,1]^k$. Next, let $1 \leq a_1 \leq a_2 \leq \cdots \leq a_{k_0}$ be real numbers and suppose $a_{k_0} > 1$. For each rectangle H_n in our original sequence we define a corresponding "shrunk" rectangle

$$h_n = \prod_{i=1}^{k_0} B(x_{n,i}, (r_n^i)^{a_i}) \times \prod_{i=k_0+1}^{k} B(x_{n,i}, r_n^i).$$

In this case we are able to prove the following.

PROPOSITION 4.8. *Let the sequences of rectangles H_n and h_n be as given above and further suppose that $|\limsup_{n\to\infty} H_n| = 1$. Then,*

$$\dim_{\mathrm{H}}\left(\limsup_{n\to\infty} h_n\right) \geq \min_{1\leq j\leq k_0}\left\{\frac{k_0 + ja_j - \sum_{i=1}^{j} a_i}{a_j} + k - k_0\right\}.$$

PROOF. Let $V = \{\mathbf{x} = (x_1, x_2, \ldots, x_k) \in [0,1]^k : x_i = 0 \text{ for all } i \geq k_0 + 1\}$. Since $|\limsup_{n\to\infty} H_n| = 1$, for almost every

$$\mathbf{b} \in \{\mathbf{x} = (x_1, x_2, \ldots, x_k) \in [0,1]^k : x_i = 0 \text{ for all } i \leq k_0\}$$

we have

$$|(V + \mathbf{b}) \cap \limsup_{n\to\infty} H_n| = 1.$$

Let us fix a $\mathbf{b}$ for which this holds and let $W = V + \mathbf{b}$. As before, $\limsup_{n\to\infty} H_n \cap W$ can be written as a sequence of k_0-dimensional balls $B_j = B(\mathbf{x}_{n_j}^{k_0}, r_{n_j}^1)$ with radii $r_{n_j}^1 (= r_{n_j}^2 = \cdots = r_{n_j}^{k_0})$ and centres $\mathbf{x}_{n_j}^{k_0} = (x_{n_j,1}, x_{n_j,2}, \ldots, x_{n_j,k_0})$. Note that $|\limsup_{j\to\infty} B_j \cap W| = 1$.

This time, for each j let

$$b_j = \prod_{i=1}^{k_0} B(x_{n_j,i}, (r_{n_j}^i)^{a_i})$$

and note that

$$\limsup_{j\to\infty} b_j \cap W = \limsup_{n\to\infty} h_n \cap W.$$

By Theorem 4.2 it follows that

$$\mathcal{H}^s(\limsup_{n\to\infty} h_n \cap W) = \infty$$

where

$$s := \min_{1 \le j \le k_0} \left\{ \frac{k_0 + j a_j - \sum_{i=1}^{j} a_i}{a_j} \right\}.$$

Since this is the case for almost every

$$\mathbf{b} \in \{\mathbf{x} = (x_1, x_2, \ldots, x_k) \in [0,1]^k : x_i = 0 \text{ for all } i \le k_0\}$$

we may use Lemma 4.6 (with $l = k - k_0$) to conclude that

$$\mathcal{H}^{s'}(\limsup_{n \to \infty} h_n) = \infty$$

where

$$s' := \min_{1 \le j \le k_0} \left\{ \frac{k_0 + j a_j - \sum_{i=1}^{j} a_i}{a_j} + k - k_0 \right\}.$$

Hence

$$\dim_{\mathrm{H}} \left(\limsup_{n \to \infty} h_n \right) \ge s',$$

as required. □

A disadvantage of using the "slicing" arguments above is that we have to impose quite strict conditions on both the original and transformed rectangles. Namely, the sides of the original rectangle which are permitted to "shrink" have to be of the same initial length (but can shrink at different rates). Meanwhile, the rest of the sides of the original rectangle are not allowed to "shrink" at all when passing to the corresponding transformed rectangle. We conclude this section by considering one more situation where all sides of the original rectangles may have different lengths and are all allowed to "shrink" in a specified manner. Let

$$H_n = \prod_{i=1}^{k} B(\mathbf{x}_{n,i}, r_n^{t_i})$$

be a sequence of rectangles in $[0,1]^k$ with $1 < t_i$ for $1 < i < k$ and $r_n \to 0$.

Let the corresponding "shrunk" rectangles be defined as

$$h_n = \prod_{i=1}^{k} B(\mathbf{x}_{n,i}, r_n^{a_i t_i}),$$

where $1 \le a_i$ for $1 \le i \le k$. Suppose without loss of generality that $1 \le a_1 t_1 \le a_2 t_2 \le \cdots \le a_k t_k$. Furthermore, suppose that

$$D := \inf \left\{ d \in \mathbb{R} : \sum_{n=1}^{\infty} r_n^d < \infty \right\}.$$

By using the "natural" covers of $\limsup_{n \to \infty} h_n$ we can get an upper bound for the Hausdorff dimension of this lim sup set; namely, we see that

$$\dim_{\mathrm{H}} \left(\limsup_{n \to \infty} h_n \right) \le \min_{1 \le j \le k} \left\{ \frac{D + j a_j t_j - \sum_{i=1}^{j} a_i t_i}{a_j t_j} \right\}. \tag{4.3}$$

PROBLEM 4.9. *Under what conditions do we get a lower bound which coincides with the upper bound given above?*

REMARK. *Throughout this section we have only considered* lim sup *sets of rectangles which are all aligned. It would also be natural to consider situations where this is not necessarily the case.*

5. Random Mass Transference Principles

It is a well known phenomenon that introducing randomness to a construction can simplify results by "smoothing" out almost impossible values in the probability space that cause problems in deterministic settings. In this section we will summarise recent progress on random analogues of the statements presented in the preceding sections. We note that the assumptions required are much weaker but with the caveat that randomness has to be introduced somewhere and precise number theoretic results cannot be recovered. The random covering sets that we will mention, as well as the random and deterministic sets we will relate to lim sup sets, have a long history of their own. While we highlight their connection to the lim sup sets mentioned in the previous sections and focus on their similarities, we note that the methods used in their proofs differ quite substantially.

We first consider a problem known as the (random) moving target problem. Let (X, μ) be a probability space, where X is a complete metric space. Let $\{B_i\}_{i\in\mathbb{N}} = \{B(x, r_i)\}_{i\in\mathbb{N}}$ be a sequence of balls centred at $x \in X$ such that $r_i \to 0$ as $i \to \infty$. We are interested in the following question.

PROBLEM 5.1. *Let $\{\widetilde{B}_i\}_{i\in\mathbb{N}} = \{B(x + a_i, r_i)\}_{i\in\mathbb{N}}$ be a sequence of balls with random centres $x + a_i$, where $a_i \in X$ are chosen independently according to the probability measure μ. Under what conditions can we deduce measure statements for the* lim sup *set $E(B_i) = \limsup_{i\to\infty} \widetilde{B}_i$?*

If $X = \mathbb{T}^1$ is the circle and μ is the uniform measure, one answer to that question should be familiar. It is the Borel–Cantelli Lemma.

LEMMA 5.2 (Borel–Cantelli Lemma). *Let $X = \mathbb{T}^1$ and let $\{B_i\}_{i\in\mathbb{N}} = \{B(x, r_i)\}_{i\in\mathbb{N}}$ be a sequence of balls centred at $x \in X$ such that $r_i \to 0$ as $i \to \infty$. Let $(a_i)_{i\in\mathbb{N}}$ be a sequence of random translations chosen independently according to the uniform measure μ. Then, we again consider $E(B_i) = \limsup_{i\to\infty} \widetilde{B}_i$ and, for almost every choice of sequence $(a_i)_{i\in\mathbb{N}}$ with respect to the product measure $\mu^{\mathbb{N}}$, we have*

$$|E(B_i)| = \begin{cases} 0 & \text{if } \sum_{i=1}^{\infty} r_i < \infty, \\ \\ 1 & \text{if } \sum_{i=0}^{\infty} r_i = \infty. \end{cases}$$

Note that the first implication, *i.e.* that the sum being finite implies zero Lebesgue measure, holds surely for any arbitrary sequence $(a_i)_{i\in\mathbb{N}}$. In particular, the a_i do not have to be chosen randomly. Using randomness though, we can make a more precise statement about the Hausdorff dimension when the lim sup set is Lebesgue null.

THEOREM 5.3 (Fan – Wu [**27**], Durand [**19**]). *Let $X = \mathbb{T}^1$ and let $\{B_i\}_{i\in\mathbb{N}}$ be a sequence of balls in X with radii r_i such that $r_i \to 0$ as $i \to \infty$. Given this sequence of radii, assume that $|E(B_i)| = 0$ for almost every sequence of uniformly chosen translations $(a_i)_{i\in\mathbb{N}} \subset \mathbb{T}^1$. Then, for almost all sequences of random translations,*

$$\dim_H E(B_i) = \min\{1, s_0\},$$

where

$$s_0 = \inf\left\{ s > 0 \; : \; \sum_{i=0}^{\infty} r_i^s < \infty \right\}. \tag{5.1}$$

Related to such statements are results in fractal geometry. We write $\mathcal{I}_0 = \{T_1, T_2, \ldots, T_N\}$ for a finite collection of contracting similarity maps on $\mathbb{R}^k$, *i.e.*

$$\|T_i(x) - T_i(y)\| = c_i\|x - y\|$$

for some $0 < c_i < 1$ for each $1 \leq i \leq N$ for all $x, y \in \mathbb{R}^k$, where $\|.\|$ is the Euclidean norm. The "best guess" for the Hausdorff dimension of the unique compact invariant attractor $F \subset \mathbb{R}^k$ satisfying $F = \bigcup_i T_i(F)$ is the *similarity dimension.* The similarity dimension is the minimum of the dimension of the ambient space and the unique exponent, s_0, satisfying the Hutchinson–Moran formula

$$\sum_{i=1}^{N} c_i^{s_0} = 1, \tag{5.2}$$

see [**35, 52**]. Its relation to (5.1) can be seen by writing F as a lim sup set

$$F = \bigcap_{l=1}^{\infty} \bigcup_{i=l}^{\infty} \bigcup_{j_1, j_2, \ldots, j_i \in \{1, \ldots, N\}} T_{j_1} \circ T_{j_2} \circ \cdots \circ T_{j_i}(\Delta),$$

where $\Delta = [-c, c]^k$ for some large enough $c \in \mathbb{R}$ such that $F \subseteq \Delta$. Notice that $\sum_j c_j^{s_0+\delta} < 1$ for any $\delta > 0$ and that $\operatorname{diam}(T_{j_1} \circ \cdots \circ T_{j_i}(\Delta)) = c_{j_1} \ldots c_{j_i} c$. So,

$$\begin{aligned}
\sum_{i=1}^{\infty} \sum_{j_1, j_2, \ldots, j_i \in \{1, \ldots, N\}} & \operatorname{diam}(T_{j_1} \circ T_{j_2} \circ \cdots \circ T_{j_i}(\Delta))^{s_0+\delta} \\
&= c^{s_0+\delta} \sum_{i=1}^{\infty} \sum_{j_1, j_2, \ldots, j_i \in \{1, \ldots, N\}} c_{j_1}^{s_0+\delta} c_{j_2}^{s_0+\delta} \ldots c_{j_i}^{s_0+\delta} \\
&= c^{s_0+\delta} \sum_{i=1}^{\infty} \left(\sum_j c_j^{s_0+\delta} \right)^i < \infty
\end{aligned} \tag{5.3}$$

using multiplicativity. Similarly, if $\delta < 0$ the sum above diverges and the similarity dimension s_0 in (5.3) coincides with the expression in (5.1). We would typically expect the similarity dimension to coincide with the Hausdorff dimension for these sets. However, this is not true in general in the deterministic setting and randomisation is one mechanism by which one can get an almost sure equality. We refer the reader to the wide literature on dimension theory of random and deterministic attractors [**23, 24, 51**], see also [**62**] for an overview of self-similar random sets.

Naturally, one is interested in higher dimensional analogues and relaxing the conditions on the covering set $E(B_i)$. Let $X = \mathbb{T}^k$ and let $\Delta \subset [0,1]^k$ have non-empty interior. Let $T_i : \mathbb{R}^k \to \mathbb{T}^k$ be a linear contraction with singular values $\sigma_1(T_i) \geq \sigma_2(T_i) \geq \cdots \geq \sigma_k(T_i)$. Recall that $\sigma_j(T_i)$ is the length of the jth longest principal semi-axis of the ellipsoid $T_i(B(\mathbf{0}, 1))$. We define the *singular value function*

$\Phi^t(T_i)$ by

$$\Phi^t(T_i) = \begin{cases} \sigma_1(T_i)\sigma_2(T_i)\ldots\sigma_n(T_i)^{t-n+1} & \text{for } n \leq t+1 < n+1 \text{ and } t < k, \\ \sigma_1(T_i)\sigma_2(T_i)\ldots\sigma_k(T_i)^t & \text{for } t \geq k. \end{cases}$$

The Hausdorff dimension of the natural lim sup set appearing in this setting is related to the behaviour of the singular value function.

THEOREM 5.4 (Järvenpää – Järvenpää – Koivusalo – Li – Suomala [**40**]). *Let $(T_i)_{i\in\mathbb{N}}$ be a sequence of maps as above with $\sigma_j(T_i) \to 0$ as $i \to \infty$ for all j. Set*

$$E(T_i) := \limsup_i (T_i(\Delta) + a_i),$$

where $a_i \in \mathbb{T}^k$ is a translation chosen independently according to the Lebesgue measure on $\mathbb{T}^k$. Then, almost surely,

$$\dim_{\mathrm{H}} E(T_i) = \inf\left\{0 < t \leq k \;:\; \sum_{i=1}^{\infty} \Phi^t(T_i) < \infty\right\}. \tag{5.4}$$

In particular, the sets can now be chosen to be rectangles, as opposed to balls. Indeed, even in the deterministic setting considered by Wang, Wu and Xu [**64**] the expression they obtain, namely (4.1), coincides with (5.4). We see this expression appearing yet again in the upper bound (4.3).

The singular value function was first used by Falconer in determining the Hausdorff dimension of self-affine sets [**22**]. Recall that a map is affine if it can be written as $\mathbf{M}x+\mathbf{v}$ for some non-singular matrix $\mathbf{M} \in \mathbb{R}^{k\times k}$ and some vector $\mathbf{v} \in \mathbb{R}^k$. Analogously to the self-similar case, if one considers the unique compact attractor F of a finite collection $\mathcal{I}$ of affine contractions, the "best guess" for the Hausdorff dimension of F is the *affinity dimension* given by the unique value $s \geq 0$ such that

$$\sum_{T\in\mathcal{I}} \Phi^s(T) = 1.$$

In the case where we are given fixed maps and randomly chosen translation vectors the Hausdorff dimension and affinity dimension coincide, see Falconer [**22**]. More recently, it was shown by Bárány, Käenmäki and Koivusalo [**3**] that one could alternatively randomise the matrices defining the maps while keeping the translation vectors fixed. The problem of determining exact conditions under which self-affine sets have Hausdorff dimension equal to affinity dimension is still open and much progress has been made towards resolving it; see a recent survey by Falconer [**25**] and [**26, 44, 54**] (and references within) for the deterministic setting, and [**30, 32, 34, 41–43, 45, 63**] for the random setting.

Dropping the linearity of the maps, T_i, Persson [**55**] proved a lower bound for the Hausdorff dimension of lim sup sets of open sets.

THEOREM 5.5 (Persson [**55**]). *Let $(A_i)_{i\in\mathbb{N}}$ be a sequence of open sets in $\mathbb{T}^k$. Let V be the Riemannian volume on $\mathbb{T}^k$ and let*

$$g_s(A_i) = \frac{|A_i|^2}{\mathcal{E}^s(A_i)}, \quad \text{where} \quad \mathcal{E}^s(A_i) = \iint_{A_i\times A_i} \frac{dV(x)\,dV(y)}{|x-y|^s}$$

is the s-energy of A_i. Then, for the lim sup *set $E(A_i)$ we obtain,*

$$\dim_{\mathrm{H}} E(A_i) \geq \inf \left\{ 0 < s \leq k \ : \ \sum_{i=1}^{\infty} g_s(A_i) < \infty \right\}.$$

Now consider the following general set up. Let U and V be open subsets of $\mathbb{R}^k$ and let $T : U \times V \to \mathbb{R}^k$ be a C^1 map such that $T(\cdot, y) : U \to \mathbb{R}^k$ and $T(x, \cdot) : V \to \mathbb{R}^k$ are diffeomorphisms for all $x \in U$ and $y \in V$. Let D_1T and D_2T be the derivatives of $T(\cdot, y)$ and $T(x, \cdot)$, respectively. Assume that

$$\|D_iT(x,y)\| \leq C_u \quad \text{and} \quad \|(D_iT(x,y))^{-1}\| \leq C_u \tag{5.5}$$

for some uniform $C_u > 0$ and all $i \in \{1, 2\}$. Let $(A_i)_{i\in\mathbb{N}}$ be a sequence of subsets of V and let $(a_i)_{i\in\mathbb{N}}$ be a sequence of points in U. The function T defines an interaction between a "generalised translation" a_i and a set A_i and embeds them without "too much distortion" into $\mathbb{R}^k$. Let $E(T, a_i, A_i) = \limsup_{i\to\infty} T(a_i, A_i)$. Note that for $T(a_i, y) = x + a_i + y$ this is equivalent to the translates setting considered above. Feng et al. [**29**] proved a (random) mass transference type statement in this general set up.

THEOREM 5.6 (Feng – Järvenpää – Järvenpää – Suomala [**29**]). *Let f be a dimension function and for each $i \in \mathbb{N}$ let $a_i \in U$ and let $A_i \subset \Delta \subset V$, where Δ is compact. Then*

$$\sum_{i=1}^{\infty} \mathcal{H}^f_\infty(A_i) < \infty \quad \textit{implies} \quad \mathcal{H}^f(E(T, a_i, A_i)) = 0.$$

Let μ be a measure on U that is not entirely singular with respect to the Lebesgue measure (see [**51**] for a definition). We denote the natural product measure on all sequences with entries in U by $\mathbb{P} = \mu^{\mathbb{N}}$ and now choose a sequence $(a_i)_{i\in\mathbb{N}}$ according to $\mathbb{P}$. Let

$$G_f(F) = \sup\{g_f(L) \ : \ L \subset F \text{ and } L \text{ is Lebesgue measurable with } |L| > 0\},$$

where g_f is the natural extension of g_s to dimension functions f,

$$g_f(A_i) = \frac{|A_i|^2}{\mathcal{E}^f(A_i)}, \quad \text{where} \quad \mathcal{E}^f(A_i) = \iint_{A_i\times A_i} \frac{dV(x)\,dV(y)}{f(|x-y|)}.$$

THEOREM 5.7 (Feng – Järvenpää – Järvenpää – Suomala [**29**]). *Suppose the same assumptions as in Theorem* 5.6. *Provided that $\mathcal{E}^f(B(0,R)) < \infty$ for all $R > 0$ and the A_i are Lebesgue measurable, then*

$$\sum_{i=1}^{\infty} G_f(A_i) = \infty \quad \textit{implies} \quad \mathcal{H}^f(E(T, a_i, A_i)) = \infty \quad \textit{for} \quad \mathbb{P}\textit{-a.e.}\ (a_i)_{i\in\mathbb{N}} \in U^{\mathbb{N}}.$$

Finally, a set $L \subset \mathbb{R}^k$ has positive Lebesgue density if

$$\liminf_{r\to 0} \frac{|L \cap B(x,r)|}{|B(x,r)|} > 0$$

for all $x \in L$.

THEOREM 5.8 (Feng – Järvenpää – Järvenpää – Suomala [**29**]). *Let f be a dimension function and recall that $V \subset \mathbb{R}^k$. Assume that $r^{-k+\varepsilon}f(r)$ is decreasing in r for some $\varepsilon > 0$. Let h be a dimension function such that $h(r) \leq f(r)^{1+\delta}$ for some $\delta > 0$ and all $r > 0$. Under the same assumptions as in Theorem* 5.6, *and* *provided that the A_i are Lebesgue measurable with positive Lebesgue density, we obtain*

$$\sum_{i=1}^{\infty} G_f(A_i) < \infty \quad \textit{implies} \quad \sum_{i=1}^{\infty} \mathcal{H}^h_\infty(A_i) < \infty.$$

As one can readily see, these latter results hold for lim sup sets of very general subsets. However, we still require positive Lebesgue density and a "nice" measure that is not singular with respect to the Lebesgue measure. Recently, Seuret [**60**] has made advances in relaxing these conditions on the measures by considering random lim sup sets with random centres chosen according to an arbitrary Borel measure and formulating their results in terms of multifractal formalism. This, in turn, was more recently extended by Ekström and Persson [**21**].

Acknowledgements. A major portion of this manuscript was prepared when ST visited DA at the University of York in December 2016. ST thanks York for their hospitality during his stay. DA would like to thank Victor Beresnevich for a number of interesting discussions prompting some of the original results presented in this article. Both authors are grateful to Victor Beresnevich, Henna Koivusalo and Sanju Velani for their helpful comments on earlier drafts of this article. They would also like to thank the referee for providing further helpful comments and thank both the referee and Stéphane Seuret for suggesting several additional references. The authors are particularly grateful to Stéphane Seuret for drawing our attention to the work of Jaffard which had previously escaped our notice.
At the time of writing DA and ST were supported, respectively, by EPSRC Doctoral Training Grants EP/M506680/1 and EP/K503162/1.

References

[1] D. Allen and S. Baker, *A General Mass Transference Principle*, Selecta Math **25** (2019), no. 3, DOI /10.1007/s00029-019-0484-9

[2] D. Allen and V. Beresnevich, *A mass transference principle for systems of linear forms and its applications*, Compos. Math. **154** (2018), no. 5, 1014–1047, DOI 10.1112/S0010437X18007121. MR3798593

[3] B. Bárány, A. Käenmäki, and H. Koivusalo, *Dimension of self-affine sets for fixed translation vectors*, J. Lond. Math. Soc. (2) **98** (2018), no. 1, 223–252, DOI 10.1112/jlms.12132. MR3847239

[4] J. Barral and S. Seuret, *Heterogeneous ubiquitous systems in $\mathbb{R}^d$ and Hausdorff dimension*, Bull. Braz. Math. Soc. (N.S.) **38** (2007), no. 3, 467–515, DOI 10.1007/s00574-007-0056-z. MR2344210

[5] V. Beresnevich, V. Bernik, M. Dodson, and S. Velani, *Classical metric Diophantine approximation revisited*, Analytic number theory, Cambridge Univ. Press, Cambridge, 2009, pp. 38–61. MR2508636

[6] V. Beresnevich, D. Dickinson, and S. Velani, *Measure theoretic laws for lim sup sets*, Mem. Amer. Math. Soc. **179** (2006), no. 846, x+91, DOI 10.1090/memo/0846. MR2184760

[7] V. Beresnevich, F. Ramírez, and S. Velani, *Metric Diophantine approximation: aspects of recent work*, Dynamics and analytic number theory, London Math. Soc. Lecture Note Ser., vol. 437, Cambridge Univ. Press, Cambridge, 2016, pp. 1–95. MR3618787

[8] V. Beresnevich and S. Velani, *A mass transference principle and the Duffin-Schaeffer conjecture for Hausdorff measures*, Ann. of Math. (2) **164** (2006), no. 3, 971–992, DOI 10.4007/annals.2006.164.971. MR2259250

[9] V. Beresnevich and S. Velani, *Schmidt's theorem, Hausdorff measures, and slicing*, Int. Math. Res. Not., posted on 2006, Art. ID 48794, 24, DOI 10.1155/IMRN/2006/48794. MR2264714

[10] V. Beresnevich and S. Velani, *Classical metric Diophantine approximation revisited: the Khintchine-Groshev theorem*, Int. Math. Res. Not. IMRN **1** (2010), 69–86, DOI 10.1093/imrn/rnp119. MR2576284

[11] V. I. Bernik and M. M. Dodson, *Metric Diophantine approximation on manifolds*, Cambridge Tracts in Mathematics, vol. 137, Cambridge Univ. Press, Cambridge, 1999. MR1727177

[12] A. S. Besicovitch, *On the sum of digits of real numbers represented in the dyadic system*, Math. Ann. **110** (1935), no. 1, 321–330, DOI 10.1007/BF01448030. MR1512941

[13] J. D. Bovey and M. M. Dodson, *The Hausdorff dimension of systems of linear forms*, Acta Arith. **45** (1986), no. 4, 337–358, DOI 10.4064/aa-45-4-337-358. MR847294

[14] S. G. Dani, M. Laurent, and A. Nogueira, *Multi-dimensional metric approximation by primitive points*, Math. Z. **279** (2015), no. 3-4, 1081–1101, DOI 10.1007/s00209-014-1404-5. MR3318261

[15] D. Dickinson and S. L. Velani, *Hausdorff measure and linear forms*, J. Reine Angew. Math. **490** (1997), 1–36, DOI 10.1515/crll.1997.490.1. MR1468922

[16] M. M. Dodson, *Hausdorff dimension, lower order and Khintchine's theorem in metric Diophantine approximation*, J. Reine Angew. Math. **432** (1992), 69–76, DOI 10.1515/crll.1992.432.69. MR1184759

[17] M. M. Dodson, B. P. Rynne, and J. A. G. Vickers, *Diophantine approximation and a lower bound for Hausdorff dimension*, Mathematika **37** (1990), no. 1, 59–73, DOI 10.1112/S0025579300012791. MR1067887

[18] R. J. Duffin and A. C. Schaeffer, *Khintchine's problem in metric Diophantine approximation*, Duke Math. J. **8** (1941), 243–255. MR0004859

[19] A. Durand, *On randomly placed arcs on the circle*, Recent developments in fractals and related fields, Appl. Numer. Harmon. Anal., Birkhäuser Boston, Inc., Boston, MA, 2010, pp. 343–351, DOI 10.1007/978-0-8176-4888-6_22. MR2743004

[20] A. Durand, *Describability via ubiquity and eutaxy in Diophantine approximation*, Ann. Math. Blaise Pascal **22** (2015), no. S2, 1–149. MR3453285

[21] F. Ekström and T. Persson, *Hausdorff dimension of random limsup sets*, J. Lond. Math. Soc. (2) **98** (2018), no. 3, 661–686, DOI 10.1112/jlms.12158. MR3893196

[22] K. J. Falconer, *The Hausdorff dimension of self-affine fractals*, Math. Proc. Cambridge Philos. Soc. **103** (1988), no. 2, 339–350, DOI 10.1017/S0305004100064926. MR923687

[23] K. Falconer, *Techniques in fractal geometry*, John Wiley & Sons, Ltd., Chichester, 1997. MR1449135

[24] K. Falconer, *Fractal geometry: Mathematical foundations and applications*, 3rd ed., John Wiley & Sons, Ltd., Chichester, 2014. MR3236784

[25] K. Falconer, *Dimensions of self-affine sets: a survey*, Further developments in fractals and related fields, Trends Math., Birkhäuser/Springer, New York, 2013, pp. 115–134, DOI 10.1007/978-0-8176-8400-6_6. MR3184190

[26] K. Falconer and T. Kempton, *The dimension of projections of self-affine sets and measures*, Ann. Acad. Sci. Fenn. Math. **42** (2017), no. 1, 473–486, DOI 10.5186/aasfm.2017.4232. MR3558544

[27] A.-H. Fan and J. Wu, *On the covering by small random intervals* (English, with English and French summaries), Ann. Inst. H. Poincaré Probab. Statist. **40** (2004), no. 1, 125–131, DOI 10.1016/S0246-0203(03)00056-6. MR2037476

[28] A.-H. Fan, J. Schmeling, and S. Troubetzkoy, *A multifractal mass transference principle for Gibbs measures with applications to dynamical Diophantine approximation*, Proc. Lond. Math. Soc. (3) **107** (2013), no. 5, 1173–1219, DOI 10.1112/plms/pdt005. MR3126394

[29] D.-J. Feng, E. Järvenpää, M. Järvenpää, and V. Suomala, *Dimensions of random covering sets in Riemann manifolds*, Ann. Probab. **46** (2018), no. 3, 1542–1596, DOI 10.1214/17-AOP1210. MR3785595

[30] J. M. Fraser and P. Shmerkin, *On the dimensions of a family of overlapping self-affine carpets*, Ergodic Theory Dynam. Systems **36** (2016), no. 8, 2463–2481, DOI 10.1017/etds.2015.21. MR3570020

[31] P. X. Gallagher, *Metric simultaneous diophantine approximation. II*, Mathematika **12** (1965), 123–127, DOI 10.1112/S0025579300005234. MR0188154

[32] D. Gatzouras and S. P. Lalley, *Statistically self-affine sets: Hausdorff and box dimensions*, J. Theoret. Probab. **7** (1994), no. 2, 437–468, DOI 10.1007/BF02214277. MR1270610

[33] A. Groshev, *A theorem on a system of linear forms*, Dokl. Akad. Nauk SSSR, **19**, (1938), 151–152, (in Russian).

[34] Y. Gui and W. Li, *Multiscale self-affine Sierpinski carpets*, Nonlinearity **23** (2010), no. 3, 495–512, DOI 10.1088/0951-7715/23/3/003. MR2586366

[35] J. E. Hutchinson, *Fractals and self-similarity*, Indiana Univ. Math. J. **30** (1981), no. 5, 713–747, DOI 10.1512/iumj.1981.30.30055. MR625600

[36] S. Jaffard, *Construction of functions with prescribed Hölder and chirp exponents*, Rev. Mat. Iberoamericana **16** (2000), no. 2, 331–349, DOI 10.4171/RMI/277. MR1809343

[37] S. Jaffard, *On lacunary wavelet series*, Ann. Appl. Probab. **10** (2000), no. 1, 313–329, DOI 10.1214/aoap/1019737675. MR1765214

[38] V. Jarník, *Diophantischen Approximationen und Hausdorffsches Mass*, Mat. Sbornik, **36**, (1929), 371–382.

[39] V. Jarník, *Über die simultanen diophantischen Approximationen* (German), Math. Z. **33** (1931), no. 1, 505–543, DOI 10.1007/BF01174368. MR1545226

[40] E. Järvenpää, M. Järvenpää, H. Koivusalo, B. Li, and V. Suomala, *Hausdorff dimension of affine random covering sets in torus*, Ann. Inst. Henri Poincaré Probab. Stat. **50** (2014), no. 4, 1371–1384, DOI 10.1214/13-AIHP556. MR3269998

[41] E. Järvenpää, M. Järvenpää, M. Wu, and W. Wu, *Random affine code tree fractals: Hausdorff and affinity dimensions and pressure*, Math. Proc. Cambridge Philos. Soc. **162** (2017), no. 2, 367–382, DOI 10.1017/S0305004116000694. MR3604920

[42] T. Jordan and N. Jurga, *Self-affine sets with non-compactly supported random perturbations*, Ann. Acad. Sci. Fenn. Math. **39** (2014), no. 2, 771–785, DOI 10.5186/aasfm.2014.3948. MR3237050

[43] T. Jordan, M. Pollicott, and K. Simon, *Hausdorff dimension for randomly perturbed self affine attractors*, Comm. Math. Phys. **270** (2007), no. 2, 519–544, DOI 10.1007/s00220-006-0161-7. MR2276454

[44] A. Käenmäki and B. Li, *Genericity of dimension drop on self-affine sets*, Statist. Probab. Lett. **126** (2017), 18–25, DOI 10.1016/j.spl.2017.02.028. MR3634575

[45] N. Luzia, *Hausdorff dimension of certain random self-affine fractals*, Stoch. Dyn. **11** (2011), no. 4, 627–642, DOI 10.1142/S0219493711003516. MR2836564

[46] A. Khintchine, *Einige Sätze über Kettenbrüche, mit Anwendungen auf die Theorie der Diophantischen Approximationen* (German), Math. Ann. **92** (1924), no. 1-2, 115–125, DOI 10.1007/BF01448437. MR1512207

[47] A. I. Khintchine, *Über die angenäherte Auflösung linearer Gleichungen in ganzen Zahlen*, Rec. Math. Soc. Moscou **32** (1925), 203–218.

[48] H. Koivusalo and M. Rams, *Mass transference principle: from balls to arbitrary shapes*, preprint (2018), arXiv:1812.08557.

[49] J. Levesley, *A general inhomogeneous Jarnik-Besicovitch theorem*, J. Number Theory **71** (1998), no. 1, 65–80, DOI 10.1006/jnth.1998.2239. MR1631030

[50] J. Levesley, C. Salp, and S. L. Velani, *On a problem of K. Mahler: Diophantine approximation and Cantor sets*, Math. Ann. **338** (2007), no. 1, 97–118, DOI 10.1007/s00208-006-0069-8. MR2295506

[51] P. Mattila, *Geometry of sets and measures in Euclidean spaces*, Cambridge Studies in Advanced Mathematics, vol. 44, Cambridge Univ. Press, Cambridge, 1995. Fractals and rectifiability. MR1333890

[52] P. A. P. Moran, *Additive functions of intervals and Hausdorff measure*, Proc. Cambridge Philos. Soc. **42** (1946), 15–23. MR0014397

[53] I. D. Morris, P. Shmerkin, *On equality of Hausdorff and affinity dimensions, via self-affine measures on positive subsystems*, preprint, (2016), arXiv:1602.08789.

[54] I. D. Morris and P. Shmerkin, *On equality of Hausdorff and affinity dimensions, via self-affine measures on positive subsystems*, Trans. Amer. Math. Soc. **371** (2019), no. 3, 1547–1582, DOI 10.1090/tran/7334. MR3894027

[55] T. Persson, *A note on random coverings of tori*, Bull. Lond. Math. Soc. **47** (2015), no. 1, 7–12, DOI 10.1112/blms/bdu087. MR3312958

[56] A. D. Pollington and R. C. Vaughan, *The k-dimensional Duffin and Schaeffer conjecture*, Mathematika **37** (1990), no. 2, 190–200, DOI 10.1112/S0025579300012900. MR1099767

[57] C. A. Rogers, *Hausdorff measures*, Cambridge Mathematical Library, Cambridge Univ. Press, Cambridge, 1998. Reprint of the 1970 original; With a foreword by K. J. Falconer. MR1692618

[58] B. P. Rynne, *Hausdorff dimension and generalized simultaneous Diophantine approximation*, Bull. London Math. Soc. **30** (1998), no. 4, 365–376, DOI 10.1112/S0024609398004536. MR1620813

[59] W. Schmidt, *A metrical theorem in diophantine approximation*, Canad. J. Math. **12** (1960), 619–631, DOI 10.4153/CJM-1960-056-0. MR0118711

[60] S. Seuret, *Inhomogeneous random coverings of topological Markov shifts*, Math. Proc. Cambridge Philos. Soc. **165** (2018), no. 2, 341–357, DOI 10.1017/S0305004117000512. MR3834004

[61] V. G. Sprindžuk, *Metric theory of Diophantine approximations*, V. H. Winston & Sons, Washington, D.C.; A Halsted Press Book, John Wiley & Sons, New York-Toronto, Ont.-London, 1979. Translated from the Russian and edited by Richard A. Silverman; With a foreword by Donald J. Newman; Scripta Series in Mathematics. MR548467

[62] S. Troscheit, *On the dimensions of attractors of random self-similar graph directed iterated function systems*, J. Fractal Geom. **4** (2017), no. 3, 257–303, DOI 10.4171/JFG/51. MR3732642

[63] S. Troscheit, *The box-counting dimensions of random box-like self-affine sets*, Indiana Univ. Math. J. **67** (2018), no. 2, 495–535, DOI 10.1512/iumj.2018.67.7295. MR3798848

[64] B.-W. Wang, J. Wu, and J. Xu, *Mass transference principle for limsup sets generated by rectangles*, Math. Proc. Cambridge Philos. Soc. **158** (2015), no. 3, 419–437, DOI 10.1017/S0305004115000043. MR3335419

Department of Mathematics, University of York, YO10 5DD, UK

Current address: School of Mathematics, University of Bristol, Bristol, BS8 1TW, United Kingdom – and – Heilbronn Institute for Mathematical Research, Bristol, United Kingdom

Email address: `demi.allen@bristol.ac.uk`

Department of Pure Mathematics, University of Waterloo, 200 University Avenue West, Waterloo, ON, N2L 3G1, Canada

Current address: Faculty of Mathematics, University of Vienna, Oskar Morgenstern Platz 1, 1090 Wien, Austria

Email address: `sascha.troscheit@univie.ac.at`

Contemporary Mathematics
Volume **731**, 2019
https://doi.org/10.1090/conm/731/14671

A measure-theoretic result for approximation by Delone sets

Michael Baake and Alan Haynes

ABSTRACT. With a view to establishing measure-theoretic approximation properties of Delone sets, we study a setup which arises naturally in the problem of averaging almost periodic functions along exponential sequences. In this setting, we establish a full converse of the Borel–Cantelli lemma. This provides an analogue of more classical problems in the metric theory of Diophantine approximation, but with the distance to the nearest integer function replaced by distance to an arbitrary Delone set.

1. Introduction

In a recent work [**2**], motivated by problems emerging in aperiodic order [**1**], we considered the problem of establishing asymptotic formulas for averages of Bohr almost periodic functions along exponential sequences of the form $(\alpha^n x)_{n\in\mathbb{N}}$, where α is a fixed real number with $|\alpha| > 1$, and $x \in \mathbb{R}$ is arbitrary. The almost everywhere convergence results which were obtained can be viewed as analogues of Birkhoff's ergodic theorem [**8**]. Their proofs relied on knowing that, for almost every $x \in \mathbb{R}$, the elements of our exponential sequence with $n \leq N$ cannot be approximated too well by elements of a given Delone set. In fact, it was precisely this occurrence of a Delone set that showed how natural the replacement of the integers by such a more general set is, and how well it blends into the realm of metric Diophantine approximation problems. Once again, systems of aperiodic order have thus suggested how one can go meaningfully beyond the standard lattice setting, while remaining firmly within the realm of physical and mathematical significance. In this short note, we return to this problem in order to more fully investigate the Diophantine approximation properties of Delone sets. Our main result is a generalization of Khintchine's theorem (even without monotonicity) to this setting (see [**4**] or [**7**] for detailed background on the metric theory of Diophantine approximation). It provides a characterization which tells us precisely when we should expect, for almost every $x \in \mathbb{R}$, to have infinitely many approximations of a certain quality.

2010 *Mathematics Subject Classification.* 11K06, 52C23.
Key words and phrases. Duffin–Schaeffer conjecture, metric number theory, Delone sets.

Recall that a Delone set is a subset Y of a metric space (X, d) for which there exist positive constants r and R such that:

(i) For any $y, y' \in Y$ with $y \neq y'$, we have that $d(y, y') \geq 2r$, and
(ii) For any $x \in X$, there exists a $y \in Y$ such that $d(x, y) \leq R$.

The supremum over all r satisfying (i) is called the packing radius of Y, and the infimum over all R satisfying (ii) is called the covering radius. See [**1**, Section 2.1] for more background on the basic theory of Delone sets.

In what follows, we suppose that Y is a Delone set in $\mathbb{R}$ with packing radius r and covering radius R. For $x \in \mathbb{R}$ and $\rho > 0$, we use $B(x, \rho)$ to denote the closed ball of radius ρ centered at x. The symbol λ denotes Lebesgue measure on $\mathbb{R}$, and $\mathrm{card}(S)$ denotes the cardinality of a set S. We also use the standard Vinogradov $\ll$ notation, so that if $f, g : D \longrightarrow \mathbb{R}$ are two functions on some common domain D, then $f \ll g$ means that there exists a constant $C > 0$ such that $|f(x)| \leq C|g(x)|$ for all $x \in D$. Our main result is the following theorem.

THEOREM. *Let $\psi : \mathbb{N} \longrightarrow [0, \infty)$ be any function, and suppose that α is a real number with $|\alpha| > 1$. If*

$$\sum_{n=1}^{\infty} \psi(n) < \infty \tag{1}$$

then, for almost every $x \in \mathbb{R}$, there are only finitely many $n \in \mathbb{N}$ and $y \in Y$ for which

$$|\alpha^n x - y| \leq \psi(n). \tag{2}$$

On the other hand, if

$$\sum_{n=1}^{\infty} \psi(n) = \infty, \tag{3}$$

then, for almost every $x \in \mathbb{R}$, there are infinitely many $n \in \mathbb{N}$ and $y \in Y$ that satisfy the above inequality.

We point out that our result amounts to a full converse of the Borel–Cantelli lemma for special collections of subsets of $\mathbb{R}$. The less trivial direction of our proof is the divergence part, in which we assume that (3) holds, and prove that the associated limsup set (the set of real numbers x for which the relevant inequalities have infinitely many solutions) has full measure. It is quite common in these types of problems to first establish a zero-full lemma (e.g. [**3**, **6**, **7**]), which proves that, whether or not (3) holds, the associated limsup set either has zero measure, or its complement does. Then all that is left to show, for the difficult part of the proof, is that the limsup set has *positive* measure. In our proof, we deviate slightly from this approach and, instead of establishing a separate zero-full lemma, we use a simple application of the Lebesgue density theorem [**5**, Theorem 3.21] to complete our proof. In essence, what we show is that, at all scales, the intersections of the individual intervals which contribute to the limsup set behave in the same way. In other words, roughly the same picture appears after zooming in on any part of the real line. Since the integers in our setup have been replaced by a completely arbitrary Delone set, this is clearly another manifestation of a behavior which should be subsumed under the theme of 'aperiodic order'.

2. Proof of main theorem

We will consider only the case when $\alpha > 1$, since the case when $\alpha < -1$ follows as an easy corollary. Let $a < b$ be real numbers and assume, with little loss of generality, that $a \geq 0$ (the cases with $a < 0$ can be dealt with by trivial modifications of our argument). For each $n \in \mathbb{N}$, define a set $\mathcal{A}_n \subseteq [a, b)$ by

$$\mathcal{A}_n = \{x \in [a, b) : |\alpha^n x - y| \leq \psi(n) \text{ for some } y \in Y\}.$$

In what follows, all constants implied by the use of the $\ll$ notation will be universal, not depending on a, b, r, or R, unless otherwise stated. Whenever implied constants do depend on some of these quantities, we will indicate this by attaching the appropriate subscripts to the $\ll$ symbol.

The proof of the first part of the theorem is a straightforward application of the convergence part of the Borel–Cantelli lemma. Write the non-negative elements of Y in increasing order as

$$y_1 < y_2 < \cdots$$

and, for each $n \in \mathbb{N}$, define

$$\mathcal{I}_n = \left\{ i \in \mathbb{N} : [a, b) \cap B\left(\frac{y_i}{\alpha^n}, \frac{\psi(n)}{\alpha^n}\right) \neq \varnothing \right\}.$$

From our hypothesis on Y we have, for all sufficiently large n (depending on a and b), that

$$\frac{(b-a)\,\alpha^n}{R} \ll \operatorname{card}(\mathcal{I}_n) \ll \frac{(b-a)\,\alpha^n}{r},$$

and it follows from this that, for n sufficiently large,

$$\lambda(\mathcal{A}_n) \ll (b-a)\,\psi(n)/r.$$

Since we are assuming that (1) holds, we conclude that almost every $x \in [a, b)$ falls in only finitely many of the sets $\mathcal{A}_n$, which is equivalent to the assertion that there are only finitely many solutions to the inequality (2). This gives the conclusion of the first part of the theorem.

The proof of the second part is slightly more complicated. Ideally, we would like to demonstrate that the sets $\mathcal{A}_n$ above are quasi-independent, or in other words that

$$\lambda(\mathcal{A}_m \cap \mathcal{A}_n) \ll_{r,R} (b-a)^{-1}\lambda(\mathcal{A}_m)\lambda(\mathcal{A}_n) \quad \text{for} \quad m \neq n.$$

Unfortunately, this is not quite true, so we need some technical modifications in our setup. First of all, choose an integer J with the property that

$$\alpha^J \geq \frac{2R(1+\frac{2}{r})}{r} + 1, \tag{4}$$

and then choose a residue class $j \in \{0, 1, \ldots, J-1\}$ modulo J with the property that

$$\sum_{n=1}^{\infty} \psi(nJ + j) = \infty.$$

Since there are only J residue classes to choose from, and since we are assuming that (3) holds, it is clear that this is possible.

Now write $\beta = \alpha^J$ and, for each $n \in \mathbb{N}$, define a set $Y^{(n)} \subseteq Y$ by

$$Y^{(n)} = Y \setminus \bigcup_{m=1}^{n-1} B(\beta^{n-m}Y, 1). \tag{5}$$

Here we are using $B(\beta^{n-m}Y, 1)$ to denote the union over $y \in Y$ of the balls $B(\beta^{n-m}y, 1)$. The reason for introducing the sets $Y^{(n)}$, which will become clearer later in the proof, is to remove the bad overlaps which occur between the sets $\mathcal{A}_n$ from the previous argument. However, we will still need to show that we have not discarded too much from Y so as to make the sum of the measures of our new sets fail to diverge. For each $X \in \mathbb{R}$, we have that

$$\operatorname{card}\{y \in Y : aX \leq y < bX\} \geq \left\lfloor \frac{(b-a)X}{R} \right\rfloor$$

and, for each $\ell \in \mathbb{N}$, we also have that

$$\operatorname{card}\{y \in \beta^{\ell} Y : aX \leq y < bX\} \leq \frac{(b-a)X}{r\beta^{\ell}} + 1.$$

Since the number of points of Y in a ball of radius 1 is bounded above by $1 + \frac{2}{r}$, it follows that, for $n, N \in \mathbb{N}$,

$$\begin{aligned}
\operatorname{card}\{y \in Y^{(n)} : aX \leq y < bX\} &\geq \left\lfloor \frac{(b-a)X}{R} \right\rfloor - \sum_{\ell=1}^{n-1} \left(\frac{(b-a)X}{r\beta^{\ell}} + 1 \right) \cdot (1 + \tfrac{2}{r}) \\
&\geq (b-a)X \left(\frac{1}{R} - \frac{1 + \frac{2}{r}}{r(\beta - 1)} \right) - 1 - n(1 + \tfrac{2}{r}) \\
&\geq \frac{(b-a)X}{2R} - 1 - n(1 + \tfrac{2}{r}). \qquad (6)
\end{aligned}$$

The final inequality here is a result of our choice of J in (4).

For $n \in \mathbb{N}$ we now define $\mathcal{A}'_n \subseteq [0,1)$, our replacement for the set $\mathcal{A}_n$ from above, by

$$\mathcal{A}'_n = \{x \in [a,b) : |\alpha^j \beta^n x - y| \leq \psi(nJ + j) \text{ for some } y \in Y^{(n)}\}.$$

From (6), and using the arguments from above, we have that, for all sufficiently large n (depending again only on a and b),

$$\lambda(\mathcal{A}'_n) \gg \frac{(b-a)\, \psi(nJ + j)}{R}.$$

For $m \neq n$, we now would like to derive an upper bound for the measure of the intersection of $\mathcal{A}'_m$ with $\mathcal{A}'_n$. With this purpose in mind, let

$$\delta = \delta(m,n) = \min \left\{ \frac{2\psi(mJ + j)}{\alpha^j \beta^m}, \frac{2\psi(nJ + j)}{\alpha^j \beta^n} \right\},$$

and

$$\Delta = \Delta(m,n) = \max \left\{ \frac{2\psi(mJ + j)}{\alpha^j \beta^m}, \frac{2\psi(nJ + j)}{\alpha^j \beta^n} \right\}.$$

Each set $\mathcal{A}'_n$ is a union of connected components, which we refer to as its component intervals. The component intervals of $\mathcal{A}'_n$ are intervals centered at points of the form $y/\alpha^j\beta^n$. To be fully accurate, at the endpoints of $[a, b)$ it may be the case that there are (at most 2) component intervals which are not of this form, but this fact is negligible in the argument we are about to give. If a component interval from $\mathcal{A}'_m$ intersects a component interval from $\mathcal{A}'_n$, the centers of these intervals must be within Δ of one another. Furthermore, the intersection of any two such intervals has measure at most δ. This translates into the upper bound

$$\lambda(\mathcal{A}'_m \cap \mathcal{A}'_n) \ll \delta \cdot \operatorname{card}\Bigg\{(y, y') \in Y^{(m)} \times Y^{(n)} \ : \ a \leq \frac{y}{\alpha^j\beta^m} < b,\ a \leq \frac{y'}{\alpha^j\beta^n} < b, \left|\frac{y}{\alpha^j\beta^m} - \frac{y'}{\alpha^j\beta^n}\right| \leq \Delta\Bigg\}, \tag{7}$$

which holds for all m and n sufficiently large. Suppose without loss of generality that $m < n$, and let us derive an estimate for the number of pairs (y, y') which we are counting on the right hand side above. If $y \in Y^{(m)}$ and $y' \in Y^{(n)}$ satisfy

$$|\beta^{n-m}y - y'| \leq \alpha^j\beta^n\Delta, \tag{8}$$

then, from the definition (5) of $Y^{(n)}$, we must also have that

$$|\beta^{n-m}y - y'| \geq 1. \tag{9}$$

It is worth pointing out that this was the reason for the introduction of the sets $Y^{(n)}$. Without the lower bound of a fixed positive constant here, the next part of the argument would not work. From Eqs. (7)-(9) we now derive that, for all m and n sufficiently large (depending on a and b),

$$\begin{aligned}\lambda(\mathcal{A}'_m \cap \mathcal{A}'_n) &\ll_{r,R} \delta \cdot \big((b-a)\,\alpha^j\beta^m\big) \cdot \big(\alpha^j\beta^n\Delta\big)\\ &\ll (b-a)\,\psi(mJ+j)\,\psi(nJ+j).\end{aligned}$$

This implies that there is a constant $K > 0$ (depending only on r and R) with the property that

$$\limsup_{N\to\infty} \left(\sum_{n\leq N} \lambda(\mathcal{A}'_n)\right)^2 \left(\sum_{m,n\leq N} \lambda(\mathcal{A}'_m \cap \mathcal{A}'_n)\right)^{-1} \geq (b-a)K. \tag{10}$$

Since the sum of the measures of the sets $\mathcal{A}'_n$ diverges, by standard arguments from probability theory (see [**7**, Lemma 2.3]), it follows that the set of x which fall in infinitely many of the sets $\mathcal{A}'_n$ has measure greater than or equal to $(b-a)K$.

For the final step of the proof we could appeal directly to [**3**, Proposition 1]. For completeness, we provide the following simple argument. Supposing still that the divergence condition (3) holds, let $\mathcal{W} \subseteq \mathbb{R}$ be the set of $x \in \mathbb{R}$ for which the inequality (2) is satisfied by infinitely many $n \in \mathbb{N}$ and $y \in Y$. If it were the case that $\lambda(\mathcal{W}^c) > 0$ then, by the Lebesgue density theorem [**5**, Theorem 3.21], we could find a point of metric density x_0 of the set $\mathcal{W}^c$. However, this would imply that

$$\lim_{\epsilon\to 0^+} \frac{\lambda(\mathcal{W} \cap B(x_0, \epsilon))}{2\epsilon} = 0,$$

which contradicts (10). Therefore we conclude that $\lambda(\mathcal{W}^c) = 0$, thereby completing the proof of our main result.

References

[1] Michael Baake and Uwe Grimm, *Aperiodic order. Vol. 1: A mathematical invitation*, Encyclopedia of Mathematics and its Applications, vol. 149, Cambridge University Press, Cambridge, 2013. with a foreword by Roger Penrose. MR3136260

[2] Michael Baake, Alan Haynes, and Daniel Lenz, *Averaging almost periodic functions along exponential sequences*, Aperiodic order. Vol. 2, Encyclopedia Math. Appl., vol. 166, Cambridge Univ. Press, Cambridge, 2017, pp. 343–362. MR3791852

[3] Victor Beresnevich, Detta Dickinson, and Sanju Velani, *Measure theoretic laws for lim sup sets*, Mem. Amer. Math. Soc. **179** (2006), no. 846, x+91, DOI 10.1090/memo/0846. MR2184760

[4] Yann Bugeaud, *Distribution modulo one and Diophantine approximation*, Cambridge Tracts in Mathematics, vol. 193, Cambridge University Press, Cambridge, 2012. MR2953186

[5] Gerald B. Folland, *Real analysis: Modern techniques and their applications*, 2nd ed., Pure and Applied Mathematics (New York), John Wiley & Sons, Inc., New York, 1999. A Wiley-Interscience Publication. MR1681462

[6] Patrick Gallagher, *Approximation by reduced fractions*, J. Math. Soc. Japan **13** (1961), 342–345, DOI 10.2969/jmsj/01340342. MR0133297

[7] Glyn Harman, *Metric number theory*, London Mathematical Society Monographs. New Series, vol. 18, The Clarendon Press, Oxford University Press, New York, 1998. MR1672558

[8] Peter Walters, *An introduction to ergodic theory*, Graduate Texts in Mathematics, vol. 79, Springer-Verlag, New York-Berlin, 1982. MR648108

Fakultät für Mathematik, Universität Bielefeld, 33501 Bielefeld, Germany
Email address: mbaake@math.uni-bielefeld.de

Department of Mathematics, University of Houston, Houston, TX, United States
Email address: haynes@math.uh.edu

Contemporary Mathematics
Volume **731**, 2019
https://doi.org/10.1090/conm/731/14672

Self-similar tilings of fractal blow-ups

M. F. Barnsley and A. Vince

ABSTRACT. New tilings of certain subsets of $\mathbb{R}^M$ are studied, tilings associated with fractal blow-ups of certain similitude iterated function systems (IFS). For each such IFS with attractor satisfying the open set condition, our construction produces a usually infinite family of tilings that satisfy the following properties: (1) the prototile set is finite; (2) the tilings are repetitive (quasiperiodic); (3) each family contains self-similar tilings, usually infinitely many; and (4) when the IFS is rigid in an appropriate sense, the tiling has no non-trivial symmetry; in particular the tiling is non-periodic.

1. Introduction

The subject of this paper is a new type of tiling of certain subsets D of $\mathbb{R}^M$. Such a domain D is a fractal blow-up (as defined in Section 3) of certain similitude iterated function systems (IFSs); see also [**3, 11**]. For an important class of such tilings it is the case that $D = \mathbb{R}^M$, as exemplified by the tiling of Figure 1 (on the right) that is based on the "golden b" tile (on the left). We are also interested, however, in situations where D has non-integer Hausdorff dimension. The left panel in Figure 2 shows the domain D, the right panel a tiling of D. These examples are explored in Section 12. In this work, tiles may be fractals; pairs of distinct tiles in a tiling are required to be non-overlapping, i.e., they intersect on a set whose Hausdorff dimension is lower than that of the individual tiles.

These tilings come in families, one family for each similitude IFS whose functions $f_1, f_2 \dots, f_N$ have scaling ratios that are integer powers $s^{a_1}, s^{a_2}, \dots, s^{a_N}$ of a single real number s and whose attractor is non-overlapping. Each such family contains, in general, an uncountable number of tilings. Each family has a finite set of prototiles.

The paper is organized as follows. Sections 2 and 3 provide background and definitions relevant to tilings and to iterated function systems. The construction of our tilings is given in Section 3. The main theorems are stated precisely in Section 3 and proved in subsequent sections. Results appear in Section 8 that define and discuss the relative and absolute addresses of tiles. These concepts, useful towards understanding the relationships between different tilings, are illustrated in Section 12. Also in Section 12 are examples of tilings of $\mathbb{R}^2$ and of a quadrant of $\mathbb{R}^2$. The Ammann (the golden b) tilings and related fractal tilings are also discussed in that section, as is a blow-up of a Cantor set.

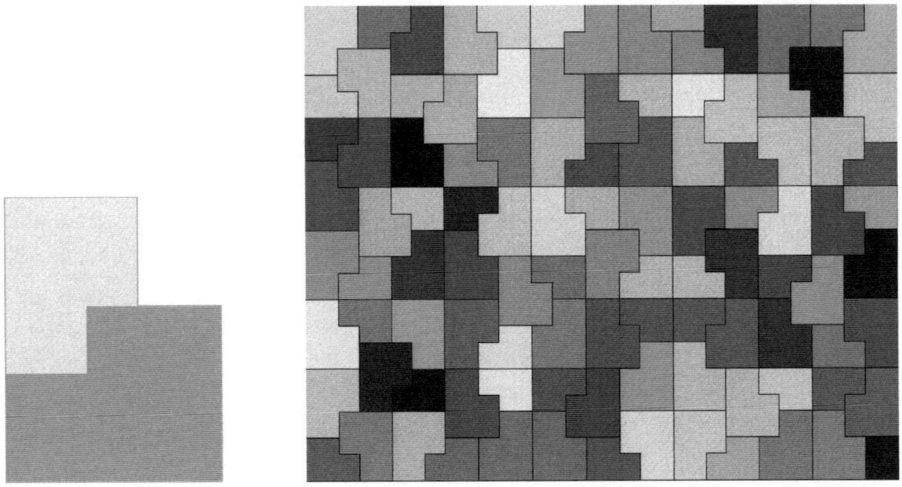

FIGURE 1. Golden b and golden b tiling.

FIGURE 2. The left image shows part of an infinite fractal blow-up D; the right image shows part of a tiling of D using a finite set of prototiles. See Section 12.

A subset P of a tiling T is called a *patch* of T if it is contained in a ball of finite radius. A tiling T is *quasiperiodic* (also called repetitive) if, for any patch P, there is a number $R > 0$ such that any ball of radius R centered at a point contained in a tile of T contains an isometric copy of P. Two tilings are *locally isomorphic* if any patch in either tiling also appears in the other tiling. A tiling T is *self-similar* if there is a similitude ψ such that $\psi(t)$ is a union of tiles in T for all $t \in T$. Such a map ψ is called a *self-similarity*.

Let $\mathcal{F}$ be a similitude IFS whose functions have scaling ratios $s^{a_1}, s^{a_2}, \dots, s^{a_N}$ as defined above. Let $[N]^*$ be the set of finite words over the alphabet $[N] := \{1, 2, \dots, N\}$ and $[N]^\infty$ be the set of infinite words over the alphabet $[N]$. Let $\widetilde{[N]}^\infty$ be the set of reversible words in $[N]^\infty$, as defined in [**4**]. For a fixed

IFS $\mathcal{F}$, our results show that:

(1) For each $\theta \in [N]^*$, our construction yields a bounded tiling, and for each $\theta \in [N]^\infty$, our construction yields an unbounded tiling. In the latter case, the tiling, denoted $\pi(\theta)$, almost always covers $\mathbb{R}^M$ when the attractor of the IFS has nonempty interior.
(2) The mapping $\theta \mapsto \pi(\theta)$, restricted to $\widetilde{[N]^\infty}$, is continuous with respect to the standard topologies on the domain and range of π.
(3) Under quite general conditions, the mapping $\theta \mapsto \pi(\theta)$ is injective.
(4) For each such tiling, the prototile set is $\{sA, s^2A, \dots, s^{a_{\max}}A\}$, where A is the attractor of the IFS and $a_{\max} = \max\{a_1, a_2, \dots, a_N\}$.
(5) The constructed tilings, in the unbounded case, are repetitive (quasiperiodic) and any two such tilings are locally isomorphic.
(6) For all $\theta \in [N]^\infty$, if θ is eventually periodic, then $\pi(\theta)$ is self-similar.
(7) If $\mathcal{F}$ is strongly rigid, then how isometric copies of a pair bounded tilings can overlap is extremely restricted: if the two tilings are such that their overlap is a subset of each, then one tiling must be contained in the other.
(8) If $\mathcal{F}$ is strongly rigid, then the constructed tilings have no non-identity symmetry. In particular, they are non-periodic.

The concept of a rigid and a strongly rigid IFS is discussed in Sections 9.

A special case of our construction (polygonal tilings, no fractals) appears in [**5**], in which we took a more recreational approach, devoid of proofs. Other references to related material are [**1**, **10**]. This work extends, but is markedly different from [**4**].

2. Tilings, Similitudes and Tiling Spaces

Given a natural number M, this paper is concerned with certain tilings of strict subsets of Euclidean space $\mathbb{R}^M$ and of $\mathbb{R}^M$ itself. A *tile* is a perfect (i.e. no isolated points) compact nonempty subset of $\mathbb{R}^M$. Fix a Hausdorff dimension $0 < D_H \leq M$. A *tiling* in $\mathbb{R}^M$ is a set of tiles, each of Hausdorff dimension D_H, such that every distinct pair is non-overlapping. Two tiles are *non-overlapping* if their intersection is of Hausdorff dimension strictly less than D_H. The *support* of a tiling is the union of its tiles. We say that a tiling tiles its support. Some examples are presented in Section 12.

A *similitude* is an affine transformation $f : \mathbb{R}^M \to \mathbb{R}^M$ of the form $f(x) = s\,O(x) + q$, where O is an orthogonal transformation and $q \in \mathbb{R}^M$ is the translational part of $f(x)$. The real number $s > 0$, a measure of the expansion or contraction of the similitude, is called its *scaling ratio*. An *isometry* is a similitude of unit scaling ratio and we say that two sets are isometric if they are related by an isometry. We write $\mathcal{E}$ to denote the group of isometries on $\mathbb{R}^M$.

The *prototile set* $\mathcal{P}$ of a tiling T is a minimal set of tiles such that every tile in T is an isometric copy of a tile in $\mathcal{P}$. The tilings constructed in this paper have a finite prototile set.

Given a tiling T we define ∂T to be the union of the set of boundaries of all of the tiles in T and we let $\rho : \mathbb{R}^M \to \mathbb{S}^M$ be the usual M-dimensional stereographic projection to the M-sphere, obtained by positioning $\mathbb{S}^M$ tangent to $\mathbb{R}^M$ at the origin. We define the distance between tilings T and T' to be

$$d_\tau(T, T') = h(\overline{\rho(\partial T)}, \overline{\rho(\partial T')})$$

where the bar denotes closure and h is the Hausdorff distance with respect to the round metric on $\mathbb{S}^M$. Let $\mathbb{K}(\mathbb{R}^M)$ be the set of nonempty compact subsets of $\mathbb{R}^M$. It is well known that d_τ provides a metric on the space $\mathbb{K}(\mathbb{R}^M)$ and that $(\mathbb{K}(\mathbb{R}^M), d_\tau)$ is a compact metric space.

This paper examines spaces consisting, for example, of $\pi(\theta)$ indexed by $\theta \in [N]^*$ with metric d_τ. Although we are aware of the large literature on tiling spaces, we do not explore the larger spaces obtained by taking the closure of orbits of our tilings under groups of isometries as in, for example, [**1**, **10**]. We focus on the relationship between the addressing structures associated with IFS theory and the particular families of tilings constructed here.

3. Definition and Properties of IFS Tilings

Let $\mathbb{N} = \{1, 2, \cdots\}$ and $\mathbb{N}_0 = \{0, 1, 2, \cdots\}$. For $N \in \mathbb{N}$, let $[N] = \{1, 2, \cdots, N\}$. Let $[N]^* = \cup_{k \in \mathbb{N}_0} [N]^k$, where $[N]^0$ is the empty string, denoted $\varnothing$.

See [**6**] for formal background on iterated function systems (IFSs). Here we are concerned with IFSs of a special form: let $\mathcal{F} = \{\mathbb{R}^M; f_1, f_2, \cdots, f_N\}$, with $N \geq 2$, be an IFS of contractive similitudes where the scaling factor of f_n is s^{a_n} with $0 < s < 1$ where $a_n \in \mathbb{N}$. There is no loss of generality in assuming that the greatest common divisor is one: $\gcd\{a_1, a_2, \cdots, a_N\} = 1$. That is, for $x \in \mathbb{R}^M$, the function $f_n : \mathbb{R}^M \to \mathbb{R}^M$ is defined by

$$f_n(x) = s^{a_n} O_n(x) + q_n$$

where O_n is an orthogonal transformation and $q_n \in \mathbb{R}^M$. It is convenient to define

$$a_{\max} = \max\{a_i : i = 1, 2, \ldots, N\}.$$

The *attractor* A of $\mathcal{F}$ is the unique solution in $\mathbb{K}(\mathbb{R}^M)$ to the equation

$$A = \bigcup_{i \in [N]} f_i(A).$$

It is assumed throughout that A obeys the open set condition (OSC) with respect to $\mathcal{F}$. As a consequence, the intersection of each pair of distinct tiles in the tilings that we construct either have empty intersection or intersect on a relatively small set. More precisely, the OSC implies that the Hausdorff dimension of A is strictly greater than the Hausdorff dimension of the *set of overlap* $\mathcal{O} = \cup_{i \neq j} f_i(A) \cap f_j(A)$. Similitudes applied to subsets of the set of overlap comprise the sets of points at which tiles may meet. See [**2**, p.481] for a discussion concerning measures of attractors compared to measures of the set of overlap.

In what follows, the space $[N]^* \cup [N]^\infty$ is equipped with a metric $d_{[N]^* \cup [N]^\infty}$ such that it becomes compact. First, define the "length" $|\theta|$ of $\theta \in [N]^* \cup [N]^\infty$ as follows. For $\theta = \theta_1 \theta_2 \cdots \theta_k \in [N]^*$ define $|\theta| = k$, and for $\theta \in [N]^\infty$ define $|\theta| = \infty$. Now define $d_{[N]^* \cup [N]^\infty}(\theta, \omega) = 0$ if $\theta = \omega$, and

$$d_{[N]^* \cup [N]^\infty}(\theta, \omega) = 2^{-\mathcal{N}(\theta, \omega)}$$

if $\theta \neq \omega$, where $\mathcal{N}(\theta,\omega)$ is the index of the first disagreement between θ and ω (and θ and ω are understood to disagree at index k if either $|\theta| < k$ or $|\omega| < k$). It is routine to prove that $([N]^* \cup [N]^\infty, d_{[N]^* \cup [N]^\infty})$ is a compact metric space.

A point $\theta \in [N]^\infty$ is *eventually periodic* if there exists $m \in \mathbb{N}_0$ and $n \in \mathbb{N}$ with $\theta_{m+i} = \theta_{m+n+i}$ for all $i \geq 1$. In this case we write $\theta = \theta_1\theta_2 \cdots \theta_m \overline{\theta_{m+1}\theta_{m+2} \cdots \theta_{m+n}}$.

For $\theta = \theta_1\theta_2 \cdots \theta_k \in [N]^*$, the following simplifying notation will be used:

$$
\begin{aligned}
f_\theta &= f_{\theta_1} f_{\theta_2} \cdots f_{\theta_k} \\
f_{-\theta} &= f_{\theta_1}^{-1} f_{\theta_2}^{-1} \cdots f_{\theta_k}^{-1} = (f_{\theta_k\theta_{k-1}\cdots\theta_1})^{-1},
\end{aligned}
$$

with the convention that f_θ and $f_{-\theta}$ are the identity function id if $\theta = \varnothing$. Likewise, for all $\theta \in [N]^\infty$ and $k \in \mathbb{N}_0$ define $\theta|k = \theta_1\theta_2 \cdots \theta_k$, and

$$
f_{-\theta|k} = f_{\theta_1}^{-1} f_{\theta_2}^{-1} \cdots f_{\theta_k}^{-1} = (f_{\theta_k\theta_{k-1}\cdots\theta_1})^{-1},
$$

with the convention that $f_{-\theta|0} = id$.

For $\sigma = \sigma_1\sigma_2 \cdots \sigma_k \in [N]^*$ and with $\{a_1, \ldots, a_N\}$ the scaling powers defined above, let

$$
e(\sigma) = a_{\sigma_1} + a_{\sigma_2} + \cdots + a_{\sigma_k} \qquad \text{and} \qquad e^-(\sigma) = a_{\sigma_1} + a_{\sigma_2} + \cdots + a_{\sigma_{k-1}},
$$

with the conventions $e(\varnothing) = e^-(\varnothing) = 0$. Let

$$
\Omega_k := \{\sigma \in [N]^* : e(\sigma) > k \geq e^-(\sigma)\}
$$

for all $k \in \mathbb{N}_0$, and note that $\Omega_0 = [N]$. We also write, in some places, $\sigma^- = \sigma_1\sigma_2 \cdots \sigma_{k-1}$ so that

$$
e^-(\sigma) = e(\sigma^-).
$$

DEFINITION 3.1. A mapping π from $[N]^* \cup [N]^\infty$ to collections of subsets of $\mathbb{R}^M$ is defined as follows. For $\theta \in [N]^*$

$$
\pi(\theta) := \{f_{-\theta} f_\sigma(A) : \sigma \in \Omega_{e(\theta)}\},
$$

and for $\theta \in [N]^\infty$

$$
\pi(\theta) := \bigcup_{k \in \mathbb{N}_0} \pi(\theta|k).
$$

Let $\mathbb{T}$ be the image of π, i.e.

$$
\mathbb{T} = \{\pi(\theta) : \theta \in [N]^* \cup [N]^\infty\}.
$$

It is a consequence of Theorem 3.2, stated below, that the elements of $\mathbb{T}$ are tilings. We refer to $\pi(\theta)$ as an *IFS tiling*, but usually drop the term "IFS". It is a consequence of the proof of Theorem 3.2, given in Section 6, that the support of $\pi(\theta)$ is what is sometimes referred to as a *fractal blow-up* [**3, 11**]. More exactly, if $F_k := f_{-\theta|k}(A)$, then

$$
\text{support}\,(\pi(\theta)) = \bigcup_{k \in \mathbb{N}_0} F_k.
$$

Thus the support of $\pi(\theta)$ is the limit of an increasing union of sets $F_0 \subseteq F_1 \subseteq F_2 \subseteq \cdots$, each similar to A.

The theorems of this paper are summarized in the rest of this section. The first two theorems, as well as a proposition in Section 8, reveal general information about the tilings in $\mathbb{T}$ without the rigidity condition that is assumed in the second two theorems. The proof of the following theorem appears in Section 6.

THEOREM 3.2. *Each set $\pi(\theta)$ in $\mathbb{T}$ is a tiling of a subset of $\mathbb{R}^M$, the subset being bounded when $\theta \in [N]^*$ and unbounded when $\theta \in [N]^\infty$. For all $\theta \in [N]^\infty$ the sequence of tilings $\{\pi(\theta|k)\}_{k=0}^{\infty}$ is nested according to*

$$\{f_i(A) : i \in [N]\} = \pi(\varnothing) \subset \pi(\theta|1) \subset \pi(\theta|2) \subset \pi(\theta|3) \subset \cdots . \tag{3.1}$$

For all $\theta \in [N]^\infty$, the prototile set for $\pi(\theta)$ is $\{s^i A : i = 1, 2, \cdots, a_{\max}\}$. Furthermore

$$\pi : [N]^* \cup \widetilde{[N]^\infty} \to \mathbb{T}$$

is a continuous map from the metric space $[N]^* \cup \widetilde{[N]^\infty}$ into the space $(\mathbb{K}(\mathbb{R}^M), d_\tau)$.

The proof of the following theorem is given in Section 7.

THEOREM 3.3. (1) *Each tiling in $\mathbb{T}$ is quasiperiodic and each pair of such tilings in $\mathbb{T}$ is locally isomorphic.*

(2) *If θ is eventually periodic, then $\pi(\theta)$ is self-similar. In fact, if $\theta = \alpha\overline{\beta}$ for some $\alpha, \beta \in [N]^*$ then $f_{-\alpha} f_{-\beta} (f_{-\alpha})^{-1}$ is a self-similarity of $\pi(\theta)$.*

In Section 9 the concept of *rigidity* of an IFS is defined. We postpone the definition because additional notation is required. There are numerous examples of rigid $\mathcal{F}$, including the golden b IFS in Section 12. The following theorem is proved in Section 9.

THEOREM 3.4. *Let $\mathcal{F}$ be strongly rigid. If $\theta, \theta' \in [N]^*$ and $E \in \mathcal{E}$ are such that $\pi(\theta) \cap E\pi(\theta')$ is a nonempty common tiling, then either $\pi(\theta) \subset E\pi(\theta')$ or $E\pi(\theta') \subset \pi(\theta)$. If $e(\theta) = e(\theta')$, then $E\pi(\theta') = \pi(\theta)$.*

A *symmetry* of a tiling is an isometry that takes tiles to tiles. A tiling is *periodic* if there exists a translational symmetry; otherwise the tiling is *non-periodic.* For example, any tiling of a quadrant of $\mathbb{R}^2$ by congruent squares is periodic. The proof of the following theorem is given in Section 10.

THEOREM 3.5. *If $\mathcal{F}$ is strongly rigid, then there does not exist any non-identity isometry $E \in \mathcal{E}$ and $\theta \in [N]^\infty$ such that $E\pi(\theta) \subset \pi(\theta)$.*

The following theorem is proved in Section 11.

THEOREM 3.6. *If $\pi(i) \cap \pi(j)$ does not tile $(\text{support } \pi(i)) \cap (\text{support } \pi(j))$ for all $i \neq j$, $i, j \in [N]$, then $\pi : [N]^* \cup [N]^\infty \to \mathbb{T}$ is one-to-one.*

4. Structure of $\{\Omega_k\}$ and Symbolic IFS Tilings

The results in this section, which will be applied later, relate to a symbolic version of the theory in this paper. The next two lemmas provide recursions for the sequence $\Omega_k := \{\sigma \in [N]^* : e(\sigma) > k \geq e^-(\sigma)\}$. In this section the square union symbol $\bigsqcup$ denotes a disjoint union.

LEMMA 4.1. *For all $k \geq a_{\max}$*

$$\Omega_k = \bigsqcup_{i=1}^{N} i\,\Omega_{k-a_i}. \tag{4.1}$$

PROOF. For all $k \in \mathbb{N}_0$ we have

$$\begin{aligned} i\,\Omega_k &= \{i\sigma : \sigma \in [N]^*, e(\sigma) > k \geq e^-(\sigma)\} \\ &= \{\omega : \omega \in [N]^*, e(\omega) > k + a_i \geq e^-(\omega), \omega_1 = i\} \\ &= \Omega_{k+a_i} \cap i[N]^*. \end{aligned}$$

It follows that

$$i\,\Omega_{k-a_i} = \Omega_k \cap i[N]^*$$

for all $k \geq a_i$, from which it follows that $\Omega_k = \bigsqcup_{i=1}^N i\Omega_{k-a_i}$ for all $k \geq a_{\max}$. □

LEMMA 4.2. *With* $\Omega_k' := \{\omega \in [N]^* : e(\omega) = k+1\}$, *we have* $\Omega_k' \subset \Omega_k$ *and*

$$\Omega_{k+1} = \{\Omega_k \backslash \Omega_k'\} \bigsqcup \left\{ \bigsqcup_{i=1}^N \Omega_k' i \right\}.$$

PROOF. (i) We first show that $\{\Omega_k \backslash \Omega_k'\} \bigsqcup \left\{ \bigsqcup_{i=1}^N \Omega_k' i \right\} \subset \Omega_{k+1}$.

Suppose $\theta \in \Omega_k \backslash \Omega_k'$. Then $e^-(\theta) \leq k < e(\theta)$ and $e(\theta) \neq k+1$. Hence $e^-(\theta) \leq k+1 < e(\theta)$ and so $\theta \in \Omega_{k+1}$.

Suppose $\theta \in \Omega_k' i$ for some $i \in [N]$. Then $\theta = \theta^- i$ where $\theta^- \in \Omega_k'$, $e^-(\theta) = e(\theta^-) = k+1$ and $e(\theta) = e(\theta^- i) = k+1+a_i$. Hence $e(\theta) > k+1 = e^-(\theta)$. Hence $e^-(\theta) \leq k+1 < e(\theta)$. Hence $\theta \in \Omega_{k+1}$.

(ii) We next show that $\Omega_{k+1} \subset \{\Omega_k \backslash \Omega_k'\} \bigsqcup \left\{ \bigsqcup_{i=1}^N \Omega_k' i \right\}$.

Let $\theta \in \Omega_{k+1}$. Then $e^-(\theta) = e(\theta^-) \leq k+1 < e(\theta)$.

If $e(\theta^-) = k+1$, then $\theta \in \Omega_k' \theta_{|\theta|} \subset \{\Omega_k \backslash \Omega_k'\} \bigsqcup \left\{ \bigsqcup_{i=1}^N \Omega_k' i \right\}$.

If $e(\theta^-) \neq k+1$, then $e(\theta^-) < k+1$. So $e(\theta^-) \leq k < k+1 < e(\theta)$; so $\theta \in \Omega_k \backslash \Omega_k' \subset \{\Omega_k \backslash \Omega_k'\} \bigsqcup \left\{ \bigsqcup_{i=1}^N \Omega_k' i \right\}$. □

For all $\theta \in [N]^*$, define $c(\theta) = \{\omega \in [N]^\infty : \omega_1\omega_2 \cdots \omega_{|\theta|} = \theta\}$. (Such sets are sometimes called *cylinder sets.*) With the metric on $[N]^\infty$ defined to be $d_0(\theta,\omega) = 2^{-\min\{k : \theta_k \neq \omega_k\}}$ for $\theta \neq \omega$, the diameter of $c(\theta)$ is $2^{-(|\theta|+1)}$. The following lemma tells us how $\{c(\theta) : \theta \in \Omega_k\}$ may be considered as a tiling of the symbolic space $[N]^\infty$.

LEMMA 4.3. *For each* $k \in \mathbb{N}_0$ *the collection of sets* $\{c(\theta) : \theta \in \Omega_k\}$ *form a partition of* $[N]^\infty$, *each part of which has diameter belonging to* $\{s^{k+1}, s^{k+2}, \ldots s^{k+a_{\max}}\}$ *where* $s = 1/2$. *That is,*

$$[N]^\infty = \bigsqcup_{\theta \in \Omega_k} c(\theta)$$

for all $k \in \mathbb{N}_0$.

PROOF. Assume that $\omega \in [N]^\infty$. There is a unique j such that $\omega|j \in \Omega_k$. Letting $\theta = w|j$ we have $\omega \in c(\theta) \subset [N]^\infty$. Therefore $[N]^\infty = \bigcup_{\theta \in \Omega_k} c(\theta)$.

Assume that $\theta, \theta' \in \Omega_k$. If $\omega \in c(\theta) \cap c(\theta')$, then by the definition of cylinder set either $\theta = \theta'$ or $|\theta| \neq |\theta'|$. However, if $|\theta| \neq |\theta'|$, then $\omega||\theta| = \theta \in \Omega_k$ and $\omega||\theta'| = \theta' \in \Omega_k$, which would contradict the uniqueness of j. Therefore $[N]^\infty = \bigsqcup_{\theta\in\Omega_k} c(\theta)$. □

5. A Canonical Sequence of Self-similar Tilings

To facilitate the proofs of the theorems stated in Section 3, another family of tilings is introduced, tilings isometric to those that are the subject of this paper. Let $A_k = s^{-k}A$, for all $k \in \mathbb{N} \cup \{-1, -2, \dots, -a_{\max}\}$, and define, for all $k \in \mathbb{N}$, a sequence of tilings T_k of A_k by $T_k = \{s^{-k} f_\sigma(A) : \sigma \in \Omega_k\}$.

The following lemma says, in particular, that T_k is a non-overlapping union of copies of T_{k-a_i} for $i \in [N]$ when $k \geq a_{\max}$, and T_k may be expressed as a non-overlapping union of copies of $T_{k-e(\omega)}$ for $\omega \in \Omega_l$ when k is somewhat larger than $l \in \mathbb{N}_0$. In this section the square union notation $\bigsqcup$ denotes a non-overlapping union.

Lemma 5.1. *For all $k \in \mathbb{N}_0$ the support of T_k is A_k. For all $\theta \in [N]^*$,*

$$\pi(\theta) = E_\theta T_{e(\theta)}$$

where E_θ is the isometry $f_{-\theta} s^{e(\theta)}$. Also

$$T_k = \bigsqcup_{i=1}^{N} E_{k,i} T_{k-a_i} \tag{5.1}$$

for all $k \geq a_{\max}$, where each of the mappings $E_{k,i} = s^{-k} \circ f_i \circ s^{k-a_i}$ is an isometry. More generally,

$$T_k = \bigsqcup_{\omega\in\Omega_l} E_{k,\omega} T_{k-e(\omega)}, \tag{5.2}$$

for all $k \geq l + a_{\max}$ and for all $l \in \mathbb{N}_0$, where each of the mappings $E_{k,\omega} = s^{-k} \circ f_\omega \circ s^{k-e(\omega)}$ is an isometry.

Proof. It is well-known that if $\mathcal{P}$ is a partition of $[N]^\infty$, then $A = \bigcup_{\omega\in\mathcal{P}} \phi(\omega)$ where $\phi : [N]^\infty \to A$ is the usual (continuous) coding map defined by $\phi(\omega) = \lim_{k\to\infty} f_{\omega|k}(x)$ for any fixed $x \in A$. By Lemma 4.3 we can choose $\mathcal{P} = \{c(\theta) : \theta \in \Omega_k\}$. Hence, the support of T_k is

$$s^{-k}\{\bigcup\{f_\sigma(A) : \sigma \in \Omega_k\}\} = s^{-k}\{\bigcup\{\phi(\omega) : \omega \in \{c(\theta) : \theta \in \Omega_k\}\}\} = s^{-k}A.$$

The expression $\pi(\theta) = E_\theta T_{e(\theta)}$ where E_θ is the isometry $f_{-\theta} s^{e(\theta)}$ follows from the definitions of $\pi(\theta)$ and T_k on taking $k = e(\theta)$.

Equation (5.1) follows from Lemma 4.1 according to these steps.

$$
\begin{aligned}
T_k &= \{s^{-k} f_\sigma(A) : \sigma \in \Omega_k\} \text{ (by definition)} \\
&= s^{-k}\{f_\sigma(A) : \sigma \in \bigsqcup_{i=1}^{N} i\Omega_{k-a_i}\} \text{ (by Lemma 4.1)} \\
&= s^{-k}\bigsqcup_{i=1}^{N}\{f_{i\sigma}(A) : \sigma \in \Omega_{k-a_i}\} \text{ (identity)} \\
&= s^{-k}\bigsqcup_{i=1}^{N} f_i(\{f_\sigma(A) : \sigma \in \Omega_{k-a_i}\}) \text{ (identity)} \\
&= \bigsqcup_{i=1}^{N} E_{k,i} T_{k-a_i} \text{ (by definition)}
\end{aligned}
$$

The function $E_{k,i} = s^{-k} \circ f_i \circ s^{k-a_i}$ is an isometry because it is a composition of three similitudes, of scaling ratios s^{-k}, s^{a_i}, and s^{k-a_i}. The proof of the last assertion is immediate: tiles meet at images under similitudes of the set of overlap $\mathcal{O} = \cup_{i \neq j} f_i(A) \cap f_j(A)$.

To prove Equation (5.2) we make use of the following identity.

$$
\Omega_k = \bigsqcup_{\omega \in \Omega_l} \omega \Omega_{k-e(\omega)} \text{ for all } l \in \mathbb{N}_0, \text{ for all } k \geq a_{\max} + l.
$$

This is proved as follows.

$$
\begin{aligned}
\Omega_k &= \{\sigma \in [N]^* : e^-(\sigma) \leq k < e^-(\sigma)\} \\
&= \{\sigma \in [N]^* : \sigma = \omega\theta \text{ for some } \omega \in \Omega_l, \theta \in \Omega_{k-e(\omega)}\} = \bigsqcup_{\omega \in \Omega_l} \omega \Omega_{k-e(\omega)}
\end{aligned}
$$

where we have used a similar observation to the start of the proof of Lemma 3, namely for any $\sigma \in \Omega_k$ with $k \geq a_{\max} + l$ there is a unique $\omega \in \Omega_l$ such that $\sigma = \omega\theta$.

We now have

$$
\begin{aligned}
T_k &= \{s^{-k} f_\sigma(A) : \sigma \in \Omega_k\} \text{ (by definition)} \\
&= \{s^{-k} f_\sigma(A) : \sigma \in \bigsqcup_{\omega \in \Omega_l} \omega \Omega_{k-e(\omega)}\} \\
&= \bigsqcup_{\omega \in \Omega_l} \{s^{-k} f_{\omega\theta}(A) : \theta \in \Omega_{k-e(\omega)}\} \\
&= \bigsqcup_{\omega \in \Omega_l} \{s^{-k} f_\omega s^{k-e(\omega)} s^{-k+e(\omega)} f_\theta(A) : \theta \in \Omega_{k-e(\omega)}\} \\
&= \bigsqcup_{\omega \in \Omega_l} E_{k,\omega} T_{k-e(\omega)} \text{ for all } k \geq a_{\max} + l. \qquad \square
\end{aligned}
$$

The following definition, formalizing the notion of an "isometric combination of tilings", will be used later, but it is convenient to place it here.

DEFINITION 5.2. Let $\{U_i : i \in \mathcal{I}\}$ be a collection of tilings. An **isometric combination of the set of tilings** $\{U_i : i \in \mathcal{I}\}$ is a tiling V that can be written in the form

$$V = \bigsqcup_{i=1}^{K} E^{(i)} U^{(i)}$$

for some $K \in \mathbb{N}$, where $E^{(i)} \in \mathcal{E}$, $U^{(i)} \in \{U_i : i \in \mathcal{I}\}$, for all $i \in \{1, 2, \ldots, K\}$.

For example, Lemma 5.1 tells us that any T_k can be written as an isometric combination of any set of tilings of the form $\{T_j, T_{j+1}, \ldots, T_{j+a_{\max}-1}\}$ when $k \geqslant j$.

PROPOSITION 1. *The sequence* $\{T_k\}$ *of tilings is self-similar in the following sense. Each of the sets in the magnified tiling* $s^{-1}T_k$ *is a union of tiles in* T_{k+1}.

PROOF. This follows at once from Lemma 4.2. The tiling T_{k+1} is obtained from T_k by applying the similitude s^{-1} and then splitting those resulting sets that are isometric to A. By splitting we mean we replace EA by $\{Ef_1(A), Ef_2(A),\ldots,Ef_N(A)\}$, see Section 9. □

6. Theorem 3.2: Existence and Continuity of Tilings

Let $A_{-\theta|k} := f_{-\theta|k}A$ for all $\theta \in [N]^\infty$. It is immediate from Definition 3.1 that the support of the tiling $\pi(\theta|k)$ is $A_{-\theta|k}$ and that $\pi(\theta|k)$ is isometric to the tiling $T_{e(k)}$ of $A_{e(k)}$. We use this fact repeatedly in the rest of this paper.

THEOREM 3.2. Each set $\pi(\theta)$ in $\mathbb{T}$ is a tiling of a subset of $\mathbb{R}^M$, the subset being bounded when $\theta \in [N]^*$ and unbounded when $\theta \in [N]^\infty$. For all $\theta \in [N]^\infty$ the sequence of tilings $\{\pi(\theta|k)\}_{k=0}^\infty$ is nested according to

$$\{f_i(A) : i \in [N]\} = \pi(\varnothing) \subset \pi(\theta|1) \subset \pi(\theta|2) \subset \pi(\theta|3) \subset \cdots . \tag{6.1}$$

For all $\theta \in [N]^\infty$, the prototile set for $\pi(\theta)$ is $\{s^i A : i = 1, 2, \cdots, a_{\max}\}$. Furthermore

$$\pi : [N]^* \cup \widetilde{[N]^\infty} \to \mathbb{T}$$

is a continuous map from the metric space $[N]^* \cup \widetilde{[N]^\infty}$ into the space $(\mathbb{K}(\mathbb{R}^M), d_\tau)$.

PROOF. Using Lemma 5.1, for $\theta = \theta_1\theta_2\cdots\theta_l \in [N]^*$ and $\theta^- = \theta_1\theta_2\cdots\theta_{l-1}$,

$$\pi(\theta) = E_\theta T_{e(\theta)} = \bigsqcup_{i=1}^{N} E_\theta E_{e(\theta),i} T_{k-a_i} \supset E_\theta E_{e(\theta),\theta_l} T_{k-a_{\theta_l}} = E_{\theta^-} T_{e(\theta^-)} = \pi(\theta^-).$$

It follows that $\{\pi(\theta|k)\}$ is an increasing sequence of tilings for all $\theta \in [N]^\infty$, as in Equation (3.1), and so converges to a well-defined limit. Since the maps in the IFS are strict contractions, their inverses are expansive, whence $\pi(\theta)$ is unbounded for all $\theta \in [N]^\infty$.

The fact that the tiles here are indeed tiles as we defined them at the start of this paper follows from three readily checked observations. (i) The tiles are nonempty perfect compact sets because they are isometric to the attractor, that is not a singleton, of an IFS of similitudes. (ii) There are only finitely many tiles that intersect any ball of finite radius. (iii) Any two tiles can meet only on a set that is contained in the image under a similitude of the set of overlap.

Next we prove that there are exactly $a_{\max}$ distinct tiles, up to isometry, in any tiling $\pi(\theta)$ for $\theta \in [N]^{\infty}$. The tiles of $\pi(\theta)$ take the form $\{f_{-\theta|k} f_{\sigma}(A) : \sigma \in \Omega_{e(\theta|k)}\}$ for some $k \in \mathbb{N}$. The mappings here are similitudes whose scaling factors are $\{s^{e(\sigma)-e(\theta|k)} : e(\sigma) - e(\theta|k) > 0 \geq e(\sigma) - e(\theta|k) - a_{|\sigma|}\}$, namely $\{s^m : m > 0 \geq m - a_{|\sigma|}\}$ for which the possible values are at most all of $\{1, 2, \ldots, a_{\max}\}$. That all of these values occur for large enough k follows from $\gcd\{a_i : i = 1, 2, \ldots, N\} = 1$.

Next we prove that $\pi : [N]^* \cup [N]^{\infty} \to \mathbb{T}$ is a continuous map from the metric space $[N]^* \cup [N]^{\infty}$ onto the space $(\mathbb{T}, d_T)$. The map $\pi|_{[N]^*} : [N]^* \to \mathbb{T}$ is continuous on the discrete part of the space $([N]^*, d_{[N]^* \cup [N]^{\infty}})$ because each point $\theta \in [N]^*$ possesses an open neighborhood that contains no other points of $[N]^* \cup [N]^{\infty}$. To show that π is continuous at points of $[N]^{\infty}$ we follow a similar method to the one in [**1**]. Let $\varepsilon > 0$ be given and let $B(R)$ be the open ball of radius R centered at the origin. Choose R so large that $h(\rho(\overline{B(R)}), \mathbb{S}^M) < \varepsilon$. This implies that if two tilings differ only where they intersect the complement of $\overline{B(R)}$, then their distance $d_{\mathcal{T}}$ apart is less than ε. But geometrical consideration of the way in which *support(*$\pi(\theta_1\theta_2\theta_3..\theta_k)$) grows with increasing k when $\theta \in \widetilde{[N]^{\infty}}$ shows that we can choose K so large that *support(*$\pi(\theta_1\theta_2\theta_3..\theta_k)) \cap \overline{B(R)}$ is constant for all $k \geq K$. It follows that

$$h(\rho(\pi(\theta_1\theta_2..\theta_k)), \rho(\pi(\theta_1\theta_2..\theta_l))) \leq \varepsilon$$

and as a consequence

$$h(\rho(\partial\pi(\theta_1\theta_2..\theta_k)), \rho(\partial\pi(\theta_1\theta_2..\theta_l))) \leq \varepsilon$$

for all $k, l \geq K$. It follows that $h(\rho(\pi(\theta)), \rho(\pi(\omega))) \leq \varepsilon$ whenever $\theta_1\theta_2..\theta_K = \omega_1\omega_2..\omega_K$. It follows that π is continuous. □

7. Theorem 3.3: When Do all Tilings Repeat the Same Patterns?

THEOREM 3.3.

(1) Each unbounded tiling in $\mathbb{T}$ is quasiperiodic and each pair of tilings in $\mathbb{T}$ is locally isomorphic.
(2) If θ is eventually periodic, then $\pi(\theta)$ is self-similar. In fact, if $\theta = \alpha\overline{\beta}$ for some $\alpha, \beta \in [N]^*$, then $f_{-\alpha} f_{-\beta} (f_{-\alpha})^{-1}$ is a self-similarity of $\pi(\theta)$.

PROOF. (1) First we prove quasiperiodicity. This is related to the self-similarity of the sequence of tilings $\{T_k\}$ mentioned in Proposition 1.

Let $\theta \in [N]^{\infty}$ be given and let P be a patch in $\pi(\theta)$. There is a $K_1 \in \mathbb{N}$ such that P is contained in $\pi(\theta|K_1)$. Hence an isometric copy of P is contained in T_{K_2} where $K_2 = e(\theta|K_1)$. Now choose $K_3 \in \mathbb{N}$ so that an isometric copy of T_{K_2} is contained in each T_k with $k \geq K_3$. That this is possible follows from the recursion (5.2) of Lemma 5.1 and $gcd\{a_i\} = 1$. In particular, $T_{K_2} \subset T_{K_3+i}$ for all $i \in \{1, 2, ..., a_{\max}\}$.

Now let $K_4 = K_3 + a_{\max}$. Then, for all $k \geq K_4$, the tiling T_k is an isometric combination of $\{T_{K_3+i} : i = 1, 2, ..., a_{\max}\}$, and each of these tilings contains a copy of T_{K_2} and in particular a copy of P.

Let $D = \max\{\|x - y\| : x, y \in A\}$ be the diameter of A. The support of T_k is $s^{-k}A$ which has diameter $s^{-k}D$. Hence *support*$(T_k) \subset B(x, 2s^{-k}D)$, the ball centered at x of radius $2s^{-k}D$, for all $x \in$ *support*(T_k). It follows that if $x \in$*support*$\pi(\theta')$

for any $\theta' \in [N]^\infty$, then $B(x, 2s^{-K_4}D)$ contains a copy of $support(T_{K_2})$ and hence a copy of P. Therefore all unbounded tilings in $\mathbb{T}$ are quasiperiodic.

In [7] Radin and Wolff define a tiling to have the local ismorphism property if for every patch P in the tiling there is some distance $d(P)$ such that every sphere of diameter $d(P)$ in the tiling contains an isometric copy of P. Above, we have proved a more general property of tilings, as defined here, of fractal blow-ups. Given P, there is a distance $d(P)$ such that each sphere of diameter $d(P)$, centered at any point belonging to the support of any unbounded tiling in $\mathbb{T}$, contains a copy of P.

(2) Let $\theta = \alpha\overline{\beta} = \alpha_1\alpha_2\cdots\alpha_l\beta_1\beta_2\cdots\beta_m\beta_1\beta_2\cdots\beta_m\beta_1\beta_2\cdots\beta_m\cdots$. We have the equivalent increasing unions

$$\pi(\theta) = \bigcup_{k\in\mathbb{N}} E_{\theta|k}T_{e(\theta|k)} = \bigcup_{j\in\mathbb{N}} E_{\theta|(l+jm)}T_{e(\theta|(l+jm))} = \bigcup_{j\in\mathbb{N}} E_{\theta|(l+jm+m)}T_{e(\theta|(l+jm+m))}$$

where, for all k,

$$E_{\theta|k} = f_{-\theta|k}s^{e(\theta|k)}.$$

We can write

$$\pi(\theta) = \bigcup_{j\in\mathbb{N}} E_{\theta|(l+jm)}T_{e(\theta|(l+jm))} = f_{-\alpha}\bigcup_{j\in\mathbb{N}} f_{-\beta}^{j}s^{e(\theta|(l+jm))}T_{e(\theta|(l+jm))},$$

and also

$$\pi(\theta) = \bigcup_{j\in\mathbb{N}} E_{\theta|(l+jm+m)}T_{e(\theta|(l+jm+m))} = f_{-\alpha}f_{-\beta}\bigcup_{j\in\mathbb{N}} f_{-\beta}^{j}s^{e(\theta|(l+jm+m))}T_{e(\theta|(l+jm+m))}.$$

Here $f_{-\beta}^{j}s^{e(\theta|(l+jm+m))}T_{e(\theta|(l+jm+m))}$ is a refinement of $f_{-\beta}^{j}s^{e(\theta|(l+jm))}T_{e(\theta|(l+jm))}$. It follows that $(f_{-\alpha}f_{-\beta})^{-1}\pi(\theta)$ is a refinement of $(f_{-\alpha})^{-1}\pi(\theta)$, from which it follows that $(f_{-\alpha})(f_{-\alpha}f_{-\beta})^{-1}\pi(\theta)$ is a refinement of $\pi(\theta)$. Therefore, every set in $(f_{-\alpha}f_{-\beta})(f_{-\alpha})^{-1}\pi(\theta)$ is a union of tiles in $\pi(\theta)$. □

8. Relative and Absolute Addresses

In order to understand how different tilings relate to one another, the notions of relative and absolute addresses of tiles are introduced. Given an IFS $\mathcal{F}$, the *set of absolute addresses* is defined to be:

$$\mathbb{A} := \{\theta.\omega : \theta \in [N]^*, \, \omega \in \Omega_{e(\theta)}, \, \theta_{|\theta|} \neq \omega_1\}.$$

Define $\widehat{\pi} : \mathbb{A} \to \{t \in T : T \in \mathbb{T}\}$ by $\widehat{\pi}(\theta.\omega) = f_{-\theta}.f_\omega(A)$. We say that $\theta.\omega$ is an *absolute address* of the tile $f_{-\theta}.f_\omega(A)$. It follows from Definition 3.1 that the map $\widehat{\pi}$ is surjective: every tile of $\{t \in T : T \in \mathbb{T}\}$ possesses at least one address. The condition $\theta_{|\theta|} \neq \omega_1$ is imposed to make cancellation unnecessary.

The *set of relative addresses* is associated with the tiling T_k of $A_k = s^{-k}A$ and is defined to be $\{.\omega : \omega \in \Omega_k\}$.

PROPOSITION 2. *There is a bijection between the set of relative addresses $\{.\omega : \omega \in \Omega_k\}$ and the tiles of T_k, for all $k \in \mathbb{N}_0$.*

PROOF. This follows from the non-overlapping union

$$A = \bigsqcup_{\omega\in\Omega_k} f_\omega(A).$$

This expression follows immediately from Lemma 4.3; see the start of the proof of Lemma 5.1. □

Accordingly, we say that $.\omega$, or equivalently $\varnothing.\omega$, where $\omega \in \Omega_k$, is *the relative address* of the tile $s^{-k}f_\omega(A)$ in the tiling T_k of A_k. Note that a tile of T_k may share the same relative address as a different tile of T_l for $l \neq k$.

Define the *set of labelled tiles* of T_k to be

$$\mathcal{A}_k = \{(.\omega, s^{-k}f_\omega(A)) : \omega \in \Omega_k\}$$

for all $k \in \mathbb{N}_0$. A key point about relative addresses is that the set of labelled tiles of T_k for $k \in \mathbb{N}$ can be computed recursively. Define

$$\mathcal{A}_k^{'} = \{(\omega, s^{-k}f_\omega(A)) \in \mathcal{A}_k : e(\omega) = k+1\} \subset \mathcal{A}_k.$$

An example of the following inductive construction is illustrated in Figure 6, and some corresponding tilings $\pi(\theta)$ labelled by absolute addresses are illustrated in Figure 7.

LEMMA 8.1. *For all $k \in \mathbb{N}_0$ we have*

$$\mathcal{A}_{k+1} = \mathcal{L}(\mathcal{A}_k \backslash \mathcal{A}_k^{'}) \cup \mathcal{M}(\mathcal{A}_k^{'})$$

where

$$\mathcal{L}(\omega, s^{-k}f_\omega(A)) = (\omega, s^{-k-1}f_\omega(A)),$$
$$\mathcal{M}(\omega, s^{-k}f_\omega(A)) = \{(\omega i, s^{-k-1}f_{\omega i}(A)) : i \in [N]\}.$$

PROOF. This follows immediately from Lemma 4.2. □

9. Strong Rigidity, Definition of "Amalgamation and Shrinking" Operation α on Tilings, and Proof of Theorem 3.4.

We begin this key section by introducing an operation, called "amalgamation and shrinking", that maps certain tilings into tilings. This leads to the main result of this section, Theorem 3.4, which, in turn, leads to Theorem 3.5.

DEFINITION 9.1. Let $T_0 = \{f_i(A) : i \in [N]\}$. The IFS $\mathcal{F}$ is said to be **rigid** if (i) there exists no non-identity isometry $E \in \mathcal{E}$ such that $T_0 \cap ET_0$ is non-empty and tiles $A \cap EA$, and (ii) there exists no non-identity isometry $E \in \mathcal{E}$ such that $A = EA$.

DEFINITION 9.2. Define $\mathbb{T}'$ to be the set of all tilings using the set of prototiles $\{s^iA : i = 1, 2, ..., a_{\max}\}$. Any tile that is isometric to $s^{a_{\max}}A$ is called a **small tile**, and any tile that is isometric to sA is called a **large tile**. We say that a tiling $P \in \mathbb{T}'$ comprises a set of **partners** if $P = ET_0$ for some $E \in \mathcal{E}$. Define $\mathbb{T}'' \subset \mathbb{T}'$ to be the set of all tilings in $\mathbb{T}'$ such that, given any $Q \in \mathbb{T}''$ and any small tile $t \in Q$, there is a set of partners of t, call it $P(t)$, such that $P(t) \subset Q$. Given any $Q \in \mathbb{T}''$ we define Q' to be the union of all sets of partners in Q.

DEFINITION 9.3. Let $\mathcal{F}$ be a rigid IFS. The amalgamation and shrinking operation $\alpha : \mathbb{T}'' \to \mathbb{T}'$ is defined by

$$\alpha Q = \{st : t \in Q \backslash Q'\} \cup \bigsqcup_{\{E \in \mathcal{E} : ET_0 \subset Q'\}} sEA.$$

LEMMA 9.4. *If $\mathcal{F}$ is rigid, the function $\alpha : \mathbb{T}'' \to \mathbb{T}'$ is well-defined and bijective; in particular for $\alpha^{-1} : \mathbb{T}' \to \mathbb{T}''$ is well defined by*

$$\alpha^{-1}(Q) = \{\alpha_Q^{-1}(q) : q \in Q\}$$

where

$$\alpha_Q^{-1}(q) = \begin{cases} s^{-1}q \text{ if } q \in Q \text{ is not a large tile} \\ s^{-1}ET_0 \text{ if } Eq \text{ is a large tile, some } E \in \mathcal{E} \end{cases}$$

PROOF. Because $\mathcal{F}$ is rigid, there can be no ambiguity with regard to which sets of tiles in a tiling are partners, nor with regard to which tiles are the partners of a given small tile. Hence $\alpha : \mathbb{T}'' \to \mathbb{T}'$ is well defined. Given any $T' \in \mathbb{T}'$ we can find a unique $Q \in \mathbb{T}''$ such that $\alpha(Q) = T'$, namely $\alpha^{-1}(Q)$ as defined in the lemma. □

LEMMA 9.5. *Let $\mathcal{F}$ be rigid and $k \in \mathbb{N}$. Then*
(i) $T_k \in \mathbb{T}''$;
(ii) $\alpha T_k = T_{k-1}$ and $\alpha^{-1}T_{k-1} = T_k$.

PROOF. As described in Lemma 8.1, T_k can constructed in a well-defined manner, starting from from T_{k-1}, by scaling and splitting, that is, by applying α^{-1}. Conversely T_{k-1} can be constructed from T_k by applying α. Statements (i) and (ii) are consequences. □

LEMMA 9.6. *If $\mathcal{F}$ is rigid, $L, M \in \mathbb{T}''$, and $L \cap M$ tiles support$(L) \cap$ support(M), then $L \cap M \in \mathbb{T}''$. Moreover,*

$$\alpha(L \cap M) = \alpha(L) \cap \alpha(M),$$

and $\alpha(L \cap M)$ tiles support$\,\alpha(L) \cap$ support$\,\alpha(M)$.

PROOF. Since $L, M \in \mathbb{T}'' \subset \mathbb{T}'$ lie in the range of α^{-1}, we can find unique $L', M' \in \mathbb{T}'$ such that

$$L = \alpha^{-1}L' \text{ and } M = \alpha^{-1}M'.$$

Note that $\alpha^{-1}(T') = \left\{\alpha^{-1}(t) : t \in T'\right\}$ for all $T' \in \mathbb{T}'$, which implies that α^{-1} commutes both with unions of disjoint tilings and also with intersections of tilings whose intersections tile the intersections of their supports. It follows that $L \cap M \in \mathbb{T}''$,

$$\begin{aligned} \alpha(L \cap M) &= \alpha(\alpha^{-1}L' \cap \alpha^{-1}M') \\ &= \alpha(\alpha^{-1}(L' \cap M')) \\ &= L' \cap M' \\ &= \alpha(L) \cap \alpha(M), \end{aligned}$$

and support $\alpha(L \cap M) =$ support $\alpha(L) \cap$ support $\alpha(M)$. □

DEFINITION 9.7. $\mathcal{F}$ is **strongly rigid** if $\mathcal{F}$ is rigid and whenever

$$i, j \in \{0, 1, 2, \ldots, a_{\max} - 1\},\ E \in \mathcal{E},\ \text{and } T_i \cap ET_j \text{ tiles } A_i \cap EA_j,$$

either $T_i \subset ET_j$ or $T_i \supset ET_j$.

Section 12 contain a few examples of strongly rigid IFSs.

LEMMA 9.8. *Let $\mathcal{F}$ be strongly rigid, $k, l \in \mathbb{N}_0$, and $E \in \mathcal{E}$.*
(i) If $ET_k \cap T_k$ is nonempty and tiles $EA_k \cap A_k$, then $E = id$.
(ii) If $EA_k \cap A_{k+l}$ is nonempty and $ET_k \cap T_{k+l}$ tiles $EA_k \cap A_{k+l}$, then $ET_k \subset T_{k+l}$.

PROOF. Suppose $ET_k \cap T_l \neq \varnothing$ and t.i.s. (tiles the intersection of supports). Without loss of generality assume $k \leq l$, for if not, then apply E^{-1}, then redefine E^{-1} as E.

Both ET_k and T_l lie in the domain of α^k, so we can apply Lemma 9.6 k times, yielding

$$\begin{aligned} \alpha^k(ET_k \cap T_l) &= s^k E s^{-k} T_0 \cap T_{l-k} \\ &:= \widetilde{E}T_0 \cap T_{l-k} \neq \varnothing, \end{aligned} \tag{9.1}$$

where $\widetilde{E}T_0 \cap T_{l-k}$ t.i.s. Now observe that by Lemma 5.1 we can write, for all $k' \geq l' + a_{\max}$,

$$T_{k'} = \bigsqcup_{\omega \in \Omega_{l'}} E_{k',\omega} T_{k'-e(\omega)} \left(= \left\{E_{k',\omega} T_{k'-e(\omega)} : \omega \in \Omega_{l'}\right\}\right),$$

where $E_{k',\omega} \in \mathcal{E}$ for all k', ω. Choosing $l' = k' - a_{\max}$ and noting that, for $\omega \in \Omega_{l'}$, we have $e(\omega) \in \{l'+1, \ldots, l' + a_{\max}\}$, and for $\omega \in \Omega_{k'-a_{\max}}$ we have that $e(\omega) \in \{k' - a_{\max} + 1, \ldots, k'\}$. Therefore $k' - e(\omega) \in \{0, 1, \ldots, a_{\max} - 1\}$ and we obtain the explicit representation

$$T_{k'} = \bigsqcup_{\omega \in \Omega_{k'-a_{\max}}} E_{k',\omega} T_{k'-e(\omega)}$$

which is an isometric combination of $\{T_0, T_1, \ldots, T_{a_{\max}-1}\}$. In particular, we can always reexpress T_{l-k} in (9.1) as isometric combination of $\{T_0, T_1, \ldots, T_{a_{\max}-1}\}$ and so there is some E' and some $T_m \in \{T_0, T_1, \ldots, T_{a_{\max}-1}\}$ with $\widetilde{E}T_0 \cap E'T_m \neq \varnothing$ and t.i.s.

By the strong rigidity assumption, this implies $\widetilde{E}T_0 \subset E'T_m$, which in turn implies $\widetilde{E}T_0 \subset T_{l-k}$ and t.i.s. Now apply α^{-k} to both sides of this last equation to obtain the conclusions of the lemma. □

THEOREM 3.4. Let $\mathcal{F}$ be strongly rigid. If $\theta, \theta' \in [N]^*$and $E \in \mathcal{E}$ are such that $\pi(\theta) \cap E\pi(\theta')$ is not empty and tiles $A_{-\theta} \cap EA_{-\theta'}$, then either $\pi(\theta) \subset E\pi(\theta')$ or $E\pi(\theta') \subset \pi(\theta)$. In this situation, if $e(\theta) = e(\theta')$, then $E\pi(\theta') = \pi(\theta)$.

PROOF. This follows from Lemma 9.8. If $\theta, \theta' \in [N]^*$and $E \in \mathcal{E}$ are such that $\pi(\theta) \cap E\pi(\theta')$ is not empty and tiles $A_{-\theta} \cap EA_{-\theta'}$, then $\theta, \theta' \in [N]^*$and $E \in \mathcal{E}$ are such that $E_\theta T_{e(\theta)} \cap EE_{\theta'} T_{e(\theta')}$ is not empty and tiles $E_\theta A_{e(\theta)} \cap EE_{\theta'} A_{e(\theta')}$, where $E_\theta = f_{-\theta} s^{e(\theta)}$ and $E_{\theta'} = f_{-\theta'} s^{e(\theta')}$ are isometries. Assume, without loss of generality, that $e(\theta) \leq e(\theta')$ and apply $E_{\theta'}^{-1} E^{-1}$ to obtain that $\theta, \theta' \in [N]^*$ and $E' = E_{\theta'}^{-1} E^{-1} E_\theta \in \mathcal{E}$ are such that $E' T_{e(\theta)} \cap T_{e(\theta')}$ is not empty and tiles $E' A_{e(\theta)} \cap A_{e(\theta')}$. By Lemma 9.8 it follows that $E' T_{e(\theta)} \subset T_{e(\theta')}$, i.e. $E_{\theta'}^{-1} E^{-1} E_\theta T_{e(\theta)} \subset T_{e(\theta')}$, i.e. $\pi(\theta) \subset E\pi(\theta')$. If also $e(\theta') \leq e(\theta)$ (i.e. $e(\theta') = e(\theta)$), then also $E\pi(\theta') \subset \pi(\theta)$. Therefore $E\pi(\theta') = \pi(\theta)$. □

10. Theorem 3.5: When is a Tiling Non-Periodic?

THEOREM 3.5. *If F is strongly rigid, then there does not exist any non-identity isometry $E \in \mathcal{E}$ and $\theta \in [N]^\infty$ such that $E\pi(\theta) \subset \pi(\theta)$.*

PROOF. Suppose there exists an isometry E such that $E\pi(\theta) = \pi(\theta)$. Then we can choose $K \in \mathbb{N}_0$ so large that $E\pi(\theta|K) \cap \pi(\theta|K) \neq \varnothing$ and $E\pi(\theta|K) \cap \pi(\theta|K)$ tiles $EA_{-\theta|K} \cap A_{-\theta|K}$. By Theorem 3.4 it follows that $E\pi(\theta|K) = \pi(\theta|K)$. This implies $EE_\theta T_{e(\theta|K)} = E_\theta T_{e(\theta|K)}$, whence, because $E_\theta T_{e(\theta|K)}$ is in the domain of $\alpha^{e(\theta|K)}$ and $\alpha^{e(\theta|K)} T_{e(\theta|K)} = T_0$, we have by Lemma 9.5

$$\begin{aligned}
&\alpha^{e(\theta|K)} EE_\theta T_{e(\theta|K)} = \alpha^{e(\theta|K)} E_\theta T_{e(\theta|K)} \\
&\quad\Longrightarrow s^{e(\theta|K)} EE_\theta s^{-e(\theta|K)} \alpha^{e(\theta|K)} T_{e(\theta|K)} = s^{e(\theta|K)} E_\theta s^{-e(\theta|K)} \alpha^{e(\theta|K)} T_{e(\theta|K)} \\
&\quad\Longrightarrow s^{e(\theta|K)} EE_\theta s^{-e(\theta|K)} T_0 = s^{e(\theta|K)} E_\theta s^{-e(\theta|K)} T_0 \\
&\quad\Longrightarrow s^{e(\theta|K)} EE_\theta s^{-e(\theta|K)} = s^{e(\theta|K)} E_\theta s^{-e(\theta|K)} \text{ (using rigidity)} \\
&\quad\Longrightarrow E = id.
\end{aligned}$$

□

It follows that if $\mathcal{F}$ is strongly rigid, then $\pi(\theta)$ is non-periodic for all θ.

11. When is $\pi : [N]^* \cup [N]^\infty \to \mathbb{T}$ invertible?

LEMMA 11.1. *For all $\mathcal{F}$ the restricted mapping $\pi|_{[N]^*.} : [N]^* \to \mathbb{T}$ is injective.*

PROOF. To simplify notation, write $\pi = \pi|_{[N]^*}$. We show how to calculate θ given $\pi(\theta)$ for $\theta \in [N]^*$. By Lemma 5.1 we have $\pi(\theta) = E_\theta T_{e(\theta)}$, where E is the isometry $f_{-\theta} s^{e(\theta)}$. Given $\pi(\theta)$, we can calculate

$$e(\theta) = \frac{\ln |A| - \ln |\pi(\theta)|}{\ln s},$$

where $|U|$ denotes the diameter of the set U.

We next show that if $E_\theta = E_{\theta'}$ for some $\theta \neq \theta'$ with $e(\theta) = e(\theta')$, then $\pi(\theta) \neq \pi(\theta')$. To do this, suppose that $E_\theta = E_{\theta'}$. This implies that $f_{-\theta} = f_{-\theta'}$ which implies $(f_{-\theta'})^{-1} f_{-\theta} = id$, which is not possible when $\theta \neq \theta'$, as we prove next. The similitude $(f_{-\theta'})^{-1} f_{-\theta}$ maps $(f_{-\theta})^{-1}(A) \subset A$ to $(f_{-\theta'})^{-1}(A) \subset A$, and these two subsets of A are distinct for all $\theta, \theta' \in [N]^*$with $\theta \neq \theta'$, as we prove next.

Let ω, ω' denote the two strings θ, θ' written in inverse order, so that $\theta \neq \theta'$ is equivalent to $\omega \neq \omega'$. First suppose $|\omega| = |\omega'| = m$ for some $m \in \mathbb{N}$. Then use

$$A = \bigsqcup_{\omega \in [N]^m} f_\omega(A),$$

which tells us that $f_\omega(A)$ and $f_{\omega'}(A)$ are disjoint. Since $(f_{-\theta'})^{-1} f_{-\theta}$ maps $(f_{-\theta})^{-1}(A) = f_\omega(A)$ to the distinct set $(f_{-\theta'})^{-1}(A) = f_{\omega'}(A)$, we must have $(f_{-\theta'})^{-1} f_{-\theta} \neq id$.

Now suppose $|\omega| = m < |\omega'| = m'$. If both strings ω and ω' agree through the first m places, then $f_\omega(A)$ is a strict subset of $f_{\omega'}^{-1}(A)$ and again we cannot have $(f_{-\theta'})^{-1} f_{-\theta} = id$. If both strings ω and ω' do not agree through the first m places, then let $p < m$ be the index of their first disagreement. Then we find that $f_\omega(A)$ is a subset of $f_{\omega|p}(A)$, while $f_{\omega'}(A)$ is a subset of the set $f_{\omega'|p}(A)$, which is disjoint from $f_{\omega|p}(A)$. Since $(f_{-\theta'})^{-1} f_{-\theta}$ maps $f_\omega(A)$ to $f_{\omega'}(A)$, we again have that $(f_{-\theta'})^{-1} f_{-\theta} \neq id$. □

We are going to need a key property of certain shifts maps on tilings, defined in the next lemma.

LEMMA 11.2. *The mappings* $S_i : \{\pi(\theta) : \theta \in [N]^l \cup [N]^\infty, l \geq a_i\} \to \mathbb{T}'$ *for* $i \in [N]$ *are well-defined by* $S_i = f_i s^{-a_i} \alpha^{a_i}$. *It is true that* $S_{\theta_1}\pi(\theta) = \pi(S\theta)$ *for all* $\theta \in [N]^l \cup [N]^\infty$ *where* $l \geq a_{\theta_1}$.

PROOF. We only consider the case $\theta \in [N]^\infty$. The case $\theta \in [N]^l$ is treated similarly. A detailed calculation, outlined next, is needed. The key idea is that $\pi(\theta)$ is broken up into a countable union of disjoint tilings, each of which belongs to the domain of α^k for all $k \leq K$ for any $K \in \mathbb{N}$. For all $K \in \mathbb{N}$ we have:

$$\pi(\theta) = E_{\theta|K} T_{e(\theta|K)} \bigsqcup \bigsqcup_{k=K}^{\infty} E_{\theta|k+1} T_{e(\theta|k+1)} \backslash E_{\theta|k} T_{e(\theta|k)}.$$

The tilings on the r.h.s. are indeed disjoint, and each set belongs to the domain of $\alpha^{e(\theta|K)}$, so we can use Lemma 9.6 applied countably many times to yield

$$S_{\theta_1}\pi(\theta) = S_{\theta_1}\left(E_{\theta|K} T_{e(\theta|K)}\right) \bigsqcup_{k=K}^{\infty} S_{\theta_1}\left(E_{\theta|k+1} T_{e(\theta|k+1)}\right) \backslash S_{\theta_1}\left(E_{\theta|k} T_{e(\theta|k)}\right).$$

Evaluating, we obtain successively

$$\begin{aligned}
S_{\theta_1}\pi(\theta) =& f_{\theta_1} s^{-a_{\theta_1}} \alpha^{a_{\theta_1}} \left(E_{\theta|K} T_{e(\theta|K)}\right) \\
& \bigsqcup_{k=K}^{\infty} f_{\theta_1} s^{-a_{\theta_1}} \alpha^{a_{\theta_1}} \left(E_{\theta|k+1} T_{e(\theta|k+1)}\right) \backslash f_{\theta_1} s^{-a_{\theta_1}} \alpha^{a_{\theta_1}} \left(E_{\theta|k} T_{e(\theta|k)}\right), \\
S_{\theta_1}\pi(\theta) =& f_{\theta_1} E_{\theta|K} s^{-a_{\theta_1}} \alpha^{a_{\theta_1}} T_{e(\theta|K)} \\
& \bigsqcup_{k=K}^{\infty} f_{\theta_1} E_{\theta|k+1} s^{-a_{\theta_1}} \alpha^{a_{\theta_1}} T_{e(\theta|k+1)} \backslash f_{\theta_1} E_{\theta|k+1} s^{-a_{\theta_1}} \alpha^{a_{\theta_1}} T_{e(\theta|k)}, \\
S_{\theta_1}\pi(\theta) =& f_{\theta_1} E_{\theta|K} s^{-a_{\theta_1}} T_{e(S\theta|K-1)} \\
& \bigsqcup_{k=K}^{\infty} f_{\theta_1} E_{\theta|k+1} s^{-a_{\theta_1}} T_{e(S\theta|k)} \backslash f_{\theta_1} E_{\theta|k} s^{-a_{\theta_1}} T_{e(S\theta|k-1)}, \\
S_{\theta_1}\pi(\theta) =& E_{S\theta|(K-1)} T_{e(S\theta|K-1)} \bigsqcup_{k=K}^{\infty} E_{S\theta|k} T_{e(S\theta|k-1)} \backslash E_{S\theta|k-1} T_{e(S\theta|k-1)} = \pi(S\theta).
\end{aligned}$$

□

THEOREM 3.6. If $\pi(i) \cap \pi(j)$ does not tile $(support\ \pi(i)) \cap (support\ \pi(j))$ for all $i \neq j$, then $\pi : [N]^* \cup [N]^\infty \to \mathbb{T}$ is one-to-one.

PROOF. The map π is one-to-one on $[N]^*$ by Lemma 11.1, so we restrict attention to points in $[N]^\infty$. If θ and θ' are such that $\theta_1 = i$ and $\theta_1' = j$, then the result is immediate because $\pi(\theta)$ contains $\pi(i)$ and $\pi(\theta')$ contains $\pi(j)$. If θ and θ' agree through their first K terms with $K \geq 1$ and $\theta_{K+1} \neq \theta'_{K+1}$, then $\pi(S^K\theta) \neq \pi(S^K\theta')$. Now apply $S_{\theta_1}^{-1} S_{\theta_2}^{-1} ... S_{\theta_K}^{-1}$ to obtain $\pi(\theta) \neq \pi(\theta')$. (We can do this last step because $S_i^{-1} = \left(f_i s^{-a_i} \alpha^{a_i}\right)^{-1} = \alpha^{-a_i} s^{a_i} f_i^{-1}$ has as its domain all of $\mathbb{T}'$ and maps $\mathbb{T}'$ into $\mathbb{T}'$.) □

12. Examples

12.1. Golden b tilings. A *golden b* $G \subset \mathbb{R}^2$ is illustrated in Figure 3. This hexagon is the only rectilinear polygon that can be tiled by a pair of differently scaled copies of itself [**8**, **9**]. Throughout this subsection the IFS is $\mathcal{F} = \{\mathbb{R}^2; f_1, f_2\}$, where

$$f_1(x,y) = \begin{pmatrix} 0 & s \\ -s & 0 \end{pmatrix} \begin{pmatrix} x \\ y \end{pmatrix} + \begin{pmatrix} 0 \\ s \end{pmatrix}, \quad f_2(x,y) = \begin{pmatrix} -s^2 & 0 \\ 0 & s^2 \end{pmatrix} \begin{pmatrix} x \\ y \end{pmatrix} + \begin{pmatrix} 1 \\ 0 \end{pmatrix},$$

where the scaling ratios s and s^2 obey $s^4 + s^2 = 1$, which tells us that $s^{-2} = \alpha^{-2}$ is the golden mean. The attractor of $\mathcal{F}$ is $A = G$. It is the union of two prototiles

FIGURE 3. A golden b is a union of two tiles, a small one and its partner, a large one. The vertices of this golden b are located at $(0,0)$ $(1,0)$ $(1,\alpha^3)$ (α^2,α^3) (α^2,α) $(0,\alpha)$ in counterclockwise order, starting at the lower left corner, where α^{-2} is the golden mean. This picture also represents a tiling $T_0 = \pi(\varnothing)$.

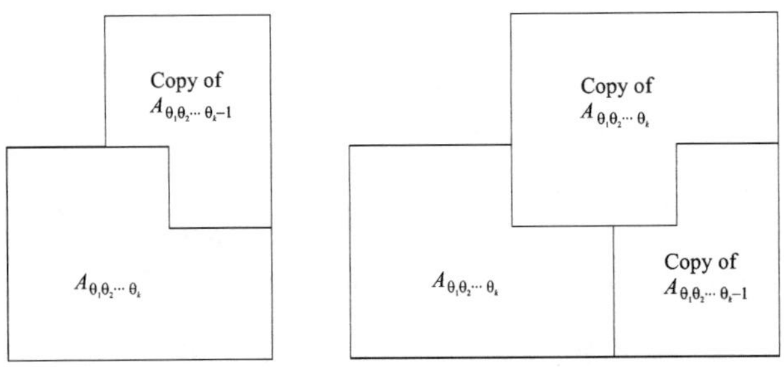

FIGURE 4. Structures of $A_{\theta_1\theta_2\cdots\theta_k 1}$ and $A_{\theta_1\theta_2\cdots\theta_k 2}$ relative to $A_{\theta_1\theta_2\cdots\theta_k}$.

$f_1(G)$ and $f_2(G)$. Copies of these prototiles are labelled L and S. In this example, note that $e(\theta) = \theta_1 + \theta_2 + \cdots + \theta_{|\theta|}$ for $\theta \in [2]^*$.

The figures in this section illustrate some earlier concepts in the context of the golden b. Using some of these figures, it is easy to check that $\mathcal{F}$ is strongly rigid, so the tilings $\pi(\theta)$ have all of the properties ascribed to them by the theorems in the earlier sections.

The relationships between $A_{\theta_1\theta_2\cdots\theta_k 1}$ and $A_{\theta_1\theta_2\cdots\theta_k 2}$ relative to $A_{\theta_1\theta_2\cdots\theta_k}$ are illustrated in Figure 4. Figure 5 illustrates some of the sets $A_{\theta_1\theta_2\theta_3..\theta_k}$ and the corresponding tilings $\pi(\theta_1\theta_2\theta_3..\theta_k)$.

In Section 8, procedures were described by which the relative addresses of tiles in $T(\theta|k)$ and the absolute addresses of tiles in $\pi(\theta|k)$ may be calculated recursively. Relative addresses for some golden b tilings are illustrated in Figure 6. Figure 7 illustrates absolute addresses for some golden b tilings.

The map $\pi : [2]^* \cup [2]^\infty \to \mathbb{T}$ is 1-1 by Theorem 3.6, because $\pi(1) \cup \pi(2)$ does not tile the interesection of the supports of $\pi(1)$ and $\pi(2)$, as illustrated in Figure 8.

We note that $\pi(\overline{21})$ is an aperiodic tiling of the upper right quadrant of $\mathbb{R}^2$.

12.2. Fractal tilings with non-integer dimension. The left hand image in Figure 9, shows the attractor of the IFS represented by the different coloured regions, there being 8 maps, and provides an example of a strongly rigid IFS. The right hand image represents the attractor of the same IFS minus one of the maps, also strongly rigid, but in this case the dimensions of the tiles is less than two and greater than one. Figure 2 (in Section 1) illustrates a part of a fractal blow up of a different but related 7 map IFS, also strongly rigid, and the corresponding tiling.

Figure 10 left shows a tiling associated with the IFS $\mathcal{F}$ represented on the left in Figure 9, while the tiling on the right is another example of a fractal tiling, obtained by dropping one of the maps of $\mathcal{F}$.

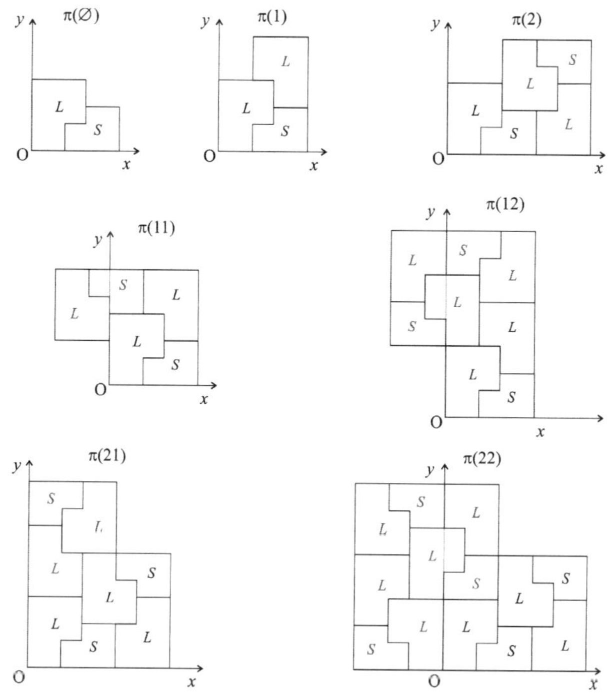

FIGURE 5. Some of the sets $A_{\theta_1\theta_2\theta_3..\theta_k}$ and the corresponding tilings $\pi(\theta_1\theta_2\theta_3..\theta_k)$. The recursive organization is such that $\pi(\varnothing) \subset \pi(\theta_1) \subset \pi(\theta_1\theta_2) \subset \cdots$ regardless of the choice $\theta_1\theta_2\theta_3.. \in \{1,2\}^\infty$.

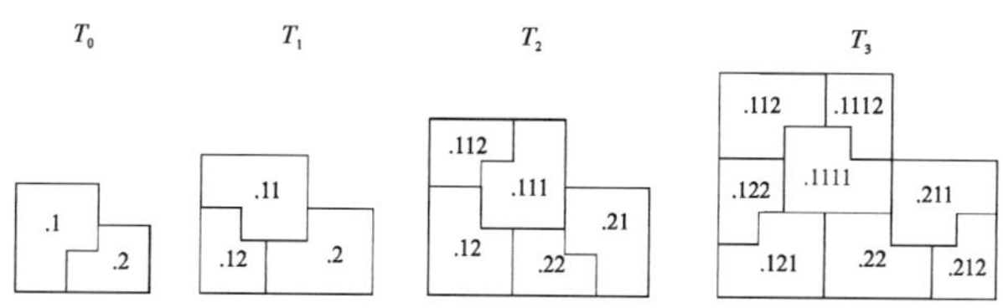

FIGURE 6. Relative addresses, the addresses of the tiles that comprise the tilings T_0, T_1, T_2, T_3 of A_0, A_1, A_2, A_3.

12.3. Tilings derived from Cantor sets. Our results apply to the case where $\mathcal{F} = \{\mathbb{R}^M; f_i(x) = s^{a_i}O_i + q_i, i \in [N]\}$ where $\{O_i, q_i : i \in [N]\}$ is fixed in a general position, the a_is are positive integers, and s is chosen small enough to ensure that the attractor is a Cantor set. In this situation the set of overlap is

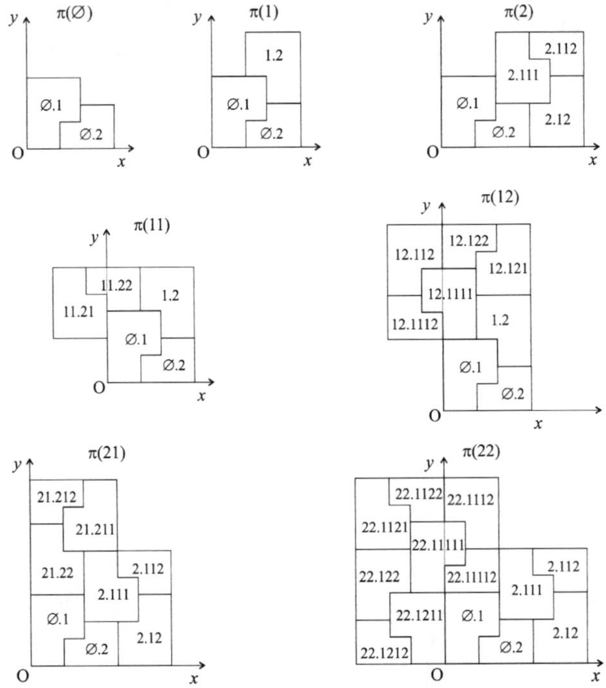

FIGURE 7. Absolute addresses associated with the golden b.

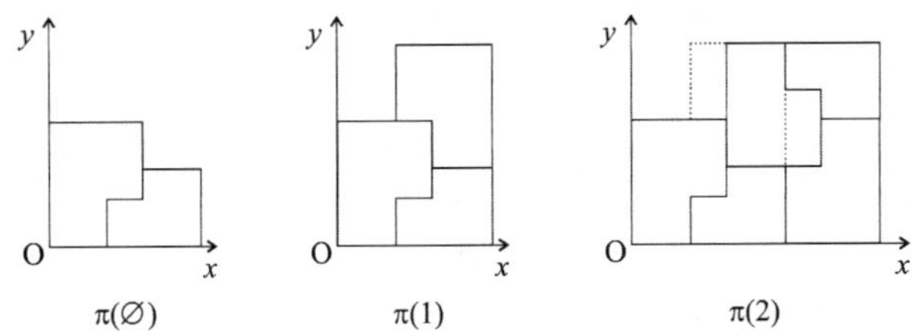

FIGURE 8. The boundaries of the tilings $\pi(\emptyset)$, $\pi(1)$, $\pi(2)$, with the parts of the boundaries of the tiles in $\pi(1)$ that are not parts of the boundaries of tiles in $\pi(2)$ superimposed in red on the rightmost image.

empty and it is to be expected that $\mathcal{F}$ is strongly rigid, in which case all tilings (by a finite set of prototiles, each a Cantor set) will be non-periodic. We can then take s to be the supremum of value such that the set of overlap is empty, to yield interesting "just touching" tilings.

Acknowledgement

We thank Louisa Barnsley for many of the illustrations. We thank a referee for helpful comments, and we thank Rodrigo Treviño for interesting discussions. We acknowledge support for this work by Australian Research Council grant DP130102738.

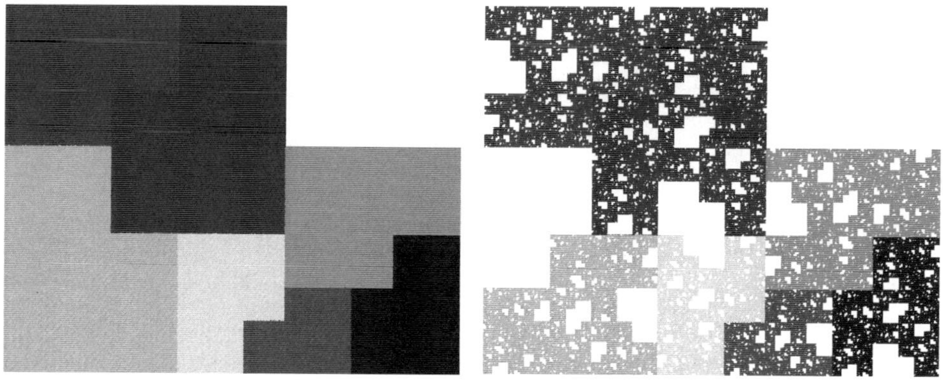

FIGURE 9. See text.

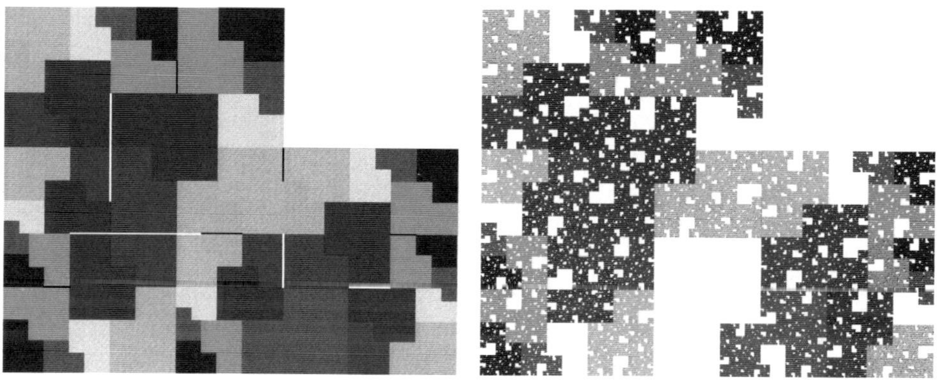

FIGURE 10. See text.

References

[1] Jared E. Anderson and Ian F. Putnam, *Topological invariants for substitution tilings and their associated C^*-algebras*, Ergodic Theory Dynam. Systems **18** (1998), no. 3, 509–537, DOI 10.1017/S0143385798100457. MR1631708

[2] Christoph Bandt, Michael Barnsley, Markus Hegland, and Andrew Vince, *Old wine in fractal bottles I: Orthogonal expansions on self-referential spaces via fractal transformations*, Chaos Solitons Fractals **91** (2016), 478–489, DOI 10.1016/j.chaos.2016.07.007. MR3551732

[3] Michael F. Barnsley and Andrew Vince, *Fast basins and branched fractal manifolds of attractors of iterated function systems*, SIGMA Symmetry Integrability Geom. Methods Appl. **11** (2015), Paper 084, 21, DOI 10.3842/SIGMA.2015.084. MR3412269

[4] Michael Barnsley and Andrew Vince, *Fractal tilings from iterated function systems*, Discrete Comput. Geom. **51** (2014), no. 3, 729–752, DOI 10.1007/s00454-014-9589-2. MR3201253

[5] Michael Barnsley and Andrew Vince, *Self-similar polygonal tiling*, Amer. Math. Monthly **124** (2017), no. 10, 905–921, DOI 10.4169/amer.math.monthly.124.10.905. MR3733299

[6] John E. Hutchinson, *Fractals and self-similarity*, Indiana Univ. Math. J. **30** (1981), no. 5, 713–747, DOI 10.1512/iumj.1981.30.30055. MR625600

[7] Charles Radin and Mayhew Wolff, *Space tilings and local isomorphism*, Geom. Dedicata **42** (1992), no. 3, 355–360, DOI 10.1007/BF02414073. MR1164542

[8] K. Scherer, A Puzzling Journey To The Reptiles And Related Animals, privately published, Auckland, New Zealand, 1987.

[9] James H. Schmerl, *Dividing a polygon into two similar polygons*, Discrete Math. **311** (2011), no. 4, 220–231, DOI 10.1016/j.disc.2010.10.021. MR2739907

[10] Lorenzo Sadun, *Tiling spaces are inverse limits*, J. Math. Phys. **44** (2003), no. 11, 5410–5414, DOI 10.1063/1.1613041. MR2014868
[11] Robert S. Strichartz, *Fractals in the large*, Canad. J. Math. **50** (1998), no. 3, 638–657, DOI 10.4153/CJM-1998-036-5. MR1629847

Australian National University, Canberra, ACT, Australia

University of Florida, Gainesville, Florida

Contemporary Mathematics
Volume **731**, 2019
https://doi.org/10.1090/conm/731/14673

Regularly varying functions, generalized contents, and the spectrum of fractal strings

Tobias Eichinger and Steffen Winter

Abstract. We revisit the problem of characterizing the eigenvalue distribution of the Dirichlet-Laplacian on bounded open sets $\Omega \subset \mathbb{R}$ with fractal boundaries. It is well-known from the results of Lapidus and Pomerance [**15**] that the asymptotic second term of the eigenvalue counting function can be described in terms of the Minkowski content of the boundary of Ω provided it exists. He and Lapidus [**7**] discussed a remarkable extension of this characterization to sets Ω with boundaries that are not necessarily Minkowski measurable. They employed so-called generalized Minkowski contents given in terms of gauge functions more general than the usual power functions. The class of valid gauge functions in their theory is characterized by some technical conditions, the geometric meaning and necessity of which is not obvious. Therefore, it is not completely clear how general the approach is and which sets Ω are covered. Here we revisit these results and put them in the context of regularly varying functions. Using Karamata theory, it is possible to get rid of most of the technical conditions and simplify the proofs given by He and Lapidus, revealing thus even more of the beauty of their results. Further simplifications arise from characterization results for Minkowski contents obtained in [**20**]. We hope our new point of view on these spectral problems will initiate some further investigations of this beautiful theory.

1. Introduction

Given a bounded open set $\Omega \subset \mathbb{R}^d$ with boundary $F := \partial\Omega$, consider the eigenvalue problem

$$(1.1)\qquad \begin{aligned} -\Delta u &= \lambda u && \text{in } \Omega, \\ u &= 0 && \text{on } F, \end{aligned}$$

where $\Delta = \sum_{i=1}^d \partial^2/\partial x_i^2$ denotes the Laplace operator. Recall that $\lambda \in \mathbb{R}$ is called an eigenvalue of (1.1), if there exists a function $u \neq 0$ in $H_0^1(\Omega)$ (the closure of

---

2010 *Mathematics Subject Classification.* 35P20, 28A80.

Key words and phrases. Minkowski content, fractal string, Laplace operator, spectral asymptotics, eigenvalue counting function, Weyl-Berry conjecture, regularly varying function, Karamata theory.

We are grateful to Günter Last for pointing us to the theory of regularly varying functions. Part of the results are based on the master's thesis of the first author. The second author was supported in part by the DFG research unit *Geometry and Physics of spatial random systems*, grant LA 965/8-2.

$C_0^\infty(\Omega)$, the space of smooth functions with compact support contained in Ω, in the Sobolev space $H_1(\Omega)$) satisfying $-\Delta u = \lambda u$ in the distributional sense.

It is well-known that the spectrum of Δ is positive and discrete, i.e. the eigenvalues of (1.1) form an increasing sequence $(\lambda_i)_{i\in\mathbb{N}}$ of strictly positive numbers with $\lambda_i \to \infty$, as $i \to \infty$. If Ω is interpreted as a vibrating membrane held fixed along its boundary, the reciprocals of the eigenvalues can be interpreted as the natural frequencies and the corresponding eigenfunctions as the natural vibrations (overtones) of the system. Much work has been devoted to the question how much geometric information about Ω can be recovered just from listening to its sound.

The famous *Weyl's law* [**23**] describes the growth of the eigenvalue counting function N, defined by

$$N(\lambda) := \#\{i \in \mathbb{N} : \lambda_i \leq \lambda\}, \quad \lambda > 0.$$

It states (originally for sufficiently smooth domains, but nowadays known to hold for arbitrary bounded open sets $\Omega \subset \mathbb{R}^d$, see [**4, 18**]) that N is asymptotically equivalent to the so-called *Weyl term* φ given by

$$\varphi(\lambda) := (2\pi)^{-d}\omega_d|\Omega|_d\lambda^{d/2}. \tag{1.2}$$

Here ω_d denotes the volume of the d-dimensional unit ball and $|\cdot|_d$ is the Lebesgue measure in $\mathbb{R}^d$. Since the dimension d of Ω as well as its volume appear in the Weyl term, we are not only able to infer the dimension of Ω from the growth of N but also its volume. For sets with sufficiently smooth boundaries, Weyl conjectured in [**24**] the existence of a second additive term of the order $(d-1)/2$ in the asymptotic expansion of N with a prefactor being determined by the surface area of the boundary F of Ω:

$$N(\lambda) = \varphi(\lambda) + c_{d-1}\mathcal{H}^{d-1}(F)\lambda^{(d-1)/2} + o(\lambda^{(d-1)/2}), \text{ as } \lambda \to \infty, \tag{1.3}$$

with a constant c_{d-1} solely depending on the dimension $d-1$ of F. Many people have contributed to the verification and generalization of Weyl's conjecture, culminating in the results of Seeley [**21, 22**], Pham The Lai [**9**] (who established that for all bounded open sets Ω with C^∞-boundary the order $\frac{d-1}{2}$ of the second term is correct) and Ivrii [**8**], who proved the correctness of the prefactor under some additional assumption (roughly, that the set of multiply reflected periodic geodesics in $\overline{\Omega}$ is of measure zero). Therefore, for sets with smooth boundaries, one can 'hear' not only the volume of the set but also the dimension and measure of its boundary.

And yet, what if the boundary is non-smooth? Motivated by experiments on wave scattering in porous media, the physicist M. V. Berry conjectured in [**1, 2**] that for general bounded open sets Ω the second term in (1.3) should be replaced by $c_{n,H}\mathcal{H}^H(F)\lambda^{H/2}$ involving Hausdorff dimension H and Hausdorff measure $\mathcal{H}^H(F)$ of the boundary F of Ω. This was disproved by Brossard and Carmona [**5**] who suggested to use Minkowski content and dimension in the second term instead. More precisely, these notions need to be considered relative to the set Ω. Recall that for any bounded set $\Omega \subset \mathbb{R}^d$, $F := \partial\Omega$ and $s \geq 0$, the *s-dimensional Minkowski content of F* (relative to Ω) is defined by

$$\mathcal{M}^s(F) = \lim_{r\searrow 0} \frac{|F_r \cap \Omega|_d}{r^{d-s}}, \tag{1.4}$$

provided the limit exists. Here $F_r := \{x \in \mathbb{R}^d : d(x, F) \leq r\}$ is the *r-parallel set* of F. We write $\underline{\mathcal{M}}^s(F)$ and $\overline{\mathcal{M}}^s(F)$ for the corresponding lower and upper limits.

The numbers $\overline{dim}_M F := \inf\{s \geq 0 : \overline{\mathcal{M}}^s(F) = 0\}$ and $\underline{dim}_M F := \inf\{s \geq 0 : \underline{\mathcal{M}}^s(F) = 0\}$ are the *upper and lower Minkowski dimension* of F, and in case these numbers coincide, the common value is known as the Minkowski dimension $\dim_M F$ of F (relative to Ω). If $\mathcal{M}^s(F)$ is both positive and finite (for some s), we say that F is *(s-dimensional) Minkowski measurable*, and if $0 < \underline{\mathcal{M}}^s(F) \leq \overline{\mathcal{M}}^s(F) < \infty$, then F is called *(s-dimensional) Minkowski nondegenerate*. Note that in this case, necessarily $\dim_M F = s$. In the above notation and also in most of the further discussion, we suppress the dependence on Ω, but of course it should be kept in mind that all these notions are considered relative to Ω here.

Lapidus [**12**] established that indeed the (upper) Minkowski dimension D of $F = \partial\Omega$ gives the correct order of growth for the second term in (1.3) (provided $\overline{\mathcal{M}}^D(F) < \infty$), and formulated the so-called *Modified Weyl-Berry conjecture* (MWB) under the assumption that F is Minkowski measurable:

$$N(\lambda) = \varphi(\lambda) - c_{d,D}\mathcal{M}^D(F)\lambda^{D/2} + o(\lambda^{D/2}), \text{ as } \lambda \to \infty. \tag{1.5}$$

Here $c_{d,D}$ is a constant depending only on d and the Minkowski dimension $D = \dim_M(F)$ of F. Lapidus and Pomerance showed in [**15**] that the MWB holds in dimension $d = 1$, and in [**16**] they constructed counterexamples, which disprove MWB in any dimension $d \geq 2$. In [**15**], they also established the following more general result, which characterizes the second order asymptotics of N for sets $\Omega \subset \mathbb{R}$ with a Minkowski nondegenerate boundary F.

Recall that (for some $a \in [0, \infty]$) two positive functions h_1, h_2 defined in some neighborhood of a are called *asymptotically similar as* $y \to a$, in symbols $h_1(y) \asymp h_2(y)$, as $y \to a$, if and only if there are positive constants $\underline{c}$, $\overline{c}$ such that $\underline{c}h_1(y) \leq h_2(y) \leq \overline{c}h_1(y)$ for all y close enough to a. Moreover, h_1 and h_2 are *asymptotically equal, as* $y \to a$, $h_1(y) \sim h_2(y)$, as $y \to a$, if and only if $\lim_{y\to a} h_1(y)/h_2(y) = 1$.

THEOREM 1.1 ([**15**, Theorem 2.4]). *Let $\Omega \subset \mathbb{R}$ be a bounded open set and $F := \partial\Omega$. Let $D \in (0, 1)$ and let $\mathcal{L} = (l_j)_{j\in\mathbb{N}}$ denote the associated fractal string. Then the following assertions are equivalent:*

(i) $0 < \underline{\mathcal{M}}^D(F) \leq \overline{\mathcal{M}}^D(F) < \infty$,
(ii) $l_j \asymp j^{-\frac{1}{D}}$, *as* $j \to \infty$,
(iii) $\sum_{j=1}^{\infty}\{l_j x\} \asymp x^D$, *as* $x \to \infty$,
(iv) $\varphi(\lambda) - N(\lambda) \asymp \lambda^{\frac{D}{2}}$, *as* $\lambda \to \infty$.

Here and throughout, $\{y\} := y - [y]$ denotes the fractional part and $[y]$ the integer part of a number $y \in \mathbb{R}$. (The connection between (iii) and (iv) is due to the general relation (6.22), see also Remark 6.7.)

Given a bounded open set $\Omega \subset \mathbb{R}^d$, unfortunately, very often its boundary $F = \partial\Omega$ is neither Minkowski measurable nor Minkowski nondegenerate. Its volume function $V_F(r) := |F_r \cap \Omega|_d$, $r > 0$ may exhibit a growth behavior which differs significantly from the behavior of the functions r^{1-D}, as $r \searrow 0$ captured by the Minkowski contents. Substituting power functions with some well-chosen more general *gauge functions* $h : (0, \infty) \to (0, \infty)$, it may be possible to understand much better the behavior of V_F at the origin. This leads to the notion of generalized Minkowski contents: For any function $h : (0, \infty) \to (0, \infty)$, the *h-Minkowski content* of F (relative to Ω), is defined by

$$\mathcal{M}(h; F) := \lim_{r\searrow 0} \frac{|F_r \cap \Omega|_d}{h(r)}, \tag{1.6}$$

provided the limit exists. We denote by $\underline{\mathcal{M}}(h; F)$ and $\overline{\mathcal{M}}(h; F)$ the corresponding lower and upper limits. Moreover, we call F *h-Minkowski measurable*, if and only if $0 < \mathcal{M}(h; F) < \infty$, and *h-Minkowski nondegenerate*, if $0 < \underline{\mathcal{M}}(h; F) \leq \overline{\mathcal{M}}(h; F) < \infty$. (Note that again all these notions are relative to the set Ω.)

Given some $\Omega \subset \mathbb{R}^d$ with boundary F and some h such that the h-Minkowski content $\mathcal{M}(h; F)$ exists, the natural question arises, whether the second order asymptotic term of the eigenvalue counting function N is still determined by the asymptotic behavior of the volume function V_F, now described by $\mathcal{M}(h; F)$.

A partial answer to this question has been given by He and Lapidus [**7**]. Introducing a class $\mathcal{G}_s$ of gauge functions (for each $s \in (0, 1)$; cf. Definition 1.3), they generalized the results in [**15**] employing h-Minkowski contents with gauge functions $h \in \mathcal{G}_s$. In particular, they obtained a generalization of (1.5) to h-Minkowski measurable boundaries as well as the following generalization of Theorem 1.1 to sets with h-Minkowski nondegenerate boundaries:

THEOREM 1.2 ([**7**, Theorem 2.7]). *Let $\Omega \subset \mathbb{R}$ be a bounded open set and $F := \partial\Omega$. Denote by $\mathcal{L} = (l_j)_{j\in\mathbb{N}}$ the associated fractal string and let $h \in \mathcal{G}_{1-D}$ for some $D \in (0, 1)$. Then the following assertions are equivalent:*

(i) $0 < \underline{\mathcal{M}}(h; F) \leq \overline{\mathcal{M}}(h; F) < \infty$,
(ii) $l_j \asymp g(j)$, *as* $j \to \infty$,
(iii) $\sum_{j=1}^{\infty} \{l_j x\} \asymp f(x)$, *as* $x \to \infty$,
(iv) $\varphi(\lambda) - N(\lambda) \asymp f(\sqrt{\lambda})$, *as* $\lambda \to \infty$,

where $g(x) := H^{-1}(1/x)$ *with* $H(x) := x/h(x)$, *and* $f(x) := 1/h(1/x)$.

The classes $\mathcal{G}_s$ of gauge functions appearing in Theorem 1.2 are defined as follows.

DEFINITION 1.3 ([**7**, Definition 2.2]). Let $s \in (0, 1)$. A function $h : (0, \infty) \to (0, \infty)$ is in $\mathcal{G}_s$, if and only if the following three conditions are satisfied:

(H1) h is a continuous, strictly increasing function with $\lim_{x\searrow 0} h(x) = 0$, $\lim_{x\searrow 0} h(x)/x = \infty$ and $\lim_{x\to\infty} h(x) = \infty$.

(H2) For any $t > 0$,
$$\lim_{x\searrow 0} \frac{h(tx)}{h(x)} = t^s,$$
where the convergence is uniform in t on any compact subset of $(0, \infty)$.

(H3) There exist some constants $\tau \in (0, 1)$, $m > 0$, x_0, $t_0 \in (0, 1]$ such that
$$\frac{h(tx)}{h(x)} \geq m t^{\tau},$$
for all $0 < x < x_0$, $0 < t < t_0$.

From the definition of the classes $\mathcal{G}_s$ it is not clear at all, how restrictive they are with respect to the family of sets Ω covered by this approach. Given Ω, it is not easy to decide, whether some suitable gauge function h exists in $\mathcal{G}_s$ (for some s) or not, such that the boundary $F = \partial\Omega$ is h-Minkowski nondegenerate or even h-Minkowski measurable. It is also not clear whether all of the conditions in the definition of $\mathcal{G}_s$ are really necessary to establish the derived connections between the eigenvalue counting function N and the volume function V_F. On the other hand, there might be a simpler class of gauge functions capable of serving the same purpose.

In the present paper we aim at shedding some light on these problems. We introduce (for each $\varrho \in \mathbb{R}$) the class $\mathcal{SR}_\varrho$ of C^1-smooth gauge functions h that are regularly varying with index ϱ, i.e. which satisfy $\lim_{x\searrow 0} \frac{h(tx)}{h(x)} = t^\varrho$ for any $t > 0$ (see Definition 2.1 and compare with condition (H2) above). Using results from Karamata theory, the theory of regularly varying functions, we establish results for the classes $\mathcal{SR}_\varrho$ completely analogous to those obtained in [**7**].

The new classes $\mathcal{SR}_\varrho$ of gauge functions have several advantages compared to the classes $\mathcal{G}_s$ in He and Lapidus [**6, 7**]. First, their definition is simpler involving only two conditions - regular variation and smoothness (and the latter will turn out to be a convenience rather than a restriction). Second, the new analogues of Theorem 1.2 and (1.5) for $\mathcal{SR}_\varrho$, stated in Theorem 2.5, hold for all $\Omega \subset \mathbb{R}$ covered by the old result. (We were hoping to extend the family of sets Ω covered, but, in fact, we will show that exactly the same family of sets is addressed by our results.) Third, using regularly varying gauge functions allows some structural insights, e.g. concerning (generalized) Minkowski measurability. Last but not least, the proofs simplify significantly when using the new class. Employing Karamata theory allows to separate very clearly the main geometric (and algebraic) arguments from pure 'asymptotic calculus'. Moreover, the differentiability allows amongst others to apply some characterization results of Minkowski contents in terms of S-contents from [**20**]. S-contents describe the asymptotic behavior of the parallel surface areas of the given set F. They have been used to simplify some parts of the proof of Theorem 1.1, see [**20**, §5]. In analogy to this approach, we add a criterion in terms of generalized S-contents, which will admit further simplification of the proofs also in this setting of general gauge functions.

For these reasons, the new classes $\mathcal{SR}_\varrho$ are certainly helpful in understanding which sets are covered by the results of He and Lapidus and which are not and whether the theory can be extended further. We see our discussion as a starting point for further investigations, paving the road for a further generalization of the theory.

2. Regularly varying functions and statement of the main results

We will now define the classes of gauge functions that we are going to use and state our main results. We start with regular variation, a notion dating back to Jovan Karamata in the 1930'ies. Nowadays, the theory of such functions is known as *Karamata theory*. While the classic theory is formulated for regular variation at ∞, see also Definition 4.1 below, we will require regular variation at 0.

DEFINITION 2.1. Let $\varrho \in \mathbb{R}$. A function $h : (0, y_0) \to (0, \infty)$ (with $y_0 > 0$, possibly ∞) is called *regularly varying (at* 0*) of index* ϱ, if and only if h is measurable and satisfies for all $t > 0$

$$\lim_{y\searrow 0} \frac{h(ty)}{h(y)} = t^\varrho.$$

We write $\mathcal{R}_\varrho$ for the family of these functions. Functions $h \in \mathcal{R}_0$ are also called *slowly varying (at* 0*)*.

Furthermore, a regularly varying function $h \in \mathcal{R}_\varrho$ is in the class $\mathcal{SR}_\varrho$ of *smoothly varying functions (at* 0*) of index* ϱ, if and only if h is C^1-smooth in a neighborhood of 0, i.e. $h \in C^1(0, y_1]$ for some $y_1 > 0$.

We first discuss the relation of the classes $\mathcal{G}_\varrho$, $\mathcal{R}_\varrho$ and $\mathcal{SR}_\varrho$. It turns out that they are asymptotically equivalent, by which we mean the following: Two classes of functions are *asymptotically equivalent*, if for any function h in one class there is an asymptotically equivalent function $\tilde{h}$ in the other class, i.e. one such that $\tilde{h}(y) \sim h(y)$, as $y \searrow 0$, and vice versa. (It is easy to see that this is an equivalence relation on classes of functions, justifying the notion.)

THEOREM 2.2. *For any $\varrho \in (0,1)$, the three classes $\mathcal{G}_\varrho$, $\mathcal{R}_\varrho$ and $\mathcal{S}_\varrho$ are asymptotically equivalent.*

It is rather obvious that any $h \in \mathcal{G}_\varrho$ is in $\mathcal{R}_\varrho$ (because of condition (H2)), similarly we have $\mathcal{SR}_\varrho \subset \mathcal{R}_\varrho$, but in both cases a reverse relation is not obvious. A proof is given in Section 4, when we discuss regularly varying functions in more detail (see Corollary 4.14 and Theorem 4.17). Theorem 2.2 is not particularly deep but very useful. It clarifies that we have some freedom in choosing gauge functions. Regular variation is the basis of all three classes, but we may impose some additional properties for our convenience without loosing generality. This is exemplified by the following simple statement saying that generalized Minkowski contents are invariant with respect to asymptotic equivalence.

PROPOSITION 2.3. *Let $F \subset \mathbb{R}^d$ be compact and let $h, \tilde{h} : (0,\infty) \to (0,\infty)$ be gauge functions such that $h(y) \sim \tilde{h}(y)$, as $y \searrow 0$. Then F is h-Minkowski nondegenerate if and only if F is $\tilde{h}$-Minkowski nondegenerate. In this case, one even has*

$$0 < \underline{\mathcal{M}}(\tilde{h}; F) = \underline{\mathcal{M}}(h; F) < \infty \quad \text{and} \quad 0 < \overline{\mathcal{M}}(\tilde{h}; F) = \overline{\mathcal{M}}(h; F) < \infty.$$

In particular, F is h-Minkowski measurable if and only if F is $\tilde{h}$-Minkowski measurable.

In connection with Proposition 2.3, Theorem 2.2 clarifies that when searching for a suitable gauge function h (for a given set F) within the class of regularly varying functions $\mathcal{R}_\varrho$ (such that F becomes h-Minkowski measurable/nondegenerate), we can as well restrict ourselves to one of the classes $\mathcal{G}_\varrho$ or $\mathcal{SR}_\varrho$. Note also that the existence of such an $h \in \mathcal{R}_\varrho$ necessarily implies that $\dim_M F$ exists and equals $d - \varrho$, cf. Remark 4.10.

From the above considerations one might get the impression that not much can be gained from reproving the results of He and Lapidus, obtained for gauge functions in $\mathcal{G}_\varrho$, in the context of the classes $\mathcal{R}_\varrho$ and $\mathcal{SR}_\varrho$. Indeed, the above results make very clear that not a single additional set Ω will be covered by the new statements. However, even though our original intention was an extension of the theory to more general sets Ω, our approach using regularly varying functions is still very useful, as it provides a more structural point of view on the studied classes of gauge functions and, more importantly, as it clarifies and simplifies many of the arguments of the long and technical proofs in [**7**]. It shows that regular variation is the essential property which makes things work.

As an interesting side result, we mention that generalized Minkowski measurability with respect to regularly varying functions can be characterized without referring to any particular gauge function:

THEOREM 2.4. *Let $\Omega \subset \mathbb{R}^d$ be a bounded open set, $F = \partial\Omega$ and $\varrho \in \mathbb{R}$. There exists a gauge function $h \in \mathcal{R}_\varrho$ for F such that F is h-Minkowski measurable, if* *and only if the volume function V_F of F (i.e., the function $r \mapsto |F_r \cap \Omega|_d$, $r > 0$) is regularly varying with index ϱ, i.e. if and only if $V_F \in \mathcal{R}_\varrho$.*

This means that the assumed regular variation of the gauge function reflects an essential property of the volume function itself. In particular, if a set is h-Minkowski measurable for some $h \in \mathcal{R}_\varrho$, then it cannot be g-Minkowski measurable for some gauge function g outside the class $\mathcal{R}_\varrho$ and vice versa. A proof of Theorem 2.4 is given in Section 4, see page 74.

Before we formulate our main result, we recall the notion of *(generalized) S-content*. For any gauge function $h : (0, y_0) \to (0, \infty)$, the *lower and upper h-S-contents* of $F := \partial\Omega$ (relative to Ω) are defined by

$$(2.1)\quad \underline{\mathcal{S}}(h; F) := \liminf_{r \searrow 0} \frac{\mathcal{H}^{d-1}(\partial F_r \cap \Omega)}{h(r)} \text{ and } \overline{\mathcal{S}}(h; F) = \limsup_{r \searrow 0} \frac{\mathcal{H}^{d-1}(\partial F_r \cap \Omega)}{h(r)},$$

respectively. If $\underline{\mathcal{S}}(h; F) = \overline{\mathcal{S}}(h; F)$, then the common value $\mathcal{S}(h; F)$ is the *h-S-content* of F. There are close general relations between (generalized) Minkowski and S-contents of a set F. In particular, for differentiable gauge functions h the h-Minkowski content $\mathcal{M}(h; F)$ and the h'-S-content $\mathcal{S}(h'; F)$ coincide whenever one of these contents exists. We refer to [**20, 25**] and Section 5 for more details. Moreover, in dimension $d = 1$ there are close relations between the (generalized) S-content of a set and the so-called string counting function of the associated fractal string, see Section 6 for details.

Another advantage of smooth gauge functions is that, for any $h \in \mathcal{SR}_{1-D}$, $D \in (0, 1)$, the following two auxiliary functions f and g are well-defined for x large enough (i.e. for all $x \in [x_0, \infty)$ for some $x_0 \geq 0$):

$$(2.2)\qquad f(x) = x\, h(1/x) \quad \text{and} \quad g(x) = H^{-1}(1/x),$$

where $H(x) = x/h(x)$, $x \in \operatorname{dom}(h)$ and H^{-1} is the inverse of H. We refer to Proposition 4.18 for details. g and f take the same role as the corresponding functions in Theorem 1.2. They characterize the decay of the lengths of the associated string (l_j), as $j \to \infty$ and of the *packing defect* $\delta(x) := \sum_{j=1}^{\infty}\{l_j x\}$, as $x \to \infty$, respectively.

THEOREM 2.5 (Main Theorem). *Let $\Omega \subset \mathbb{R}$ be a bounded open set and $F := \partial\Omega$. Assume $\dim_M F = D \in (0, 1)$ and denote by $\mathcal{L} = (l_j)_{j\in\mathbb{N}}$ the associated fractal string. Let $h \in \mathcal{SR}_{1-D}$ and let f and g be given as in (2.2).*

I. (*Two-sided bounds.*) *The following assertions are equivalent:*
  (i) $0 < \underline{\mathcal{M}}(h; F) \leq \overline{\mathcal{M}}(h; F) < \infty$,
  (ii) $0 < \underline{\mathcal{S}}(h'; F) \leq \overline{\mathcal{S}}(h'; F) < \infty$,
  (iii) $l_j \asymp g(j)$, *as* $j \to \infty$,
  (iv) $\sum_{j=1}^{\infty}\{l_j x\} \asymp f(x)$, *as* $x \to \infty$,
  (v) $\varphi(\lambda) - N(\lambda) \asymp f(\sqrt{\lambda})$, *as* $\lambda \to \infty$.

II. (*Minkowski measurability.*) *The following assertions are equivalent:*
  (vi) *F is h-Minkowski measurable,*
  (vii) *F is h'-S measurable,*
  (viii) $l_j \sim L\, g(j)$, *as $j \to \infty$ for some positive $L > 0$.*

Under these latter assertions, h-Minkowski content, h'-S-content of F and the constant L are connected by the relation

$$\mathcal{M}(h;F) = \mathcal{S}(h';F) = \frac{2^{1-D}L^D}{1-D}. \tag{2.3}$$

Moreover, these assertions imply

$$N(\lambda) = \varphi(\lambda) - c_{1,D}\mathcal{M}(h;F)f(\sqrt{\lambda}) + o(f(\sqrt{\lambda})), \qquad \text{as } \lambda \to \infty, \tag{2.4}$$

where $c_{1,D} = 2^{D-1}\pi^{-D}(1-D)(-\zeta(D))$ with ζ being the Riemann zeta function.

Note that Theorem 2.5 comprises the classical results of Lapidus and Pomerance [**15**, cf. Theorems 2.1-2.4] for sets Ω with Minkowski measurable/nondegenerate boundaries as a special case (by choosing the gauge function $h(y) = y^{1-D}$). It parallels the results of He and Lapidus in [**7**, Theorems 2.4-2.7] obtained for gauge functions h of the classes $\mathcal{G}_{1-D}$. Our proofs bring the generalized theory of He and Lapidus in some parts closer back to the original arguments in the proofs of [**15**] (in some other parts, however, we will use results on S-contents from [**20**]).

REMARK 2.6. One may wonder, whether also a converse to the last assertion in Theorem 2.5 holds, i.e. whether the existence of an asymptotic second term for N as in (2.4) (assumed to hold for some f given in terms of some h as in (2.2)) implies the h-Minkowski measurability of the boundary F. He and Lapidus have addressed this question (for $h \in \mathcal{G}_{1-D}$). In [**7**, Theorem 2.9] they showed in particular that such a converse fails for all dimensions D, for which the Riemann zeta function has a zero on the line $\Re(s) = D$, generalizing thus earlier results of Lapidus and Maier [**13, 14**] and showing a connection of this question to the Riemann hypothesis. More precisely, for such D the converse fails for any $h \in \mathcal{G}_{1-D}$ which is differentiable and satisfies $yh'(y)/h(y) \geq \mu$ for some $\mu > 0$ (and all y). Since the differentiability implies $h \in \mathcal{SR}_{1-D}$, we can conclude from our Lemma 4.16 below that $\lim_{y\searrow 0} yh'(y)/h(y) = 1-D$, implying in particular that $yh'(y)/h(y) \geq (1-D)/2 =: \mu$ for all y in a suitable neighborhood of 0. Hence this additional hypothesis is not necessary in [**7**, Theorem 2.9], or in other words, this statement directly applies to any $h \in \mathcal{SR}_{1-D}$. We are optimistic that our methods might also help to simplify the proof of this statement, but we have not yet tried to do this.

REMARK 2.7. In the present paper, we have restricted ourselves to subsets Ω of $\mathbb{R}$. In [**7**], also some results in the higher dimensional case are obtained, i.e. for bounded open sets $\Omega \in \mathbb{R}^d$, $d > 1$. It is known that the MWB conjecture fails in this case and therefore these results are necessarily of a weaker nature. More precisely, under the assumption that the upper h-Minkowski content of $F := \partial\Omega$ is finite (for some suitable gauge function h), an upper bound for the second asymptotic term of N is derived in terms of the function f (determined by h as in (2.2)). Even though the assumptions on the gauge function h are slightly different in these higher dimensional results, it may be worth to revisit them in the light of the methods used here.

Although we have clarified significantly the (one-dimensional) theory of He and Lapidus by showing that regular variation is the driving force behind these results (while the other conditions can be omitted or follow automatically), we think that this is not the end of the story. Several examples of sets Ω and suitable associated gauge functions h have been discussed in [**7**, cf. §7 and the Appendix], to which the results apply. Since all occurring gauge functions h are differentiable and thus in the classes $\mathcal{SR}_{1-D}$, they can also serve as examples for Theorem 2.5 (and to apply the theorem, it is not necessary to check any of the technical conditions (H1)-(H3), only the regular variation). However, there might exist sets Ω, for which regularly varying gauge functions are not suitable but for which nevertheless results similar to Theorem 2.5 hold. It would be desirable to have such an example or to prove that it does not exist. In either case, we believe that the concept of regular variation – and its various generalizations, e.g. *de Haan theory*, cf. [**3**] – will be useful for such an attempt.

REMARK 2.8. The results of Lapidus and Pomerance [**15, 16**] and in particular the discovered connections to the Riemann zeta function have initiated an extensive study of fractal strings which led to the theory of complex dimensions, see [**17**] and the references therein. Nowadays these spectral problems and many related questions, including Minkowski measurability, are studied with the help of various zeta functions, which allow to derive explicit formulas for functions such as the the eigenvalue counting function or the tube volume. We have not made any efforts to relate our results to this zeta function theory. It seems, however, that most results in this theory do not cover sets that are generalized Minkowski measurable (but not Minkowski measurable) or generalized Minkowski nondegenerate (but not Minkowski nondegenerate). An exception is the recent preprint [**11**], where in Section 5 a class of h-Minkowski measurable sets (in $\mathbb{R}^d$) is discussed with gauge functions of the form $h(t) = t^{d-D}(\log t^{-1})^m$, $m > 0$. This is part of the recent strong efforts to extend parts of the theory to higher dimensions, see the monograph [**10**]. In the language of [**10**, §6.1], where a classification scheme for subsets of $\mathbb{R}^d$ is proposed in terms of the behavior of their tube functions, the boundaries F studied here and in [**7**] fall into the class of *weakly degenerate sets*. There are also some connections between gauge functions of regular variation as discussed here and the weaker notion of gauge functions of slow growth/slow decay defined in [**10**, §6.1].

The remaining parts of the paper are organized as follows. After some preliminary considerations concerning asymptotic similarity in Section 3, we recall in Section 4 some results from Karamata theory and discuss various useful consequences. On the way, we will prove the asymptotic equivalence of the classes $\mathcal{G}_\varrho$, $\mathcal{R}_\varrho$ and $\mathcal{SR}_\varrho$ (Theorem 2.2) as well as Theorem 2.4. Sections 5 and 6 are devoted to the proof of Theorem 2.5. In Section 5, we will treat the 'geometric part' and establish the connection between Minkowski contents, S-contents and the growth behaviour of the lengths of the associated fractal string, while in Section 6, we finally establish the connection between the geometry of Ω and its spectral properties. An overview over the various steps of the proof is given in the diagrams on page 93.

3. Some preliminaries on asymptotic equivalence and similarity

In this section we collect some basic (and well-known) facts about asymptotic similarity $\asymp$ and asymptotic equivalence $\sim$ of functions. Recall the definitions from page 65. Here we formulate all facts for asymptotic similarity/equivalence at $a = 0$, but for the case $a = \infty$ they hold analogously.

Consider the family $\mathcal{F}$ of real valued positive functions f defined on a right neighborhood of 0, i.e. functions $f : (0, y_0) \to (0, \infty)$ for some $y_0 > 0$. Observe that $\sim$ and $\asymp$ are equivalence relations on $\mathcal{F}$. In particular, both relations are transitive, i.e. if $f_1 \asymp f_2$ and $f_2 \asymp f_3$ then $f_1 \asymp f_3$, as $y \searrow 0$ and similarly for $\sim$. Moreover, since asymptotic equivalence implies asymptotic similarity, we also have the following implication:

$$f_1(y) \asymp f_2(y) \text{ and } f_2(y) \sim f_3(y) \text{ implies } f_1(y) \asymp f_3(y), \text{ as } y \searrow 0. \tag{3.1}$$

We collect some useful rules for $\asymp$ and $\sim$. Some further rules will be stated later on for functions with extra properties such as regular variation or smoothness (see e.g. Lemma 4.5 or 4.7).

LEMMA 3.1. *Let $f_1, f_2, h, g_1, g_2 \in \mathcal{F}$. Then, as $y \searrow 0$,*

(i) *$f_1 \sim f_2/h$, if and only if $h \cdot f_1 \sim f_2$;*
(ii) *if $f_1 \sim f_2$ and $g_1 \sim g_2$, then $f_1 \cdot g_1 \sim f_2 \cdot g_2$ and $f_1/g_1 \sim f_2/g_2$;*
(iii) *if $f_1 \sim f_2 \sim g_1 \sim g_2$, then $f_1 + f_2 \sim g_1 + g_2$.*

The assertions hold analogously with $\sim$ replaced by $\asymp$.

As a first simple application, we provide a proof of Proposition 2.3. Note that a statement similar to Proposition 2.3 could be formulated for S-contents.

PROOF OF PROPOSITION 2.3. The first assertion is a direct consequence of (3.1) applied to $f_1 = V_F$, $f_2 = h$ and $f_3 = \tilde{h}$. To see the equality of the upper and lower generalized Minkowski contents, let $\epsilon \in (0, 1)$. Then there is a $y_0 > 0$ such that

$$(1-\epsilon)h(y) < \tilde{h}(y) < (1+\epsilon)h(y)$$

for all $y \in (0, y_0]$. Hence we obtain

$$0 < \frac{V_F(y)}{(1+\epsilon)h(y)} \le \frac{V_F(y)}{\tilde{h}(y)} \le \frac{V_F(y)}{(1-\epsilon)h(y)} < \infty$$

and the assertion follows by taking upper and lower limits in this relation as $y \searrow 0$ and then letting $\epsilon \searrow 0$. The last assertion is a direct consequence of the transitivity of $\sim$. $\square$

4. Regularly varying functions and Karamata theory

In this section we discuss functions of regular variation in more detail. In our exposition we concentrate on those results which turned out to be useful for our purposes and which prepare the proof of our main result. On the way we will also prove Theorems 2.2 and 2.4. While the classic theory is formulated for regular variation at ∞ (see Definition 4.1 below), we will focus on *regular variation at* 0 (cf. Definition 2.1). We will reformulate some of the classic results of Karamata theory for this case (which is an easy exercise due to Remark 4.2). For the versions at ∞ (which we will also use occasionally) and further details we refer to the monograph [**3**].

DEFINITION 4.1. Let $\varrho \in \mathbb{R}$. A function $f : (x_0, \infty) \to (0, \infty)$ (with $x_0 \ge 0$) is called *regularly varying (at ∞) of index ϱ*, if and only if f is measurable and satisfies for all $t > 0$

$$\lim_{x\to\infty} \frac{f(tx)}{f(x)} = t^{\varrho}.$$

We write $\mathcal{R}_\varrho[\infty]$ for the family of all such functions. $f \in \mathcal{R}_0[\infty]$ are also called *slowly varying (at ∞)*.

Note that, by Definition 2.1, the domain $\mathrm{dom}(h)$ of any $h \in \mathcal{R}_\varrho$ is an interval of the form $(0, y_0)$ (a right neighborhood of zero). h can easily be extended to a function on $(0, \infty)$ without affecting the regular variation property at 0. It is easy to see that one can even extend h to $(0, \infty)$ in such a way that properties like continuity, differentiability or strict monotonicity (which h may have) are preserved. (Of course, if $\lim_{y \to y_0} h(y) = \infty$, then one has to redefine h near y_0 for this extension.) Similarly, any function $f \in \mathcal{R}_\varrho[\infty]$ (with domain $\mathrm{dom}(f) = (x_0, \infty)$) can be extended to the whole positive axis.

As pointed out above, regularly varying functions (at ∞) are well-studied, see e.g. [**3**]. The following simple observation allows to transfer virtually all results known for regular variation at ∞ to regular variation at 0.

REMARK 4.2. $h \in \mathcal{R}_\varrho$ if and only if the function $\tilde{h}$ defined by $\tilde{h}(x) = h(1/x)$, $1/x \in \mathrm{dom}(h)$ is in $\mathcal{R}_{-\varrho}[\infty]$. Indeed, if the domain of $h \in \mathcal{R}_\varrho$ is $(0, y_0)$ then the domain of $\tilde{h}$ is the interval $(1/y_0, \infty)$ and for any $t > 0$, one obtains

$$\lim_{x \to \infty} \frac{\tilde{h}(tx)}{\tilde{h}(x)} = \lim_{x \to \infty} \frac{h(1/(tx))}{h(1/x)} = \lim_{y \searrow 0} \frac{h(t^{-1}y)}{h(y)} = t^{-\varrho},$$

implying $\tilde{h} \in \mathcal{R}_{-\varrho}[\infty]$.

REMARK 4.3. Observe that condition (H2) in the definition of the classes $\mathcal{G}_\varrho$ is regular variation at 0 plus an extra uniformity requirement in the variable t. We will see below in Theorem 4.6 that this local uniformity is automatically satisfied for regularly varying functions. In [**7**], (H2) is called a *homogeneity property*. Indeed, any regularly varying function $h \in \mathcal{R}_\varrho$ is *asymptotically homogeneous of degree* ϱ in the following sense: for any $t > 0$,

$$h(ty) = h(y)(t^\varrho + o(1)), \text{ as } y \searrow 0.$$

This relation is equivalently described by $h(ty) \sim t^\varrho h(y)$, as $y \searrow 0$.

The following simple observation characterizes regular variation in terms of slow variation, compare also with the Characterization Theorem [**3**, Theorem 1.4.1]:

LEMMA 4.4. *Let $\varrho \in \mathbb{R}$. For any $h \in \mathcal{R}_\varrho$, there is a slowly varying function $\ell \in \mathcal{R}_0$ such that $h(y) = y^\varrho \ell(y)$ for any $y \in \mathrm{dom}(h)$.*

PROOF. Define ℓ by $\ell(y) := h(y)/y^\varrho$, $y \in \mathrm{dom}(h)$ and observe that, for any $t > 0$, $\ell(ty)/\ell(y) \to 1$, as $y \searrow 0$ implying $\ell \in \mathcal{R}_0$. □

Our next observation is that the classes $\mathcal{R}_\varrho$ are stable with respect to asymptotic equivalence. Any function g asymptotically equivalent to some regularly varying function h is already regularly varying itself. In contrast, asymptotic similarity $\asymp$ does not preserve regular variation in general.

LEMMA 4.5. *Let $h \in \mathcal{R}_\varrho$ and let $g : (0, y_0) \to (0, \infty)$ be a measurable function such that $h(y) \sim g(y)$, as $y \searrow 0$. Then $g \in \mathcal{R}_\varrho$, i.e. g is regularly varying at 0 with the same index ϱ.*

PROOF. The function z defined by $z(y) := \frac{g(y)}{h(y)}$ for all $y \in \mathrm{dom}(h) \cap \mathrm{dom}(g)$ is measurable and converges to 1 as $y \searrow 0$. Therefore,

$$\frac{g(ty)}{g(y)} = \frac{h(ty)}{h(y)} \frac{z(ty)}{z(y)} \to t^\varrho, \text{ as } y \searrow 0, \tag{4.1}$$

for all $t > 0$. Hence g is regularly varying of index ϱ. □

This simple observation will be used frequently in the proofs later on. It is also the key to Theorem 2.4, which we prove now.

PROOF OF THEOREM 2.4. For the forward implication, by assumption, there exists some $h \in \mathcal{R}_\varrho$ and some $M > 0$, such that $\mathcal{M}(h; F) = M$, which means that $\lim_{r \searrow 0} \frac{V_F(r)}{M\,h(r)} = 1$ or, equivalently $V_F(r) \sim M\,h(r)$, as $r \searrow 0$. Since any (positive) constant multiple of a function in $\mathcal{R}_\varrho$ is still in $\mathcal{R}_\varrho$, it follows from Lemma 4.5, that $V_F \in \mathcal{R}_\varrho$. The reverse implication is transparent by noting that one can choose $h := V_F$. □

The Uniform Convergence Theorem (UCT) is one of the central results of Karamata theory. We formulate a version for functions $h \in \mathcal{R}_\varrho$. It can easily be derived from the corresponding statement for functions in $\mathcal{R}_\varrho[\infty]$, cf. Remark 4.2.

THEOREM 4.6 (UCT, cf. [**3**, Theorem 1.5.2] for a version at ∞). *Let $\varrho \in \mathbb{R}$ and $h \in \mathcal{R}_\varrho$. Then $\frac{h(ty)}{h(y)} \to t^\varrho$, as $y \searrow 0$ converges uniformly in t on any compact subset of $(0, \infty)$.*

Theorem 4.6 implies in particular that condition (H2) is satisfied for any regularly varying function $h \in \mathcal{R}_\varrho$. As a further consequence of Theorem 4.6, we may now prove the following result on asymptotic similarity/equivalence of regularly varying functions. It will be extremely useful in the proofs later on.

LEMMA 4.7. *Let $\kappa \in \mathbb{R}$ and $g \in \mathcal{R}_\kappa$. Let $f_1, f_2 : (0, \infty) \to (0, \infty)$ such that $\lim_{y \searrow 0} f_2(y) = 0$ and assume $f_1(y) \asymp f_2(y)$, as $y \searrow 0$. Then*

$$g(f_1(y)) \asymp g(f_2(y)), \quad \text{as } y \searrow 0. \tag{4.2}$$

Moreover, if $g \in \mathcal{R}_0$ or $f_1(y) \sim f_2(y)$, as $y \searrow 0$, then

$$g(f_1(y)) \sim g(f_2(y)), \quad \text{as } y \searrow 0. \tag{4.3}$$

Completely analogous relations hold, if $\lim_{y \searrow 0} f_2(y) = \infty$ and $g \in \mathcal{R}_\kappa[\infty]$, or if $f_1(x) \asymp f_2(x)$, as $x \to \infty$.

PROOF. By assumption, there are positive constants y_0, c_1, c_2 such that $c(y) := \frac{f_1(y)}{f_2(y)} \in [c_1, c_2]$ for all $y \in (0, y_0)$. Assume first that $\kappa \geq 0$ and fix some $\varepsilon \in (0, c_1^\kappa)$. By the Uniform Convergence Theorem 4.6, the convergence $g(tx)/g(x) \to t^\kappa$ (as $x \searrow 0$) is uniform in t on the compact interval $[c_1, c_2]$, i.e. there exists some $\delta > 0$ such that $|g(tx)/g(x) - t^\kappa| \leq \varepsilon$ for each $0 < x < \delta$ and each $t \in [c_1, c_2]$. Therefore,

$$\frac{g(f_1(y))}{g(f_2(y))} = \frac{g(c(y) f_2(y))}{g(f_2(y))} \leq c(y)^\kappa + \varepsilon \leq c_2^\kappa + \varepsilon =: c_2',$$

for each $y \in (0, y_0)$ such that $f_2(y) < \delta$. Observe that the hypothesis on f_2 implies there exists some $y_1 = y_1(\delta) > 0$ such that the $f_2(y) < \delta$ is true for all $0 < y < y_1$. In a similar way one obtains that the expression $g(f_1(y))/g(f_2(y))$ is bounded from below by $c_1' := c_1^\kappa - \varepsilon$ for all $y \in (0, y_1)$. This shows (4.2) for the case $\kappa \geq 0$.

For $\kappa < 0$, the only difference is that now the function $t \mapsto t^\kappa$ is decreasing. Fixing some $\varepsilon \in (0, c_2^\kappa)$, a similar argument as in the previous case gives the upper bound $c_2' := c_1^\kappa + \varepsilon$ and the lower bound $c_1' := c_2^\kappa - \varepsilon$ for the expression $g(f_1(y))/g(f_2(y))$ for all $y \in (0, y_1)$. This completes the proof of the first assertion.

In case $\kappa = 0$ the constants used above are $c_1' = 1 - \varepsilon$ and $c_2' = 1 + \varepsilon$, where the $\varepsilon > 0$ was fixed. Observing that ε can be chosen as close to zero as we wish, assertion (4.3) follows for $g \in \mathcal{R}_0$. Assertion (4.3) is also valid under the hypothesis

$f_1 \sim f_2$, since in this case there is for any $\varepsilon > 0$ some y_0 such that $(1-\varepsilon) \leq f_1(y)/f_2(y) \leq (1+\varepsilon)$ for all $y \in (0, y_0]$. This implies

$$\frac{g((1-\varepsilon)f_2(y))}{g(f_2(y))} \leq \frac{g(f_1(y))}{g(f_2(y))} \leq \frac{g((1+\varepsilon)f_2(y))}{g(f_2(y))} \to (1+\varepsilon)^\kappa,$$

as $y \searrow 0$. Similarly, the left expression converges to $(1-\varepsilon)^\kappa$, as $y \searrow 0$. Now (4.3) follows by letting ε tend to 0.

The arguments for the case $\lim_{y\searrow 0} f_2(y) = \infty$ and $g \in \mathcal{R}_\kappa[\infty]$, and for $f_1(x) \asymp f_2(x)$, as $x \to \infty$ are completely analogous. □

A second central result of Karamata theory is the following Representation Theorem, which provides an integral representation for slowly varying functions. Again, we formulate it for slow variation at 0.

THEOREM 4.8 (Representation Theorem, cf. [**3**, Theorem 1.3.1]). *A function ℓ is slowly varying at 0, if and only if it has a representation*

$$\ell(y) = c(y) \exp\left(\int_y^a \frac{\varepsilon(u)}{u} du\right), \; y \in (0, a), \tag{4.4}$$

for some $a > 0$, where c and ε are measurable functions with $\lim_{y\searrow 0} c(y) = C \in (0, \infty)$ and $\lim_{y\searrow 0} \varepsilon(y) = 0$.

PROOF. Applying [**3**, Theorem 1.3.1] to the function $\tilde{\ell} \in \mathcal{R}_0[\infty]$ given by $\tilde{\ell}(x) := \ell(1/x)$, yields the existence of some positive constants $\tilde{a}$ and C and some functions $\tilde{c}$ and $\tilde{\varepsilon}$ with $\tilde{c}(x) \to C$ and $\tilde{\varepsilon}(x) \to 0$ as $x \to \infty$ such that

$$\tilde{\ell}(x) = \tilde{c}(x) \exp\left(\int_{\tilde{a}}^x \frac{\tilde{\varepsilon}(u)}{u} du\right), \; x \geq \tilde{a}.$$

Therefore, we get for any $0 < y < a := 1/\tilde{a}$,

$$\ell(y) = \tilde{\ell}(1/y) = \tilde{c}(1/y) \exp\left(\int_{\tilde{a}}^{1/y} \frac{\tilde{\varepsilon}(u)}{u} du\right) = c(y) \exp\left(\int_{1/a}^{1/y} \frac{\tilde{\varepsilon}(u)}{u} du\right),$$

where $c(y) := \tilde{c}(1/y)$ satisfies obviously $c(y) \to C$ as $y \searrow 0$. The representation (4.4) follows now with the substitution $u \mapsto 1/u$ and setting $\varepsilon(y) := \tilde{\varepsilon}(1/y)$. □

Note that the representation of ℓ given by Theorem 4.8 is not unique, for we may add to ε any function δ with $\lim_{y\searrow 0} \delta(y) = 0$ (and adjust c accordingly). Moreover, the upper bound a of the integration interval in (4.4) may be shifted arbitrarily in $\mathrm{dom}(\ell)$, again by adjusting the function c. If $a > 1$, for instance, we have

$$\ell(y) = c(y) \exp\left(\int_y^a \frac{\varepsilon(u)}{u} du\right) = \tilde{c}(y) \exp\left(\int_y^1 \frac{\varepsilon(u)}{u} du\right), \tag{4.5}$$

where

$$\tilde{c}(y) := c(y) \exp\left(\int_1^a \frac{\varepsilon(u)}{u} du\right).$$

Combining Lemma 4.4 with the Representation Theorem 4.8, we obtain an analogous representation for regularly varying functions: Since for $h \in \mathcal{R}_\varrho$, $\varrho \neq 0$, there is some $\ell \in \mathcal{R}_0$ such that $h(y) = y^\varrho \ell(y)$, there must exist functions c and ε (with $c(y) \to C \in (0, \infty)$ and $\varepsilon(y) \to 0$, as $y \searrow 0$) such that

$$h(y) = y^\varrho \ell(y) = c(y) \exp\left(\varrho \cdot \log y + \int_y^1 \varepsilon(u) \frac{du}{u}\right) = c(y) \exp\left(\int_y^1 -\varrho + \varepsilon(u) \frac{du}{u}\right)$$

From this representation it is easy to conclude the following asymptotic behaviour.

PROPOSITION 4.9 (cf. [**3**, Proposition 1.5.1]). *Let $\varrho \neq 0$ and $h \in \mathcal{R}_\varrho$. Then*

$$h(y) \to \begin{cases} 0, & \text{if } \varrho > 0 \\ \infty, & \text{if } \varrho < 0 \end{cases}, \quad \text{as } y \searrow 0.$$

REMARK 4.10. Let $\Omega \subset \mathbb{R}^d$ be a bounded open set and $F = \partial\Omega$. Assume that F is h-Minkowski nondegenerate for some $h \in \mathcal{R}_{d-D}$, i.e. $0 < \underline{\mathcal{M}}(h; F) \leq \overline{\mathcal{M}}(h; F) < \infty$. Then the Minkowski dimension of F exists and equals D. Indeed, by Lemma 4.4, we can write h as $h(r) = r^{d-D}\ell(r)$ for some $\ell \in \mathcal{R}_0$ and so, for any $\gamma > 0$,

$$\mathcal{M}^{D+\gamma}(F) = \lim_{r \searrow 0} \frac{V_F(r)}{r^{d-D-\gamma}} = \lim_{r \searrow 0} \frac{V_F(r)}{r^{d-D}\ell(r)} r^\gamma \ell(r) \leq \overline{\mathcal{M}}(h; F) \lim_{r \searrow 0} r^\gamma \ell(r).$$

Since $r \mapsto r^\gamma \ell(r)$ is regularly varying with index $\gamma > 0$, by Proposition 4.9, the last limit vanishes, implying $\mathcal{M}^{D+\gamma}(F) = 0$ and thus $\overline{\dim}_M F \leq D$. Similarly, we get $\mathcal{M}^{D+\gamma}(F) \geq \underline{\mathcal{M}}(h; F) \lim_{r \searrow 0} r^\gamma \ell(r)$, where the last limit is now $+\infty$ for $\gamma < 0$, implying $\underline{\dim}_M F \geq D$. This shows $\dim_M F = D$. Therefore, clearly, the index of any suitable regularly varying gauge function h for F is necessarily equal to its (Minkowski) co-dimension $d - D$. We point out that this does not imply that for every F with dimension $\dim_M F = D$ there exists a suitable gauge function $h \in \mathcal{R}_{d-D}$.

The following fact will be useful in the sequel, e.g. in Theorem 4.19 below. It is another immediate consequence of the Representation Theorem.

COROLLARY 4.11 (cf. [**3**, Corollary 1.4.2]). *Let $\varrho \in \mathbb{R}$ and $h \in \mathcal{R}_\varrho$. Then there is $x_0 > 0$ such that the functions h and $\frac{1}{h}$ are locally bounded and locally integrable on $(0, x_0]$.*

The Representation Theorem allows to show that any regularly varying function $h \in \mathcal{R}_\varrho$ of index $\varrho \in (0, 1)$ already satisfies condition (H3) (cf. Definition 1.3). This is another important step towards the asymptotic equivalence of the classes $\mathcal{G}_D$ and $\mathcal{SR}_D$.

PROPOSITION 4.12. *Let $\varrho \in (0, 1)$ and $h \in \mathcal{R}_\varrho$. Then h satisfies hypothesis (H3), i.e. there are constants $m > 0$, $\tau \in (0, 1)$, $y_0, t_0 \in (0, 1]$ such that*

$$\frac{h(ty)}{h(y)} \geq m t^\tau,$$

for all $0 < y < y_0$ and $0 < t < t_0$.

PROOF. By Lemma 4.4, $h(y) = y^\varrho \ell(y)$ for some $\ell \in \mathcal{R}_0$, and, by the Representation Theorem 4.8, we can write ℓ as in equation (4.4). Therefore, for any $t, y \in (0, 1)$,

$$\frac{\ell(y)}{\ell(ty)} = \frac{c(y) \cdot \exp\left(\int_y^1 \frac{\varepsilon(u)}{u} du\right)}{c(ty) \cdot \exp\left(\int_{ty}^1 \frac{\varepsilon(u)}{u} du\right)} = \frac{c(y)}{c(ty)} \exp\left(-\int_{ty}^y \varepsilon(u) \frac{du}{u}\right), \tag{4.6}$$

where c, ε are measurable functions such that $c(y) \to C$ (for some $C \in (0,\infty)$) and $\varepsilon(y) \to 0$ as $y \searrow 0$. Let $\delta > 0$ such that $C - \delta > 0$ and let $\gamma > 0$ such that $\varrho + \gamma < 1$. Since, $c(y) \to C$ as $y \searrow 0$, there is $y_0 \in (0,1)$ such that $|c(y) - C| < \delta$ for all $y \in (0, y_0)$. Moreover, since $\varepsilon(y) \to 0$, there is also some $y_1 \in (0, y_0]$ such that $|\varepsilon(y)| < \gamma$ for any $y \in (0, y_1)$. Therefore, we obtain for any $y \in (0, y_1)$,

$$\frac{\ell(y)}{\ell(ty)} \leq \frac{C+\delta}{C-\delta} \exp\left(\int_{ty}^{y} |\varepsilon(u)| \frac{du}{u}\right) \leq M \exp\left(\int_{ty}^{y} \gamma \frac{du}{u}\right) = M t^{-\gamma}, \tag{4.7}$$

where we have set $M := \frac{C+\delta}{C-\delta}$. This implies $\ell(ty)/\ell(y) \geq M^{-1} t^{\gamma}$ and thus

$$\frac{h(ty)}{h(y)} = t^{\varrho} \frac{\ell(ty)}{\ell(y)} \geq M^{-1} t^{\varrho+\gamma},$$

for all $y \in (0, y_1)$ and all $t \in (0,1)$. Thus the assertion follows for $m := M^{-1}$ and $\tau := \varrho + \gamma$, which by the choice of γ is strictly less than 1. □

Up to now, we have seen that any regularly varying function $h \in \mathcal{R}_\varrho$ with $\varrho \in (0,1)$ satisfies conditions (H2) and (H3), by Theorem 4.6 and Proposition 4.12, respectively. Concerning condition (H1), it is easy to see from Proposition 4.9 that such h also satisfy $\lim_{y\searrow 0} h(y) = 0$ and $\lim_{y\searrow 0} h(y)/y = \infty$. The condition $\lim_{y\to\infty} h(y) = \infty$ is not always satisfied but it is not very relevant. One can easily extend or redefine h on some interval $[x_0, \infty)$ bounded away from zero to meet this condition without affecting the asymptotic properties of h at 0. On the other hand, regularly varying functions need neither be continuous nor strictly increasing. It will be our aim now to clarify that for each $h \in \mathcal{R}_\varrho$ there is an asymptotically equivalent function with these two properties. It turns out that there are even smooth representatives for h.

Smoothly varying functions. We will now discuss the classes $\mathcal{SR}_\varrho$ in more detail. We will show in particular the asymptotic equivalence of the classes $\mathcal{G}_\varrho$ (cf. Definition 1.3) and $\mathcal{SR}_\varrho$, see Theorem 4.17. Recall that for any $\varrho \in \mathbb{R}$, a regularly varying function $h \in \mathcal{R}_\varrho$ is in the class $\mathcal{SR}_\varrho$, if and only if there is some $y_0 > 0$ such that $h \in C^1(0, y_0]$. Similarly, we write $\mathcal{SR}_\varrho[\infty]$ for the class of functions that are smoothly varying at ∞.

By the following result, we may assume slowly varying functions $\ell \in \mathcal{R}_0$ to be smooth modulo asymptotic equivalence, implying that the classes $\mathcal{SR}_0$ and $\mathcal{R}_0$ are asymptotically equivalent.

THEOREM 4.13 (cf. [**3**, Theorem 1.3.3]). *Let $\ell \in \mathcal{R}_0$. Then there is a smoothly varying function $\ell_1 \in \mathcal{SR}_0$ asymptotically equivalent to ℓ (i.e. $\ell_1(y) \sim \ell(y)$, as $y \searrow 0$). ℓ_1 can be chosen in such a way that all derivatives of the function p_1 defined by $p_1(x) := \log(\ell_1(e^{-x})), e^{-x} \in \mathrm{dom}(\ell_1)$ vanish asymptotically, i.e., for all $n \in \mathbb{N}$,*

$$p_1^{(n)}(x) \to 0, \quad \text{as } x \to \infty.$$

Combining Theorem 4.13 with Lemma 4.4, we obtain the following result, which establishes in particular that the classes $\mathcal{R}_\varrho$ and $\mathcal{SR}_\varrho$ are asymptotically equivalent for any $\varrho \in \mathbb{R}$.

COROLLARY 4.14. *Let $\varrho \in \mathbb{R}$ and $h \in \mathcal{R}_\varrho$. Then there is a smoothly varying function $h_1 \in \mathcal{SR}_\varrho$ asymptotically equivalent to h (i.e. with $h_1(y) \sim h(y)$, as $y \searrow 0$).*

h_1 can be chosen in such a way that the function p_1 defined by $p_1(x) := \log(h_1(e^{-x}))$, $x \in \mathrm{dom}(h_1)$ satisfies

$$\lim_{x\to\infty} p_1'(x) = -\varrho \qquad \textit{and} \qquad \lim_{x\to\infty} p_1^{(n)}(x) = 0$$

for each $n \in \mathbb{N}, n \geq 2$.

PROOF. By Lemma 4.4, the function h can be written as $h(y) = y^\varrho \ell(y)$ for some $\ell \in \mathcal{R}_0$. Theorem 4.13 implies the existence of a smoothly varying function $\ell_1 \in \mathcal{SR}_0$ asymptotically equivalent to ℓ. Defining h_1 by $h_1(y) := y^\varrho \ell_1(y), y > 0$, we obtain

$$p_1(x) = \log(h_1(e^{-x})) = \log(e^{-\varrho x}\ell_1(e^{-x})) = -\varrho x + \log(\ell_1(e^{-x})).$$

By Theorem 4.13, all derivatives of the function $x \mapsto \log(\ell_1(e^{-x})), x > 0$ tend to zero, as $x \to \infty$, which yields the assertion. □

The next statement characterizes the asymptotic behavior of the the derivative of a smoothly varying function. It is not only needed in the proof of the subsequent proposition but it will also be extremely useful in the proofs in Section 5. The expression $E_h(y) := \frac{y\,h'(y)}{h(y)}$ is known as the *elasticity* of a function h at y.

LEMMA 4.15. *For any $\varrho \in \mathbb{R}$ and any function $h \in \mathcal{SR}_\varrho$, the elasticity of h at y converges to ϱ as $y \searrow 0$, i.e.*

$$\lim_{y\searrow 0} \frac{yh'(y)}{h(y)} = \varrho. \tag{4.8}$$

For $\varrho \neq 0$, this can be rephrased as $h'(y) \sim \varrho\, h(y)/y$, as $y \searrow 0$.

PROOF. It is enough to show the assertion for slowly varying functions. Indeed, for $h \in \mathcal{SR}_\varrho$, there is, by Lemma 4.4, some function $\ell \in \mathcal{SR}_0$ such that $h(y) = y^\varrho \ell(y)$ on $\mathrm{dom}(h)$ and therefore

$$E_h(y) = \frac{y\,h'(y)}{h(y)} = \frac{y\,(y^\varrho \ell(y))'}{y^\varrho \ell(y)} = \frac{\varrho y^\varrho \ell(y) + y^{\varrho+1}\ell'(y)}{y^\varrho \ell(y)} = \varrho + \frac{y\,\ell'(y)}{\ell(y)} = \varrho + E_\ell(y).$$

That is, the elasticities of h and ℓ differ pointwise by ϱ and so in particular, as $y \searrow 0$, $E_h(y) \to \varrho$ if and only if $E_\ell(y) \to 0$.

So let $h \in \mathcal{SR}_0$. For any fixed $y \in \mathrm{dom}(h)$, we apply the linear substitution $u(t) = yt - y$ in the differential quotient of h at y and obtain

$$\begin{aligned}
\lim_{y\searrow 0}\lim_{u\searrow 0} \frac{h(y+u) - h(y)}{u}\frac{y}{h(y)} &= \lim_{y\searrow 0}\lim_{t\to 1} \frac{h(ty) - h(y)}{y(t-1)h(y)} y \\
&= \lim_{t\to 1}\frac{1}{t-1}\lim_{y\searrow 0}\frac{h(ty) - h(y)}{h(y)} = 0,
\end{aligned}$$

where in the last expression we have changed the order of the limits. This is justified, since, by the Uniform Convergence Theorem 4.6, the inner limit in this last expression vanishes uniformly in t on any compact interval. (So choose one containing 1 in its interior.) It follows that also the outer limit (as $t \to 1$) of this expression exists (and equals 0), implying that it is safe to interchange the order of the limits. □

Our next observation is that smoothly varying functions are monotone near zero. One of the consequences is that the derivatives of smoothly varying functions are again regularly varying.

PROPOSITION 4.16. *Let $\varrho \neq 0$ and $h \in \mathcal{SR}_\varrho$. If $\varrho > 0$, then h is strictly increasing in some right neighborhood of 0, and if $\varrho < 0$, then h is strictly decreasing in some right neighborhood of 0. In particular, the function $\varrho^{-1}h'$ (when restricted to this neighborhood) is in the class $\mathcal{R}_{\varrho-1}$.*

PROOF. Let $\varrho \neq 0$ and $h \in \mathcal{SR}_\varrho$. Fix some $\epsilon \in (0,1)$ and let $(0, y_0)$ be an interval in which h is differentiable. Then (4.8) in Lemma 4.15 implies that there is some $y_1 \in (0, y_0]$ such that for each $y \in (0, y_1)$,

$$(1-\epsilon)\varrho h(y)/y < h'(y) < (1+\epsilon)\varrho h(y)/y.$$

For $\varrho > 0$, the expression on the left is strictly positive, which means in particular that h' is strictly positive in the interval $(0, y_1)$. Hence the function h is strictly increasing in $(0, y_0)$. For $\varrho < 0$, the expression on the right is strictly negative, which implies that h' is strictly negative in $(0, y_1)$. Hence h is strictly decreasing in $(0, y_1)$. For the last assertion note that $\varrho^{-1}h'$ is positive in the interval $(0, y_1)$. Moreover, by Lemma 4.15, we have $\varrho^{-1}h' \sim h(y)/y$, as $y \searrow 0$. Therefore, $\varrho^{-1}h' \in \mathcal{R}_{\varrho-1}$ follows from Lemma 4.7, since the function $y \mapsto h(y)/y$ is in $\mathcal{R}_{\varrho-1}$. This completes the proof. □

We have gathered all the results necessary in order to show the asymptotic equivalence of the classes $\mathcal{SR}_\varrho$ and $\mathcal{G}_\varrho$.

THEOREM 4.17. *Let $\varrho \in (0,1)$. For any function $h \in \mathcal{G}_\varrho$, there is a smoothly varying function $\tilde{h} \in \mathcal{SR}_\varrho$ asymptotically equivalent to h, and vice versa. Hence, the classes $\mathcal{G}_\varrho$ and $\mathcal{SR}_\varrho$ are asymptotically equivalent.*

PROOF. Let $\varrho \in (0,1)$ and $h \in \mathcal{G}_\varrho$. By condition (H2), h is regularly varying (at 0), and therefore, by Corollary 4.14, there exists some function $\tilde{h} \in \mathcal{SR}_\varrho$ such that $\tilde{h}(y) \sim h(y)$, as $y \searrow 0$.

To show the converse, let $\tilde{h} \in \mathcal{SR}_\varrho$. Then $\tilde{h}$ satisfies hypothesis (H2) by the Uniform Convergence Theorem 4.6. Furthermore, $\tilde{h}$ satisfies (H3) by Proposition 4.12. It remains to show that the function $\tilde{h}$ satisfies hypothesis (H1) modulo asymptotic equivalence.

The function $\tilde{h}$ is by definition C^1-smooth and therefore in particular continuous on some interval $(0, y_0]$ for some $y_0 > 0$. By Proposition 4.16, $\tilde{h}$ is also strictly increasing on the interval $(0, y_1]$, for some $y_1 \in (0, a]$. Therefore, $\tilde{h}$ can now easily be extended or redefined on the interval (y_1, ∞) in such a way that it becomes continuous and strictly increasing on $(0, \infty)$ and satisfies $\lim_{y\to\infty} \tilde{h}(y) = \infty$ (e.g. by setting $h(y) := (y - y_1) + \tilde{h}(y_1)$ for $y > y_1$ and $h(y) := \tilde{h}(y)$ on $(0, y_1]$). Note that this will not affect the asymptotic properties at 0, i.e. for the new function h, we have $h(y) \sim \tilde{h}(y)$ as $y \searrow 0$ and $h \in \mathcal{R}_\varrho$. In particular, h still satisfies (H1) and (H2). By the Lemma 4.4, we have the representation $h(y) = y^\varrho \ell(y)$ for some $\ell \in \mathcal{R}_0$ and so it is easy to see from Proposition 4.9 that

$$\lim_{y\searrow 0} \frac{h(y)}{y} = \lim_{y\searrow 0} y^{\varrho-1}\ell(y) = \infty \quad \text{and} \quad \lim_{y\searrow 0} h(y) = 0,$$

since $\varrho \in (0,1)$. Therefore, we have found a function h, asymptotically equivalent to $\tilde{h}$, satisfying all the conditions of the class $\mathcal{G}_\varrho$. This completes the proof. □

Combining Corollary 4.14 and Theorem 4.17, we conclude that for any $\varrho \in (0,1)$ the three classes $\mathcal{R}_\varrho$, $\mathcal{SR}_\varrho$ and $\mathcal{G}_\varrho$ are asymptotically equivalent, hence Theorem 2.2

is proved. Regular variation is indeed the essential property which makes things work. Therefore, it is only natural to expect that in Theorem 1.2 the class $\mathcal{G}_{1-D}$ can be substituted with the class $\mathcal{R}_{1-D}$. However, in order to do this, one needs to clarify that the auxiliary functions f and g associated in Theorem 1.2 to any $h \in \mathcal{G}_{1-D}$ have a well defined counterpart for any $h \in \mathcal{R}_{1-D}$. This could be achieved by using the concept of generalized asymptotic inverses (see e.g. [**3**, §1.5.7]), which, however, would impose additional technical difficulties. The latter can be avoided by restricting to smoothly varying functions, which, due to the asymptotic equivalence of the classes $\mathcal{R}_{1-D}$ and $\mathcal{SR}_{1-D}$, is rather a convenience than a restriction. In fact, we will not only see that the functions f and g are well-defined for any $h \in \mathcal{SR}_{1-D}$, but also that they inherit from h regular variation and smoothness.

For this purpose we let $D \in (0,1)$ and $h \in \mathcal{SR}_{1-D}$ (with $1-D \in (0,1)$). We define the function H by

$$H(y) := y/h(y), y \in \mathrm{dom}(h)$$

and observe that $H \in \mathcal{SR}_D$. By Proposition 4.16, H is strictly increasing on some interval $(0, y_0)$ and therefore it can be inverted on this interval. The inverse function H^{-1} is then well-defined on the interval $(0, H(y_0))$ (and may be extended beyond this interval in some arbitrary way, if needed). H^{-1} inherits the properties of smoothness and strict monotonicity from H. Moreover, observing that for any $z \in (0, H(y_0))$ there is a unique $y \in (0, y_0)$ such that $z = H(y)$ (and that z depends continuously on y), we obtain for any $t > 0$,

$$\lim_{z \searrow 0} \frac{H^{-1}(tz)}{H^{-1}(z)} = \lim_{y \searrow 0} \frac{H^{-1}(tH(y))}{H^{-1}(H(y))} = \lim_{y \searrow 0} \frac{H^{-1}(H(t^{1/D}y))}{H^{-1}(H(y))} = t^{1/D},$$

where we have used for the second equality that $H \in \mathcal{R}_D$. Hence, we have shown that H^{-1} is smoothly varying with index $1/D$, i.e. $H^{-1} \in \mathcal{SR}_{1/D}$. We are now ready to define the functions f and g.

PROPOSITION 4.18. *Let $D \in (0,1)$ and $h \in \mathcal{SR}_{1-D}$. Then there is some $x_0 \geq 0$ such that the functions g and f are well-defined for any $x \in [x_0, \infty)$ by*

$$g(x) := H^{-1}(1/x) \quad \text{and} \quad f(x) := \frac{1}{H(1/x)} = xh(1/x),$$

and C^1-smooth on this interval. Moreover, $g \in \mathcal{SR}_{-1/D}[\infty]$ and $f \in \mathcal{SR}_D[\infty]$.

PROOF. By the considerations above, H^{-1} is well-defined on $(0, H(y_0)]$, where y_0 is chosen such that h (and thus H) is strictly increasing and C^1-smooth on $(0, y_0]$. Letting $x_g := 1/H(y_0)$, we have $1/x \in (0, H(y_0)]$ for any $x \in [x_g, \infty)$, and so $H^{-1}(1/x)$ and thus $g(x)$ are well-defined and smooth on $[x_g, \infty)$. Since $H^{-1} \in \mathcal{R}_{1/D}$, it follows directly from Remark 4.2 that $g \in \mathcal{R}_{-1/D}[\infty]$.

Similarly, f is well-defined and C^1 on $[x_f, \infty)$ for $x_f := 1/y_0$ whenever H is C^1 on $(0, y_0]$. Moreover, $y \mapsto 1/H(y)$ is regularly varying (at 0) with index $-D$, since H has index D. Thus, again by Remark 4.2, we obtain $f \in \mathcal{SR}_D[\infty]$. (The actual assertion of the statement is satisfied for $x_0 := \max\{x_g, x_f\}$, however, we will not need a common interval for g and f in our applications.) □

In the proof of Theorem 1.2 given in [**7**], hypothesis (H3) is a technical assumption in order to leisurely apply Lebesgue dominated convergence. Karamata theory, however, allows to circumvent this kind of reasoning due to, among others, the following powerful result, known as Karamata's Theorem. As we will only need a version for functions regularly varying at ∞, we directly restate the corresponding result in [**3**] for this class of functions.

THEOREM 4.19 (Karamata's Theorem; direct half [**3**, Theorem 1.5.11]). *Let $\varrho \in \mathbb{R}$ and $f \in \mathcal{R}_\varrho[\infty]$. Let further $x_0 > 0$ be such that f is locally bounded in $[x_0, \infty)$, cf. Cor. 4.11. Then the following assertions hold.*

(*i*) *For any $\sigma \geq -(\varrho+1)$,*

$$x^{\sigma+1} f(x) \Big/ \int_{x_0}^{x} u^\sigma f(u) du \to \sigma + \varrho + 1, \text{ as } x \to \infty.$$

(*ii*) *For any $\sigma < -(\varrho+1)$ (and for $\sigma = -(\varrho+1)$ if $\int_\cdot^\infty u^{-(\varrho+1)} f(u) du < \infty$),*

$$x^{\sigma+1} f(x) \Big/ \int_{x}^{\infty} u^\sigma f(u) du \to -(\sigma + \varrho + 1), \text{ as } x \to \infty.$$

One of the most interesting consequences of Karamata's Theorem 4.19 in the context of regularly varying gauge functions is that we may take slowly varying functions $\ell \in \mathcal{R}_0[\infty]$ out of integrals in the following fashion:

$$\int_x^\infty u^\varrho \ell(u) du \sim \ell(x) \int_x^\infty u^\varrho du, \text{ as } x \to \infty, \tag{4.9}$$

whenever $\varrho < -1$ (and this is the case we will need). Indeed, this follows easily from part (ii) of Karamata's Theorem. Moreover, this theorem provides the following useful relations. We point out that statements similar to (i) and (ii) below have been proved and used by He and Lapidus for functions related to the classes $\mathcal{G}_\varrho$, see [**7**, Proposition 3.2 and Lemma 3.3].

PROPOSITION 4.20. *Let $\varrho < -1$ and $g \in \mathcal{R}_\varrho[\infty]$. Then*

(i) $\displaystyle\int_x^\infty g(u) du \sim -\frac{1}{\varrho+1} x g(x)$, *as $x \to \infty$;*

(ii) *for any $t > 0$,*

$$\frac{\int_{tx}^\infty g(u) du}{\int_x^\infty g(u) du} \to t^{\varrho+1}, \text{ as } x \to \infty,$$

where the convergence is uniform in t on any compact subset of $(0, \infty)$;

(iii) $\displaystyle\sum_{j=k}^\infty g(j) \sim -\frac{1}{\varrho+1} k g(k)$, *as $k \to \infty$.*

PROOF. Assertion (i) follows directly from Karamata's Theorem 4.19 (ii) for $\sigma = 0$.

For a proof of (ii) observe that the function $x \mapsto xg(x), x \in \mathrm{dom}(g)$ is regularly varying (at ∞) with index $\varrho + 1$. Hence, by Lemma 4.5 and assertion (i), the function $G(x) := \int_x^\infty g(u) du$, $x \in \mathrm{dom}(g)$ is in $\mathcal{R}_{\varrho+1}[\infty]$. Therefore, the convergence $G(tx)/G(x) \to t^{\varrho+1}$, as $x \to \infty$ is obvious and the uniformity follows from the Uniform Convergence Theorem 4.6 (version at ∞).

It remains to prove (iii). By Lemma 4.4, there is some $\ell \in \mathcal{R}_0[\infty]$ such that $g(x) = x^\varrho \ell(x)$, $x \in \mathrm{dom}(g)$. Define $\tilde{g}$ by $\tilde{g}(x) := g(j)$, for $x \in [j, j+1) \cap \mathrm{dom}(g)$ and $j \in \mathbb{N}$, and $\tilde{\ell}$ by $\tilde{\ell}(x) := \tilde{g}(x)/x^\varrho$, $x \in \mathrm{dom}(g)$. We claim that $\tilde{g} \in \mathcal{R}_\varrho[\infty]$ or, equivalently, $\tilde{\ell} \in \mathcal{R}_0[\infty]$. To prove this, by Lemma 4.5, it suffices to show that $\tilde{\ell}(x) \sim \ell(x)$, as $x \to \infty$. First observe that, for any $x \in [j, j+1) \cap \mathrm{dom}(g)$, $\tilde{\ell}(x) = g(j)/x^{\varrho}$ and thus, since $x \mapsto x^{\varrho}$ is decreasing,

$$\ell(j) = \frac{g(j)}{j^{\varrho}} \leq \tilde{\ell}(x) \leq \frac{g(j)}{(j+1)^{\varrho}} = \left(\frac{j}{j+1}\right)^{\varrho} \ell(j).$$

Since $j = [x]$ and thus $j/(j+1) \to 1$ as $x \to \infty$, we conclude $\tilde{\ell}(x) \sim \ell([x])$, as $x \to \infty$. Finally, we note that $\ell([x]) \sim \ell(x)$, as $x \to \infty$, since $\ell \in \mathcal{R}_0[\infty]$. Indeed, letting $t_x := [x]/x$ and noting that $1/2 \leq 1 - 1/x \leq t_x \leq 1$ for any $x \geq 2$, the Uniform Convergence Theorem 4.6 (applied to ℓ for t on the compact interval $[1/2, 1]$) yields

$$\lim_{x\to\infty} \frac{\ell([x])}{\ell(x)} = \lim_{x\to\infty} \frac{\ell(t_x\, x)}{\ell(x)} = 1.$$

This completes the proof of our claim that $\tilde{g} \in \mathcal{R}_{\varrho}[\infty]$. Now assertion (iii) follows directly by applying (i) to $\tilde{g}$, since, for any $k \in \mathbb{N} \cap \mathrm{dom}(g)$,

$$\sum_{j=k}^{\infty} g(j) = \sum_{j=k}^{\infty} \tilde{g}(j) = \int_k^{\infty} \tilde{g}(u)du. \qquad \square$$

5. The geometric part of the proof

In this section we discuss the equivalence of the first three assertions in Theorem 2.5 as well as that of assertions (vi), (vii) and (viii). The direct proofs of (i) $\Leftrightarrow$ (iii) and (vi) $\Leftrightarrow$ (viii) given in [**7**] are rather technical, cf. [**7**, Theorems 3.4, 3.8, 3.10 and 4.1]. Using characterization results for Minkowski contents in terms of S-contents from [**19, 20**] does not only allow to add another equivalent criterion to each of the two parts of Theorem 2.5, but also to simplify the proofs significantly. Instead of proving (i) $\Leftrightarrow$ (iii) directly, we will establish (i) $\Leftrightarrow$ (ii) and (ii) $\Leftrightarrow$ (iii) separately. Similarly we will show (vi) $\Leftrightarrow$ (vii) and (vii) $\Leftrightarrow$ (viii). The (generalized) S-contents, which describe the behavior of the boundary measure of the parallel sets, provide thus an extremely useful connection between Minkowski contents (volume of the parallel sets) and the growth of the lengths in the associated fractal string.

5.1. S-contents vs. Minkowski contents. Our first aim is to verify the equivalences (i) $\Leftrightarrow$ (ii) and (vi) $\Leftrightarrow$ (vii) in Theorem 2.5. It turns out that they can be derived essentially from the results in [**20**]. Therefore, we will not reprove the equivalence here, but rather explain the minor modifications necessary in the relevant results of [**20**] to cover our present situation. In fact, we can establish this equivalence for any bounded open set $\Omega \subset \mathbb{R}^d$ and any gauge function $h \in \mathcal{SR}_{d-D}, D \in (0, d)$. There is no need to restrict to subsets of $\mathbb{R}$ for this result.

THEOREM 5.1. *Let $\Omega \subset \mathbb{R}^d$ be bounded and $F = \partial\Omega$. Let $h \in \mathcal{SR}_{d-D}$ for some $D \in [0, d)$. Then the following assertions are equivalent:*

(i) $0 < \underline{\mathcal{M}}(h; F) \leq \overline{\mathcal{M}}(h; F) < \infty$,
(ii) $0 < \underline{\mathcal{S}}(h'; F) \leq \overline{\mathcal{S}}(h'; F) < \infty$.

In particular, these assertions imply $\dim_M F = D$.

REMARK 5.2. Recall that the (generalized) Minkowski contents and S-contents appearing in the statement above as well as in Theorem 5.3 below are defined relative to the set Ω, cf. (1.6) and (2.1). The results from [**20**] that we are going to use in the proofs are formulated for the 'full' contents (with the set Ω in the definitions (1.6) and (2.1) omitted). However, due the fact that Ω is metrically associated with its boundary F and that therefore the volume function $r \mapsto V_F(r) = |F_r \cap \Omega|_d$ is a Kneser function, all the results in [**20**] hold literally for the relative contents used here. For more details we refer to the discussion of relative contents in [**25**].

PROOF. In view of Remark 5.2, the stated equivalence follows essentially by combining [**20**, Theorem 3.2] (where the easier implication (ii) $\Rightarrow$ (i) is established) with [**20**, Theorem 3.4] (where the reverse implication is obtained). While in [**20**, Theorem 3.2] h is assumed to be differentiable with derivative h' being non-zero in some right-neighborhood of 0, an assumption met for any $h \in \mathcal{SR}_{d-D}$ due to Proposition 4.16, there are additional assumptions in [**20**, Theorem 3.4]: h is assumed to be of the form $h(y) = y^{d-D}g(y)$ with g being non-decreasing and

$$\limsup_{y \searrow 0} \frac{yh'(y)}{h(y)} < \infty.$$

The latter assumption is satisfied due to Lemma 4.15, which says that the above limit exists and equals $d-D$ for $h \in \mathcal{SR}_{d-D}$. Moreover, by Lemma 4.4, h is clearly of the form $h(y) = y^{d-D}g(y)$ for some $g \in \mathcal{SR}_0$. But the slowly varying factor g is not necessarily non-decreasing. However, inspecting the proof of [**20**, Theorem 3.4] it is rather easy to see that 'non-decreasing' can be replaced by 'slowly varying' in this statement. The monotonicity of g is only used once in the proof of [**20**, Proposition 3.3] to ensure that $g(ar)$ is bounded from below by $g(r)$ for some $a > 1$ and all sufficiently small $r > 0$. If $g \in \mathcal{R}_0$ and $\epsilon > 0$, then we can certainly find some $r_1 > 0$ such that $g(ar) \geq (1-\epsilon)g(r)$ for any $r \in (0, r_1)$ and this suffices to extend the argument in the proof of [**20**, Proposition 3.3] to functions $g \in \mathcal{R}_0$. Since [**20**, Theorem 3.4] is essentially a direct application of [**20**, Proposition 3.3], also this statement extends to functions $g \in \mathcal{SR}_0$. For the assertion $\dim_M F = D$ see Remark 4.10. This completes the proof. $\square$

Also the following statement is a direct consequence of a result in [**20**] and the elasticity properties derived in Lemma 4.15. It establishes the equivalence (vi) $\Leftrightarrow$ (vii) in Theorem 2.5, i.e. the equality of the h-Minkowski content and the h'-S-content.

THEOREM 5.3 (cf. [**20**, Theorem 3.7]). *Let $\Omega \subset \mathbb{R}^d$ be bounded and $F = \partial\Omega$. Let $h \in \mathcal{SR}_{d-D}$ for some $D \in [0, d)$. Suppose $M > 0$. Then*

$$\mathcal{M}(h; F) = M, \quad \text{if and only if} \quad \mathcal{S}(h'; F) = M.$$

Moreover, in this case $\dim_M F = D$.

PROOF. By Lemma 4.4, h can be written in the form $h(y) = g(y)y^{d-D}$ for some $g \in \mathcal{SR}_0$ and by Lemma 4.15, we have $\lim_{y \to 0} g'(y)y/g(y) = 0$. Therefore, taking into account Remark 5.2, the hypothesis of [**20**, Theorem 3.7] is satisfied except that the function g is now slowly varying and not necessarily non-decreasing. However, we can argue similarly as in the proof of Theorem 5.1 above that the forward implication in [**20**, Theorem 3.7] follows essentially from [**20**, Proposition 3.6], which is a refinement of [**20**, Proposition 3.3] and in which the estimates can easily be modified to work for slowly varying g. The reverse implication in [**20**, Theorem 3.7] (which is the reverse implication in Theorem 5.3) is a direct consequence of [**20**, Theorem 3.2], which can be applied since, by Proposition 4.16,

h' is non-zero in some right neighborhood of 0. Finally, for a proof of $\dim_M F = D$ see Remark 4.10. $\square$

5.2. S-Contents and fractal strings. Recall that in dimension $d = 1$ we can associate to any bounded open set $\Omega \subset \mathbb{R}$ its fractal string $\mathcal{L} = (l_j)_{j\in\mathbb{N}}$, encoding the lengths l_j of the connected components I_j of Ω. They appear in $\mathcal{L}$ in non-increasing order, i.e. $l_1 \geq l_2 \geq l_3 \geq \ldots$, and according to their multiplicities. We will discuss now the relations between S-contents and the asymptotic growth of the lengths l_j, and establish in particular the equivalences $(ii) \Leftrightarrow (iii)$ and $(vii) \Leftrightarrow (viii)$ of Theorem 2.5. Note that (for $F = \partial\Omega$) the '*surface area*' of $\partial F_\varepsilon \cap \Omega$ in the definition of the S-content reduces in dimension $d = 1$ to the counting measure $\mathcal{H}^0$. Therefore, the key idea behind these two relations is the following simple geometric observation (everything else is '*asymptotic calculus*'): All small intervals I_j of Ω with length $l_j \leq 2\varepsilon$ are completely covered by F_ε and do not contribute to the boundary ∂F_ε, while each of the large intervals ($l_j > 2\varepsilon$) contributes exactly two points. That is, for any $j \in \mathbb{N}$ and $\varepsilon \in [l_j/2, l_{j-1}/2)$, we have

$$\mathcal{H}^0(\partial F_\varepsilon \cap \Omega) = 2(j-1). \tag{5.1}$$

Recall from Proposition 4.18 that for $D \in (0,1)$ and $h \in \mathcal{SR}_{1-D}$, the function g is given by $g(x) := H^{-1}(1/x)$, where H^{-1} is the inverse of $H(y) := y/h(y)$.

THEOREM 5.4. *Let $\Omega \subset \mathbb{R}$ be open and bounded, $F := \partial\Omega$ and $\mathcal{L} = (l_j)_{j\in\mathbb{N}}$ the associated fractal string. Let $D \in (0,1)$ and $h \in \mathcal{SR}_{1-D}$. Then the following assertions are equivalent:*

(i) $0 < \underline{\mathcal{S}}(h'; F) \leq \overline{\mathcal{S}}(h'; F) < \infty$,
(ii) $l_j \asymp g(j)$, *as* $j \to \infty$.

PROOF. We first reformulate both assertions using 'asymptotic calculus'. Then their equivalence will follow easily from (5.1).

On the one hand, assertion (i) can be rewritten as $\mathcal{H}^0(\partial F_\varepsilon \cap \Omega) \asymp h'(\varepsilon)$, as $\varepsilon \searrow 0$, which, by Lemma 4.15, is equivalent to $\mathcal{H}^0(\partial F_\varepsilon \cap \Omega) \asymp h(\varepsilon)/\varepsilon$, as $\varepsilon \searrow 0$.

On the other hand, since $g(x) = H^{-1}(1/x)$ and $H \in \mathcal{SR}_D$, we can apply H to both sides of (ii) to see that, by Lemma 4.7, (ii) is equivalent to

$$H(l_j) = l_j/h(l_j) \asymp 1/j, \text{ as } j \to \infty.$$

By Lemma 3.1, this can also be written as $j \asymp h(l_j)/l_j$, as $j \to \infty$. Using the asymptotic homogeneity of $h \in \mathcal{SR}_{1-D}$ (which implies $h(y) \sim 2^{1-D}h(y/2)$, as $y \searrow 0$, cf. Remark 4.3) and recalling that positive constants do not matter in '$\asymp$'-relations, this is equivalently given by $2j \asymp h(l_j/2)/(l_j/2)$, as $j \to \infty$. Since $(j-1)/j \to 1$, as $j \to \infty$, we can replace j by $j-1$ on the left. Therefore, the statement of the theorem is equivalent to

$$\mathcal{H}^0(\partial F_\varepsilon \cap \Omega) \asymp \frac{h(\varepsilon)}{\varepsilon}, \text{ as } \varepsilon \searrow 0, \quad \text{iff} \quad 2(j-1) \asymp \frac{h(r_j)}{r_j}, \text{ as } j \to \infty, \tag{5.2}$$

where $r_j := l_j/2$, $j \in \mathbb{N}$. (r_j is the 'inradius' of an interval of length l_j.)

The forward implication in (5.2) is obvious: since (5.1) implies in particular that $\mathcal{H}^0(\partial F_{r_j} \cap \Omega) = 2(j-1)$ for each $j \in \mathbb{N}$, we just need to plug in the sequence $(r_j)_{j\in\mathbb{N}}$ for ε in the left assertion.

For a proof of the reverse implication in (5.2), assume the right hand side holds, which means, there are constants c_1, c_2 such that $c_1 \leq 2(j-1)/(h(r_j)/r_j) \leq c_2$ for each j. Since the function $\varepsilon \mapsto h(\varepsilon)/\varepsilon$ is in $\mathcal{SR}_{-D}$, it is strictly decreasing in some right neighborhood $(0,\varepsilon_0)$ of 0, cf. Proposition 4.16. Moreover, $\lim_{\varepsilon\searrow 0} h(\varepsilon)/\varepsilon = +\infty$. Therefore, for any sufficiently large $j\in\mathbb{N}$ (such that $r_{j-1}<\varepsilon_0$) and any $\varepsilon\in[r_j,r_{j-1})$, the quotient $\frac{\mathcal{H}^0(\partial F_\varepsilon\cap\Omega)}{h(\varepsilon)/\varepsilon}$ is bounded from above and below by

$$c_1 \le \frac{2(j-1)}{h(r_j)/r_j} \le \frac{\mathcal{H}^0(\partial F_\varepsilon\cap\Omega)}{h(\varepsilon)/\varepsilon} \le \frac{2(j-1)}{h(r_{j-1})/r_{j-1}} \le c_2 + \frac{2}{h(r_{j-1})/r_{j-1}},$$

where the last summand on the right vanishes as $j\to\infty$. But this implies the left assertion in (5.2), completing the proof of Theorem 5.4. □

A very similar argument allows to establish the equivalence of generalized S-measurability of $F=\partial\Omega$ and generalized 'L-measurability' of the associated fractal string, i.e. the equivalence of the assertions (vii) and (viii) in Theorem 2.5.

THEOREM 5.5. *Let $F\subset\mathbb{R}$ be a compact set and let $\mathcal{L}=(l_j)_{j\in\mathbb{N}}$ be the associated fractal string. Let $D\in(0,1)$ and $h\in\mathcal{SR}_{1-D}$. Let $L>0$. Then the following assertions are equivalent:*

(i) *F is h'-S measurable (i.e. $0<\mathcal{S}(h',F)<\infty$) with $\mathcal{S}(h',F)=S:=\frac{2^{1-D}L^D}{1-D}$,*
(ii) *$l_j\sim L\cdot g(j)$, as $j\to\infty$.*

PROOF. We follow the line of the argument in the proof of Theorem 5.4 and start by reformulating the statement. In view of (5.2) and using again $r_j:=l_j/2$, our first claim is that the relation (i)⇔(ii) is equivalent to the following equivalence:

(5.3)
$$\mathcal{H}^0(\partial F_\varepsilon\cap\Omega)\sim 2^{1-D}L^D\frac{h(\varepsilon)}{\varepsilon},\ \text{as } \varepsilon\searrow 0,\ \text{ iff }\ 2(j-1)\sim 2^{1-D}L^D\frac{h(r_j)}{r_j},\ \text{as } j\to\infty.$$

Indeed, assertion (i) in Theorem 5.5 means $\mathcal{H}^0(\partial F_\varepsilon\cap\Omega)\sim S\cdot h'(\varepsilon)$, as $\varepsilon\searrow 0$ and, by Lemma 4.15, $h'(\varepsilon)$ can be replaced by $(1-D)h(\varepsilon)/\varepsilon$ (since $h\in\mathcal{SR}_{1-D}$), showing the equivalence of (i) and the left assertion in (5.3). (Recall that $(1-D)S=2^{1-D}L^D$.) Similarly, by applying H to both sides of assertion (ii) and recalling that $g(x)=H^{-1}(1/x)$, we can infer from Lemma 4.7 that (ii) is equivalent to $H(L^{-1}l_j)=L^{-1}l_j/h(L^{-1}l_j)\sim 1/j$, as $j\to\infty$. Taking into account the homogeneity of h (cf. Remark 4.3) and Lemma 3.1, this can be rephrased as $j\sim L^D h(l_j)/l_j$, as $j\to\infty$, which is easily seen to be equivalent to the right assertion in (5.3), using again the homogeneity of h. This completes the proof of the above claim. It is therefore sufficient to prove the equivalence in (5.3).

The forward implication in (5.3) is again obvious from (5.1), by plugging in the sequence $(r_j)_{j\in\mathbb{N}}$ for ε in the left assertion. The reverse implication in (5.3) requires now a slightly refined argument. Assume the right hand side holds. Then, for each $\delta>0$, there is some $j_0=j_0(\delta)$ such that

$$1-\delta\le\frac{2(j-1)}{c\cdot h(r_j)/r_j}\le 1+\delta$$

for each $j\ge j_0$, where $c:=2^{1-D}L^D$. Since the function $\varepsilon\mapsto h(\varepsilon)/\varepsilon$ is in $\mathcal{SR}_{-D}$, it is strictly decreasing in some right neighborhood $(0,\varepsilon_0)$ of 0, cf. Proposition 4.16. Therefore, for any sufficiently large $j\in\mathbb{N}$ (such that $j\ge j_0$ and $r_{j-1}/2<\varepsilon_0$) and any $\varepsilon \in [r_j, r_{j-1})$, the quotient $\frac{\mathcal{H}^0(\partial F_\varepsilon \cap \Omega)}{c \cdot h(\varepsilon)/\varepsilon}$ is bounded from above and below by

$$1-\delta \leq \frac{2(j-1)}{c \cdot h(r_j)/r_j} \leq \frac{\mathcal{H}^0(\partial F_\varepsilon \cap \Omega)}{c \cdot h(\varepsilon)/\varepsilon} \leq \frac{2(j-1)}{c \cdot h(r_{j-1})/r_{j-1}} \leq 1+\delta+\frac{2}{c \cdot h(r_{j-1})/r_{j-1}},$$

where again the last summand on the right vanishes as $j \to \infty$. Since the argument works for any $\delta > 0$, the left assertion in (5.3) follows, completing the proof of Theorem 5.5. □

6. The 'spectral' part of the proof of Theorem 2.5

In this section, we will finally establish the connection between the geometric and the spectral properties of Ω, i.e. in particular the equivalence of the assertions (iii) and (iv) in part I of Theorem 2.5 and the validity of (2.4) in part II. Because of the results in the previous section, we can use a combination of the assertions (i)-(iii) of Theorem 2.5 to conclude the validity of (iv) – we will use (ii) and (iii). In contrast, it is enough to show that (iv) implies at least one of the assertions (i)-(iii).

It will be convenient now to use the *string counting function* J defined by

$$J(\varepsilon) := \max\{j \mid l_j > \varepsilon\}, \tag{6.1}$$

for any fractal string $\mathcal{L} = (l_j)_{j \in \mathbb{N}}$, counting the number of lengths l_j in $\mathcal{L}$ that are strictly larger than ε. Since

$$2J(2\varepsilon) = \mathcal{H}^0(\partial F_\varepsilon \cap \Omega), \tag{6.2}$$

for any $\varepsilon > 0$, it is easy to see that the asymptotics of $\mathcal{H}^0(\partial F_\varepsilon \cap \Omega)$ (described by the S-content) determines the asymptotics of $J(\varepsilon)$, as $\varepsilon \searrow 0$, and vice versa.

PROPOSITION 6.1. *Under the hypothesis of Theorem 2.5, any of the assertions (i),(ii) and (iii) in Theorem 2.5 is equivalent to*

$$J(\varepsilon) \asymp h(\varepsilon)/\varepsilon, \quad \textit{as } \varepsilon \searrow 0. \tag{6.3}$$

Similarly, any of the assertions (vi), (vii) or (viii) in Theorem 2.5 is equivalent to

$$J(\varepsilon) \sim L^D h(\varepsilon)/\varepsilon, \quad \textit{as } \varepsilon \searrow 0, \tag{6.4}$$

where L is the constant in (viii). In particular, this implies $J \in \mathcal{R}_{-D}$.

PROOF. Due to the results of the previous section, it suffices to relate (6.3) to assertion (ii) and (6.4) to (vii), which is easy to do with the help of (6.2). For the first equivalence recall from the proof of Theorem 5.4 (see (5.2)) that assertion (ii) in Theorem 2.5 is equivalent to $\mathcal{H}^0(\partial F_\varepsilon \cap \Omega) \asymp h(\varepsilon)/\varepsilon$, as $\varepsilon \searrow 0$ and apply (6.2).

For the second stated equivalence, recall from the proof of Theorem 5.5 (see (5.3)) that assertion (vii) is equivalent to $\mathcal{H}^0(\partial F_\varepsilon \cap \Omega) \sim 2^{1-D} L^D h(\varepsilon)/\varepsilon$ as $\varepsilon \searrow 0$. Therefore, by (6.2), $2J(2\varepsilon) \sim 2^{1-D} L^D h(\varepsilon)/\varepsilon$, as $\varepsilon \searrow 0$. Substituting 2ε by ε and taking into account the asymptotic homogeneity of h (of degree $1-D$), we obtain

$$J(\varepsilon) \sim 2^{-D} L^D \frac{h(\varepsilon/2)}{\varepsilon/2} \sim 2^{1-D} L^D \frac{h(\varepsilon)(1/2)^{1-D}}{\varepsilon} = L^D \frac{h(\varepsilon)}{\varepsilon}, \text{ as } \varepsilon \searrow 0,$$

which completes the proof of (6.4). Finally, since $\varepsilon \mapsto h(\varepsilon)/\varepsilon$ is regularly varying with index $-D$, $J \in \mathcal{R}_{-D}$ follows from (6.4) by Lemma 4.5. □

The following observation will turn out to be very useful in the proof of the implication (iii)⇒(iv).

LEMMA 6.2. *Assume that assertion (iii) of Theorem 2.5 holds. Then*

$$\sum_{j>J(2\varepsilon)} l_j \asymp h(\varepsilon), \quad as\ \varepsilon \searrow 0.$$

PROOF. Due to the assumption $l_j \asymp g(j)$ as $j \to \infty$, we find positive constants $j_0, \underline{\alpha}, \overline{\alpha}$ such that $l_j/g(j) \in [\underline{\alpha}, \overline{\alpha}]$ for all $j \geq j_0$. Fix $\varepsilon_0 > 0$ small enough such that $J(2\varepsilon_0) \geq j_0$. Then we have for any $0 < \varepsilon \leq \varepsilon_0$

$$\underline{\alpha} \sum_{j>J(2\varepsilon)} g(j) \leq \sum_{j>J(2\varepsilon)} l_j \leq \overline{\alpha} \sum_{j>J(2\varepsilon)} g(j).$$

This shows

$$\sum_{j>J(2\varepsilon)} l_j \asymp \sum_{j>J(2\varepsilon)} g(j), \quad \text{as } \varepsilon \searrow 0. \tag{6.5}$$

Since $J(2\varepsilon) \to \infty$ as $\varepsilon \searrow 0$ and $g \in \mathcal{SR}_{-1/D}$ (with $-1/D < -1$), we infer from Proposition 4.20 (iii) that

$$\sum_{j>J(2\varepsilon)} g(j) \sim \frac{D}{1-D} J(2\varepsilon) g(J(2\varepsilon)), \quad \text{as } \varepsilon \searrow 0.$$

Using (6.3), it follows from Lemma 4.7 and the definition of g that the right hand side is asymptotically similar to $\frac{h(\varepsilon)}{\varepsilon} g(\frac{h(\varepsilon)}{\varepsilon}) = h(\varepsilon)$, as $\varepsilon \searrow 0$. Combining this with (6.5), the assertion of the lemma follows. □

Now we are ready to reformulate and prove the implication (iii) ⇒ (iv) in Theorem 2.5. It is convenient to employ the function

$$\tilde{\delta}(2\varepsilon) := \delta(\frac{1}{2\varepsilon}) - \sum_{j=1}^{\infty} \{\frac{l_j}{2\varepsilon}\}, \quad \varepsilon > 0. \tag{6.6}$$

THEOREM 6.3. *Let $D \in (0,1)$ and $h \in \mathcal{SR}_{1-D}$. Let $(l_j)_{j\in\mathbb{N}}$ be a fractal string with $l_j \asymp g(j)$, as $j \to \infty$. Then*

$$\tilde{\delta}(2\varepsilon) \asymp \frac{h(\varepsilon)}{\varepsilon}, \quad as\ \varepsilon \searrow 0, \tag{6.7}$$

or, equivalently, $\delta(x) \asymp f(x)$, as $x \to \infty$, where f and g are given as in Proposition 4.18.

PROOF. The equivalence of the two assertions is obvious from the substitution $x = 1/2\varepsilon$ and the definition of f. Therefore, it suffices to prove (6.7). We split $\tilde{\delta}$ as follows

$$\tilde{\delta}(2\varepsilon) = \sum_{j=1}^{\infty} \{\frac{l_j}{2\varepsilon}\} = \sum_{j>J(2\varepsilon)} \{\frac{l_j}{2\varepsilon}\} + \sum_{j\leq J(2\varepsilon)} \{\frac{l_j}{2\varepsilon}\}. \tag{6.8}$$

Now observe that for $j > J(2\varepsilon)$, we have $l_j/2\varepsilon < 1$ and thus $\{l_j/2\varepsilon\} = l_j/2\varepsilon$. Therefore, we can employ Lemma 6.2 to the first sum on the right and infer that

$$\sum_{j>J(2\varepsilon)} \{\frac{l_j}{2\varepsilon}\} = \frac{1}{2\varepsilon} \sum_{j>J(2\varepsilon)} l_j \asymp \frac{h(\varepsilon)}{2\varepsilon}, \quad \text{as } \varepsilon \searrow 0.$$

Since $0 \le \{y\} < 1$, we infer for the second sum on the right of (6.8) that, for any $\varepsilon > 0$,

$$0 \le \sum_{j \le J(2\varepsilon)} \{\frac{l_j}{2\varepsilon}\} \le J(2\varepsilon),$$

and by (6.3) in Proposition 6.1, $J(2\varepsilon)$ is bounded above by $c \cdot h(\varepsilon)/\varepsilon$ for some $c > 0$. Combining the estimates of both sums, we conclude that $\tilde{\delta}(2\varepsilon)$ is bounded from above and below by some constant multiple of $h(\varepsilon)/\varepsilon$ for all sufficiently small $\varepsilon > 0$, proving assertion (6.7). □

Our next aim is to show that h-Minkowski measurability (i.e. any of the assertions (vi), (vii), (viii) in Theorem 2.5) implies the exact asymptotic second term of the eigenvalue counting function as stated in (2.4). For the main argument we follow closely the idea of the proof in [**15**] for the case of Minkowski measurable sets (which is also used in [**7**]) to split the sum $\delta(x)$ in a very special way into three summands and then estimate each of them separately. In the estimation part Karamata theory turns out to be very useful again, allowing a simpler argument as in [**7**]. Our first step is a refinement and generalization of Lemma 6.2 above.

LEMMA 6.4. *Assume that assertion (viii) of Theorem* 2.5 *holds. Then, for any* $k \in \mathbb{N}$,

$$\sum_{j > J(k\varepsilon)} l_j \sim \frac{DL^D}{1-D} k^{1-D}\, h(\varepsilon), \quad \text{as } \varepsilon \searrow 0.$$

PROOF. Due to the assumption $l_j \sim L\,g(j)$ as $j \to \infty$, we find for any $\gamma > 0$ some $j_0 > 0$ such that $l_j/(L\,g(j)) \in [1-\gamma, 1+\gamma]$ for all $j \ge j_0$. Given $k \in \mathbb{N}$, fix $\varepsilon_0 = \varepsilon_0(k) > 0$ small enough such that $J(k\varepsilon_0) \ge j_0$. Then we have for any $0 < \varepsilon \le \varepsilon_0$

$$(1-\gamma)L \sum_{j > J(k\varepsilon)} g(j) \le \sum_{j > J(k\varepsilon)} l_j \le (1+\gamma)L \sum_{j > J(2\varepsilon)} g(j).$$

Since we can find such j_0 and ε_0 for any $\gamma > 0$, we infer

$$\sum_{j > J(k\varepsilon)} l_j \sim L \sum_{j > J(k\varepsilon)} g(j), \qquad \text{as } \varepsilon \searrow 0,$$

for any fixed $k \in \mathbb{N}$. Since $J(k\varepsilon) \to \infty$ as $\varepsilon \searrow 0$ and $g \in \mathcal{SR}_{-1/D}$ (with $-1/D < -1$), we infer from Proposition 4.20 (iii) that, for any $k \in \mathbb{N}$,

$$\sum_{j > J(k\varepsilon)} g(j) \sim \frac{D}{1-D} J(k\varepsilon) g(J(k\varepsilon)), \qquad \text{as } \varepsilon \searrow 0.$$

Applying (6.3) and taking into account the definition of g (cf. Proposition 4.18) and Lemma 4.7, we infer

$$J(k\varepsilon)g(J(k\varepsilon)) \sim L^D \frac{h(k\varepsilon)}{k\varepsilon} g\left(L^D \frac{h(k\varepsilon)}{k\varepsilon}\right) \sim L^{D-1} \frac{h(k\varepsilon)}{k\varepsilon} g\left(\frac{h(k\varepsilon)}{k\varepsilon}\right) = L^{D-1}\, h(k\varepsilon),$$

as $\varepsilon \searrow 0$. Combining the last three estimates and using the homogeneity of h, the assertion of the lemma follows. □

Now we are ready for the main step in the proof of an exact asymptotic second term of N. We will show that (viii) in Theorem 2.5 implies an exact estimate for the packing defect δ.

THEOREM 6.5. *Let $D \in (0,1)$ and $h \in \mathcal{SR}_{1-D}$. Let $(l_j)_{j\in\mathbb{N}}$ be a fractal string with $l_j \sim Lg(j)$, as $j \to \infty$. Then*

$$\delta(x) \sim -\zeta(D) L^D f(x), \qquad \text{as } x \to \infty, \tag{6.9}$$

where g and f are given as in Proposition 4.18.

PROOF. For fixed $k \in \mathbb{N}$, we split δ as follows

$$\delta(x) = \sum_{j>J(1/x)} \{l_j x\} + \sum_{q=2}^{k} \sum_{j=J(q/x)+1}^{J((q-1)/x)} \{l_j x\} + \sum_{j \le J(k/x)} \{l_j x\} \tag{6.10}$$

Now observe that $J(q/x) < j \le J((q-1)/x)$ implies $[l_j x] = q-1$, and therefore the second sum equals

$$\begin{aligned} \sum_{q=2}^{k} \sum_{j=J(q/x)+1}^{J((q-1)/x)} (l_j x - (q-1)) &= \sum_{j=J(k/x)+1}^{J(1/x)} l_j x - \sum_{q=2}^{k} (q-1)\left(J\left(\frac{q-1}{x}\right) - J\left(\frac{q}{x}\right)\right) \\ &= x \sum_{j=J(k/x)+1}^{J(1/x)} l_j - \sum_{q=1}^{k-1} J\left(\frac{q}{x}\right) + (k-1) J\left(\frac{k}{x}\right). \end{aligned}$$

Combining the first sum of this last expression with the first sum in (6.10), in which all terms satisfy $l_j x < 1$ implying $\{l_j x\} = l_j x$, we obtain

$$\delta(x) = x \sum_{j>J(k/x)} l_j + \left(kJ\left(\frac{k}{x}\right) - \sum_{q=1}^{k-1} J\left(\frac{q}{x}\right)\right) + \sum_{j \le J(k/x)} \{l_j x\} - J\left(\frac{k}{x}\right),$$

for each $k \in \mathbb{N}$ and each $x > 0$. We will see in a moment that the first two terms

$$A := x \sum_{j>J(k/x)} l_j \quad \text{and} \quad B := kJ\left(\frac{k}{x}\right) - \sum_{q=1}^{k-1} J\left(\frac{q}{x}\right)$$

contribute asymptotically to $\delta(x)$ as $x \to \infty$ for any $k \in \mathbb{N}$, while the remainder term

$$C := \sum_{j \le J(k/x)} \{l_j x\} - J\left(\frac{k}{x}\right) = \sum_{j \le J(k/x)} (\{l_j x\} - 1)$$

will vanish as $k \to \infty$. (Note that A, B, C depend on x and k.) Up to here we followed the argument of He and Lapidus in the proof of [**7**, Theorem 4.3] (which is similar to the one in the proof of [**15**, Theorem 4.2]), the estimates are now derived in an easier way using Karamata theory.

First, by Lemma 6.4, we obtain for A:

$$A \sim \frac{DL^D}{1-D} k^{1-D} x h(1/x) = L^D \frac{D}{1-D} k^{1-D} f(x), \qquad \text{as } x \to \infty.$$

Second, since $J \in \mathcal{R}_{-D}$ by Proposition 6.1, the homogeneity implies in particular that $J(q/x) \sim q^{-D} J(1/x)$, as $x \to \infty$, for any $q > 0$ and together with (6.4) we infer that

$$B \sim \left(k^{1-D} - \sum_{q=1}^{k-1} q^{-D}\right) J\left(\frac{1}{x}\right) \sim L^D \left(k^{1-D} - \sum_{q=1}^{k-1} q^{-D}\right) f(x), \qquad \text{as } x \to \infty.$$

Combining the estimates for A and B we conclude that, for each $k \in \mathbb{N}$,

$$\frac{A+B}{L^D f(x)} \to \frac{1}{1-D} k^{1-D} - \sum_{q=1}^{k-1} q^{-D} = w_k(D) + \frac{1}{1-D} \text{ as } x \to \infty, \tag{6.11}$$

where the function w_k is defined for each $s \in \mathbb{C}$ by

$$w_k(s) := \int_1^k (t^{-s} - [t]^{-s}) dt \quad \left(= -\frac{1}{1-s} + \frac{1}{1-s} k^{1-s} - \sum_{q=1}^{k-1} q^{-s}, s \neq 1 \right).$$

The functions w_k are entire and, as $k \to \infty$, they converge (uniformly on any compact subset of $\Re(s) > 0$) to the function

$$w(s) := \int_1^\infty (t^{-s} - [t]^{-s}) dt.$$

w is analytic in $\Re(s) > 0$ and known to satisfy the relation $w(s) = -1/(1-s) - \zeta(s)$ for $\Re(s) > 0$, see e.g. [**15**, eq. (2.3)]. In particular, this implies

$$w_k(D) + \frac{1}{1-D} \to -\zeta(D), \text{ as } k \to \infty. \tag{6.12}$$

It remains to estimate C. The obvious relation $-1 \leq \{l_j x\} - 1 < 0$ implies $-J(k/x) \leq C \leq 0$ for any $k \in \mathbb{N}$. Using again Proposition 6.1, we infer that $J(k/x) \sim k^{-D} J(1/x) \sim L^D k^{-D} f(x)$, as $x \to \infty$. Hence the expression $-L^D k^{-D} f(x)$ is essentially a lower bound for C. More precisely,

$$-k^{-D} = \lim_{x\to\infty} \frac{-J(k/x)}{L^D f(x)} \leq \liminf_{x\to\infty} \frac{C}{L^D f(x)} \leq \limsup_{x\to\infty} \frac{C}{L^D f(x)} \leq 0. \tag{6.13}$$

Combining now the estimates for C in (6.13) and $A+B$ in (6.11), and recalling that $\delta(x) = A + B + C$, we infer that, for each $k \in \mathbb{N}$,

$$w_k(D) + \frac{1}{1-D} - k^{-D} \leq \liminf_{x\to\infty} \frac{\delta(x)}{L^D f(x)} \leq \limsup_{x\to\infty} \frac{\delta(x)}{L^D f(x)} = w_k(D) + \frac{1}{1-D}$$

Letting now $k \to \infty$, the left and right expression both converge to $-\zeta(D)$, which completes the proof of the theorem. □

It remains to show that assertion (iv) in Theorem 2.5 implies at least one of the assertions (i) – (iii) (and therefore all of them by the results in Section 5). Due to the equivalence of (6.3) with (ii), the following statement is essentially the implication (iv) ⇒ (ii).

THEOREM 6.6. *Let $D \in (0,1)$ and $h \in \mathcal{SR}_{1-D}$. Let $\mathcal{L} = (l_j)_{j\in\mathbb{N}}$ be a fractal string such that $\delta(x) \asymp f(x)$, as $x \to \infty$. Then $J(1/x) \asymp f(x)$, as $x \to \infty$.*

PROOF. By assumption, there are positive constants a_1, a_2 and x_0 such that

$$a_1 f(x) \leq \delta(x) \leq a_2 f(x), \tag{6.14}$$

for all $x \geq x_0$.
Part 1: We first derive a lower bound for J. Fix some small $\epsilon_0 > 0$ and choose some integer $k \geq 2$ with

$$k > \left(\frac{2a_2(1+\epsilon_0)}{a_1} \right)^{\frac{1}{1-D}}, \text{ i.e. such that } \quad k^D < \frac{a_1}{2a_2(1+\epsilon_0)} k.$$

Since $f \in \mathcal{SR}_D[\infty]$, there is an $x_1 \geq x_0 > 0$ such that

$$k^D(1-\epsilon_0) \leq \frac{f(kx)}{f(x)} \leq k^D(1+\epsilon_0) < \frac{a_1}{2a_2}k, \tag{6.15}$$

for all $x \geq x_1$, where the last inequality is due to the choice of k.

We split the packing defect $\delta(x)$ into two sums as follows

$$\delta(x) = \sum_{\{l_j x\}<k^{-1}} \{l_j x\} + \sum_{\{l_j x\}\geq k^{-1}} \{l_j x\} =: U(x) + V(x). \tag{6.16}$$

Observe that for any $\gamma > 0$ such that $k\{\gamma\} < 1$, we have $k\{\gamma\} = \{k\gamma\}$. Indeed, since $k[\gamma] \in \mathbb{N}$, we get

$$1 > k\{\gamma\} = \{k\{\gamma\}\} = \{k(\gamma - [\gamma])\} = \{k\gamma - k[\gamma]\} = \{k\gamma\}.$$

Applying this to the sum U, we obtain for $x \geq x_1$,

$$kU(x) = \sum_{\{l_j x\}<k^{-1}} \{l_j kx\} < \delta(kx) \overset{(6.14)}{\leq} a_2 f(kx) \overset{(6.15)}{\leq} \frac{a_1}{2} k f(x) \overset{(6.14)}{\leq} \frac{1}{2} k\delta(x).$$

Thus $U(x) \leq \frac{1}{2}\delta(x)$, which together with $U(x) + V(x) = \delta(x)$ implies that

$$V(x) \geq \frac{1}{2}\delta(x)$$

for all $x \geq x_1$. On the other hand, since every summand in V is less than 1, we have

$$V(x) \leq \sum_{\{l_j x\}\geq k^{-1}} 1 \leq \#\{j : \{l_j x\} \geq k^{-1}\} \leq \#\{j : l_j x \geq k^{-1}\} = J\left(\frac{1}{kx}\right).$$

Combining both estimates, we obtain for $x \geq x_1$

$$J\left(\frac{1}{kx}\right) \geq \frac{1}{2}\delta(x) \geq \frac{1}{2} a_1 f(x),$$

which provides a lower bound on $J(1/x)$ in terms of $f(x)$ as desired. Substituting kx by x and using again the homogeneity (6.15), we conclude that

$$J\left(\frac{1}{x}\right) \geq \frac{a_1}{2} f(x/k) \geq \frac{a_1}{2} k^{-D}(1-\epsilon_0) f(x), \tag{6.17}$$

for all $x \geq kx_1$.

Part 2: In order to derive an upper bound for J, we first prove the following estimate: there are positive constants c, x_2 such that, for each $x \geq x_2$,

$$J\left(\frac{1}{x}\right) - J\left(\frac{2}{x}\right) \leq c f(x). \tag{6.18}$$

For a proof of (6.18) consider the interval $I := (J(\frac{2}{x}), J(\frac{1}{x})]$. Let $J_\sigma \subset I$ denote the subset of integers j in I with $\{l_j x\} \geq \frac{1}{2}$, and $J_\kappa \subset I$ denote the set of those integers j in I with $\{l_j x\} < \frac{1}{2}$. We write $\sigma := \#J_\sigma$ and $\kappa := \#J_\kappa$ for the corresponding numbers of elements. Then $\sigma + \kappa = J(\frac{1}{x}) - J(\frac{2}{x})$. On the one hand, we see that

$$\delta(x) = \sum_{j\in\mathbb{N}} \{l_j x\} \geq \sum_{j\in J_\sigma} \{l_j x\} \geq \sum_{j\in J_\sigma} \frac{1}{2} = \frac{1}{2}\sigma. \tag{6.19}$$

On the other hand, $J(\frac{2}{x}) < j \leq J(\frac{1}{x})$ implies $\frac{2}{x} > l_j \geq \frac{1}{x}$, or equivalently $2 > l_j x \geq 1$. Yet for $j \in J_\kappa$, we have $\{l_j x\} \leq \frac{1}{2}$, which together yields $l_j x \in [1, \frac{3}{2}]$. Hence $\{l_j x\} \leq \{\frac{l_j x}{2}\} \in [\frac{1}{2}, \frac{3}{4}]$. Therefore, we obtain

$$\delta\left(\frac{x}{2}\right) = \sum_{j\in\mathbb{N}} \left\{l_j \frac{x}{2}\right\} \geq \sum_{j\in J_\kappa} \{l_j x\} \geq \sum_{j\in J_\kappa} \frac{1}{2} = \frac{1}{2}\kappa. \tag{6.20}$$

Combining inequalities (6.19) and (6.20), and taking into account (6.14), we obtain

$$J\left(\frac{1}{x}\right) - J\left(\frac{2}{x}\right) = \sigma + \kappa \leq 2\delta(x) + 2\delta\left(\frac{x}{2}\right) \leq 2a_2 f(x) + 2a_2 f\left(\frac{x}{2}\right). \tag{6.21}$$

Since $f \in \mathcal{R}_D[\infty]$, the homogeneity property implies in particular that we can find positive constants $c_2 < 1$ and x_2 such that

$$f(x/2) < c_2 f(x),$$

for all $x \geq x_2$. (Without loss of generality, we may assume that $x_2 \geq x_0$.) Applying this to (6.21) yields

$$J\left(\frac{1}{x}\right) - J\left(\frac{2}{x}\right) \leq 2a_2(1+c_2)f(x),$$

for each $x \geq x_2$, proving assertion (6.18) for the constant $c := 2a_2(1+c_2)$.

Now we use (6.18) to derive an upper bound for J. For $x \geq 2x_2$ let $m = m(x)$ be the unique integer such that $2^m x_2 < x \leq 2^{m+1}x_2$. Note that, by the choice of m, we have $2^m/x \geq 1/2x_2$, which implies $J(2^m/x) \leq J(1/2x_2) =: j_0$. Writing $J(1/x)$ as a telescope sum and applying (6.18), we infer

$$\begin{aligned} J(1/x) &= \sum_{k=0}^{m-1} \left(J(2^k/x) - J(2^{k-1}/x)\right) + J(2^m/x) \leq \sum_{k=0}^{m-1} cf\left(x/2^k\right) + j_0 \\ &\leq c\sum_{k=0}^{m-1} c_2^k f(x) + j_0 \leq cf(x)\sum_{k=0}^{\infty} c_2^k + j_0 = \frac{c}{1-c_2} f(x) + j_0, \end{aligned}$$

for any $x \geq 2x_2$, where the convergence of the geometric series is ensured, since $c_2 < 1$. This gives an upper bound on $J(1/x)$ in terms of $f(x)$ and completes the proof. (Note that $f(x) \to \infty$, as $x \to \infty$.) □

To complete the proof of Theorem 2.5, we recall that there is a direct connection between the packing defect δ and the string counting function N valid for any bounded open set $\Omega \subset \mathbb{R}$ independent of any additional assumptions on the growth behavior. It is given by the equation

$$\varphi(\lambda) - N(\lambda) = \delta(\sqrt{\lambda}/\pi), \quad \lambda > 0, \tag{6.22}$$

cf. e.g. [**15**, eq. (2.2)] and see also Remark 6.7 below. It implies the equivalence (iv) $\Leftrightarrow$ (v) in part I of Theorem 2.5 as well as the equivalence of assertion (6.9) in Theorem 6.5 with assertion (2.4) in part II of Theorem 2.5. Therefore, by Theorem 6.5, the assertions (vi)-(viii) imply indeed (2.4). This completes the proof.

REMARK 6.7. Equation (6.22) is seen as follows. For an open interval $I = (a, b)$ of length $l = b-a$, consider the Laplace operator d^2/dy^2. Under Dirichlet boundary conditions (i.e. $u(a) = u(b) = 0$) the eigenvalues are $\lambda_k = (\pi/l)^2 k^2$, $k \in \mathbb{N}$. Hence, for the eigenvalue counting function $N(I;\lambda)$ of I, we have

$$N(I;\lambda) = \#\{k \in \mathbb{N} : \lambda_k \leq \lambda\} = \#\{k \in \mathbb{N} : k \leq l\sqrt{\lambda}/\pi\} = [l\sqrt{\lambda}/\pi].$$

Therefore, we get for the eigenvalue counting function N of Ω (consisting of the disjoint open intervals I_j of lengths l_j, $j \in \mathbb{N}$)

$$N(\lambda) = \sum_{j=1}^{\infty} N(I_j; \lambda) = \sum_{j=1}^{\infty} [l_j \sqrt{\lambda}/\pi] = \sum_{j=1}^{\infty} [l_j x],$$

where $x := \sqrt{\lambda}/\pi$. Since $|\Omega|_1 = \sum_{j=1}^{\infty} l_j$, it follows in particular that

$$\varphi(\lambda) - N(\lambda) = \sqrt{\lambda}/\pi \sum_{j=1}^{\infty} l_j - \sum_{j=1}^{\infty} [l_j x] = \sum_{j=1}^{\infty} l_j x - \sum_{j=1}^{\infty} [l_j x] = \delta(x).$$

Summary of the proof of Theorem 2.5. The following diagram gives an overview over the various steps of the proof of part I of Theorem 2.5. It also shows the central role of the S-content (or the string counting function J) for the proof.

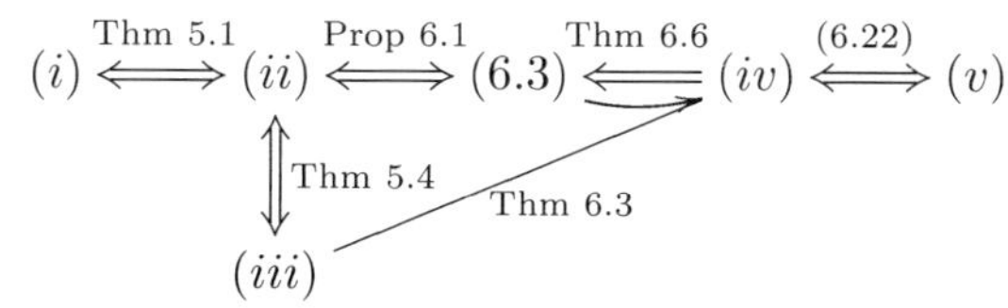

Our proof of part II of Theorem 2.5 has a similar structure:

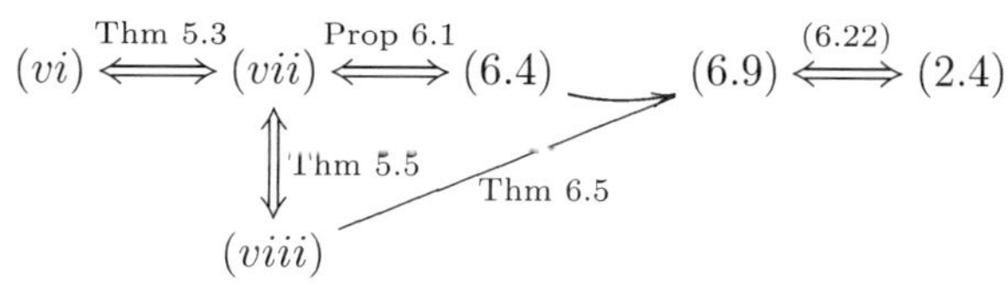

Relation (2.3) between the contents and L is obvious from the definition of the constant S in Theorem 5.5 and Theorem 5.3. □

References

[1] M. V. Berry, *Distribution of modes in fractal resonators*, Structural stability in physics (Proc. Internat. Symposia Appl. Catastrophe Theory and Topological Concepts in Phys., Inst. Inform. Sci., Univ. Tübingen, Tübingen, 1978), Springer Ser. Synergetics, vol. 4, Springer, Berlin, 1979, pp. 51–53, DOI 10.1007/978-3-642-67363-4_7. MR556688

[2] M. V. Berry, *Some geometric aspects of wave motion: wavefront dislocations, diffraction catastrophes, diffractals*, Geometry of the Laplace operator (Proc. Sympos. Pure Math., Univ. Hawaii, Honolulu, Hawaii, 1979), Proc. Sympos. Pure Math., XXXVI, Amer. Math. Soc., Providence, R.I., 1980, pp. 13–28. MR573427

[3] N. H. Bingham, C. M. Goldie, and J. L. Teugels, *Regular variation*, Encyclopedia of Mathematics and its Applications, vol. 27, Cambridge University Press, Cambridge, 1989. MR1015093

[4] M. Š. Birman and M. Z. Solomjak, *Spectral asymptotics of nonsmooth elliptic operators. I, II* (Russian), Trudy Moskov. Mat. Obšč. **27** (1972), 3–52; ibid. 28 (1973), 3–34. MR0364898

[5] Jean Brossard and René Carmona, *Can one hear the dimension of a fractal?*, Comm. Math. Phys. **104** (1986), no. 1, 103–122. MR834484

[6] Christina Q. He and Michel L. Lapidus, *Generalized Minkowski content and the vibrations of fractal drums and strings*, Math. Res. Lett. **3** (1996), no. 1, 31–40, DOI 10.4310/MRL.1996.v3.n1.a3. MR1393380

[7] Christina Q. He and Michel L. Lapidus, *Generalized Minkowski content, spectrum of fractal drums, fractal strings and the Riemann zeta-function*, Mem. Amer. Math. Soc. **127** (1997), no. 608, x+97, DOI 10.1090/memo/0608. MR1376743

[8] V. Ja. Ivriĭ, *The second term of the spectral asymptotics for a Laplace-Beltrami operator on manifolds with boundary* (Russian), Funktsional. Anal. i Prilozhen. **14** (1980), no. 2, 25–34. MR575202

[9] Phạm The Lại, *Meilleures estimations asymptotiques des restes de la fonction spectrale et des valeurs propres relatifs au laplacien* (French), Math. Scand. **48** (1981), no. 1, 5–38. MR621413

[10] Michel L. Lapidus, Goran Radunović, and Darko Žubrinić, *Zeta functions and complex dimensions of relative fractal drums: theory, examples and applications*, Dissertationes Math. **526** (2017), 105, DOI 10.4064/dm757-4-2017. MR3764036

[11] ———, *Minkowski measurability criteria for compact sets and relative fractal drums in euclidean spaces*, Preprint (2016), arXiv:1609.04498.

[12] Michel L. Lapidus, *Fractal drum, inverse spectral problems for elliptic operators and a partial resolution of the Weyl-Berry conjecture*, Trans. Amer. Math. Soc. **325** (1991), no. 2, 465–529, DOI 10.2307/2001638. MR994168

[13] Michel L. Lapidus and Helmut Maier, *Hypothèse de Riemann, cordes fractales vibrantes et conjecture de Weyl-Berry modifiée* (French, with English summary), C. R. Acad. Sci. Paris Sér. I Math. **313** (1991), no. 1, 19–24. MR1115940

[14] Michel L. Lapidus and Helmut Maier, *The Riemann hypothesis and inverse spectral problems for fractal strings*, J. London Math. Soc. (2) **52** (1995), no. 1, 15–34, DOI 10.1112/jlms/52.1.15. MR1345711

[15] Michel L. Lapidus and Carl Pomerance, *The Riemann zeta-function and the one-dimensional Weyl-Berry conjecture for fractal drums*, Proc. London Math. Soc. (3) **66** (1993), no. 1, 41–69, DOI 10.1112/plms/s3-66.1.41. MR1189091

[16] Michel L. Lapidus and Carl Pomerance, *Counterexamples to the modified Weyl-Berry conjecture on fractal drums*, Math. Proc. Cambridge Philos. Soc. **119** (1996), no. 1, 167–178, DOI 10.1017/S0305004100074053. MR1356166

[17] Michel L. Lapidus and Machiel van Frankenhuijsen, *Fractal geometry, complex dimensions and zeta functions*, Springer Monographs in Mathematics, Springer, New York, 2006. Geometry and spectra of fractal strings. MR2245559

[18] Guy Métivier, *Valeurs propres de problèmes aux limites elliptiques irrégulières* (French), Bull. Soc. Math. France Suppl. Mém. **51–52** (1977), 125–219. MR0473578

[19] Jan Rataj and Steffen Winter, *On volume and surface area of parallel sets*, Indiana Univ. Math. J. **59** (2010), no. 5, 1661–1685, DOI 10.1512/iumj.2010.59.4165. MR2865426

[20] Jan Rataj and Steffen Winter, *Characterization of Minkowski measurability in terms of surface area*, J. Math. Anal. Appl. **400** (2013), no. 1, 120–132, DOI 10.1016/j.jmaa.2012.10.059. MR3003969

[21] R. Seeley, *A sharp asymptotic remainder estimate for the eigenvalues of the Laplacian in a domain of* $\mathbf{R}^3$, Adv. in Math. **29** (1978), no. 2, 244–269, DOI 10.1016/0001-8708(78)90013-0. MR506893

[22] R. Seeley, *An estimate near the boundary for the spectral function of the Laplace operator*, Amer. J. Math. **102** (1980), no. 5, 869–902, DOI 10.2307/2374196. MR590638

[23] Hermann Weyl, *Das asymptotische Verteilungsgesetz der Eigenwerte linearer partieller Differentialgleichungen (mit einer Anwendung auf die Theorie der Hohlraumstrahlung)* (German), Math. Ann. **71** (1912), no. 4, 441–479, DOI 10.1007/BF01456804. MR1511670

[24] H. Weyl, *Über die Abhängigkeit der Eigenschwingungen einer Membran und deren Begrenzung* (German), J. Reine Angew. Math. **141** (1912), 1–11, DOI 10.1515/crll.1912.141.1. MR1580843

[25] Steffen Winter, *Localization results for Minkowski contents*, J. Lond. Math. Soc. 99 (2019) no. 2, 553-582, DOI:10.1112/jlms.12180 arXiv:1610.03117.

Technische Universität Berlin, Service-centric Networking, Telekom Innovation Laboratories, Ernst-Reuter-Platz 7, 10587 Berlin, Germany

Karlsruhe Institute of Technology, Institute of Stochastics, Englerstr. 2, 76131 Karlsruhe, Germany

Contemporary Mathematics
Volume **731**, 2019
https://doi.org/10.1090/conm/731/14674

Dimensions of limit sets of Kleinian groups

Kurt Falk

ABSTRACT. In this paper we give a brief survey of results on dimension gaps for limit sets of geometrically infinite Kleinian groups. We concentrate on an important notion from geometric group theory, amenability, as a criterion for the existence of such gaps.

1. Dimensions and invariants in conformal dynamics

One goal often encountered in mathematics is to prove equality of independently defined invariants for large classes of a certain mathematical object. An instance of this in conformal dynamics has been the attempt to show that the critical exponent of the Poincaré series associated to a conformal dynamical system , e.g. a Kleinian group or a rational map on the Riemann sphere, coincides with the Hausdorff dimension of the corresponding limit set or Julia set, respectively. In this paper we shall give a brief survey on results exploring under what circumstances several invariants associated with a Kleinian group, namely the critical exponent, the Hausdorff and Minkowski dimensions of the limit set and the so-called convex core entropy, coincide or not. Since we will be dealing mainly with *Kleinian groups*, i.e. discrete, torsion-free subgroups of the group of orientation preserving isometries of $(n+1)$-dimensional hyperbolic space $\mathbb{H}^{n+1}$, $n \in \mathbb{N}$, let us first explain briefly what these invariants are. The *limit set* $L(G)$ of a Kleinian group G is the set of accumulation points of some and thus any orbit Gx of G, $x \in \mathbb{H}^{n+1}$, and is a subset of the boundary $\partial\mathbb{H}^{n+1}$ of $\mathbb{H}^{n+1}$ due to the discreteness of G. The *Poincaré series* of G is defined as a Dirichlet series

$$P(x, y, s) := \sum_{g \in G} e^{-s\, d(x, gy)},$$

where d is the hyperbolic metric, $x, y \in \mathbb{H}$ are arbitrary but fixed and are often chosen to be 0 when working in the Poincaré unit ball model of hyperbolic geometry, and s is a real parameter. The abscissa of convergence of the series reduces for real values of s to the *critical exponent* of G, which can be expressed in two different

2010 *Mathematics Subject Classification.* Primary 30F40, Secondary 37F30.
The author was supported by the NSF conference grant *Fractal Geometry and Complex Dimensions*, a meeting celebrating the 60th birthday of Prof. Michel Lapidus.

ways:

$$\begin{aligned}\delta(G) &:= \inf\{s>0 : P(x,y,s)<\infty\}\\ &= \limsup_{R\to\infty} \frac{\log \#(B(x,R)\cap Gy)}{R}.\end{aligned}$$

By a theorem of Roblin [**49**], the upper limit above is in fact a limit. G is called *of convergence type* if $P(x,y,\delta(G))<\infty$, and *of divergence type* otherwise. It is not too difficult to see that for *non-elementary* Kleinian groups, i.e. groups which are not generated by only one isometry, the critical exponent is positive and less or equal to n, the dimension of the boundary of hyperbolic space. The *convex hull of the limit set* $H(G)$ of a Kleinian group G is defined to be the closed metric convex hull of the union of all geodesics in $\mathbb{H}^{n+1}$ with both ends in the limit set $L(G)$. The group G acts discontinuously not only on $\mathbb{H}$, but also on $H(G)$, and the quotient $C(M_G) := H(G)/G$ is called the *convex core* of the hyperbolic manifold $M_G := \mathbb{H}^{n+1}/G$. Equivalently, $C(M_G)$ is the closed metric convex hull of the union of all closed geodesics in M_G, thus being the region in M_G where closed (geodesic) orbits occur. A limit point $\xi \in L(G)$ is called *radial* if the projection to M_G of some infinite geodesic ray in $\mathbb{H}$ towards ξ returns infinitely often to some compact subset of M_G. The set of radial limit points of G is called the *radial limit set* of G and is denoted by $L_r(G)$. A somewhat non-standard notion is that of the *convex core entropy* $h_c(M_G)$ [**24**] of a hyperbolic manifold M_G which is defined analogously to the classical volume entropy, namely as the exponential growth rate of the volume of a growing ball intersected with $H(G)$ (see Section 4). For more details on Kleinian groups and Hyperbolic Geometry we refer to [**4**], [**36**], [**46**], [**37**], [**33**] and [**39**]. We shall assume that the reader is familiar with standard notions from Fractal Geometry like the Hausdorff and the Minkowski dimension.

While some early results of Patterson's [**41**] (see also [**44**]) suggested that equality of the critical exponent of a Kleinian group and the Hausdorff dimension of its limit set should hold in fairly general circumstances, it was also Patterson [**42**], [**43**] (see also [**3**] and [**25**]) who gave the first examples of Kleinian groups G for which $\delta(G)$ is strictly less than $\dim_H(L(G))$. These are groups *of the first kind*, i.e. so that $L(G)$ is the whole boundary of $\mathbb{H}^{n+1}$, which obviously have limit sets of full dimension n. The existence of such examples raises the natural question of what the 'dynamical meaning' of such a *dimension gap*, i.e. a positive difference between the Hausdorff dimension of the limit set and the critical exponent, might be. The answer was given only later by a theorem of Bishop and Jones [**8**] who showed that, for all non-elementary Kleinian groups G,

$$\dim_{\mathrm{H}}(L_r(G)) = \delta(G).$$

Thus, the dimension gap is in fact a gap between the conservative part and the rest of the dynamics within the convex core of the corresponding hyperbolic manifold. This becomes very clear when considering the following. If we fix one point $o \in C(M_G)$ we get a very suggestive 1-1 correspondence between the dynamical behaviour of geodesic rays originating at o and the type of point at infinity the lifts of these geodesic rays to $\mathbb{H}$ end in: geodesic movements staying in $C(M_G)$ correspond to limit points of G and geodesic movements returning infinitely often to some compact set (and thus necessarily staying in $C(M_G)$) correspond to radial limit points of G (see also Figure 1).

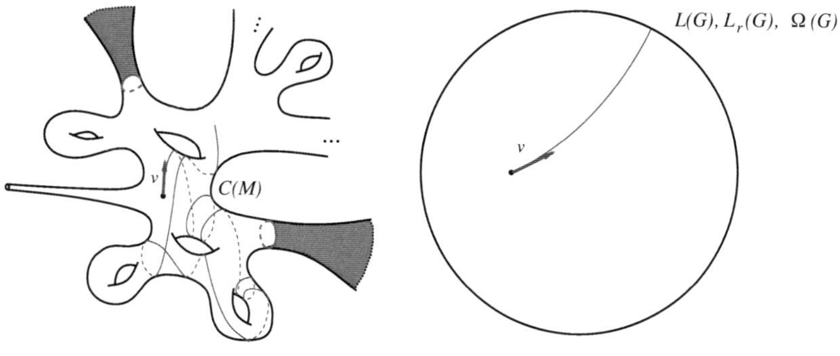

FIGURE 1. Types of dynamics in a hyperbolic manifold.

From a modern perspective it is clear that the dimension gap showcased by Bishop and Jones' theorem above can only occur for geometrically infinite groups, as shown by the classical result of Beardon and Maskit [**5**] which states that a Kleinian group G is geometrically finite if and only if $L(G) = L_r(G) \dot{\cup} L_p(G)$. Here, $L_p(G)$ denotes the set of *bounded parabolic fixed points* of G, i.e. those parabolic fixed points for which any invariant horosphere projects to a region whose intersection with the convex core has finite hyperbolic volume. Also, recall that the group G and the manifold M_G are called *geometrically finite* if, for some and thus all $\varepsilon > 0$, the ε-neighbourhood of $C(M_G)$ has finite hyperbolic volume, and *geometrically infinite* otherwise. Furthermore, G is called *convex cocompact* if it acts cocompactly on $H(G)$ or, equivalently, if $C(M_G)$ is compact. Before coming to the main theme of this paper, namely the existence of a dimension gap, it is important to list the known situations in which equality of the critical exponent and the Hausdorff dimension of the limit set occurs. For non-elementary geometrically finite Kleinian groups, it is well-known that $\delta(G) = \dim_{\mathrm{H}}(L(G))$ (see the original work [**60**] by Sullivan or Nicholls' book [**39**] for more details), and that $\dim_{\mathrm{H}}(L(G)) = \dim_{\mathrm{M}}(L(G))$, where $\dim_{\mathrm{M}}$ denotes the *box counting* or *Minkowski dimension* (see Stratmann and Urbański [**59**]). For Kleinian groups acting on the 3-dimensional hyperbolic space $\mathbb{H}^3$, Bishop [**6**], [**7**] has shown $\delta(G) = \dim_{\mathrm{M}}(L(G))$ if the group is non-elementary and analytically finite (i.e. with associated hyperbolic manifold having finitely many Riemann surfaces of finite type as boundary components) and if its limit set has null 2-dimensional measure (see also [**8**]). Taking into consideration the solution of the Ahlfors measure conjecture (which states that the limit set of a finitely generated Kleinian group acting on $\mathbb{H}^3$ is either the whole boundary or has null Lebesgue measure), following by work of Canary [**16**] from the Tameness Conjecture (which states that any complete hyperbolic 3-manifold with finitely generated fundamental group is topologically tame, i.e. homeomorphic to the interior of a compact 3-manifold) proven in [**2**] and [**15**], the above results imply that all finitely generated non-elementary Kleinian groups acting on $\mathbb{H}^3$ satisfy $\delta(G) = \dim_{\mathrm{M}}(L(G))$.

Having said this, let us return to the problem of identifying Kleininan groups with a dimension gap. It is the seminal work of Brooks [**13**] that brings a very important notion from geometric group theory into the picture: amenability. While fractal dimensions of limit sets were not his main motivation, but rather classical considerations in Riemannian geometry concerning the role played by the bottom of the L^2-spectrum of a Riemannian manifold in its large scale geometry, which he first investigated in [**11**], Brooks was of course aware of the fact that his results from [**13**] actually generalise earlier observations on critical exponents of Kleinian groups by Rees [**47**],[**48**]. Brooks [**13**] essentially shows that as long as $\delta(G) > n/2$ for a convex cocompact Kleinian group G acting on $\mathbb{H}^{n+1}$, we have for any normal subgroup N of G that

$$\delta(N) = \delta(G) \iff G/N \text{ amenable.}$$

Using Bishop and Jones' observation above, namely that $\dim_{\mathrm{H}}(L_r(G)) = \delta(G)$ for any non-elementary Kleinian group, this can be easily seen to produce an entire new class of groups with dimension gap [**24**] (see Example 2.3 for more details). It is worth noting here that there is a purely algebraic analogon [**18**], [**27**] of Brooks' theorem which predates his result, but of which he does not seem to have been aware. We discuss all these ideas at some length in Section 2.

Since the connection between amenability and the question of existence of dimension gaps becomes evident through the work of Brooks, it is obvious that any successful attempt at generalising his result will produce new examples of Kleinian groups with dimension gap. Let us first mention the one approach which is closest in method and spirit to the original. Tapie [**63**] takes Brooks' idea further by considering copies of a Riemannian manifold with boundary that can be glued together to a Riemannian manifold without boundary following the pattern of an infinite graph. Under a certain analytic condition, namely that the building block manifold admits an eigenfunction to the bottom of its L^2-spectrum which 'extends well' to a function on the glued manifold, Tapie shows that the amenability of the skeleton graph is equivalent to the bottoms of the L^2-spectra of the building block and the glued manifold coinciding. The one aspect which also goes beyond Brooks' considerations is the fact that Tapie gives an estimate for the size of the gap in terms of the associated isoperimetric constants. For more details on Tapie's work we refer the reader to his paper [**63**], and also to the interesting survey [**51**] by Roblin and Tapie.

Next, we delve into the realm of symbolic dynamics. In an attempt to drop the condition on $\delta(G)$ in the statement of Brooks' theorem, which is in place essentially due to Brooks' method of proof, Stadlbauer [**57**] formulated and solved the problem in terms of symbolic dynamics or, more precisely, topological Markov chains. He showed that under certain circumstances, the Gurevič pressures of a topological Markov chain and a group extension of this chain by some countable group coincide precisely when the group is amenable. This translates to Kleinian groups and their limit sets via the encoding procedure for the dynamics of the geodesic flow given by Bowen and Series [**10**] and developed further by Stadlbauer and Stratmann [**56**],[**58**]. Stadlbauer showed that while the assumption $\delta(G) > n/2$ can be dropped, another condition on G needs to be satisfied, namely, instead of G being convex cocompact, one needs to assume that it is *essentially free* [**58**], i.e. that all relations of G are due to parabolic elements. It is also interesting to note that Stadlbauer in fact recovers the algebraic version of the dimension gap problem [**18**], [**27**] mentioned above. Details on Stadlbauer's work are to be found in Section 3. We also refer the reader to the closely related work of Jaerisch [**29**], [**30**] and [**31**], where the radial limit set of a normal subgroup of the free group (with at least two generators) w.r.t. a graph directed Markov system is introduced, and it is shown that the Hausdorff dimensions of the radial limit sets of the normal subgroup and the free group coincide if and only if the quotient is amenable. The most general result along these lines has been obtained by Dougall and Sharp [**22**] (see also [**21**]) who, just as Stadlbauer before them, generalise the pioneering work of Kesten [**34**], [**35**] on the amenability of groups to the the context of *pinched Hadamard manifolds*, i.e. connected, simply connected and geodesically complete Riemannian manifolds with sectional curvatures bounded between negative constants. Their main result states that if G is a convex cocompact group of isometries of a pinched Hadamard manifold and N some normal subgroup of G, then the coincidence of the critical exponents of G and N is equivalent to the amenability of G/N and to the coincidence of the exponential growth rates of the length spectra of closed geodesics of the manifolds associated with G and N. For more details we refer to Section 3. During the refereeing process for this paper, a new preprint [**19**] by Coulon, Dal'bo and Sambusetti appeared on the arXiv, in which the authors extend Stadlbauer's approach to proving Brooks' theorem to the context of Gromov hyperbolic groups acting properly cocompactly on CAT(-1)-spaces.

Independently of Tapie's afore-mentioned approach [**63**] to generalise Brooks' theorem, a similar attempt at a generalisation was given in [**24**], with more emphasis on the dynamical aspects of the problem. Just as in [**63**], the convex cocompact group G is eliminated from the picture allowing for the study of hyperbolic surfaces given by infinitely generated Fuchsian groups (i.e. Kleinian groups acting on hyperbolic 2-space), which, in contrast to [**63**], are *not* made of identical copies of a surface with boundary. In [**24**] the critical exponent $\delta(G)$ is replaced by the convex core entropy which actually turns out to always coincide with the upper Minkowski dimension of the limit set. This sets it apart from the classical notion of topological entropy, since Otal and Peigné [**40**] have generalised earlier work of Sullivan [**61**] to show that even in pinched negative curvature the topological entropy of the geodesic flow coincides with the critical exponent. After formulating the main results which relate the amenability of the graph associated with a so-called *pants decomposition* of the hyperbolic surface of infinite type under scrutiny to the existence of a dimension gap (see Section 4 for more details), we go on to split up the main question into two questions. First, when is the critical exponent strictly smaller than the convex core entropy, and second, when does the Hausdorff dimension of the limit set coincide with the convex core entropy of the surface. At the end of Section 4 we discuss conditions under which the latter question has a positive answer. Finally, in Section 5 we discuss some open problems and possible strategies to attack these two questions.

Acknowledgements. The author would like to thank the referee for the very useful comments and corrections which helped improve this survey significantly.

2. Brooks' Theorem; the role of amenability

In a series of papers, most notably [**11**] and [**13**], Brooks considers the fundamental question of the behaviour of the bottom λ_0 of the L^2-spectrum under Riemannian coverings, i.e. local isometries $M_2 \to M_1$, where M_1 is a complete Riemannian manifold. He resolved this question first [**11**] for the case that M_1 is compact, thus making $\lambda_0(M_1) = 0$, and M_2 is its universal cover, and then [**13**] for the more general case that $\pi_1(M_2)$ is normal in $\pi_1(M_1)$. Here, the notion of amenability plays a crucial role. While it is not too difficult to show that if $\pi_1(M_1)/\pi_1(M_2)$ is amenable, then $\lambda_0(M_1) = \lambda_0(M_2)$ (see also [**55**] and [**50**] for different points of view on this implication), the main difficulty consists in showing the converse.

Let us give a brief reminder on the notion of amenability of (countably generated) discrete groups. J. von Neumann defined such a group G to be *amenable*, if there exists an *left-invariant mean* on $L^\infty(G,\mathbb{R})$. i.e. a non-negative bounded linear functional $\mu : L^\infty(G,\mathbb{R}) \to \mathbb{R}$ so that $\mu(1) = 1$ and $\mu(\tau_g f) = \mu(f)$ for all $f \in L^\infty(G,\mathbb{R})$, where, for each $g \in G$, $\tau_g : L^\infty(G,\mathbb{R}) \to L^\infty(G,\mathbb{R})$ is defined by $\tau_g f(h) := f(g^{-1}h)$. Later, Følner [**26**] gave the following very useful characterisation in the finitely generated case. A finitely generated group G is amenable if and only if there exists a sequence (F_n) of finite sets in G so that $\bigcup_n F_n = G$ and, for all $g \in G$,

$$\lim_{n\to\infty} \frac{\# gF_n \triangle F_n}{\# F_n} = 0,$$

where $\triangle$ refers to the symmetric difference. The condition which is most intuitive for our purposes is the following (see discussion of amenability in [**11**]). A finitely generated group G is amenable if and only if there exists a sequence (F_n) of finite sets in G so that

$$\lim_{n\to\infty} \frac{\#\partial F_n}{\# F_n} = 0,$$

where ∂F_n is the boundary of F_n in the Cayley graph of G.

THEOREM 2.1 ([**13**]). *Let M_1 be a $(n+1)$-dimensional, $n \in \mathbb{N}$, hyperbolic manifold given by a convex cocompact Kleinian group which satisfies $\lambda_0(M_1) < n^2/4$, and consider some normal cover M_2 of M_1. Then,*

$$\lambda_0(M_1) = \lambda_0(M_2) \iff \pi_1(M_1)/\pi_1(M_2) \text{ amenable.}$$

Note that this generalises previous results by Rees [**47**] [**48**] who considered the special case where $\pi_1(M_1)/\pi_1(M_2)$ is isomorphic to $\mathbb{Z}^d$ for some $d \in \mathbb{N}$, i.e. is abelian. For the proof of the more involved implication, Brooks employs a method of Sullivan [**62**] and Agmon [**1**] to shift the bottom of the L^2-spectrum of a Riemannian manifold M to 0; one uses a positive harmonic function ϕ corresponding to the eigenvalue λ of the Laplace-Beltrami operator Δ_M on M to define the new operator

$$P := \frac{1}{\phi}(\Delta_M - \lambda) \circ [\phi],$$

where $[\phi]$ denotes the multiplication operator by ϕ. P has the same spectrum as Δ_M, just shifted by λ, and is self-adjoint w.r.t. the volume form of M multiplied by ϕ^2. Then Brooks follows the line of thought from his earlier work [**11**],[**12**] where he showed that, for a compact Riemannian manifold M, the amenability of $\pi_1(M)$ is equivalent to the bottom of the L^2-spectrum of the universal cover $\widetilde{M}$ being 0. Assuming $\lambda_0(\widetilde{M}) = 0$, one considers a sequence (f_k) of smooth functions with compact support on $\widetilde{M}$ whose Rayleigh quotients satisfy

$$\frac{\int_{\widetilde{M}} |\nabla f_k|^2}{\int_{\widetilde{M}} |f_k|^2} \longrightarrow 0.$$

Then one uses Cheeger's method [**17**] to produce hypersurfaces S_k separating $\widetilde{M}$ into a bounded component $\mathrm{int}(S_k)$ and an unbounded one, so that the corresponding isoperimetric ratios satisfy

$$\frac{\mathrm{area}(S_k)}{\mathrm{vol}(\mathrm{int}(S_k))} \longrightarrow 0.$$

This is done by choosing for each k a finite union H_k of translates of some fundamental domain for the action of $\pi_1(M)$ so that H_k contains the support of f_k, and then applying Cheeger's inequality to deduce the existence of the hypersurfaces $S_k \subset H_k$ with the desired property. After replacing the hypersurfaces S_k by integral currents T_k, $k \in \mathbb{N}$, of uniformly bounded mean curvature with isoperimetric ratio less than or equal to that of the corresponding S_k, one is then able to apply the following characterisation of amenability [**11**, Proposition 2]. The fundamental group of M is amenable if and only if for any $\varepsilon > 0$ there is a union H_ε of fundamental domains of the action of $\pi_1(M)$ on $\widetilde{M}$ which satisfies the isoperimetric inequality

$$\frac{\operatorname{area}(\partial H_\varepsilon)}{\operatorname{vol}(H_\varepsilon)} < \varepsilon.$$

Of course these ideas from [**11**] need adjustment in order to prove Theorem 2.1 and we refer the reader to [**13**] for further details.

It is of course clear that while Brooks formulates and proves his theorem in terms of the bottom of the L^2-spectrum, the statement can be reinterpreted for our purposes in terms of critical exponents of Kleinian groups via the following well-known correspondence due to Sullivan in the form presented here. At this point it is interesting to note that Roblin and Tapie [**52**] have generalised this result for manifolds with pinched negative curvature.

THEOREM 2.2 ([**23**], [**41**], [**62**]). *For any Kleinian group G acting on $\mathbb{H}^{n+1}$, $n \geq 1$, with critical exponent $\delta = \delta(G)$ we have*

$$\lambda_0(M) = \begin{cases} \dfrac{n^2}{4}, & \text{if } \delta \leq \dfrac{n}{2} \\ \delta(n-\delta), & \text{if } \delta \geq \dfrac{n}{2} \end{cases}$$

Thus, Brooks' theorem can be formulated as follows, using non-amenability instead of amenability in order to emphasize the existence of a dimension gap. Let G be a convex cocompact group acting on $\mathbb{H}^{n+1}$ so that $\delta(G) > n/2$ and let $N \lhd G$ be a non-trivial normal subgroup. Then,

$$\delta(N) < \delta(G) \iff G/N \text{ not amenable.}$$

Now consider a non-trivial normal subgroup N of a convex cocompact Kleinian group G so that $\delta(G) > n/2$ and G/N is non-amenable; apply Brooks' Theorem, together with Bishop and Jones' observation [**8**] described in the first section and the fact that G is convex cocompact, to obtain

$$\dim_{\mathrm{H}}(L_r(N)) = \delta(N) < \delta(G) = \dim_{\mathrm{H}}(L(G)) = \dim_{\mathrm{H}}(L(N)).$$

The last equality is obvious since N being a non-trivial normal subgroup of G implies that $L(N) = L(G)$. This means that, for the hyperbolic manifold $\mathbb{H}^{n+1}/N$, we do have a dimension gap. Here is a class of examples of such normal subgroups N. The basic idea is to consider two Schottky groups whose sets of generators are 'geometrically far apart' and conjugate the generators of one of the groups by the elements of the other, thus obtaining an infinite generating set for a normal subgroup of the free product of the initial Schottky groups. These examples are not to be confused with Peigné's class of geometrically infinite Kleinian groups of divergence type with finite Bowen-Margulis measure [**45**].

EXAMPLE 2.3. ([**25**]) Let G_0 and G_1 be Schottky groups and assume G_0 is freely generated by hyperbolic isometries $g_1, \ldots, g_k$, and that G_1 is freely generated by more than one hyperbolic isometry. Consider the (open) fundamental domains F_0 resp. F_1 defined by their generators and the corresponding choice of hyperplanes. Furthermore, assume that these fundamental domains have disjoint complements in hyperbolic space: $F_0^{\mathbf{c}} \cap F_1^{\mathbf{c}} = \emptyset$. Put $N := ker(\varphi)$, where $\varphi : G \to G_1$ is the canonical group homomorphism. Thus, $0 \to N \to G \to G_1 \to 0$ is a short exact sequence and

$$N = \langle hg_ih^{-1} : i = 1, ..., k, h \in G_1 \rangle.$$

Furthermore, N is the normal subgroup of G generated by G_0 in G, and $G/N \cong G_1$. Clearly, G_1 is not amenable since it is freely generated by at least two generators.

As mentioned in the first section, there are also other classes of (infinitely generated) Kleinian groups with critical exponent strictly smaller than the Hausdorff dimension of their limit set, e.g. [**42**], [**43**] (see also [**25**]). The examples in [**43**] are constructed so that the limit set has full Hausdorff dimension but the critical exponent can be chosen to be arbitrarily close to 0. In the normal subgroup case, however, the gap cannot get 'too large':

THEOREM 2.4 ([**25**]). *If N is a non-trivial normal subgroup of a Kleinian group G, then*

$$\delta(N) \geq \frac{\delta(G)}{2}.$$

This result has been refined using different techniques [**50**], [**32**], where it is shown in addition to the above that if G is of divergence type, then the inequality is strict. The inequality also turns out to be sharp, as shown in [**9**].

We finish this section by discussing an algebraic analogon of Brooks' theorem and Theorem 2.4 due to Cohen [**18**] and Grigorchuk [**27**]. Let G be a finitely generated group and let $|\cdot|$ denote the word length w.r.t. some finite generating set. Let

$$\gamma(G) := \limsup_{n\to\infty} \sqrt[n]{\#\{g \in G : |g| \leq n\}} = \lim_{n\to\infty} \sqrt[n]{\#\{g \in G : |g| \leq n\}}.$$

The fact that the limit exists is a theorem [**18**]. In order to emphasize the analogy to the geometric results, we modify slightly the definition of *cogrowth* or *entropic dimension* as in [**18**], [**27**], by saying that the *entropy* of G is given by

$$h(G) := \log \gamma(G) = \lim_{n\to\infty} \frac{\log \#\{g \in G : |g| \leq n\}}{n}.$$

Now, if G is the free group with k generators, then it is not difficult to see that

$$\gamma(G) = 2k - 1 \quad \text{and thus} \quad h(G) = \log(2k-1),$$

and we have the following result which predates the other theorems described in this section.

THEOREM 2.5 ([**18**], [**27**]). *If G is the free group in k generators, and $N \triangleleft G$ some non-trivial normal subgroup, then*

$$\frac{h(G)}{2} < h(N) \leq h(G).$$

Equality in the second inequalitiy occurs precisely when G/N is amenable.

3. Group extensions of topological Markov chains

In this section we discuss the generalisation of Brooks' Theorem 2.1 given by Stadlbauer [**57**] who extends Brooks' ideas to the realm of symbolic dynamics or, more precisely, topological Markov chains, and then employs the thermodynamical formalism and methods from the theory of random walks on infinite graphs to obtain his main result. We also state the so far most general form of Brooks' result obtained by Dougall and Sharp [**22**] using these methods in the context of Hadamard manifolds.

We shall introduce some key notions in symbolic dynamics and then formulate the main theorems of the section. For more details we refer the reader to the original articles [**57**] and [**22**]. The one-sided *topological Markov chain* (Σ_A, θ) with countable *alphabet* I consists of the symbolic space

$$\Sigma_A := \{(w_k)_{k\in\mathbb{N}} : w_k \in I \text{ and } a_{w_k w_{k+1}} = 1 \ \forall k \in \mathbb{N}\}$$

where $A := (a_{ij})_{i,j\in I}$ is an adjacency matrix so that $a_{ij} \in \{0,1\}$ and $\sum_j a_{ij} > 0$ for all $i \in I$, and the *shift map*

$$\theta : \Sigma_A \to \Sigma_A, \ (w_1 w_2 \dots) \mapsto (w_2 w_3 \dots).$$

This is a topological dynamical system with topology given by *cylinder sets* $[w]$, where w is a finite admissible word, for which the notions of *topological transitivity* and *mixing* make sense. Also, one can define the so-called *big images and preimages (b.i.p.) property* [**54**] which in fact coincides with the notion of *finite irreducibility* as introduced in [**38**]. Given a strictly positive, continuous *potential function* $\varphi : \Sigma_A \to \mathbb{R}^+$ of bounded variation, one can define the *Gurevič pressure* [**53**] of $(\Sigma_A, \theta, \varphi)$

$$P(\theta, \varphi) := \limsup_{n\to\infty} \frac{1}{n} \log Z_n^a$$

as the exponential growth rate of certain *partition functions* Z_a^n, $a \in I$, w.r.t. the shift map and the potential function. If $\log\varphi$ is Hölder continuous and $(\Sigma_A, \theta, \varphi)$ topologically mixing, then a variational principle holds [**53**]. For a finite admissible word v of length $n \in \mathbb{N}$ and $f : \Sigma_A \to \mathbb{C}$, let $f \circ \tau_v : \Sigma_A \to \mathbb{C}$ be given by $x \mapsto \mathbf{1}_{\varphi^n([v])}(x) \cdot f(\tau_v(x))$, where $\mathbf{1}.$ denotes the characteristic function of a set. Using this, define the *Ruelle* or *transfer operator* by

$$L_\varphi f := \sum_v (\varphi \circ \tau_v) \cdot (f \circ \tau_v),$$

where the sum is taken over all admissible words of length 1 and f is in some appropriate complex function space such that the possibly infinite sum is well defined.

We now come to the analogue of normal covers in the context of topological Markov chains, namely their group extensions. Let (Σ_A, θ) be a topological Markov chain, G a countable discrete group and $\psi : \Sigma_A \to G$ so that ψ is constant on the cylinder sets in (Σ_A, θ) associated to admissible words of length 1. With $X := \Sigma_A \times G$ endowed with the product topology, we define the *group extension* (X, T) of (Σ_A, θ) by

$$T : X \to X, \ \ (x, g) \mapsto (\theta x, g\,\psi(x)).$$

As it turns out, (X, T) is also a topological Markov chain. Furthermore, any potential $\varphi : \Sigma_A \to \mathbb{R}^+$ lifts to a potential $\hat{\varphi} : X \to \mathbb{R}$ by simply setting $\hat{\varphi}(x, g) := \varphi(x)$, inverse branches of T are given by $\hat{\tau}_v(x, g) := (\tau_v(x), g\,\psi(v)^{-1})$ for all admissible finite words v, the Gurevič pressure $P(T, \hat{\varphi})$ is defined in the same way as $P(\theta, \varphi)$ and, finally, the Ruelle operator for (X, T) is defined by

$$\hat{L}_{\hat{\varphi}}(\xi, g) := \sum_v \hat{\varphi} \circ \hat{\tau}_v(\xi, \mathrm{id}_G) \cdot f \circ \hat{\tau}_v(\xi, g),$$

where $\xi \in [v]$ with v admissible of length 1, and $g \in G$.

The core idea in Stadlbauer's work is to generalise Kesten's [**34**] characterisation of amenability of groups to group extensions of topological Markov chains. Kesten shows that a probability measure on a countable group which satisfies a certain symmetry condition defines an operator on the ℓ^2-space of the group whose spectral radius is equal to 1 if and only if the group is amenable. While this theorem holds 'fibre-wise' for group extensions of topological Markov chains, it becomes necessary for the Markov chain to be *(weakly) symmetric* in a certain sense in order to formulate and prove an analogous result for its group extensions. (Again, we refer the interested reader to [**57**] for further details.)

THEOREM 3.1 ([**57**]). *Consider the topologically transitive and symmetric group extension (X, T) of the topologically mixing topological Markov chain (Σ_A, θ) by a discrete group G, and a weakly symmetric potential φ so that $P(\theta, \varphi) < \infty$. Then,*

$$G \text{ amenable} \implies P(T, \hat{\varphi}) = P(\theta, \varphi).$$

While this result can still be seen as a more or less direct consequence of Kesten's theorem mentioned above, the proof of the converse is way more involved and relies on a very careful analysis of the action of the Ruelle operator on the embedding of $\ell^2(G)$ into a certain subspace of the space of continuous functions on X.

THEOREM 3.2 ([**57**]). *Consider a topological Markov chain (Σ_A, θ) with the b.i.p.-property, a Hölder continuous potential φ with $\|L_\varphi(1)\| < \infty$, and a topologically transitive group extension (X, T) by the countable group G. Then,*

$$P(T, \hat{\varphi}) = P(\theta, \varphi) \implies G \text{ amenable}.$$

Now how does this relate to conformal dynamics? In other words, how do the above theorems generalise Brooks' Theorem 2.1 about the critical exponents of a Kleinian group G and a normal subgroup N of G? The idea is to use the so-called *Bowen-Series map* [**10**], generalised by Stadlbauer and Stratmann [**56**],[**58**] to fit the present situation, in order to encode the dynamics of the geodesic flow on $\mathbb{H}^{n+1}/G$ and $\mathbb{H}^{n+1}/N$ into a group extension of a topological Markov chain. More precisely, under certain circumstances it is possible to give a coding of a radial limit point ξ of G by keeping count not only of which fundamental domains a geodesic ray towards ξ intersects, but also by which side of that fundamental domain the ray enters it. The alphabet of the Markov chain is given by the collection of sides of a fundamental domain. The condition needed on G is for it to be *essentially free* [**58**], meaning that all relations in any presentation of G are due to parabolic elements of G. Thus, all finitely generated Fuchsian and all Schottky groups are essentially free. Also, note that all essentially free groups are geometrically finite but not vice versa. For instance, convex cocompact surface groups acting on hyperbolic 3-space are not necessarily essentially free. One then applies Theorem 3.1 and Theorem 3.2 to the group extension induced by N, relates the Gurevič pressures to the critical exponent of the underlying Kleinian groups, and obtains the following.

THEOREM 3.3 ([**57**]). *Let G be an essentially free Kleinian group acting on $\mathbb{H}^{n+1}$ and let $N \lhd G$ be a non-trivial normal subgroup. Then,*

$$\delta(N) = \delta(G) \iff G/N \text{ amenable.}$$

Note that while the condition on G to be essentially free is more restrictive than for it to be convex cocompact, the condition on the critical exponent of G from Brooks' Theorem 2.1 has disappeared.

As mentioned in the beginning of this section, we now turn to the more recent work of Dougall and Sharp [**22**]. In order to state their main result, we need some preparations. If X is a pinched Hadamard manifold and G a discrete group of isometries of X, then the quotient manifold $M := X/G$ also has pinched negative curvature. The critical exponent $\delta(G)$ of G is defined just as in the case of Kleininan groups. With $\mathcal{C}(M)$ denoting the countably infinite set of closed geodesics in M, and $l(\gamma)$ denoting the length of a closed geodesic γ in M, one defines

$$h(M) := \lim_{T\to\infty} \frac{\log \#\{\gamma \in \mathcal{C}(M) : l(\gamma) \leq T\}}{T}$$

to be the exponential growth rate of the length spectrum of M (see [**22**] for details).

THEOREM 3.4 ([**22**]). *Let G be a convex cocompact group of isometries of a pinched Hadamard manifold X and let N be some normal subroup of G. Then the following conditions are equivalent:*

(i) $\delta(G) = \delta(N)$;
(ii) $h(X/G) = h(X/N)$;
(iii) *G/N is amenable.*

4. The convex core entropy

This section deals mainly with yet another partial generalisation of Brooks' Theorem 2.1, as given in [**24**]. As in [**63**], the idea is to consider Kleinian groups which are *not* normal subgroups of some other 'well-behaved' (e.g. convex cocompact or even geometrically finite) group G. While Brooks' result identifies the (non-) amenability of G/N as the criterion for the existence of a dimension gap between $\delta(N)$ and $\dim_H(L(N))$, where N is some normal subgroup of G, we now remove G from the picture and replace G/N by a 'combinatorial skeleton' of the hyperbolic manifold given by N. Thus, the main question becomes what to replace $\delta(G)$ with, as dealing directly with $\dim_H(L(N))$ seems rather difficult. Recall that $\dim_H(L(N))$ coincides with $\delta(G)$ in the case where $N \lhd G$ and G is convex cocompact.

We shall look at two ways to find the desired replacement for $\delta(G)$; first, by generalising a classical notion from dynamics, the volume entropy, and second, by considering an 'extended Poincaré series' of N together with its critical exponent. We shall then see in Theorem 4.1 that the two approaches in fact lead to the same invariant.

Recall that for a compact Riemannian manifold (X, g) the volume entropy is defined as

$$\lim_{R\to\infty} \frac{\log \operatorname{vol}_g B_R(z)}{R},$$

where $B_R(z)$ is the ball of radius R centred at some $z \in \widetilde{X}$, with $\widetilde{X}$ the universal cover of X, and vol_g is the volume element induced by the Riemannian metric

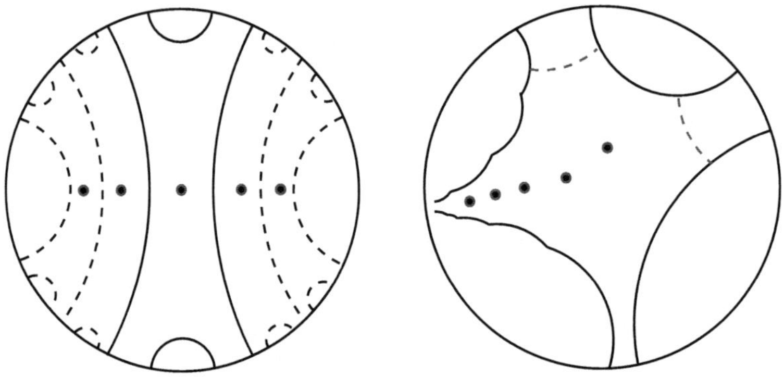

FIGURE 2. Real and 'fake' orbit points.

g; for any (not necessarily compact) hyperbolic manifold $\mathbb{H}^{n+1}/G$, with G some Kleinian group, the volume entropy just equals n. It is interesting to mention that this changes dramatically when looking at manifolds of negative pinched curvature. Dal'bo, Peigné, Picaud and Sambusetti [**20**] have shown that in this case there are examples of discrete groups of isometries whose critical exponent is strictly smaller than the volume entropy of the corresponding manifold.

For a closed set Λ in $\partial\mathbb{H}$, we define the *convex core entropy* [**24**] as

$$h_c(\Lambda) := \limsup_{R\to\infty} \frac{\log \operatorname{vol}(B_R(z) \cap H_\varepsilon(\Lambda))}{R},$$

where $B_R(z)$ is the ball of centre z and radius R in $\mathbb{H}^{n+1}$, and $H_\varepsilon(\Lambda)$ is the ε-neighbourhood (for some $\varepsilon > 0$) of the convex hull $H(\Lambda)$ of Λ in $\mathbb{H}^{n+1}$. For a hyperbolic manifold $M_G = \mathbb{H}^{n+1}/G$, we define the convex core entropy $h_c(M_G)$ to be $h_c(L(G))$.

Note that the definition is independent of the choice of $z \in \mathbb{H}^{n+1}$ and of a positive constant $\varepsilon > 0$. The latter is seen below in Theorem 4.1. The simplest reason why we need to take the ε-neighbourhood is that, in the degenerate case such as a Fuchsian group viewed as a Kleinian group, the volume of the convex hull is zero and there is no use considering its volume entropy.

Let us now turn to the idea of an 'extended Poincaré series' and its critical exponent. We shall look at certain discrete sets which are also known under other names like 'ε-nets', 'uniformly discrete sets' or 'Delone sets'. Roughly speaking, one 'fills up' the convex hull of some closed set in the boundary of hyperbolic space with uniformly distributed 'fake' orbit points and then considers the resulting series. We call a discrete set $X = \{x_i\}_{i=1}^{\infty}$ in the convex hull $H(\Lambda) \subset \mathbb{H}^{n+1}$ *uniformly distributed* if the following two conditions are satisfied:

(i) there exists a constant $M < \infty$ such that, for every point $z \in H(\Lambda)$, there is some $x_i \in X$ such that $d(x_i, z) \leq M$;
(ii) there exists a constant $m > 0$ such that any distinct points x_i and x_j in X satisfy $d(x_i, x_j) \geq m$.

Note that Λ is thus the limit set of X in the sense that $\Lambda = \overline{X} \setminus X$. The first natural example of such a uniformly distributed set in an convex hull is of course any orbit of some (convex) cocompact Kleinian group inside the convex hull of its limit set. Now, for a uniformly distributed set X, we define [**24**] the *Poincaré series* of X

with exponent $s > 0$ and reference point $z \in \mathbb{H}^{n+1}$ by

$$P^s(X, z) := \sum_{x \in X} e^{-s\, d(x,z)}.$$

The *critical exponent* of X is

$$\Delta = \Delta(X) := \inf\{s > 0 \mid P^s(X, z) < \infty\} = \limsup_{R \to \infty} \frac{\log \#(B_R(z) \cap X)}{R}.$$

The Poincaré series of X or, more simply, X itself is said to be of *convergence type* if $P^\Delta(X, z) < \infty$, and of *divergence type* otherwise.

As it turns out, not only do the convex core entropy $h_c(\Lambda)$ and the critical exponent $\Delta(X)$ of some uniformly distributed set X in the convex hull of Λ coincide, but they are also equal to the upper Minkowski dimension $\overline{\dim}_M(\Lambda)$ of Λ, since both $\Delta(X)$ and $\overline{\dim}_M(\Lambda)$ are exponential growth rates of the number of balls of decreasing size necessary to cover Λ.

THEOREM 4.1 ([**24**]). *Given a closed set $\Lambda \subset \mathbb{S}^n$, we have for any uniformly distributed set X in the convex hull $H(\Lambda)$*

$$\Delta(X) = h_c(\Lambda) = \overline{\dim}_M(\Lambda).$$

When $\Lambda = L(G)$ is the limit set of some non-elementary Kleinian group G, we thus have

$$\delta(G) \leq \dim_H(L(G)) \leq h_c(L(G)) = \overline{\dim}_M(L(G)).$$

In light of this result one may want to revisit the initial question of when $\delta(G) < \dim_H(L(G))$ holds for a Kleinian group G, as we now have two gaps to tend to. We will need some preparations in order to adress these new questions. For now we shall restrict to the 2-dimensional case, i.e. to $\mathbb{H}^2$, since this simplifies the arguments relating the 'combinatorial' isoperimetric constant to the classical one, which are necessary in [**11**], [**12**] and [**13**]. This will allow us to give the desired generalisation of Brooks' Theorem 2.1 to a situation where G is dropped altogether and one considers the amenability of some other object than G/N.

Every hyperbolic surface S has a not necessarily uniquely determined *pants decomposition* $\mathcal{P} = \mathcal{P}(S)$, in the sense that there is a (possibly infinite) collection of disjoint simple closed geodesics of S such that the complement consists of pieces homeomorphic to a disk with two holes called *pairs of pants*. Metrically, pants could be relatively compact in S, or could have punctures or funnels, i.e. boundary at infinity. If $\mathcal{P}$ is a pants decomposition for S, then the associated graph $\mathcal{G} = \mathcal{G}(\mathcal{P})$ is defined as the set of pants with edges between pants whenever these are adjacent via some boundary simple closed geodesic.

A hyperbolic surface S has *bounded geometry* if the set of all lengths of closed geodesics in S is uniformly bounded away from 0. We say that S has *strongly bounded geometry* if the convex core $C(S)$ admits a pants decomposition possibly including degenerate pairs of pants such that there are constants $C \geq c > 0$ with the property that the lengths of geodesic boundary components of each pair of pants are bounded from above by C and from below by c.

The notion of amenability generalises to graphs in the following way. A connected graph $\mathcal{G} = (V, E)$ is *amenable* if

$$\inf_K \frac{\#\partial K}{\#K} = 0,$$

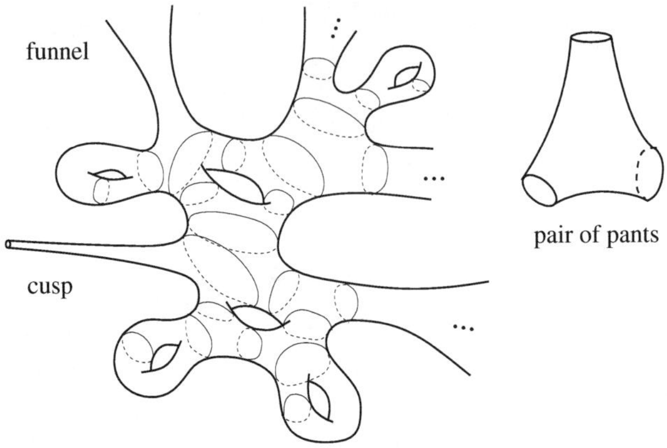

FIGURE 3. The pants decomposition.

where the infimum is taken over all finite connected subgraphs K of $\mathcal{G}$. $\mathcal{G}$ is *strongly amenable* if, for some fixed vertex $v \in V$,

$$\lim_{m\to\infty} \inf_{K^{(m)}} \frac{\#\partial K^{(m)}}{\#K^{(m)}} = 0,$$

where the infimum is taken over all finite connected subgraphs $K^{(m)}$ with $\#K^{(m)} \leq m$ and $K^{(m)} \ni v$. $\mathcal{G}$ is *uniformly strongly amenable* if the above convergence as $m \to \infty$ is uniform independently of the choice of $v \in V$. A connected graph is *transitive* if its autorphism group acts transitively on the set of edges. Note that for transitive graphs the above notions of amenability coincide. Cayley graphs of finitely generated groups are transitive, as such a group always acts transitively on its own Cayley graph.

Next, we consider the notion of an isoperimetric constant which exists both in hyperbolic geometry and for graphs. As a direct consequence of the definition, a graph is non-amenable if and only if

$$\sup_K \frac{\#K}{\#\partial K} < \infty,$$

where the supremum is taken over all finite connected subgraphs K of $\mathcal{G}$. We shall refer to $\sup_K(\#K/\#\partial K)$ as the *isoperimetric constant* of $\mathcal{G}$. For convenience we define the *isoperimetric constant* or *Cheeger constant* of a hyperbolic surface S as

$$h(S) := \sup_W \frac{A(W)}{\ell(\partial W)},$$

where the supremum is taken over all relatively compact domains $W \subset S$ with smooth boundary ∂W. Here, $A(W)$ denotes the hyperbolic area of A and $\ell(\partial W)$ the hyperbolic length of the boundary. Note that this is the reciprocal of the 'usual' isoperimetric constant, as defined and used for instance in [**13**]. The next classical result establishes a relationship between $h(S)$ and the bottom of the L^2-spectrum of S, denoted as before by λ_0. The first inequality is also employed by Brooks in [**11**] and [**13**].

THEOREM 4.2 ([**17**], [**14**]). *For every hyperbolic surface S we have*

$$\frac{1}{4\,h(S)^2} \leq \lambda_0(S) \leq \frac{B}{h(S)}$$

for some universal constant $B > 0$.

For hyperbolic surfaces S, the constant B does not depend on S, only on the dimension 2. In the original source [**14**] it depends on a bound for the Ricci curvature of the considered manifold, and we refer the interested reader to Buser's work [**14**] for details on the general case.

In order to obtain the desired partial generalisation of Brooks' theorem, we now only need to put the puzzle pieces together; relate the isoperimetric constants

$$h(S) = \sup_W \frac{A(W)}{\ell(\partial W)} \qquad \text{and} \qquad \sup_K \frac{\#K}{\#\partial K}$$

of S and $\mathcal{G}$, respectively, to each other, thus showing that the non-amenability of $\mathcal{G}$ implies that the Cheeger constant of S is finite. Theorem 4.2 then implies that $\lambda_0(S) > 0$ and thus $\delta(G) < 1$. This is the core of the argument yielding the following result, and is in fact closely related to the methods used in [**11**] and [**13**].

THEOREM 4.3 ([**24**]). *Consider a hyperbolic surface of infinite type S with strongly bounded geometry, given by the Fuchsian group G; further, consider the graph $\mathcal{G}$ associated to some arbitrary pants decomposition of S. Then we have the following implications:*

(i) *If $\mathcal{G}$ is uniformly strongly amenable, then either both $\delta(G)$ and $\dim_H(L(G))$ are striclty less than 1, or they are both equal to 1.*
(ii) *If $\mathcal{G}$ is non-amenable, then $\delta(G) < 1$.*

In particular, if $S = \mathbb{H}^2/G$ and $\mathcal{G}$ are as in Theorem 4.3, and if we assume that $\mathcal{G}$ is transitive and that $\dim_H(L(G)) = h_c(S) = 1$, then $\mathcal{G}$ is non-amenable if and only if $\delta(G) < 1$. Note that the methods used in [**24**] do not allow for a formulation of Theorem 4.3 in which 1 is replaced by $\dim_H(L(G))$. However, these considerations and the statement of Theorem 4.3 are reason enough to give the following not too surprising conjecture.

CONJECTURE 4.4. Let $S = \mathbb{H}^2/G$ be a hyperbolic surface of infinite type with strongly bounded geometry, and $\mathcal{G}$ be the graph associated to some arbitrary pants decomposition of S. Then, $\mathcal{G}$ is non-amenable if and only if $\delta(G) < h_c(S)$.

Note that we formulate the conjecture in terms of the convex core entropy h_c, and *not* the Hausdorff dimension of the limit set, since this seems to be a more approachable problem. It does, however, leave the question open of when $\dim_H(L(G))$ and $h_c(S)$ coincide. In order to address it, we first introduce a certain type of 'homogeneity condition'. For the remainder of this section we return to the general case of $(n+1)$-dimensional hyperbolic space.

We say that a uniformly distributed set X is of *bounded type* [**24**] if there exists a constant $\rho \geq 1$ such that

$$\frac{\#(X \cap B_R(x))}{\#(X \cap B_R(o))} \leq \rho$$

for every $x \in X$ and for every $R > 0$. Here we assume that the origin $o \in \mathbb{H}^{n+1}$ belongs to X. Clearly, any orbit of a convex cocompact Kleinian group is of bounded type. Also, any normal subgroup of a Schottky group as in Example 2.3 admits a uniformly distributed set of bounded type within the convex hull of its limit set, namely the orbit of the Schottky group.

We shall now see that if the bounded type condition is satisfied by a uniformly distributed set, then the Hausdorff and box dimensions of the corresponding limit set coincide.

THEOREM 4.5 ([**24**]). *Let Λ be a closed subset in $\partial\mathbb{H}^{n+1}$ and assume that there is a uniformly distributed set X of bounded type in the convex hull $H(\Lambda)$. Then the Δ-dimensional Hausdorff measure of Λ is positive. In particular,* $\dim_{\mathrm{H}}(\Lambda) \geq \Delta(X)$, *and hence we have that*

$$\dim_{\mathrm{H}}(\Lambda) = \Delta(X) = h_{\mathrm{c}}(\Lambda) = \underline{\dim}_{\mathrm{M}}(\Lambda) = \overline{\dim}_{\mathrm{M}}(\Lambda).$$

Furthermore, the Poincaré series of X is of divergence type, i.e. $P^{\Delta}(X,z) = \infty$.

A moment's consideration shows that the bounded type condition for uniformly distributed sets in fact prevents orbits of geometrically finite Kleinian groups with parabolic elements from being examples of such sets. This is clearly not a satisfactory state of affairs, so we need to adapt the definition slightly. We say that a uniformly distributed set X is of *weakly bounded type* [**24**] if there exist a constant $\rho \geq 1$ and a family of mutually disjoint horoballs $\{D_p\}_{p\in\Phi}$ in $\mathbb{H}^{n+1}$ with set of tangency points $\Phi \subset \partial\mathbb{H}^{n+1}$ such that

$$\frac{\#(X \cap B_R(x))}{\#(X \cap B_R(o))} \leq \rho$$

for every $x \in X'$ and for every $R > 0$. Here $X' = X \cap (\mathbb{H}^{n+1} \setminus \bigcup_{p\in\Phi} D_p)$ and we assume that the origin $o \in \mathbb{H}^{n+1}$ belongs to X'. It is not difficult to see that any orbit of a geometrically finite Kleinian group with parabolic elements is an elementary example of such a X'. Moreover, the simplest examples of non-geometrically finite hyperbolic manifolds which admit uniformly distributed sets of weakly bounded type are given by infinite normal covers of geometrically finite manifolds.

In this situation one can show the following [**24**]. Suppose that a uniformly distributed set X in the convex hull of $\Lambda \subset \mathbb{S}^n$ is of weakly bounded type and the *Patterson measure* μ_z, $z \in \mathbb{H}^{n=1}$, *associated with* X has no atom on the tangency points $p \in \Phi \subset \partial\mathbb{H}^{n+1}$ of the horoballs $\{D_p\}_{p\in\Phi}$ appearing in the definition of weak boundedness. (For more details on μ_z see [**24**].) Then the packing dimension $\dim_{\mathrm{P}}(\Lambda)$ of Λ coincides with $\Delta(X) = h_c(\Lambda) = \overline{\dim}_{\mathrm{M}}(\Lambda)$. Moreover, the Poincaré series of X is of divergence type, i.e. $P^{\Delta}(X,z) = \infty$. In particular, let G be a non-elementary Kleinian group acting on $\mathbb{H}^{n+1}$ all of whose parabolic fixed points are bounded. If the limit set $L(G)$ of G is of weakly bounded type for the set Φ of all parabolic fixed points of G, i.e. its convex hull contains a uniformly distributed set X of weakly bounded type, then

$$\dim_{\mathrm{P}}(L(G)) = h_c(L(G)) = \Delta(X) \quad \text{and} \quad P^{\Delta}(X,z) = \infty.$$

5. An outlook; open problems

Given some (non-elementary) Kleinian group G, it makes sense to split up the question whether a gap $\delta(G) < \dim_H(L(G))$ exists into the following two questions:

(i) When is $\dim_{\mathrm{H}}(L(G)) = h_{\mathrm{c}}(L(G))$? More generally, as we have seen that $h_{\mathrm{c}}(L(G)) = \dim_{\mathrm{M}}(L(G))$, when is $\dim_H(L(G)) = \dim_{\mathrm{M}}(L(G))$?

(ii) When is $\delta(G) < h_c(L(G))$?

About (i): So far there exist only sufficient conditions for the general equality

$$\dim_H(\Lambda) = \dim_M(\Lambda), \quad \Lambda \text{ closed},$$

which usually amount to some sort of suitable 'homogeneity' of Λ. Our bounded type condition on uniformly distributed sets is such a 'homogeneity condition', not directly on Λ but on its convex hull inside hyperbolic space. However, there is no useful characterisation for this equality, albeit many concrete examples of $\dim_H(\Lambda) < \dim_M(\Lambda)$, which employ some sort of 'inhomogeneity'. This should not exist in conformal dynamical systems (or at least in some large class of such systems), which means that conjecturing

$$\dim_H(L(G)) = h_c(L(G))$$

for all (or 'most') non-elementary Kleinian groups is not completely unrealistic.

About (ii): This appears to be more likely to be answered than the direct question of whether $\delta(G) < \dim_H(L(G))$, since $h_c(L(G)) = \Delta(X)$ for some uniformly distributed X within $H(L(G))$ and $\delta(G)$ are 'similarly defined' invariants and may thus be easier to put in relation to each other. There are at least two possible approaches. The first would be the attempt to generalise Brooks' approach, as exercised by Tapie [**63**] or in [**24**]. The problem with the method in [**24**] is that it appears difficult to generalise the pants decomposition of surfaces along totally geodesic pieces, i.e. simple closed geodesics, to higher dimensions, where there is no analogous decomposition method with totally geodesic separating submanifolds of codimension 1.

The other approach would be attempting to generalise Stadlbauer's method [**57**] of proof for Brooks' theorem . The problem here is that translating the geometric problem from hyperbolic manifolds to the setting of Markov chains works via the Bowen-Series map, which has only been meaningfully defined for rather restricted classes of Kleinian groups.

References

[1] Shmuel Agmon, *On positivity and decay of solutions of second order elliptic equations on Riemannian manifolds*, Methods of functional analysis and theory of elliptic equations (Naples, 1982), Liguori, Naples, 1983, pp. 19–52. MR819005

[2] I. Agol, 'Tameness of hyperbolic 3-manifolds', preprint, arXiv:math/0405568v1 [math.GT].

[3] Tohru Akaza and Harushi Furusawa, *The exponent of convergence of Poincaré series on some Kleinian groups*, Tôhoku Math. J. (2) **32** (1980), no. 3, 447–452, DOI 10.2748/tmj/1178229604. MR590041

[4] Alan F. Beardon, *The geometry of discrete groups*, Graduate Texts in Mathematics, vol. 91, Springer-Verlag, New York, 1983. MR698777

[5] Alan F. Beardon and Bernard Maskit, *Limit points of Kleinian groups and finite sided fundamental polyhedra*, Acta Math. **132** (1974), 1–12, DOI 10.1007/BF02392106. MR0333164

[6] Christopher J. Bishop, *Minkowski dimension and the Poincaré exponent*, Michigan Math. J. **43** (1996), no. 2, 231–246, DOI 10.1307/mmj/1029005460. MR1398152

[7] Christopher J. Bishop, *Geometric exponents and Kleinian groups*, Invent. Math. **127** (1997), no. 1, 33–50, DOI 10.1007/s002220050113. MR1423024

[8] Christopher J. Bishop and Peter W. Jones, *Hausdorff dimension and Kleinian groups*, Acta Math. **179** (1997), no. 1, 1–39, DOI 10.1007/BF02392718. MR1484767

[9] Petra Bonfert-Taylor, Katsuhiko Matsuzaki, and Edward C. Taylor, *Large and small covers of a hyperbolic manifold*, J. Geom. Anal. **22** (2012), no. 2, 455–470, DOI 10.1007/s12220-010-9204-6. MR2891734

[10] Rufus Bowen and Caroline Series, *Markov maps associated with Fuchsian groups*, Inst. Hautes Études Sci. Publ. Math. **50** (1979), 153–170. MR556585

[11] Robert Brooks, *The fundamental group and the spectrum of the Laplacian*, Comment. Math. Helv. **56** (1981), no. 4, 581–598, DOI 10.1007/BF02566228. MR656213

[12] Robert Brooks, *The spectral geometry of the Apollonian packing*, Comm. Pure Appl. Math. **38** (1985), no. 4, 359–366, DOI 10.1002/cpa.3160380402. MR792395

[13] Robert Brooks, *The bottom of the spectrum of a Riemannian covering*, J. Reine Angew. Math. **357** (1985), 101–114, DOI 10.1515/crll.1985.357.101. MR783536

[14] Peter Buser, *A note on the isoperimetric constant*, Ann. Sci. École Norm. Sup. (4) **15** (1982), no. 2, 213–230. MR683635

[15] Danny Calegari and David Gabai, *Shrinkwrapping and the taming of hyperbolic 3-manifolds*, J. Amer. Math. Soc. **19** (2006), no. 2, 385–446, DOI 10.1090/S0894-0347-05-00513-8. MR2188131

[16] Richard D. Canary, *Ends of hyperbolic 3-manifolds*, J. Amer. Math. Soc. **6** (1993), no. 1, 1–35, DOI 10.2307/2152793. MR1166330

[17] Jeff Cheeger, *A lower bound for the smallest eigenvalue of the Laplacian*, Problems in analysis (Papers dedicated to Salomon Bochner, 1969), Princeton Univ. Press, Princeton, N. J., 1970, pp. 195–199. MR0402831

[18] Joel M. Cohen, *Cogrowth and amenability of discrete groups*, J. Funct. Anal. **48** (1982), no. 3, 301–309, DOI 10.1016/0022-1236(82)90090-8. MR678175

[19] Rémi Coulon, Françoise Dal'Bo, and Andrea Sambusetti, *Growth gap in hyperbolic groups and amenability*, Geom. Funct. Anal. **28** (2018), no. 5, 1260–1320, DOI 10.1007/s00039-018-0459-6. MR3856793

[20] Françoise Dal'Bo, Marc Peigné, Jean-Claude Picaud, and Andrea Sambusetti, *On the growth of nonuniform lattices in pinched negatively curved manifolds*, J. Reine Angew. Math. **627** (2009), 31–52, DOI 10.1515/CRELLE.2009.010. MR2494912

[21] R. Dougall, 'Critical exponents of normal subgroups, the spectrum of group extended transfer operators, and Kazhdan distance', preprint, arXiv:1702.06115 [math.DS].

[22] Rhiannon Dougall and Richard Sharp, *Amenability, critical exponents of subgroups and growth of closed geodesics*, Math. Ann. **365** (2016), no. 3-4, 1359–1377, DOI 10.1007/s00208-015-1338-1. MR3521093

[23] J. Elstrodt, 'Die Resolvente zum Eigenwertproblem der automorphen Formen in der hyperbolischen Ebene', I, *Math. Ann.* **203** (1973), 295–300; II, *Math. Z.* **132** (1973), 99–134; III, *Math. Ann.* **208** (1974), 99–132.

[24] Kurt Falk and Katsuhiko Matsuzaki, *The critical exponent, the Hausdorff dimension of the limit set and the convex core entropy of a Kleinian group*, Conform. Geom. Dyn. **19** (2015), 159–196, DOI 10.1090/ecgd/279. MR3351952

[25] Kurt Falk and Bernd O. Stratmann, *Remarks on Hausdorff dimensions for transient limit sets of Kleinian groups*, Tohoku Math. J. (2) **56** (2004), no. 4, 571–582. MR2097162

[26] Erling Følner, *On groups with full Banach mean value*, Math. Scand. **3** (1955), 243–254, DOI 10.7146/math.scand.a-10442. MR0079220

[27] R. I. Grigorchuk, *Symmetrical random walks on discrete groups*, Multicomponent random systems, Adv. Probab. Related Topics, vol. 6, Dekker, New York, 1980, pp. 285–325. MR599539

[28] J. H. Hubbard, *Teichmüller theory and Applications to Geometry Topology and Dynamics; Volume 1: Teichmüller theory*, Matrix Editions, New York, 2006.

[29] Johannes Jaerisch, *Conformal fractals for normal subgroups of free groups*, Conform. Geom. Dyn. **18** (2014), 31–55, DOI 10.1090/S1088-4173-2014-00263-X. MR3175016

[30] Johannes Jaerisch, *Fractal models for normal subgroups of Schottky groups*, Trans. Amer. Math. Soc. **366** (2014), no. 10, 5453–5485, DOI 10.1090/S0002-9947-2014-06095-9. MR3240930

[31] Johannes Jaerisch, *Group-extended Markov systems, amenability, and the Perron-Frobenius operator*, Proc. Amer. Math. Soc. **143** (2015), no. 1, 289–300, DOI 10.1090/S0002-9939-2014-12237-4. MR3272754

[32] Johannes Jaerisch, *A lower bound for the exponent of convergence of normal subgroups of Kleinian groups*, J. Geom. Anal. **25** (2015), no. 1, 298–305, DOI 10.1007/s12220-013-9427-4. MR3299281

[33] Michael Kapovich, *Hyperbolic manifolds and discrete groups*, Progress in Mathematics, vol. 183, Birkhäuser Boston, Inc., Boston, MA, 2001. MR1792613

[34] Harry Kesten, *Symmetric random walks on groups*, Trans. Amer. Math. Soc. **92** (1959), 336–354, DOI 10.2307/1993160. MR0109367

[35] Harry Kesten, *Full Banach mean values on countable groups*, Math. Scand. **7** (1959), 146–156, DOI 10.7146/math.scand.a-10568. MR0112053

[36] Bernard Maskit, *Kleinian groups*, Grundlehren der Mathematischen Wissenschaften [Fundamental Principles of Mathematical Sciences], vol. 287, Springer-Verlag, Berlin, 1988. MR959135

[37] Katsuhiko Matsuzaki and Masahiko Taniguchi, *Hyperbolic manifolds and Kleinian groups*, Oxford Mathematical Monographs, The Clarendon Press, Oxford University Press, New York, 1998. Oxford Science Publications. MR1638795

[38] R. Daniel Mauldin and Mariusz Urbański, *Gibbs states on the symbolic space over an infinite alphabet*, Israel J. Math. **125** (2001), 93–130, DOI 10.1007/BF02773377. MR1853808

[39] Peter J. Nicholls, *The ergodic theory of discrete groups*, London Mathematical Society Lecture Note Series, vol. 143, Cambridge University Press, Cambridge, 1989. MR1041575

[40] Jean-Pierre Otal and Marc Peigné, *Principe variationnel et groupes kleiniens* (French, with English and French summaries), Duke Math. J. **125** (2004), no. 1, 15–44, DOI 10.1215/S0012-7094-04-12512-6. MR2097356

[41] S. J. Patterson, *The limit set of a Fuchsian group*, Acta Math. **136** (1976), no. 3-4, 241–273, DOI 10.1007/BF02392046. MR0450547

[42] S. J. Patterson, *Some examples of Fuchsian groups*, Proc. London Math. Soc. (3) **39** (1979), no. 2, 276–298, DOI 10.1112/plms/s3-39.2.276. MR548981

[43] S. J. Patterson, *Further remarks on the exponent of convergence of Poincaré series*, Tohoku Math. J. (2) **35** (1983), no. 3, 357–373, DOI 10.2748/tmj/1178228995. MR711352

[44] S. J. Patterson, *Lectures on measures on limit sets of Kleinian groups*, Analytical and geometric aspects of hyperbolic space (Coventry/Durham, 1984), London Math. Soc. Lecture Note Ser., vol. 111, Cambridge Univ. Press, Cambridge, 1987, pp. 281–323. MR903855

[45] Marc Peigné, *On the Patterson-Sullivan measure of some discrete group of isometries*, Israel J. Math. **133** (2003), 77–88, DOI 10.1007/BF02773062. MR1968423

[46] John G. Ratcliffe, *Foundations of hyperbolic manifolds*, Graduate Texts in Mathematics, vol. 149, Springer-Verlag, New York, 1994. MR1299730

[47] Mary Rees, *Checking ergodicity of some geodesic flows with infinite Gibbs measure*, Ergodic Theory Dynamical Systems **1** (1981), no. 1, 107–133. MR627791

[48] Mary Rees, *Divergence type of some subgroups of finitely generated Fuchsian groups*, Ergodic Theory Dynamical Systems **1** (1981), no. 2, 209–221. MR661820

[49] Thomas Roblin, *Sur la fonction orbitale des groupes discrets en courbure négative* (French, with English and French summaries), Ann. Inst. Fourier (Grenoble) **52** (2002), no. 1, 145–151. MR1881574

[50] Thomas Roblin, *Un théorème de Fatou pour les densités conformes avec applications aux revêtements galoisiens en courbure négative* (French, with French summary), Israel J. Math. **147** (2005), 333–357, DOI 10.1007/BF02785371. MR2166367

[51] Thomas Roblin and Samuel Tapie, *Exposants critiques et moyennabilité* (French), Géométrie ergodique, Monogr. Enseign. Math., vol. 43, Enseignement Math., Geneva, 2013, pp. 61–92, DOI 10.1002/9781118557754.ch3. MR3220551

[52] Thomas Roblin and Samuel Tapie, *Critical exponent and bottom of the spectrum in pinched negative curvature*, Math. Res. Lett. **22** (2015), no. 3, 929–944, DOI 10.4310/MRL.2015.v22.n3.a15. MR3350111

[53] Omri M. Sarig, *Thermodynamic formalism for countable Markov shifts*, Ergodic Theory Dynam. Systems **19** (1999), no. 6, 1565–1593, DOI 10.1017/S0143385799146820. MR1738951

[54] Omri Sarig, *Existence of Gibbs measures for countable Markov shifts*, Proc. Amer. Math. Soc. **131** (2003), no. 6, 1751–1758, DOI 10.1090/S0002-9939-03-06927-2. MR1955261

[55] Richard Sharp, *Critical exponents for groups of isometries*, Geom. Dedicata **125** (2007), 63–74, DOI 10.1007/s10711-007-9137-9. MR2322540

[56] M. Stadlbauer, 'The Bowen-Series map for some free groups', Dissertation, University of Göttingen 2002. (Preprint in *Math. Gottingensis* **5** (2002), 1–53.)

[57] Manuel Stadlbauer, *An extension of Kesten's criterion for amenability to topological Markov chains*, Adv. Math. **235** (2013), 450–468, DOI 10.1016/j.aim.2012.12.004. MR3010065

[58] Manuel Stadlbauer and Bernd O. Stratmann, *Infinite ergodic theory for Kleinian groups*, Ergodic Theory Dynam. Systems **25** (2005), no. 4, 1305–1323, DOI 10.1017/S014338570400104X. MR2158407

[59] B. Stratmann and M. Urbański, *The box-counting dimension for geometrically finite Kleinian groups*, Fund. Math. **149** (1996), no. 1, 83–93. MR1372359

[60] Dennis Sullivan, *The density at infinity of a discrete group of hyperbolic motions*, Inst. Hautes Études Sci. Publ. Math. **50** (1979), 171–202. MR556586

[61] Dennis Sullivan, *Entropy, Hausdorff measures old and new, and limit sets of geometrically finite Kleinian groups*, Acta Math. **153** (1984), no. 3-4, 259–277, DOI 10.1007/BF02392379. MR766265

[62] Dennis Sullivan, *Related aspects of positivity in Riemannian geometry*, J. Differential Geom. **25** (1987), no. 3, 327–351. MR882827

[63] S. Tapie, 'Graphes, moyennabilité et bas du spectre de variétés topologiquement infinies', preprint, arXiv:1001.2501 [math.DG].

Mathematisches Seminar, Christian-Albrechts-Universität zu Kiel, Ludewig-Meyn-Strasse 4, 24118 Kiel, Germany

Email address: `falk@math.uni-kiel.de`

Contemporary Mathematics
Volume **731**, 2019
https://doi.org/10.1090/conm/731/14675

The spectral operator and resonances

Machiel van Frankenhuijsen

ABSTRACT. The spectral operator and fractal strings were developed to understand the Riemann hypothesis through the relationship between geometric and spectral oscillations. Following Mark Kac [**10**], Lapidus asked in [**13, 14**] if one can hear the shape of a fractal string, and found a remarkable connection with the location of the zeros of $\zeta(s)$. In [**21**], Lapidus and Pomerance showed that spectral oscillations can only arise from geometric oscillations. Then, Lapidus and Maier constructed in [**19**] an example of a fractal string with geometric oscillations but no oscillations in the spectrum. The resulting vision was worked out by Lapidus and the author in [**22**–**29**]. As a test for this philosophy, we introduced the class of generalized Cantor strings, the oscillations of which resonate with each other, and therefore remain audible in the spectrum. This proved that the Riemann zeta function does not vanish at any vertical arithmetic progression. In this exposition, we emphasize the more recent work of Herichi and Lapidus [**5**–**8**], who have analyzed the spectral operator in detail, discovering a close connection between invertibility and the universality of $\zeta(s)$ in the right half of the critical strip. Their findings suggest new directions in this approach to understanding the Riemann hypothesis.

1. Introduction

There have been very few convincing heuristic arguments to explain the validity of the Riemann hypothesis. One approach is through the study of the spectrum of fractal strings. In 1991, Lapidus and Maier discovered that the frequencies (or spectrum) of a fractal string are connected with its geometry via the zeros of the Riemann zeta function. Together with the work of Lapidus and Pomerance this led, in the collaboration of Lapidus with the author, to a theory of complex dimensions and a tentative definition of the spectral operator.

The theory of the spectral operator for fractal strings was developed and made rigorous in the work of Herichi and Lapidus in 2012 (see [**5**–**8**]). Their work shows a connection between the invertibility of the spectral operator and universality of the Riemann zeta function in the right half of the critical strip, the property that every nonvanishing power series with radius of convergence $1/4$ can be approximated by translates $z \mapsto \zeta(z + 3/4 + it)$.

We illustrate the philosophy of the connection between the geometry and the spectrum (the direct and inverse spectral problem) in the question of the existence of zeros in arithmetic progression of $\zeta(s)$, which can be successfully solved with this approach. We also discuss what might still be missing to explain the Riemann hypothesis in this framework.

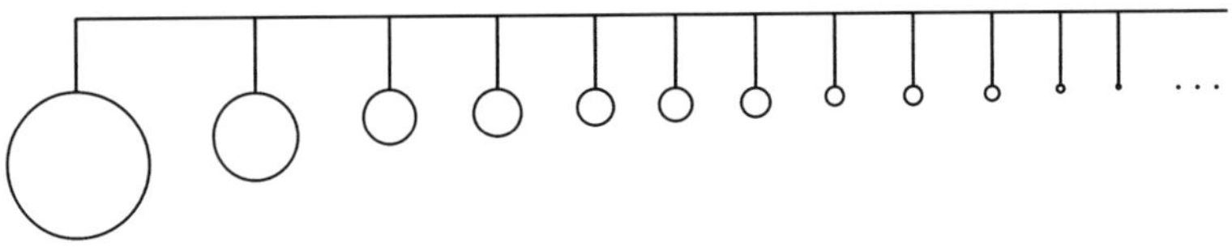

FIGURE 1. A fractal string.

This paper is an elaboration of the talk given at the California Polytechnic State University in San Luis Obispo on June 27, 2016, on the occasion of the sixtieth birthday of Michel Lapidus in the conference *Fractal Geometry and Complex Dimensions*. This conference, organized by Erin Pearse, John Rock, and Tony Samuel, brought together many researchers of fractal geometry, many of them students of Michel. I want to thank him here for our friendship and his enthusiastic support of and interest in the many people making discoveries in fractal geometry.

2. Fractal strings

A fractal string is a one-dimensional drum with a fractal boundary. More formally, a fractal string is defined as a decreasing sequence of "lengths". We take a slightly different point of view and consider an unbounded increasing sequence of curvatures (Figure 1),

$$0 < c_1 \le c_2 \le c_3 \le \dots.$$

A circle of curvature c corresponds to a circumference of (or proportional to) $l = 1/c$, and we assume that the total length is finite,

$$L = l_1 + l_2 + l_3 + \dots \quad \text{is finite.} \tag{1}$$

The geometry of a fractal string is described by the counting function of the curvatures,

$$N(x) = \sum_{n:\, c_n \le x} 1 = \sum_{n:\, l_n \ge 1/x} 1.$$

This function counts the number of curvatures, or inverse lengths, up to x.

REMARK 2.1. The functions N and c are essentially inverses of each other: if $c_{n+1} > c_n$, then

$$N(c_n) = n,$$

and for all positive x,

$$c_{N(x)} \le x,$$

with equality if $x = c_n$ for some n. Note that N is increasing and takes only integer values.

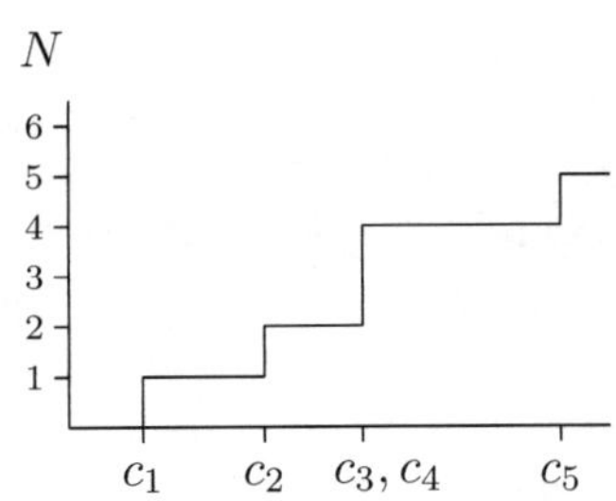

2.1. Explicit formulas and the geometric zeta function. The fractal string $\mathcal{T}$, and its zeta function $\zeta_{\mathcal{T}}$, are called *languid* if $\zeta_{\mathcal{T}}$ satisfies two growth conditions **L1**, along horizontal lines $\sigma + iT_n$, and **L2**, along a vertical "screen" $S(t) + it$ contained in the half-plane $[\mathrm{Re} < D]$[1] (see Definition 5.2 in [**28**] or [**29**]):

L1: $\zeta_{\mathcal{T}}(\sigma + iT_n) = O\big((1 + |T_n|^{\kappa}\big)$ for all $\sigma > S(T_n)$,

L2: $\zeta_{\mathcal{T}}(S(t) + it) = O\big(|t|^{\kappa}\big)$ for all $|t| \geq 1$.

Every fractal string that we consider here is languid, but for strings that are not languid, an explicit formula does not always exist.

For a languid string, the counting function of the curvatures is described by an explicit formula

$$N(x) = \sum\nolimits_{\omega} r_{\omega} \frac{x^{\omega}}{\omega} + O(x^0),$$

for certain (complex) exponents ω, called the *complex dimensions*. These complex dimensions, and the associated coefficients r_{ω}, are determined by the *geometric zeta function*

$$\zeta_{\mathcal{T}}(s) = \sum_{n=1}^{\infty} c_n^{-s} = \sum_{n=1}^{\infty} l_n^s. \tag{2}$$

This series converges for $s = 1$ by (1) and has an abscissa of convergence D, called the Minkowski dimension of $\mathcal{T}$. The series (2) defines $\zeta_{\mathcal{T}}(s)$ for $\mathrm{Re}\, s > D$. If this function has an analytic continuation, we find a singularity at $s = D$ and possibly other singularities. If the singularities of $\zeta_{\mathcal{T}}$ are simple poles then

$$r_{\omega} - \mathrm{res}(\zeta_{\mathcal{T}}, \omega),$$

the residue at each pole ω. In general, the terms are more complicated (see [**28**, Theorem 6.1; **29**]).

Another important invariant that can be read off from the zeta function is the total length,

$$L = \zeta_{\mathcal{T}}(1).$$

From the formula

$$\zeta_{\mathcal{T}}(s) = s \int_0^{\infty} N(x) x^{-s-1}\, dx \tag{3}$$

(to be derived below) one sees that the analytic continuation of $\zeta_{\mathcal{T}}$ to $\mathrm{Re}\, s \leq D$ is intimately connected with the existence of an asymptotic expansion for N, about which initially it is only known that $N(x) = O\big(x^{D+\varepsilon}\big)$ for all $\varepsilon > 0$.

One way to derive (3) is by introducing the Heaviside function

$$H(x) = \begin{cases} 0 & \text{if } x < 0, \\ 1 & \text{if } x > 0, \end{cases}$$

and noting that

$$s \int_0^{\infty} H(x - c) x^{-s-1}\, dx = c^{-s}$$

[1]We use the notation $[P]$ for the set $\{x \colon P(x)\}$. Thus, $[\mathrm{Re} < D]$ denotes the half-plane of complex numbers $\{s \in \mathbb{C} \colon \mathrm{Re}\, s < D\}$.

for $\operatorname{Re} s > 0$. The formula then follows by adding the integrals for $N(x)$, since the geometric counting function can be expressed as $\sum_{n=1}^{\infty} H(x - c_n)$.

3. The spectrum of a fractal string

The frequencies of a circle of curvature c are $c,\ 2c,\ 3c, \dots$, and the counting function of the frequencies is $\nu(N)$, given by

$$\nu(N)(x) = \sum_{k,n:\, kc_n \leq x} 1 = N(x) + N(x/2) + N(x/3) + \dots. \tag{4}$$

We call this the *spectral operator*. For the sum in (4) to be finite, it is necessary and sufficient that $N(x) = 0$ in a neighborhood of 0.

In terms of the complex dimensions, the explicit formula for the spectral counting function is

$$\nu(N)(x) = Lx + \sum\nolimits_{\omega} \zeta(\omega) r_\omega \frac{x^\omega}{\omega} + O(x^0), \tag{5}$$

where for each ω, $\zeta(\omega)$ is the value of the Riemann zeta function at the complex dimension. Thus, the spectral operator adds the *Weyl-term* Lx, and each term of the explicit formula of N is multiplied by $\zeta(\omega)$ in the explicit formula for $\nu(N)$.

REMARK 3.1. The terms with

$$x^\omega = x^\sigma \big(\cos(t \log x) + i \sin(t \log x)\big)$$

for nonreal $\omega = \sigma + it$ are called *oscillatory terms* because x^ω is multiplicatively periodic, with the same argument at x as at $xe^{2\pi/t}$. The period is determined by the imaginary part t, and the value

$$\mathbf{p} = \frac{2\pi}{\operatorname{Im} \omega} = \frac{2\pi}{t}$$

is called the *oscillatory period*. The amplitude of these oscillations varies with x^σ.

Fractal strings for which $N(x)$ does not have an asymptotic expansion in terms with multiplicative oscillations usually have a zeta function without poles. See (8) and Remark 4.2 below for an example.

3.1. The direct spectral problem. In 1990 (see [**21**]), Lapidus and Pomerance discovered that if the geometry of a fractal string has no oscillations, then the spectrum has no oscillations either:

$$\text{If } N(x) = Mx^D + o(x^D), \text{ then } \nu(N)(x) = Lx + \zeta(D)Mx^D + o(x^D). \tag{6}$$

Here, the coefficient M, if it exists, is the D-dimensional *Minkowski content* of the fractal string.

REMARK 3.2. The factor $\zeta(D)$ is negative, and it is useful to write the formula in (6) as

$$\nu(N)(x) = Lx - (-\zeta(D))Mx^D + o(x^D)$$

to emphasize the fact that the counting function of the spectrum is bounded by the Weyl term. This is generally true for the eigenvalues of the Laplacian with Dirichlet boundary conditions. For Neumann boundary conditions[2] on the other hand, the spectral counting function is usually bounded from below by the Weyl term.

---

[2]Karl Gottfried Neumann (1832-1925).

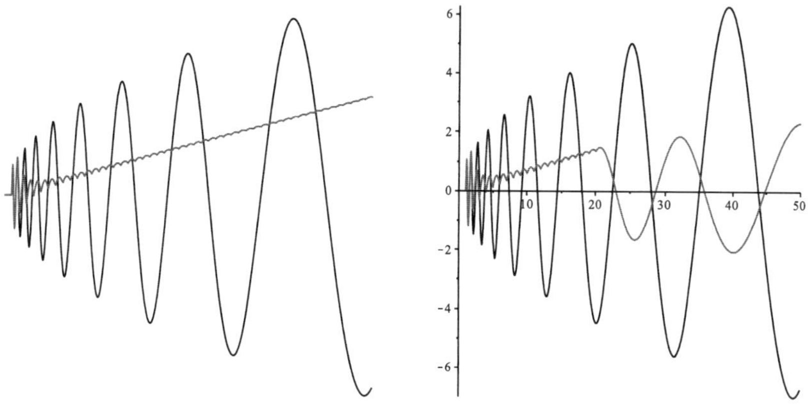

FIGURE 2. $\rho = 1/2 + 14.134725142\,i$; $\nu(N)$ truncated at the 20th term.

3.2. The inverse spectral problem. In 1991 (see [**18, 19**]), Lapidus and Maier discovered that the inverse problem of recovering the geometric counting function from the spectrum,

$$\text{If } \nu(N)(x) = Lx + Cx^D + o\big(x^D\big), \text{ then } N(x) = (C/\zeta(D))x^D + o\big(x^D\big)\ ?$$

holds only if the Riemann zeta function does not vanish on the line $\operatorname{Re} s = D$.

Thus you can hear irregularities in the geometry of a fractal string of dimension D, in particular, geometric oscillations as in Remark 3.1, if and only if $\zeta(s) \neq 0$ for $\operatorname{Re} s = D$.

REMARK 3.3. More precisely, the error term $o\big(x^D\big)$ above should be replaced by the slightly more restrictive $O\big(x^{D-\varepsilon}\big)$ for some $\varepsilon > 0$. The methods of [**27–29**] allow more general error terms.

3.3. Can you hear the shape of a fractal string? From the above, it follows that you can hear the Minkowski dimension D of a fractal string (by (6), since $\zeta(D) \neq 0$), and you can hear if the string is Minkowski measurable if and only if the Riemann zeta function does not vanish on the vertical line [Re $= D$]. In their paper, Lapidus and Maier construct a fractal string with geometric oscillations that disappear in the spectrum (see [**18, 19**]).

In essence, the example of Lapidus–Maier is as follows: Take the geometric counting function

$$N(x) = \begin{cases} \operatorname{Im}\big(x^\rho\big) & \text{for } x \geq 1, \\ 0 & \text{for } x \leq 1, \end{cases} \tag{7}$$

for a zero $\rho = \sigma + it$ of the Riemann zeta function. Thus

$$N(x) = \frac{x^\rho - x^{\bar\rho}}{2i} = x^\sigma \sin(t \log x), \qquad \text{for } x \geq 1,$$

is the explicit formula for the counting function, and the corresponding "fractal string" would have two complex conjugate complex dimensions (and no Minkowski

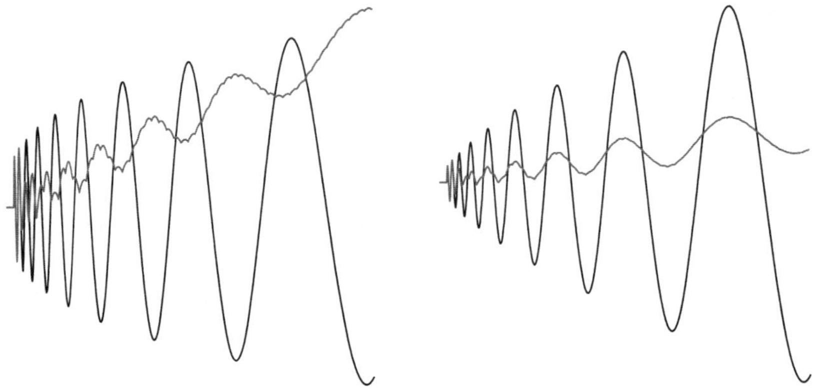

FIGURE 3. $1/3 + 14.134725142\,i$ and $2/3 + 14.134725142\,i$.

dimension). The oscillations in the geometry disappear in the spectrum by (5). The spectral operator ν gives a function essentially without oscillations,

$$\nu(N)(x) = Lx + O(x^0), \qquad \text{where } L = \int_1^\infty \operatorname{Im}(x^\rho)x^{-2}\,dx.$$

This is illustrated in the left diagram of Figure 2. The value $\rho = \frac{1}{2} + i\,14.1\ldots$ is a zero of $\zeta(s)$, and we see multiplicative oscillations in this diagram for the geometric counting function. The spectral counting function in the same diagram has no oscillations. (Note that $\nu(N)$ has a corner at each integer due to the nondifferentiability of $N(x)$ at $x = 1$.)

The second diagram on the right shows that all terms in $\nu(N)$ are necessary: A sum of twenty terms $N(x) + N(x/2) + \cdots + N(x/20)$ only gives $\nu(N)(x)$ for $x \leq 20$ due to the fact that $N(x) = 0$ for $x \leq 1$. The large oscillations for $x > 20$ are cancelled by the next terms $N(x/21) + \ldots$ in the sum for $\nu(N)$. This is reminiscent of the fact that to have Weil-positivity, all Euler factors need to be included, as explained in [**4**, formula (1.18), p. 96].

The diagrams in the next Figure 3 illustrate two geometric counting functions constructed in the same way, but for two values of ρ that are not zeros of the Riemann zeta function. We see that in that case, the geometry and the spectrum of the corresponding fractal strings have similar oscillations. In particular, the geometric oscillations of both strings remain audible in the spectrum.

REMARK 3.4. The 'counting function' (7) is not the geometric counting function of a fractal string because $N(x)$ is not increasing and does not jump by integer amounts at curvature values c_n. Lapidus and Maier obtain an actual counting function (and thus creating a fractal string of Minkowski dimension $D = 1/2$) by first adding Ax^σ for large enough A so that the function becomes increasing ($A = 2|\rho|$ is the least value that can be used), and then defining

$$N(x) = \left[x^\sigma(A + \sin(t\log x)\right] \qquad (\rho = \sigma + it),$$

where $[y]$ denotes the integer part of y. The curvatures c_n are now defined by the locations of the jumps of this function, as in Remark 2.1. The explicit formula for N is

$$N(x) = Ax^\sigma + \frac{1}{2i}x^\rho - \frac{1}{2i}x^{\bar\rho} + O(x^0).$$

The Minkowski dimension of the resulting fractal string is $D = \sigma$. By Section 4.4 below, it is not known if there exist such strings (that are not Minkowski measurable but the spectrum has no oscillations) of dimension $\sigma \neq 1/2$.

EXAMPLE 3.5 (**The spectral operator and the prime number theorem**). As is well-known (see [**3**, **9**], for example), the function that counts the prime powers with a weight of $\log p$ has an explicit formula in terms of the zeros of the Riemann zeta function,

$$\psi(x) = \sum_{p^k \leq x} \log p = x - \sum_\rho \frac{x^\rho}{\rho} - \frac{1}{2}\log(1 - x^{-2}),$$

where ρ runs over all zeros of $\zeta(s)$ with $\operatorname{Re}\rho > 0$. Applying the spectral operator, we see that this is another example where jumps in a geometric counting function disappear,

$$\nu(\psi)(x) = x \log x + O(1).$$

4. The Riemann zeta function

The zeta function of Riemann was perhaps first studied by Euler (1707-1783). It is defined for $\operatorname{Re} s > 1$ by the series

$$\zeta(s) = 1 + \frac{1}{2^s} + \frac{1}{3^s} + \dots.$$

Thus $\zeta(1)$ is the divergent harmonic series, and $\zeta(2) = \pi^2/6$, the result that made Euler famous in 1735. Riemann (1826-1866) was the first to consider this series for complex values of s and used it to heuristically study the prime counting function in [**33**] using the Euler product (see Section 4.1).

There are several methods to find the analytic continuation of $\zeta(s)$ to $\operatorname{Re} s \leq 1$. One method is to use "summation by parts" (Abel's summation formula) to rewrite a sum $\sum_{n=1}^N a_n$ as $Na_{N+1} + \sum_{n=1}^N (a_n - a_{n+1})n$, to obtain, for $\operatorname{Re} s > 1$,

$$\zeta(s) = s\int_0^\infty [x]\, x^{-s-1} dx. \tag{8}$$

Then, writing the integer part $[x]$ as the difference of x and its fractional part, $[x] = x - \{x\}$, we obtain a formula that is valid for $\operatorname{Re} s > 0$, and we see that the pole at $s = 1$ is simple with residue 1.

REMARK 4.1. The next step is to write (8) as

$$\frac{1}{s-1} + \frac{1}{2} - s\int_1^\infty (\{x\} - 1/2)x^{-s-1}\, dx.$$

Thus we see the first Bernoulli polynomial

$$B_1(x) = \{x\} - 1/2.$$

Integrating by parts gives a formula for $\zeta(s)$ that converges for $\operatorname{Re} s > -1$, and integrating by parts n times gives a formula involving the n-th Bernoulli polynomial that converges for $\operatorname{Re} s > 1 - n$. This gives a way to compute the Riemann zeta function at any complex value. It also gives the values $\zeta(-1), \zeta(-3), \ldots$, and the zeros at the negative even values $s = -2, -4, -6, \ldots$.

This is called Euler's method for summing a series. Also the values at the positive even integers, the values of $\zeta(2), \zeta(4), \ldots$, were known to Euler. With this information, one could make a guess at the functional equation (see Section 4.2).

REMARK 4.2. Comparing (8) with (3), we see that the Riemann zeta function is the geometric zeta function of the fractal string with curvatures $c_n = n$. The counting function of this string is $N(x) = [x]$ and its dimension is $D = 1$. It has no other complex dimensions and $N(x)$ does not have multiplicatively periodic terms. Note that $N(x) = x - \{x\} = x + O(1)$ and $\{x\}$ is additively periodic.

4.1. The Euler product. The Euler product expresses analytically that every natural number has a unique factorization in prime powers. Let

$$\zeta_p(s) = \frac{1}{1 - p^{-s}}$$

be the Euler factor for the prime number p. Then for $\operatorname{Re} s > 1$,

$$\zeta(s) = \prod\nolimits_p \zeta_p(s) = \frac{1}{1 - 2^{-s}} \times \frac{1}{1 - 3^{-s}} \times \frac{1}{1 - 5^{-s}} \times \ldots,$$

a product over all prime numbers. Since none of the factors vanishes for $\operatorname{Re} s > 1$, it follows that $\zeta(s)$ has no zeros in the half-plane $[\operatorname{Re} > 1]$.

By the functional equation (see Section 4.2), it follows that $\zeta(s)$ also does not vanish for $\operatorname{Re} s < 0$, except at $-2, -4, -6, \ldots$, the points where $\zeta_{\mathbb{R}}(s)$ (see (10) below) has poles. Thus the zeros of the completed zeta function $\zeta_{\mathbb{Z}}(s)$ (again, see (10)) are located inside $0 \leq \operatorname{Re} s \leq 1$. It is even known that no zeros can lie on $[\operatorname{Re} = 1]$ or $[\operatorname{Re} = 0]$, and certain "zero free regions" that imply that a zero must lie at least a certain minimal distance from these lines.

REMARK 4.3. Unlike the series for $\zeta(s)$, there is no "summation by parts" method known to extend the Euler product to $\operatorname{Re} s \leq 1$, even though such a transformation of the Euler product would be of great interest. On the other hand, the Euler product for the spectral operator in the next section converges for $0 \leq D \leq 1$.

4.1.1. *The Euler product for the spectral operator.* The spectral operator is meaningful for fractal strings of dimension between 0 and 1. Equivalently, the spectral operator ν acts on functions $N(x)$ of order $O(x^D)$ for $0 \leq D \leq 1$.

For a prime number p, let

$$\nu_p(N)(x) = N(x/p) + N(x/p^2) + N(x/p^3) + \ldots.$$

The spectral operator is the infinite composition of these operators,

$$\nu = \prod\nolimits_p \nu_p.$$

In the right space (Hilbert space or in general, normed space), this composition converges in the operator norm, see Section 5.

The differential operator $\partial = d/dt$ acts like multiplication by h when acting on e^{ht}. In the philosophy of noncommutative geometry, this means that ∂ should be regarded as *every* complex number, and in the right Hilbert space, $\operatorname{spec}(\partial)$ can be an unbounded vertical line. The shift is defined as $S_h f(t) = f(t-h)$. By the Taylor series and the operational calculus,

$$S_h f(t) = e^{-h\partial} f(t).$$

Thus we see that ∂ is the infinitesimal generator of the semigroup of shifts.

Making a change of variables, we write $f(t) = N(e^t)$. Then $N(x/2)$ corresponds to $N(e^t/2) = f(t - \log 2) = 2^{-\partial} f(t)$, and in general, again by the operational calculus,

$$a(f)(t) = f(t) + f(t - \log 2) + f(t - \log 3) + \cdots = \zeta(\partial) f(t). \tag{9}$$

Thus we have the following commutative diagram

$$\begin{array}{ccc} N & \longrightarrow & \nu(N) \\ \big\downarrow (e^t) & & \big\downarrow (e^t) \\ f & \longrightarrow & a(f) \end{array}$$

where $a = \zeta(\partial)$. Introducing

$$a_p(f)(t) = f(t) + f(t - \log p) + f(t - 2\log p) + \cdots = \zeta_p(\partial) f(t),$$

the Euler product for the spectral operator in additive variables is

$$a(f) = \left(\prod\nolimits_p a_p\right)(f).$$

In Section 5, we introduce the Hilbert space of Herichi and Lapidus where this composition converges in the operator norm. By the operational calculus, we can express the Euler product for the spectral operator as

$$\zeta(\partial) = \prod\nolimits_p \zeta_p(\partial).$$

4.1.2. *Universality.* By the work of Laurinčikas [**30**], a nonvanishing analytic function in the disc $[|z| < 1/4]$ without zeros can be approximated arbitrarily closely by translates

$$z \mapsto \zeta(3/4 + it + z)$$

for certain real values of t. And clearly, since these approximations can be made as close as we please, we find an infinite sequence of such t values.

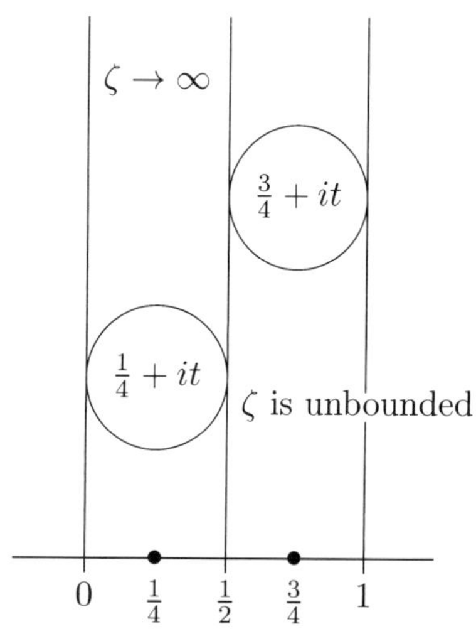

A consequence of universality is that the image $\zeta(c + it)$ is dense in $\mathbb{C}$ for every vertical line $[\mathrm{Re} = c]$, $1/2 < c < 1$.

The function that is approximated must be nonvanishing because its logarithm is approximated by the Euler sum

$$\sum\nolimits_p -\log\left(1 - p^{-s}\right).$$

Only for $c = \operatorname{Re} s > 1/2$ is the sequence p^{-c} square summable, but not summable itself. This allows one to use an argument with projections in Hilbert space, using that the prime numbers are multiplicatively independent. On the other hand, universality does not hold in the left half of the critical strip. (We refer to [**30**] for complete details.)

4.2. The functional equation. The Euler product can be completed with one more factor to obtain a function satisfying the so-called functional equation. This was discovered by Riemann around 1858, but it was already essentially known to Euler (probably around 1740, see Remark 4.1). It is surprising that Euler never formulated it:

$$\zeta_{\mathbb{Z}}(1-s) = \zeta_{\mathbb{Z}}(s),$$

where

$$\zeta_{\mathbb{Z}}(s) = \zeta_{\mathbb{R}}(s)\zeta(s) \quad \text{and} \quad \zeta_{\mathbb{R}}(s) = \pi^{-s/2}\Gamma(s/2). \tag{10}$$

The functional equation, and the Euler product, was eventually fully explained in Tate's thesis [**35**].

4.3. Growth on vertical lines. Clearly, $\zeta(D+it)$ is bounded for every $D > 1$. In other words,

$$\zeta(D+it) = O\big(t^0\big) \quad \text{for every } D > 1.$$

By the functional equation (since by Stirling's formula, the growth of $\zeta_{\mathbb{R}}(s)$ can be accurately estimated on vertical lines),

$$\zeta(D+it) = O\big(t^{1/2-D}\big) \quad \text{for every } D < 0.$$

The Lindelöf hypothesis says that $\zeta(D+it)$ grows like $t^{\mu(D)}$ for the simplest possible μ:

$$\zeta(D+it) = O\big(t^{\mu(D)+\varepsilon}\big) \quad \text{for all } \varepsilon > 0,$$

where $\mu(D)$ is the convex function

$$\mu(D) = \begin{cases} 0 & \text{for } D \geq 1/2, \\ 1/2 - D & \text{for } D \leq 1/2. \end{cases}$$

From general properties of analytic functions, it is known that μ must be a convex function. It follows that $\mu(D) \leq (1-D)/2$ for $0 \leq D \leq 1$. It is also known, for example, that $\mu(1/2) \leq 1/6$. The Lindelöf hypothesis would follow if it were known that $\mu(1/2) = 0$. It also follows from the Riemann hypothesis, to be discussed next.

4.4. The Riemann hypothesis. Whenever one finds a zero of $\zeta_{\mathbb{Z}}$, it turns out that it lies on the line $[\operatorname{Re} = 1/2]$. This experimental fact is called the Riemann hypothesis. If the Euler product could be suitably extended to $[\operatorname{Re} > 1/2]$, the Riemann hypothesis would follow.

See also [**1, 2, 4, 15, 37**] for other ideas for the Riemann hypothesis.

REMARK 4.4 (**Connes' Approach**). In [**1**, **2**], Alain Connes studies the shift on the real line,

$$S_h f(t) = f(t-h) = e^{-h\partial} f(t).$$

The shift is the action of Frobenius on a curve over a finite field. Connes' heuristic argument combines Weil's proof and Selberg's Trace Formula for the spectrum of the Laplacian on a hyperbolic curve. See also [**37**] for Connes' approach for curves over a finite field.

We have seen that in the context of fractal strings, the shift defines the spectral operator on an additive space.

5. The spectral operator

As we have seen in (9), the spectral operator is expressed as $a = \zeta(\partial)$, using the infinitesimal generator ∂ of the additive shift. Herichi and Lapidus choose a Hilbert space to restrict the spectrum of ∂ to be a line. Let $\mathcal{H}_c$ be the Hilbert space of functions on $(0, \infty)$ such that the norm $\|f\|$ is finite, where $\|f\|$ is defined by the relation

$$\|f\|^2 = \int_0^\infty |f(t)|^2 e^{-2ct}\, dt.$$

In this Hilbert space,

$$\operatorname{spec}(\partial) = \{s\colon \operatorname{Re} s = c\},$$

a vertical line in the critical strip.

By universality (which only holds in the right half of the critical strip, see Section 4.1.2 and [**11**]), the spectrum of $\zeta(\partial)$ is dense in $\mathbb{C}$ for $1/2 < c < 1$. It follows that the operator a (and ν) is not invertible: either the Riemann hypothesis is false, and for some $c > 1/2$ the line $[\mathrm{Re} = c]$ crosses a zero of ζ, or else 0 lies in the closure of the image of ζ of every vertical line in the right half $[1/2 < \mathrm{Re} < 1]$, hence the inverse of $\zeta(\partial)$ is unbounded.

It is of interest to consider invertibility in the Hilbert space $\mathcal{H}_c$ for all c, and not just the right half of the critical strip. For $0 < c < 1/2$, the image of a vertical line in the left half of the critical strip is a curve converging to infinity if the Riemann hypothesis holds, in which case ν is invertible. But if the Riemann hypothesis does not hold then the spectral operator would not be invertible on lines that cross a zero of ζ or on which zeros accumulate.

For $c > 1$, the image $\zeta([\mathrm{Re} = c])$ is shaped like a disc around 1 (but it is not known if the closure of the image contains an open set for any c). For large c (certainly for $c \geq 2$), this disc does not come close to 0, and for c approaching 1, this disc becomes unbounded.

Note that $c > 1$ corresponds to $D > 1$, which is not a dimension of a fractal string. Also $c < 0$ is not geometrically meaningful. In that case, $\zeta([\mathrm{Re} = c])$ is a curve going to infinity.

Thus the three interesting cases of the spectral operator are the dimensions $0 < D < 1/2$, dimensions $1/2 < D < 1$, and the 'critical dimension' $D = 1/2$. If the Riemann hypothesis holds, then ν is noninvertible for $D \geq 1/2$, but the reason for this noninvertibility is different for the critical dimension then for $D > 1/2$. The spectral operator would be invertible for $D < 1/2$. Herichi and Lapidus call this a *phase transition* at dimension $1/2$.

REMARK 5.1. Herichi and Lapidus also consider quasi-invertibility and almost invertibility of the spectral operator (see [**6**] for details). In general, invertibility implies quasi-invertibility, which in turn implies almost invertibility. Also, the Riemann hypothesis holds if and only if the spectral operator is quasi-invertible for every $c \neq 1/2$.

6. Zeros in arithmetic progression

One of the early successes of the philosophy of the invertibility of the spectral operator was the study of zeros in arithmetic progressions. We explain here the approach to see how it could apply to the Riemann hypothesis.

In [**23**] (see also [**27–29**]), Lapidus and the author considered generalized Cantor strings. These are defined to have curvatures a^n with multiplicity b^n. If b is not an integer then this is not an actual string, but the argument does not require this. The geometric counting function is

$$N(x) = \sum_{a^n \leq x} b^n.$$

It is important that N is an increasing function: it is piecewise constant and jumps by $b^n = x^D$ at $x = a^n$.

EXAMPLE 6.1. The case $a = 3$, $b = 2$ is the Cantor string. Another interesting case is $b = \sqrt{a}$, which gives all Cantor strings of Minkowski dimension $D = 1/2$. In general,

$$b = a^D$$

for a Cantor string of Minkowski dimension D, and the oscillatory period determines

$$a = e^{2\pi/\mathbf{p}}.$$

The direct computation of the spectral counting function from the geometry is straightforward: $\nu(N)(x) = N(x) + N(x/2) + \ldots$. Since N is increasing and jumps by x^D at $x = a^n$, $\nu(N)(x)$ jumps by at least x^D at $x = a^n$. Thus both N and $\nu(N)$ have oscillations of order x^D.

The explicit formula for the geometric counting function is essentially a Fourier series computation,

$$N(x) = \sum_{n=-\infty}^{\infty} \frac{x^{D+in\mathbf{p}}}{D + in\mathbf{p}} + O(x^0).$$

By the general theory of the spectral operator and explicit formulas,

$$\nu(N)(x) = Lx + \sum_{n=-\infty}^{\infty} \zeta(D + in\mathbf{p}) \frac{x^{D+in\mathbf{p}}}{D + in\mathbf{p}} + O(x^0).$$

Note that the convergence of the spectral counting function becomes worse as D becomes smaller. If the Lindelöf hypothesis holds, then the convergence for $D > 1/2$ is about as good as for N itself. That is, the convergence is conditional, and the infinite sum must be interpreted as $\lim_{M\to\infty} \sum_{n=-M}^{M}$.

If $\zeta(D + in\mathbf{p})$ were to vanish for all $n \neq 0$, then $\nu(N)(x)$ would asymptotically be equal to

$$\nu(N)(x) = Lx - (-\zeta(D)) \frac{x^D}{D} + O(1),$$

and there would be no oscillations and no jumps. This contradiction shows that ζ does not have zeros in arithmetic progression, other than $-2, -4, -6, \ldots$, the poles of $s \cdot \Gamma(s/2)$.

Thus the direct computation of the spectrum shows that there are jumps (of size x^D at least), and the explicit formula then shows that $\zeta(D + in\mathbf{p})$ must be nonzero for at least one value of n. In fact, since only infinitely many oscillations can produce enough resonance to create a discontinuity, we even conclude that $\zeta(D + in\mathbf{p})$ must be nonzero for infinitely many n.

REMARK 6.2. The same reasoning applies to many other zeta functions, namely those for which the corresponding spectral operator leads to an increasing function with large jumps.

REMARK 6.3. In Section 3.3, we have seen that it is possible that $\nu(N)$ does not have oscillations even if N has them. For Cantor strings, on the other hand, the oscillations have frequencies exactly at the multiples of $\mathbf{p}$, and they resonate with each other to such an amount that they create jumps. And jumps cannot disappear in the spectrum.

We thank the referee for pointing out the work of Li and Radziwiłł [**31**], and the earlier work of Martin and Ng [**32**]. In that work, the authors obtain an estimate for the number of zeros of $\zeta(s)$ in a shifted arithmetic progression $1/2 + i(an + b)$, and show that $\zeta(s) \neq 0$ in such a progression at least $1/3$ of the time. It is pointed out in [**32**, §1] that their methods also apply to vertical arithmetic progressions in $[\mathrm{Re} = \sigma]$ for $\sigma \neq 1/2$, but that the result is superseded by the known density estimates for zeros off the critical line. Finally, the results of [**31**] provide an estimate for the length of a finite progression of zeros, as in the next section.

6.1. Finite progressions. For finite progressions of zeros of $\zeta(s)$ the same idea works, but instead of jumps, we look at steep parts of the graph of a 'geometric counting function.' For this reason, x must remain relatively small, and a simple order estimate does not suffice anymore. Instead, an explicit error term, and a good estimate of the error term is needed.

The appropriate "truncated Cantor strings" are defined by the explicit formula for the geometric counting function (see [**36**]),

$$N(x) = \sum_{n=-N}^{N} \left(1 - \frac{|n|}{N}\right) \frac{x^{D+in\mathbf{p}} - 1}{D + in\mathbf{p}}, \qquad \text{for } x \geq 1.$$

This can be computed to yield a direct formula for $N(x)$ that involves the classical Féjér kernel of Fourier analysis. This kernel is positive, thus showing that $N(x)$ is increasing.

The spectrum is given by the explicit formula

$$\nu(N)(x) = L(x-1) + \sum_{n=-N}^{N} \zeta(D + in\mathbf{p}) \left(1 - \frac{|n|}{N}\right) \frac{x^{D+in\mathbf{p}} - 1}{D + in\mathbf{p}} + O(1). \tag{11}$$

The direct computation of the spectrum yields that $\nu(N)$ is increasing. It also yields the $O(1)$-term in (11):

$$O(1) = \int_0^\infty (\{xa^t\} - \{a^t\}) K_N(t) b^{-t}\, dt.$$

In this integral, we see additive and logarithmic oscillations together, from the kernel K_N and from $\{xa^t\} - \{a^t\}$, respectively. Since $\{xy\} - \{y\}$ is zero on average, it is likely that this integral is very small, and can be estimated independently of $\mathbf{p}$. If that could be done, then the length of a progression of zeros could be bounded independently of $\mathbf{p}$ as well. For D close to 1, this estimate would become even better, and this would yield the best possible result that there are no progressions of length 1 close to [Re $= 1$]. Thus one would obtain a good zero free region, of the type $\zeta(D + i\mathbf{p}) \neq 0$ for $D > 1 - \varepsilon$.

REMARK 6.4. See [**36**] for more details. In that paper, we were only able to estimate the error term (the $O(1)$-term above) by a function that increases linearly with $\mathbf{p}$, thus proving that for arithmetic progressions of zeros on the critical line,

$$\zeta(1/2 + in\mathbf{p}) \neq 0 \quad \text{for some } n \text{ with } 0 < |n| < 13\mathbf{p}.$$

For arithmetic progressions of zeros to the right of the critical line (for $1/2 < D < 1$), we were able to show that

$$\zeta(D + in\mathbf{p}) \neq 0 \quad \text{for some } n \text{ with } 0 < |n| < 60\mathbf{p}^{\frac{1}{D}-1} \log \mathbf{p}.$$

Note that this last bound for the length of an arithmetic progression of zeros of $\zeta(s)$ becomes better the closer D is to 1.

7. Resonances

We have seen in Section 6 that very specific resonances, namely those that create jumps or approximations of jumps in the geometric counting function, do not disappear in the spectrum, i.e., "can be heard." This led, in the work of Lapidus and the author, to speculations about the strongest possible result that could be obtained.

7.1. Two zeros on a line. If the Riemann zeta function has two zeros on a vertical line, $\zeta(D+i\mathbf{p}) = 0$ and $\zeta(D+i\mathbf{q}) = 0$, then we can construct a fractal string whose explicit formula has terms that disappear in the spectrum. On the other hand, the geometric oscillations resonate with each other precisely if $\mathbf{p}$ and $\mathbf{q}$ are related by integer multiples. The simplest case is $\mathbf{q} = 2\mathbf{p}$, and if one could deduce by a direct computation that the spectrum will have oscillations of period either $\mathbf{p}$ or $2\mathbf{p}$, one would have shown (even on the line $D = 1/2$) that if $\zeta(D + i\mathbf{p}) = 0$ then $\zeta(D + 2i\mathbf{p}) \neq 0$.

More generally, the oscillatory periods resonate with each other if $m\mathbf{q} = n\mathbf{p}$ for some integers m and n, and thus fractal strings provide a strategy to show, for example, that if $\zeta(1/2 + i\alpha) = 0$ and $\zeta(1/2 + i\beta) = 0$, then α/β is irrational.

Thus, fractal strings provide a heuristic reason that the zeros of $\zeta(s)$ with $\operatorname{Re} s > 0$ should have rationally independent imaginary parts. This heuristic reason depends on the more general heuristic that geometric resonances remain audible in the spectrum.

REMARK 7.1. The same ideas apply to zeros of more general Dirichlet series, see [**29**, Section 11.2].

The harder problem of zeros on a horizontal line was already formulated by Lapidus around 2000: If one could show, by a direct argument, that oscillatory terms of the form $x^{D_1+i\mathbf{p}}$ and $x^{D_2+i\mathbf{p}}$ cannot both disappear from the spectrum, then one could prove the Riemann hypothesis. In this case one is helped by the fact that the corresponding frequencies are the same, but the corresponding amplitudes (namely, x^{D_1} and x^{D_2}, where $D_1 \neq D_2$) only reinforce each other by an infinitesimal amount as $x \to \infty$.

Even though resonances play an important role in the elaboration of these ideas, there is no good heuristic reason that resonances cannot disappear. The following conjecture is formulated with an aim to discover the heuristic reasons for its truth.

CONJECTURE 7.2. *Geometric resonances remain audible in the spectrum as oscillations.*

A consequence would be that only oscillations can disappear, but only in Minkowski dimension $D = 1/2$.

A closer investigation of the theory of the spectral operator that has been developed by Herichi, Lapidus, and the author, reveals that the present theory of the spectral operator uses the Euler product of the Riemann zeta function in a way that is convergent for $0 \leq D \leq 1$. On the other hand, the functional equation of $\zeta_{\mathbb{Z}}(s)$ does not play a role, and closely related, there is no counterpart of the archimedean factor $\zeta_{\mathbb{R}}(s)$ in the definition of the spectral operator.

With a view to explaining the Riemann hypothesis, Lapidus and the author have developed a theory of dual strings, and Lapidus, Lu, and the author have developed a theory of p-adic fractal strings [**16**, **17**] with a view to constructing an adelic completion. Such a theory would provide a way to incorporate all important properties of the Riemann zeta function (its Euler product, the functional equation, and the gamma factor) in this approach to the Riemann hypothesis.

7.2. Two research questions. As in the work of Herichi and Lapidus, we propose the following research question:

Explain the phase transition at $D = 1/2$.

In [**6**–**8**], the difference in behavior of the spectral operator for $D > 1/2$ and $D < 1/2$ is called a phase transition. What is the nature of this phase transition?

Since this question may be close to asking for a proof of the Riemann hypothesis, we conclude with a more modest research question:

Is there a theory of dual strings that reflects the functional equation?

Conjecturally in such a theory, self-dual strings have dimension $D \geq 1/2$, and if $D > 1/2$, the geometric oscillations in dimension D resonate with the ones in dimension $1 - D$.

References

[1] Alain Connes, *Trace formula in noncommutative geometry and the zeros of the Riemann zeta function*, Selecta Math. (N.S.) **5** (1999), no. 1, 29–106, DOI 10.1007/s000290050042. MR1694895

[2] Alain Connes, *Noncommutative geometry and the Riemann zeta function*, Mathematics: frontiers and perspectives, Amer. Math. Soc., Providence, RI, 2000, pp. 35–54. MR1754766

[3] H. M. Edwards, *Riemann's zeta function*, Dover Publications, Inc., Mineola, NY, 2001. Reprint of the 1974 original [Academic Press, New York; MR0466039 (57 #5922)]. MR1854455

[4] Haran, S., *On Riemann's zeta function*, in: Dynamical, Spectral, and Arithmetic Zeta Functions (M. L. Lapidus, M. van Frankenhuijsen, eds.), Contemporary Mathematics **290**, Amer. Math. Soc., Providence, RI, 2001, 93–112.

[5] Hafedh Herichi and Michel L. Lapidus, *Fractal complex dimensions, Riemann hypothesis and invertibility of the spectral operator*, Fractal geometry and dynamical systems in pure and applied mathematics. I. Fractals in pure mathematics, Contemp. Math., vol. 600, Amer. Math. Soc., Providence, RI, 2013, pp. 51–89, DOI 10.1090/conm/600/11948. MR3203399

[6] Hafedh Herichi and Michel L. Lapidus, *Riemann zeros and phase transitions via the spectral operator on fractal strings*, J. Phys. A **45** (2012), no. 37, 374005, 23, DOI 10.1088/1751-8113/45/37/374005. MR2970522

[7] Herichi, H., Lapidus, M., *Fractal Strings, the Spectral Operator and the Riemann Hypothesis: Zeta Values, Riemann Zeros, Phase Transitions and Quantized Universality*, Research Memoir.

[8] Herichi, H., Lapidus, M., *Spectral Operator and Convergence of its Euler Product in the Critical Strip*, preprint.

[9] A. E. Ingham, *The distribution of prime numbers*, Cambridge Mathematical Library, Cambridge University Press, Cambridge, 1990. Reprint of the 1932 original; With a foreword by R. C. Vaughan. MR1074573

[10] Mark Kac, *Can one hear the shape of a drum?*, Amer. Math. Monthly **73** (1966), no. 4, 1–23, DOI 10.2307/2313748. MR0201237

[11] Justas Kalpokas and Jörn Steuding, *On the value-distribution of the Riemann zeta-function on the critical line*, Mosc. J. Comb. Number Theory **1** (2011), no. 1, 26–42. MR2948324

[12] Michel L. Lapidus, *Fractal drum, inverse spectral problems for elliptic operators and a partial resolution of the Weyl-Berry conjecture*, Trans. Amer. Math. Soc. **325** (1991), no. 2, 465–529, DOI 10.2307/2001638. MR994168

[13] M. L. Lapidus, *Vibrations of fractal drums, the Riemann hypothesis, waves in fractal media and the Weyl-Berry conjecture*, Ordinary and partial differential equations, Vol. IV (Dundee, 1992), Pitman Res. Notes Math. Ser., vol. 289, Longman Sci. Tech., Harlow, 1993, pp. 126–209. MR1234502

[14] Michel L. Lapidus, *Fractals and vibrations: can you hear the shape of a fractal drum?*, Fractals **3** (1995), no. 4, 725–736, DOI 10.1142/S0218348X95000643. Symposium in Honor of Benoit Mandelbrot (Curaçao, 1995). MR1410291

[15] Michel L. Lapidus, *In search of the Riemann zeros: Strings, fractal membranes and noncommutative spacetimes*, American Mathematical Society, Providence, RI, 2008. MR2375028

[16] Michel L. Lapidus, Lũ' Hùng, and Machiel van Frankenhuijsen, *Minkowski measurability and exact fractal tube formulas for p-adic self-similar strings*, Fractal geometry and dynamical systems in pure and applied mathematics. I. Fractals in pure mathematics, Contemp. Math., vol. 600, Amer. Math. Soc., Providence, RI, 2013, pp. 161–184, DOI 10.1090/conm/600/11949. MR3203402

[17] Lapidus, M., Lu, H., van Frankenhuijsen, M., *Minkowski Dimension and Explicit Tube Formulas for p-Adic Fractal Strings.*

[18] Michel L. Lapidus and Helmut Maier, *Hypothèse de Riemann, cordes fractales vibrantes et conjecture de Weyl-Berry modifiée* (French, with English summary), C. R. Acad. Sci. Paris Sér. I Math. **313** (1991), no. 1, 19–24. MR1115940

[19] Michel L. Lapidus and Helmut Maier, *The Riemann hypothesis and inverse spectral problems for fractal strings*, J. London Math. Soc. (2) **52** (1995), no. 1, 15–34, DOI 10.1112/jlms/52.1.15. MR1345711

[20] Michel L. Lapidus and Carl Pomerance, *Fonction zêta de Riemann et conjecture de Weyl-Berry pour les tambours fractals* (French, with English summary), C. R. Acad. Sci. Paris Sér. I Math. **310** (1990), no. 6, 343–348. MR1046509

[21] Michel L. Lapidus and Carl Pomerance, *The Riemann zeta-function and the one-dimensional Weyl-Berry conjecture for fractal drums*, Proc. London Math. Soc. (3) **66** (1993), no. 1, 41–69, DOI 10.1112/plms/s3-66.1.41. MR1189091

[22] Lapidus, M., van Frankenhuijsen, M., *Complex dimensions of fractal strings and explicit formulas for geometric and spectral zeta functions*, preprint IHÉS/M/97/34.

[23] Lapidus, M., van Frankenhuijsen, M., *Complex dimensions and oscillatory phenomena, with applications to the geometry of fractal strings and to the critical zeros of zeta functions*, preprint IHÉS/M/97/38.

[24] Michel L. Lapidus and Machiel van Frankenhuysen, *Complex dimensions of fractal strings and oscillatory phenomena in fractal geometry and arithmetic*, Spectral problems in geometry and

arithmetic (Iowa City, IA, 1997), Contemp. Math., vol. 237, Amer. Math. Soc., Providence, RI, 1999, pp. 87–105, DOI 10.1090/conm/237/1710790. MR1710790

[25] Michel L. Lapidus and Machiel van Frankenhuysen, *Complex dimensions of fractal strings and oscillatory phenomena in fractal geometry and arithmetic*, Spectral problems in geometry and arithmetic (Iowa City, IA, 1997), Contemp. Math., vol. 237, Amer. Math. Soc., Providence, RI, 1999, pp. 87–105, DOI 10.1090/conm/237/1710790. MR1710790

[26] Michel L. Lapidus and Machiel van Frankenhuijsen, *Fractality, self-similarity and complex dimensions*, Fractal geometry and applications: a jubilee of Benoît Mandelbrot. Part 1, Proc. Sympos. Pure Math., vol. 72, Amer. Math. Soc., Providence, RI, 2004, pp. 349–372. MR2112111

[27] Michel L. Lapidus and Machiel van Frankenhuysen, *Fractal geometry and number theory*, Birkhäuser Boston, Inc., Boston, MA, 2000. Complex dimensions of fractal strings and zeros of zeta functions. MR1726744

[28] Michel L. Lapidus and Machiel van Frankenhuijsen, *Fractal geometry, complex dimensions and zeta functions*, Springer Monographs in Mathematics, Springer, New York, 2006. Geometry and spectra of fractal strings. MR2245559

[29] Michel L. Lapidus and Machiel van Frankenhuijsen, *Fractal geometry, complex dimensions and zeta functions*, 2nd ed., Springer Monographs in Mathematics, Springer, New York, 2013. Geometry and spectra of fractal strings. MR2977849

[30] Antanas Laurinčikas, *Limit theorems for the Riemann zeta-function*, Mathematics and its Applications, vol. 352, Kluwer Academic Publishers Group, Dordrecht, 1996. MR1376140

[31] Xiannan Li and Maksym Radziwiłł, *The Riemann zeta function on vertical arithmetic progressions*, Int. Math. Res. Not. IMRN **2** (2015), 325–354, DOI 10.1093/imrn/rnt197. MR3340323

[32] Greg Martin and Nathan Ng, *Nonzero values of Dirichlet L-functions in vertical arithmetic progressions*, Int. J. Number Theory **9** (2013), no. 4, 813–843, DOI 10.1142/S1793042113500140. MR3060861

[33] Riemann, B., *Ueber die Anzahl der Primzahlen unter einer gegebenen Grösse*, in [**34**, p. 145] and translated in [**3**, p. 299].

[34] Riemann, B., *Gesammelte Werke*, Teubner, Leipzig, 1892 (reprinted by Dover Books, New York, 1953).

[35] J. T. Tate, *Fourier analysis in number fields, and Hecke's zeta-functions*, Algebraic Number Theory (Proc. Instructional Conf., Brighton, 1965), Thompson, Washington, D.C., 1967, pp. 305–347. MR0217026

[36] Machiel van Frankenhuijsen, *Arithmetic progressions of zeros of the Riemann zeta function*, J. Number Theory **115** (2005), no. 2, 360–370, DOI 10.1016/j.jnt.2005.01.002. MR2180508

[37] Machiel van Frankenhuijsen, *The Riemann hypothesis for function fields*, London Mathematical Society Student Texts, vol. 80, Cambridge University Press, Cambridge, 2014. Frobenius flow and shift operators. MR3468729

Email address: vanframa@uvu.edu

Department of Mathematics, Utah Valley University, 800 West University Parkway, Orem, Utah, 84058-5999

Contemporary Mathematics
Volume **731**, 2019
https://doi.org/10.1090/conm/731/14676

Measure-geometric Laplacians for discrete distributions

M. Kesseböhmer, T. Samuel, and H. Weyer

ABSTRACT. In 2002 Freiberg and Zähle introduced and developed a harmonic calculus for measure-geometric Laplacians associated to continuous distributions. We show their theory can be extended to encompass distributions with finite support and give a matrix representation for the resulting operators. In the case of a uniform discrete distribution we make use of this matrix representation to explicitly determine the eigenvalues and the eigenfunctions of the associated Laplacian.

1. Introduction

Motivated by the fundamental theorem of calculus, and based on the works of Feller [**Fel57**] and Kac and Kreĭn [**KKn58**], given an atomless Borel probability measure μ supported on a compact subset of $\mathbb{R}$, Freiberg and Zähle [**FZ02**] introduced a measure-geometric approach to define a first order differential operator ∇^{μ} and a second order differential operator $\Delta^{\mu,\mu} := \nabla^{\mu} \circ \nabla^{\mu}$, with respect to μ. In the case that μ is the Lebesgue measure, it was shown that ∇^{μ} coincides with the weak derivative. Moreover, a harmonic calculus for $\Delta^{\mu,\mu}$ was developed and, when μ is a self-similar measure supported on a Cantor set, the authors proved the eigenvalue counting function of $\Delta^{\mu,\mu}$ is comparable to the square-root function. In [**KSW16**] for continuous measures the exact eigenvalues and eigenfunctions were obtained and it was shown the eigenvalues do not depend on the given measure. Arzt [**Arz15**] has also considered the Kreĭn-Feller operator $\Delta^{\mu,\Lambda} := \nabla^{\mu} \circ \nabla^{\Lambda}$, where μ denotes a continuous Borel probability measure and Λ denotes the Lebesgue measure, see [**Fre05, Fuj87**] for further results in this direction.

Here, we show this framework can be extended to included purely atomic measures μ. Unlike in the case when one has a measure with a continuous distribution function (see for instance [**FZ02, KSW16**]), we prove the operators ∇^{μ} and $\Delta^{\mu,\mu}$ are no longer symmetric. To circumvent this problem, we consider the operator ∇^{μ}, its adjoint $(\nabla^{\mu})^{*}$ and define the μ-Laplacian to be $\Delta^{\mu} = -(\nabla^{\mu})^{*} \circ \nabla^{\mu}$. We give matrix representations for these operators, noting they coincide with the normalised Laplacian matrix of a cycle graph [**Big93**] and resemble a discretisation of a one-dimensional Laplacian on a non-uniform grid [**LeV07**]. Further, we discuss properties of the eigenvalues and eigenfunctions of the operator Δ^{μ}. In particular,

2010 *Mathematics Subject Classification.* 35P20; 42B35; 47G30.
Key words and phrases. measure-geometric Laplacians; spectral asymptotics.

we show the eigenfunctions for distributions with finite support are not necessarily of the form $f^{\mu}_{\kappa}(\cdot) := \sin(\pi\kappa F_{\mu}(\cdot))$ or $g^{\mu}_{\kappa}(\cdot) := \cos(\pi\kappa F_{\mu}(\cdot))$, for $\kappa \in \mathbb{R}\backslash\{0\}$ and where F_{μ} denotes the distribution function of μ. This differs from the case of continuous distributions, see [**FZ02, KSW16, Zäh05**]. Additionally, in the case that μ is a uniform discrete probability distribution we explicitly determine the eigenvalues and eigenfunctions of Δ^{μ}.

Outline. In Section 2 we present necessary definitions and basic properties of ∇^{μ}, $(\nabla^{\mu})^{*}$ and Δ^{μ} and give matrix representations for these operators. In Section 3 we prove general results concerning the spectral properties of Δ^{μ}. We conclude with Section 4, where explicit computations are carried out when μ is a uniform discrete probability distribution.

2. Definitions and analytic properties of Δ^{μ}

Set $\mathcal{I} := [0,1]$ and let δ_z denote the Dirac-measure at z, for some fixed $z \in \mathcal{I}$. Let μ denote the probability measure $\mu := \sum_{i=1}^{N} \alpha_i \delta_{z_i}$, where $N \in \mathbb{N}$, $0 \leq z_1 < z_2 < \cdots < z_N < 1$ and $\alpha_i > 0$, for $i \in \{1, \dots, N\}$. We denote the set of real-valued square-integrable functions on $\mathcal{I}$ by $\mathfrak{L}^2_{\mu} = \mathfrak{L}^2_{\mu}(\mathcal{I})$, we define $\mathcal{N}_{\mu}(\mathcal{I})$ to be set of $\mathfrak{L}^2_{\mu}$-functions which are constant zero μ-almost everywhere, and we let $L^2_{\mu} = L^2_{\mu}(\mathcal{I}) := \mathfrak{L}^2_{\mu}(\mathcal{I}) \setminus \mathcal{N}_{\mu}(\mathcal{I})$. The latter space is a finite-dimensional inner product space with inner product $\langle \cdot, \cdot \rangle$ given by

$$\langle f, g \rangle = \langle f, g \rangle_{\mu} := \sum_{i=1}^{N} \alpha_i f(z_i) g(z_i).$$

We define the set of *μ-differentiable functions on $\mathcal{I}$ with periodic boundary conditions* by

(1)

$$\mathscr{D}^1_{\mu} = \mathscr{D}^1_{\mu}(\mathcal{I}) := \Big\{ f \in \mathfrak{L}^2(\mu) \colon \text{there exists } f' \in L^2_{\mu} \text{ such that } f(0) = f(1) \text{ and } f(x) = f(0) + \int \mathbb{1}_{[0,x)} f' \, \mathrm{d}\mu \text{ for all } x \in \mathcal{I} \Big\},$$

where we understand $[0,0) = \emptyset$. Note, the function f' defined in (1) is unique in L^2_{μ}. Since $f(0) = f(1) = f(0) + \int \mathbb{1}_{[0,1)} f' \, \mathrm{d}\mu$, it follows that

$$\int \mathbb{1}_{[0,1)} f' \, \mathrm{d}\mu = 0. \tag{2}$$

For $f \in \mathscr{D}^1_{\mu}$ and f' as in (1), the operator $\nabla^{\mu} \colon \mathscr{D}^1_{\mu} \to L^2_{\mu}$ defined by $\nabla^{\mu} f := f'$ is called the *μ-derivative*. Linearity of the integral yields ∇^{μ} is linear on $\mathscr{D}^1_{\mu}$. As μ is a linear combination of Dirac measures, we can reformulate the defining equation of $\nabla^{\mu} f$ given in (1) by

$$f(x) = f(0) + \sum_{\substack{i \in \{1,\dots,N\} \\ z_i < x}} \alpha_i \nabla^{\mu} f(z_i), \tag{3}$$

where $f \in \mathscr{D}^1_{\mu}$ and $x \in \mathcal{I}$. Thus, if μ is a Dirac measure, that is $N = 1$, then (2) becomes $\alpha_1 \nabla^{\mu} f(z_1) = 0$. Hence, from (3), it follows that $\mathscr{D}^1_{\mu}$ is the set of constant functions. In other words, the operator ∇^{μ} is the null-operator, and so, from here on we assume $N \geq 2$.

The periodic boundary conditions and (3) together imply that a function $f \in \mathscr{D}_\mu^1$ is piecewise constant; namely, $f|_{[0,z_1]\cup(z_N,1]}$ and $f|_{(z_i,z_{i+1}]}$ are constant, for $i \in \{1,\dots,N-1\}$. Therefore, f is uniquely determined by the vector $(f(z_1),\dots,f(z_N))^\top$, and thus, there exists an $N \times N$-matrix A with

$$A(f(z_1),\dots,f(z_N))^\top = (\nabla^\mu f(z_1),\dots,\nabla^\mu f(z_N))^\top.$$

From (3) and the fact that $f(1) = f(0) = f(z_1)$, we have

$$\nabla^\mu f(z_N) = \frac{f(z_1) - f(z_N)}{\alpha_N} \quad \text{and} \quad \nabla^\mu f(z_n) = \frac{f(z_{n+1}) - f(z_n)}{\alpha_n},$$

for $n \in \{1,\dots,N-1\}$, and hence,

$$A = \begin{pmatrix} -\alpha_1^{-1} & \alpha_1^{-1} & 0 & \cdots & 0 & 0 & 0 \\ 0 & -\alpha_2^{-1} & \alpha_2^{-1} & \cdots & 0 & 0 & 0 \\ 0 & 0 & -\alpha_3^{-1} & \cdots & 0 & 0 & 0 \\ \vdots & \vdots & \vdots & \ddots & \vdots & \vdots & \vdots \\ 0 & 0 & 0 & \cdots & -\alpha_{N-2}^{-1} & \alpha_{N-2}^{-1} & 0 \\ 0 & 0 & 0 & \cdots & 0 & -\alpha_{N-1}^{-1} & \alpha_{N-1}^{-1} \\ \alpha_N^{-1} & 0 & 0 & \cdots & 0 & 0 & -\alpha_N^{-1} \end{pmatrix}.$$

Since $\sum_{i=1}^N \alpha_i g^2(z_i) < \infty$, for all $g \in \mathscr{D}_\mu^1$, there exists a natural embedding $\pi\colon \mathscr{D}_\mu^1 \to L_\mu^2$. In fact, from the matrix representation given above it follows that $\pi(\mathscr{D}_\mu^1) = L_\mu^2$. In other words, every equivalence class of L_μ^2 has a μ-differentiable representative, and so, from here on we will not distinguish between $\mathscr{D}_\mu^1$ and $\pi(\mathscr{D}_\mu^1)$.

Notice, A is not self-adjoint, and A^2 is self-adjoint if and only if $N = 2$ and $\alpha_1 = \alpha_2$. Hence, the operators ∇^μ and $\Delta^\mu := \nabla^\mu \circ \nabla^\mu$ are not in general self-adjoint. To obtain a self-adjoint operator we follow the program of Kigami [**Kig93**, **Kig01**, **Kig03**], and Kigami and Lapidus [**KL01**], and use the bilinear form $\mathcal{E}$ defined by

$$\mathcal{E}(f,g) = \mathcal{E}^\mu(f,g) := \langle \nabla^\mu f, \nabla^\mu g \rangle,$$

for $f,g \in \mathscr{D}_\mu^1$. We refer to $\mathcal{E}$ as the *μ-energy form.*

THEOREM 2.1. *The μ-energy form $\mathcal{E}$ is a Dirichlet form.*

PROOF. The μ-energy form is bilinear since the inner product is bilinear, ∇^μ is linear and every equivalence class of L_μ^2 has a μ-differentiable representative. The symmetry and the non-negativity of $\mathcal{E}$ follow from the properties of the inner product. For every $f \in \mathscr{D}_\mu^1$, the function $\hat{f}\colon \mathcal{I} \to \mathbb{R}$, defined by $\hat{f}(x) := \min(\max(f(x),0),1)$, belongs to $\mathscr{D}_\mu^1$, and as

$$|\hat{f}(z_{i+1}) - \hat{f}(z_i)| \le |f(z_{i+1}) - f(z_i)| \quad \text{and} \quad |\hat{f}(z_1) - \hat{f}(z_N)| \le |f(z_1) - f(z_N)|,$$

it follows that $\mathcal{E}(\hat{f},\hat{f}) \le \mathcal{E}(f,f)$. The properties of $\langle\cdot,\cdot\rangle$ and $\mathcal{E}$ yield $\mathscr{D}_\mu^1$ equipped with $\langle\cdot,\cdot\rangle_{\mathcal{E}} := \langle\cdot,\cdot\rangle + \mathcal{E}(\cdot,\cdot)$ is an inner product space. Now, for every Cauchy sequence $(f_n)_{n\in\mathbb{N}}$ in $(\mathscr{D}_\mu^1, \langle\cdot,\cdot\rangle_{\mathcal{E}})$, we have both $(f_n)_{n\in\mathbb{N}}$ and $(\nabla^\mu f_n)_{n\in\mathbb{N}}$ are Cauchy-sequences in L_μ^2. Hence, there exist $\tilde{f}_0, \tilde{f}_1 \in L_\mu^2$ with $\lim_{n\to\infty}\|f_n - \tilde{f}_0\| = 0$ and $\lim_{n\to\infty}\|\nabla^\mu f_n - \tilde{f}_1\| = 0$, where $\|f\|^2 := \langle f,f\rangle$, for $f \in L_\mu^2$. Since ∇^μ is linear, it is continuous, and so $\nabla^\mu \tilde{f}_0 = \tilde{f}_1$. This implies that $\lim_{n\to\infty} f_n = \tilde{f}_0 \in \mathscr{D}_\mu^1$, with respect to $\langle\cdot,\cdot\rangle_{\mathcal{E}}$. □

Notice, since the analysis reduces to a finite dimensional vector space, the μ-energy form is also a graph energy form as treated in [**KL12**].

We say that $f \in \mathscr{D}_\mu^1$ belongs to $\mathscr{D}_\mu^2 = \mathscr{D}_\mu^2(\mathcal{I})$, if there exists a $h \in L_\mu^2$, necessarily unique, such that $\mathcal{E}(f,g) = -\langle h, g\rangle$, for all $g \in \mathscr{D}_\mu^1$. We define the *μ-Laplacian* to be the operator $\Delta^\mu \colon \mathscr{D}_\mu^2 \to L_\mu^2$ given by $\Delta^\mu f := h$. Indeed, for an arbitrary $g \in \mathscr{D}_\mu^1$, we observe

$$(4) \qquad \langle \nabla^\mu f, \nabla^\mu g\rangle = -\langle \Delta^\mu f, g\rangle, \quad \text{and thus,} \quad \Delta^\mu = -\left(\nabla^\mu\right)^* \circ \nabla^\mu.$$

With this, we conclude $B := -A^\top A$ is a matrix representation of Δ^μ; in fact, for $N = 2$,

$$(5) \qquad B = \begin{pmatrix} -\alpha_1^{-2} - \alpha_2^{-2} & \alpha_1^{-2} + \alpha_2^{-2} \\ \alpha_1^{-2} + \alpha_2^{-2} & -\alpha_1^{-2} - \alpha_2^{-2} \end{pmatrix},$$

and, for $N \geq 3$, we have B is the $N \times N$-matrix

$$\begin{pmatrix} -\alpha_N^{-2} - \alpha_1^{-2} & \alpha_1^{-2} & 0 & \cdots & 0 & 0 & \alpha_N^{-2} \\ \alpha_1^{-2} & -\alpha_1^{-2} - \alpha_2^{-2} & \alpha_2^{-2} & \cdots & 0 & 0 & 0 \\ 0 & \alpha_2^{-2} & -\alpha_2^{-2} - \alpha_3^{-2} & \cdots & 0 & 0 & 0 \\ \vdots & \vdots & \vdots & \ddots & \vdots & \vdots & \vdots \\ 0 & 0 & 0 & \cdots & -\alpha_{N-3}^{-2} - \alpha_{N-2}^{-2} & \alpha_{N-2}^{-2} & 0 \\ 0 & 0 & 0 & \cdots & \alpha_{N-2}^{-2} & -\alpha_{N-2}^{-2} - \alpha_{N-1}^{-2} & \alpha_{N-1}^{-2} \\ \alpha_N^{-2} & 0 & 0 & \cdots & 0 & \alpha_{N-1}^{-2} & -\alpha_{N-1}^{-2} - \alpha_N^{-2} \end{pmatrix}.$$

THEOREM 2.2. *The operator Δ^μ is linear, self-adjoint and non-positive.*

PROOF. Linearity follows from linearity of ∇^μ and bilinearity of $\mathcal{E}$. Self-adjointness is a consequence of symmetry of $\mathcal{E}$. An application of (4) yields $\langle \Delta^\mu f, f\rangle = -\langle \nabla^\mu f, \nabla^\mu f\rangle \leq 0$, and hence, Δ^μ is non-positive. □

The above demonstrates one can view the operator Δ^μ as a Laplacian matrix of a weighted cycle graph, see [**Big93, Chu97**]. Indeed, the analytic results from spectral graph theory may be carried over to our setting, which would certainly be interesting to investigate and could lead to further results concerning Δ^μ. Moreover, the matrix B resembles the matrices appearing in the finite difference methods used in numerical analysis of ordinary differential equations[1], see [**LeV07**]. We also observe the matrix representation of the operator Δ^μ only depends on the order and the weighting of the atoms of μ and is independent on the distances between z_i and z_j, for all $i, j \in \{1, \dots, N\}$.

3. Spectral properties of Δ^μ

By definition of the matrix B, to find the eigenvalues and eigenfunctions of Δ^μ, it suffices to compute the eigenvalues and eigenvectors of B.

PROPOSITION 3.1. *If λ is an eigenvalue of Δ^μ, then $\lambda \in \mathbb{R}$ and $2 \min_{i \in \{1,\dots,N\}} B_{i,i} \leq \lambda \leq 0$.*

PROOF. Theorem 2.2 together with the fact that B is self-adjoint with entries in $\mathbb{R}$ yields all eigenvalues are non-positive real numbers. Since the spectral norm

[1]We would like to thank Paul Choboter, Maik Gröger, Jens Rademacher and Alfred Schmidt for bringing this to our attention.

is bounded above by the column-sum norm, given an eigenvalue λ of B, it follows that

$$|\lambda| \leq 2\max\{|B_{i,i}| \colon i \in \{1, \dots, N\}\}. \qquad \square$$

PROPOSITION 3.2. *The operator Δ^{μ} has a simple eigenvalue at $\lambda = 0$ where the corresponding eigenfunction is the constant function with value 1.*

PROOF. A direct calculation reveals $B(1, \dots, 1)^{\top} = (0, \dots, 0)^{\top}$, and hence, we have $\lambda = 0$ is an eigenvalue of Δ^{μ} where the constant function with value 1 is the corresponding eigenfunction. A row reduced echelon form of A is an upper triangular matrix with a single zero on the diagonal, and so $\operatorname{rank}(A) = N - 1$. Combining this with the fact that $\operatorname{rank}(B) = \operatorname{rank}(A^{\top}A) = \operatorname{rank}(A)$ it follows that the eigenvalue $\lambda = 0$ of B is simple. $\square$

We observe, for $N = 2$, that the eigenvalues of Δ^{μ} are $\lambda_0 = 0$ and $\lambda_1 = -2(\alpha_1^{-2} + \alpha_2^{-2})$ with corresponding eigenfunctions,

$$f_0(x) = 1, \text{ for all } x \in [0,1], \quad \text{and} \quad f_1(x) = \begin{cases} 1 & \text{for } x \in [0, z_1] \cup (z_2, 1], \\ -1 & \text{otherwise.} \end{cases}$$

This follows since λ_0 and λ_1 are eigenvalues of the matrix B, given in (5), with corresponding eigenvectors $v^{(0)} = (1,1)^{\top}$ and $v^{(1)} = (1,-1)^{\top}$, respectively. In particular, we see that, in this case, the lower bound in Theorem 3.1 is sharp.

Different to the case of continuous distributions, the eigenfunctions for distributions with finite support are not necessarily of the form

$$f_{\kappa}^{\mu}(x) = \sin(\pi\kappa F_{\mu}(x)) \quad \text{or} \quad g_{\kappa}^{\mu}(x) = \cos(\pi\kappa F_{\mu}(x)),$$

for $x \in [0,1]$ and $\kappa \in \mathbb{R} \setminus \{0\}$, where F_{μ} denotes the distribution function of μ, which we now address in the following paragraph.

Let m_1 and m_2 denote two positive real numbers with $3m_1 + 3m_2 = 1$, and set $r = m_2/m_1$. Consider the discrete distribution $\mu = \sum_{i=1}^{6} \alpha_i \delta_{z_i}$, where $0 < z_1 < z_2 < \cdots < z_6 < 1$, $\alpha_1 = \alpha_3 = \alpha_5 = m_1$ and $\alpha_2 = \alpha_4 = \alpha_6 = m_2$. A direct calculation shows the eigenvalues for the matrix representation of Δ^{μ} are

$$\lambda_0 = 0, \qquad \lambda_1 = \lambda_5 = -(m_1^{-2} + m_2^{-2}) + \sqrt{m_1^{-4} + m_2^{-4} - m_1^{-2}m_2^{-2}},$$

$$\lambda_3 = -2(m_1^{-2} + m_2^{-2}), \quad \lambda_2 = \lambda_4 = -(m_1^{-2} + m_2^{-2}) - \sqrt{m_1^{-4} + m_2^{-4} - m_1^{-2}m_2^{-2}},$$

with corresponding eigenvectors

$$\begin{aligned}
v^{(0)} &= (1,1,1,1,1,1)^{\top}, \\
v^{(1)} &= \left(r^2, \sqrt{1+r^4-r^2}, 1-r^2, -\sqrt{1+r^4-r^2}, -1, 0\right)^{\top}, \\
v^{(2)} &= \left(r^2, -\sqrt{1+r^4-r^2}, 1-r^2, \sqrt{1+r^4-r^2}, -1, 0\right)^{\top}, \\
v^{(3)} &= (1,-1,1,-1,1,-1)^{\top}, \\
v^{(4)} &= \left(\sqrt{1+r^4-r^2}, 1-r^2, -\sqrt{1+r^4-r^2}, r^2, 0, -1\right)^{\top}, \\
v^{(5)} &= \left(\sqrt{1+r^4-r^2}, r^2-1, -\sqrt{1+r^4-r^2}, -r^2, 0, 1\right)^{\top}.
\end{aligned}$$

For the eigenvalues with multiplicity two, namely $\lambda_1 = \lambda_5$ and $\lambda_2 = \lambda_4$, notice that the sets of tuples

$$S_{1,5} := \left\{ \left(v_1^{(1)}, v_1^{(5)}\right), \ldots, \left(v_6^{(1)}, v_6^{(5)}\right) \right\} \text{ and } S_{2,4} := \left\{ \left(v_1^{(2)}, v_1^{(4)}\right), \ldots, \left(v_6^{(2)}, v_6^{(4)}\right) \right\}$$

determine the same ellipse, namely,

$$\sqrt{1 + r^{-4} - r^{-2}}\,\left(x^2 + y^2 - 1\right) = \left(2 - r^{-2}\right) xy,$$

which is non-axisymmetric; see Figure 2 for an example and compare with Theorem 4.1 and Corollary 4.2, where the analogous set of tuples lie on the unit circle. This latter property demonstrates that the eigenspace for $\lambda_1 = \lambda_5$ (or $\lambda_2 = \lambda_4$) is not spanned by $\{f^{\mu}_{\kappa_1}, g^{\mu}_{\kappa_2}\}$ for any $\kappa_1, \kappa_2 \in \mathbb{R}$. Moreover, in this explicit case, setting

$$w^{(\kappa)} := (\sin(\pi\kappa F_\mu(z_1)), \ldots, \sin(\pi\kappa F_\mu(z_6)))^\top,$$
$$u^{(\kappa)} := (\cos(\pi\kappa F_\mu(z_1)), \ldots, \cos(\pi\kappa F_\mu(z_6)))^\top,$$

a direct calculation shows $Bw^{(\kappa)} \neq \lambda_i w^{(\kappa)}$ and $Bu^{(\kappa)} \neq \lambda_i u^{(\kappa)}$, for all $\kappa \in \mathbb{R} \setminus \{0\}$ and $i \in \{0, \ldots, 5\}$. This latter result also holds when replacing $F_\mu(z_i)$ by $\widetilde{F}_\mu(z_i) := F_\mu(z_i - \varepsilon)$, for a fixed $\varepsilon \in (0, \min\{z_1, \min\{z_{i+1} - z_i \colon i \in \{1, \ldots, 5\}\}\})$. For further details and examples, we refer the reader to [**Wey18**].

For the case $m_1 = 1/4$ and $m_2 = 1/12$, in Figure 1 the corresponding eigenfunctions of the operator Δ^μ are sketched and Figure 2 illustrates the point plots of $S_{1,5}$ and $S_{2,4}$ together with the corresponding ellipse.

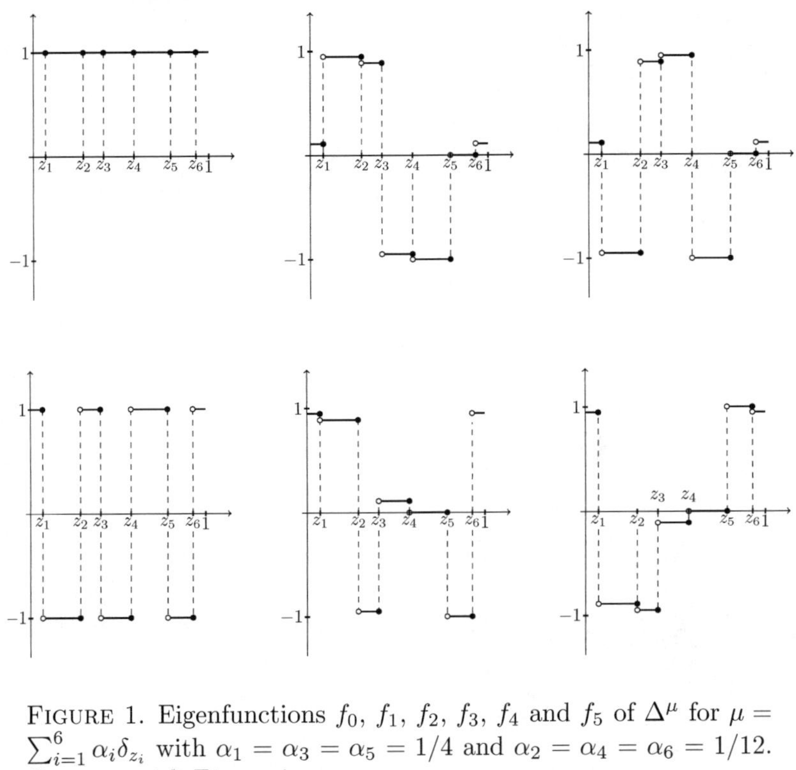

FIGURE 1. Eigenfunctions f_0, f_1, f_2, f_3, f_4 and f_5 of Δ^μ for $\mu = \sum_{i=1}^{6} \alpha_i \delta_{z_i}$ with $\alpha_1 = \alpha_3 = \alpha_5 = 1/4$ and $\alpha_2 = \alpha_4 = \alpha_6 = 1/12$. Compare with Figure 4.

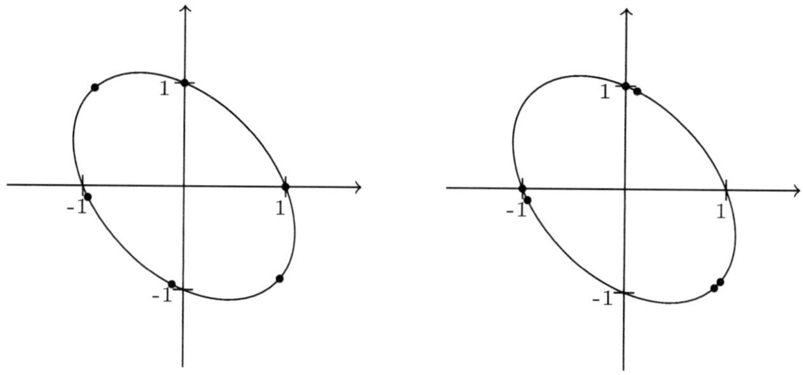

FIGURE 2. Point plot of $S_{1,5}$ (left) and point plot of $S_{2,4}$ (right) together with the curve given by $\sqrt{73}(x^2+y^2-1) = -7xy$.

4. Uniform discrete probability distributions

Here, we consider the case when μ is a uniform discrete probability distribution with $N \geq 3$, namely $\alpha_i = N^{-1}$ for all $i \in \{1, \ldots, N\}$, in which case,

$$B = \begin{pmatrix} -2N^2 & N^2 & 0 & \cdots & 0 & 0 & N^2 \\ N^2 & -2N^2 & N^2 & \cdots & 0 & 0 & 0 \\ 0 & N^2 & -2N^2 & \cdots & 0 & 0 & 0 \\ \vdots & \vdots & \vdots & \ddots & \vdots & \vdots & \vdots \\ 0 & 0 & 0 & \cdots & -2N^2 & N^2 & 0 \\ 0 & 0 & 0 & \cdots & N^2 & -2N^2 & N^2 \\ N^2 & 0 & 0 & \cdots & 0 & N^2 & -2N^2 \end{pmatrix}.$$

The following theorem reveals the spectrum of this matrix, and in the case that N is even, the result gives a second example for which the lower bound in Theorem 3.1 is sharp.

THEOREM 4.1. *The eigenvalues of the matrix B given directly above are of the form*

$$\lambda_l = -2N^2 + 2N^2 \cos(2\pi l/N)$$

with corresponding eigenvectors

$$v^{(l)} = \Big(1, \exp(2\pi \mathbf{i}\, l/N), \exp(2\pi \mathbf{i}\, 2l/N), \ldots, \exp(2\pi \mathbf{i}\,(N-1)l/N)\Big)^{\top},$$

for $l \in \{0, \ldots, N-1\}$.

PROOF. Set $m_1 = -2N^2, m_2 = m_N = N^2$ and $m_i = 0$, for $i \in \{3, \ldots, N-1\}$. The identity $Bv = \lambda v$ can be formulated as a system of N difference equations of the form

$$\sum_{k=1}^{N-j} m_k v_{k+j} + \sum_{k=N-j+1}^{N} m_k v_{k-N+j} = \lambda v_{j+1}, \tag{6}$$

where $j \in \{0, \ldots, N-1\}$ and $v = (v_1, \ldots, v_N)$. To obtain the eigenvalue λ_l, we follow the ansatz $v_k^{(l)} = \varphi_l^{k-1}$, where $\varphi_l^k := \exp(2\pi \mathbf{i}\, kl/N)$, for $k, l \in \{0, \ldots, N-1\}$. Substituting this into (6), and using the facts that $\varphi_l^{-N} = \varphi_l^0 = 1$ and $\varphi_l^j \neq 0$, for all $j, l \in \{0, \ldots, N-1\}$, we obtain $\lambda_l = \sum_{k=1}^N m_k \varphi_l^{k-1}$. Hence, for $l \in \{0, \ldots, N-1\}$, we have $Bv^{(l)} = \lambda_l v^{(l)}$ and

$$\begin{aligned}\lambda_l &= -2N^2 + N^2 \exp(2\pi \mathbf{i}\, l/N) + N^2 \exp(2\pi \mathbf{i}\, l(N-1)/N) \\ &= -2N^2 + 2N^2 \cos(2\pi l/N). \end{aligned}$$ □

COROLLARY 4.2. *The eigenvalues of the operator* Δ^μ *are*

$$\lambda_l = -2N^2 + 2N^2 \cos(2\pi l/N),$$

for $l \in \{0, \ldots, N-1\}$, *with corresponding eigenfunctions* $f_l \in \mathscr{D}_\mu^2$, *where*

(1) f_0 *is the constant function with value* 1,

and, for $j \in \{1, \ldots, N-1\}$,

(2) $f_l|_{[0,z_1]\cup(z_N,1]} = 0$ *and* $f_l|_{(z_j,z_{j+1}]} = \operatorname{Im}(\exp(2\pi \mathbf{i}\, jl/N))$, *for* $0 < l < N/2$, *and*

(3) $f_l|_{[0,z_1]\cup(z_N,1]} = 1$ *and* $f_l|_{(z_j,z_{j+1}]} = \operatorname{Re}(\exp(2\pi \mathbf{i}\, jl/N))$, *for* $N/2 \leq l \leq N-1$.

(See Figures 3 *and* 4.*)*

In the situation of Corollary 4.2, we have for N tending to infinity, that the pure point measures converge weakly to the Lebesgue measure, and indeed, the eigenfunctions of the discrete Laplacians converge uniformly to (appropriately rescaled) cosine and sine functions. Moreover, the eigenvalue approach $-(2\pi k)^2$; the corresponding eigenvalues of the classical weak Laplacian.

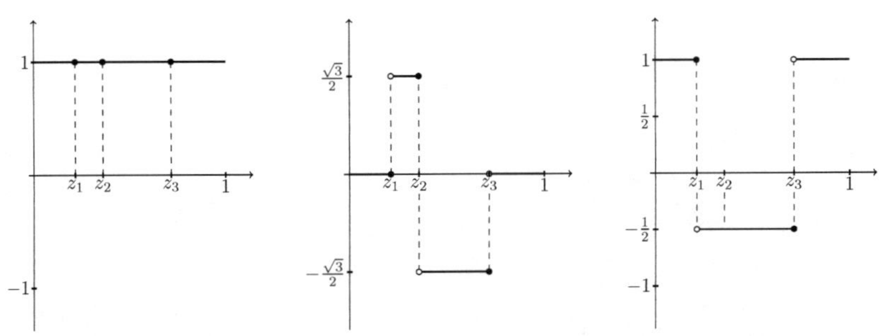

FIGURE 3. Eigenfunctions f_0, f_1 and f_2 of Δ^μ with corresponding eigenvalues $\lambda_0 = 0$, $\lambda_1 = -27$ and $\lambda_2 = -27$, for μ a uniform discrete probability distribution with $N = 3$.

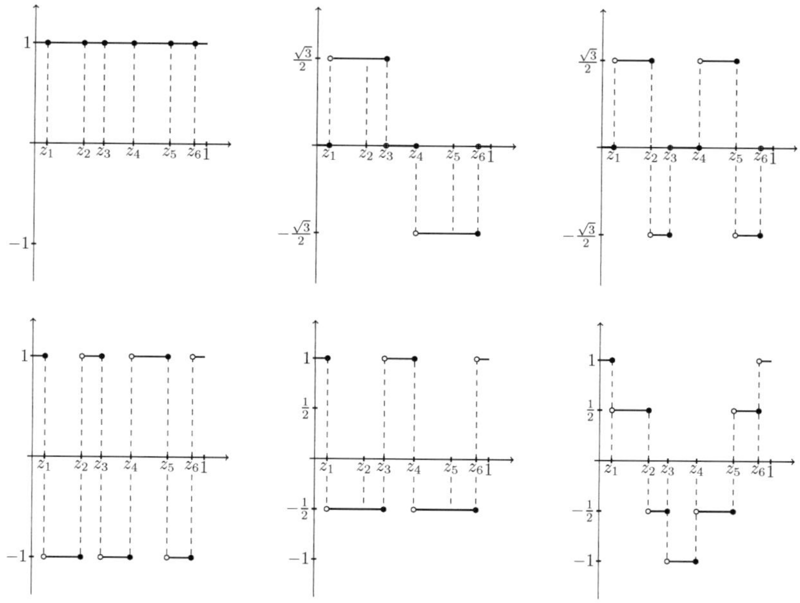

FIGURE 4. Eigenfunctions f_0, f_1, f_2, f_3, f_4 and f_5 of Δ^μ with corresponding eigenvalues $\lambda_0 = 0$, $\lambda_1 = -36$, $\lambda_2 = -108$, $\lambda_3 = -144$, $\lambda_4 = -108$ and $\lambda_5 = -36$, for μ a uniform discrete probability distribution with $N = 6$; compare with Figure 1.

References

[Arz15] Peter Arzt, *Measure theoretic trigonometric functions*, J. Fractal Geom. **2** (2015), no. 2, 115–169, DOI 10.4171/JFG/18. MR3353090

[Big93] Norman Biggs, *Algebraic graph theory*, 2nd ed., Cambridge Mathematical Library, Cambridge University Press, Cambridge, 1993. MR1271140

[Chu97] Fan R. K. Chung, *Spectral graph theory*, CBMS Regional Conference Series in Mathematics, vol. 92, Published for the Conference Board of the Mathematical Sciences, Washington, DC; by the American Mathematical Society, Providence, RI, 1997. MR1421568

[Fel57] William Feller, *Generalized second order differential operators and their lateral conditions*, Illinois J. Math. **1** (1957), 459–504. MR0092046

[Fre05] Uta Freiberg, *Spectral asymptotics of generalized measure geometric Laplacians on Cantor like sets*, Forum Math. **17** (2005), no. 1, 87–104, DOI 10.1515/form.2005.17.1.87. MR2110540

[Fuj87] Takahiko Fujita, *A fractional dimension, self-similarity and a generalized diffusion operator*, Probabilistic methods in mathematical physics (Katata/Kyoto, 1985), Academic Press, Boston, MA, 1987, pp. 83–90. MR933819

[FZ02] U. Freiberg and M. Zähle, *Harmonic calculus on fractals—a measure geometric approach. I*, Potential Anal. **16** (2002), no. 3, 265–277, DOI 10.1023/A:1014085203265. MR1885763

[Kig93] Jun Kigami, *Harmonic calculus on p.c.f. self-similar sets*, Trans. Amer. Math. Soc. **335** (1993), no. 2, 721–755, DOI 10.2307/2154402. MR1076617

[Kig01] Jun Kigami, *Analysis on fractals*, Cambridge Tracts in Mathematics, vol. 143, Cambridge University Press, Cambridge, 2001. MR1840042

[Kig03] Jun Kigami, *Harmonic analysis for resistance forms*, J. Funct. Anal. **204** (2003), no. 2, 399–444, DOI 10.1016/S0022-1236(02)00149-0. MR2017320

[KKn58] I. S. Kac and M. G. Kreĭn, *Criteria for the discreteness of the spectrum of a singular string* (Russian), Izv. Vysš. Učebn. Zaved. Matematika **1958** (1958), no. 2 (3), 136–153. MR0139804

[KL01] Jun Kigami and Michel L. Lapidus, *Self-similarity of volume measures for Laplacians on p.c.f. self-similar fractals*, Comm. Math. Phys. **217** (2001), no. 1, 165–180, DOI 10.1007/s002200000326. MR1815029

[KL12] Matthias Keller and Daniel Lenz, *Dirichlet forms and stochastic completeness of graphs and subgraphs*, J. Reine Angew. Math. **666** (2012), 189–223, DOI 10.1515/CRELLE.2011.122. MR2920886

[KSW16] M. Kesseböhmer, T. Samuel, and H. Weyer, *A note on measure-geometric Laplacians*, Monatsh. Math. **181** (2016), no. 3, 643–655, DOI 10.1007/s00605-016-0906-0. MR3552804

[LeV07] Randall J. LeVeque, *Finite difference methods for ordinary and partial differential equations*, Society for Industrial and Applied Mathematics (SIAM), Philadelphia, PA, 2007. Steady-state and time-dependent problems. MR2378550

[Wey18] H. Weyer, *Measure-geometric laplacians on the real line*, Ph.D. thesis, Universität Bremen, 2018.

[Zäh05] M. Zähle, *Harmonic calculus on fractals—a measure geometric approach. II*, Trans. Amer. Math. Soc. **357** (2005), no. 9, 3407–3423, DOI 10.1090/S0002-9947-05-03854-7. MR2146630

Fachbereich 3 Mathematik, Universität Bremen, Bibliothekstr. 1, 28359 Bremen, Germany.

Email address: mhk@math.uni-bremen.de.

Mathematics Department, California Polytechnic State University, San Luis Obispo, CA, USA.

Email address: ajsamuel@calpoly.edu

Fachbereich 3 Mathematik, Universität Bremen, Bibliothekstr. 1, 28359 Bremen, Germany.

Email address: hendrik@math.uni-bremen.de

Contemporary Mathematics
Volume **731**, 2019
https://doi.org/10.1090/conm/731/14677

An overview of complex fractal dimensions: from fractal strings to fractal drums, and back

Michel L. Lapidus

ABSTRACT. Our main goal in this long survey article is to provide an overview of the theory of complex fractal dimensions and of the associated geometric or fractal zeta functions, first in the case of fractal strings (one-dimensional drums with fractal boundary), in Section 2, and then in the higher-dimensional case of relative fractal drums and, in particular, of arbitrary bounded subsets of Euclidean space of $\mathbb{R}^N$, for any integer $N \geq 1$, in Section 3.

Special attention is paid to discussing a variety of examples illustrating the general theory rather than to providing complete statements of the results and their proofs, for which we refer to the author's previous (joint) books mentioned in the paper.

Finally, in an epilogue (Section 4), entitled "From quantized number theory to fractal cohomology", we briefly survey aspects of related work (motivated in part by the theory of complex fractal dimensions) of the author with H. Herichi (in the real case) [**HerLap1**], along with [**Lap8**], and with T. Cobler (in the complex case) [**CobLap1**], respectively, as well as in the latter part of a book in preparation by the author, [**Lap10**].

To my daughter, Julie Anne Myriam Lapidus, who has recently embarked on a journey fraught with perilous and frightening dangers, the ultimate struggle for the continuation of her own beautiful and young life.
You are my scintillating treasure and will always remain my untarnished jewel.
With all my love.

2010 *Mathematics Subject Classification.* Primary 11M06, 11M41, 28A12, 28A80, 30D10, 30B50, 35P20, 42B20, 45Q05, 81Q20; Secondary 30D30, 37C30, 37C45, 40A10, 42B35, 42B37, 44A10, 45Q05, 81R40.

Key words and phrases. Fractal strings and drums, relative fractal drums (RFDs), complex dimensions, fractal zeta functions, distance and tube zeta functions, relative fractal zeta functions, fractal tube formulas, Minkowski dimension and content, Minkowski measurability criteria, fractality, unreality, Riemann zeta function, Riemann hypothesis, direct and inverse problems for fractal strings, quantized number theory, infinitesimal shift, spectral operator, regularized determinants, fractal cohomology.

Work partially supported by the US National Science Foundation under the research grants DMS-0707524 and DMS-1107750, as well as by the Burton Jones Endowed Chair Fund while he has been the holder of the F. Burton Jones Endowed Chair in Pure Mathematics at the University of California, Riverside, since July 2017.

CONTENTS

1. Introduction

Our main goal in this research expository article is to provide an overview of the theory of complex dimensions and of the associated geometric or fractal zeta functions, first in the case of fractal strings (one-dimensional drums with fractal boundary) or essentially equivalently, for compact subsets of the real line (see §2), and then (see §3), in the higher-dimensional case of relative fractal drums and, in particular, of bounded subsets in Euclidean space $\mathbb{R}^N$, for any integer $N \geq 1$.

Special attention is paid to providing a variety of examples (especially, in higher dimensions) illustrating the general theory rather than to stating precise theorems in their greatest generality or providing their proofs (or even a sketch thereof). For a rigorous and quite detailed account of the theory, we refer instead the interested reader to the books by the author and Machiel van Frankenhuijsen, [**Lap-vF4**] for the case of fractal strings (i.e., $N = 1$) and the recent book by the author, Goran Radunović and Darko Žubrinić, [**LapRaZu1**], when $N \geq 1$ is arbitrary.

In the epilogue (see §4), we survey related work, motivated in part by the theory of complex dimensions of fractal strings, on quantized number theory (in the real setting) and the Riemann hypothesis (joint with Hafedh Herichi; see the papers [HerLap2–5] and the forthcoming book [**HerLap1**], along with [**Lap8**]), as well as on quantized number theory (in the complex setting), regularized determinants and "fractal cohomology" (see [CobLap1–2], joint with Tim Cobler, along with the author's book in preparation, [**Lap10**], of which the present paper is both the seed and a significantly condensed version).

Except at the very end of this introduction, we now focus on §2 and §3 in the rest of §1. For a brief overview of the contents of §4, we refer the interested reader to the overall introduction of §4 and to the beginning of each of the subsections of §4 (namely, §§4.1–4.4), along with part *(iii)* towards the end of this section.

Complex dimensions provide a natural way to extract the information about the oscillatory nature of fractal objects. This is done via generalized explicit formulas (extending to this setting Riemann's original explicit formula for the prime number counting function, see §2.3).

In this paper, we have chosen to emphasize a type of explicit formulas called *fractal tube formulas* and which enable us to express the volume, $V(\varepsilon)$, of the ε-neighborhoods of the given fractal set A or, more generally, of the given relative fractal drum (A,Ω) (RFD, in short) in $\mathbb{R}^N$,[1] as an extended power series in ε with exponents the underlying complex codimensions and (normalized) coefficients (in the case of simple poles) the residues of the associated geometric or fractal zeta function.

For example, under suitable assumptions (and still in the case of simple poles), we have the following (pointwise or distributional) exact fractal tube formula for a given bounded set A (or, more generally, RFD (A,Ω)) in $\mathbb{R}^N$ (up to a possible error term, which can be estimated explicitly):

$$V(\varepsilon)=\sum_{\omega\in\mathcal{D}}c_\omega\frac{\varepsilon^{N-\omega}}{N-\omega}, \tag{1.2}$$

where $c_\omega:=\operatorname{res}(\zeta_A,\omega)$ (or, more generally, $c_\omega:=\operatorname{res}(\zeta_{A,\Omega},\omega)$) for each $\omega\in\mathcal{D}$, and where $\mathcal{D}=\mathcal{D}_A$ (respectively, $\mathcal{D}=\mathcal{D}_{A,\Omega}$) denotes the set of complex dimensions of A (respectively, of the RFD (A,Ω)), viewed as a multiset (i.e., a set with finite integer multiplicities). Here, ζ_A (respectively, $\zeta_{A,\Omega}$) is the *distance zeta function* of A (respectively, of (A,Ω)) defined (in the case of the bounded set A, for example) for all $s\in\mathbb{C}$ with $Re(s)$ sufficiently large by the Lebesgue (and hence, absolutely convergent) integral

$$\zeta_A(s)=\int_{A_\delta}d(x,A)^{s-N}dx, \tag{1.3}$$

for some fixed $\delta>0$ whose specific value is unimportant from the point of view of the theory of complex dimensions.

More specifically, (1.3) holds for all $s\in\mathbb{C}$ with $Re(s)>\overline{D}$, where $\overline{D}$ is the (upper) Minkowski dimension of A, and this lower bound is optimal. In other words, the abscissa of convergence of ζ_A coincides with $\overline{D}$; this is one of the first basic results of the theory (see part (a) of §3.3.1).

---

[1] An RFD in $\mathbb{R}^N$ is a pair (A,Ω), with $A\subseteq\mathbb{R}^N$, Ω open in $\mathbb{R}^N$ and $\Omega\subseteq A_{\delta_1}$, for some $\delta_1>0$, where for any $\varepsilon>0$,

$$A_\varepsilon:=\{x\in\mathbb{R}^N: d(x,A)<\varepsilon\} \tag{1.1}$$

is the ε-neighborhood of A and $d(x,A)$ denotes the Euclidean distance from $x\in\mathbb{R}^N$ to A. Also, we let $V(\varepsilon)=V_A(\varepsilon):=|A_\varepsilon|$ (respectively, $V(\varepsilon)=V_{A,\Omega}(\varepsilon):=|A_\varepsilon\cap\Omega|$) in the case of a bounded set A (respectively, RFD (A,Ω)) in $\mathbb{R}^N$. Here and thereafter, $|\cdot|=|\cdot|_N$ denotes the N-dimensional volume (or Lebesgue measure in $\mathbb{R}^N$).

The (visible) *complex dimensions* of A are defined as the poles of the meromorphic continuation (if it exists) of ζ_A to some given connected open neighborhood of the vertical line $\{Re(s) = \overline{D}\}$ (or, equivalently, of the closed right half-plane $\{Re(s) \geq \overline{D}\}$, since ζ_A is holomorphic on the open right half-plane $\{Re(s) > \overline{D}\}$; see part (*b*) of §3.3.1).[2]

In particular, if $\overline{D}$ itself is a pole of ζ_A (under mild conditions, it is always a nonremovable singularity of ζ_A), then it is a complex dimension having the largest possible real part.

Provided $\overline{D} < N$ (i.e., if $\overline{D} \neq N$ since we always have $\overline{D} \leq N$), all of the above results and definitions extend to another useful fractal zeta function, called the *tube zeta function* of A and denoted by $\widetilde{\zeta}_A$.[3] The fractal zeta functions ζ_A and $\widetilde{\zeta}_A$ are connected via a functional equation (see (3.15)), which implies that the (visible) complex dimensions of A can be defined indifferently via either ζ_A or $\widetilde{\zeta}_A$. Furthermore, the fractal tube formula (1.2) has a simple counterpart expressed in terms of the residues of $\widetilde{\zeta}_A$ (instead of those of ζ_A) evaluated at the complex dimensions of A. Namely, up to a possible error term which can be estimated explicitly,

$$V(\varepsilon) = \sum_{\omega \in \mathcal{D}} d_\omega \varepsilon^{N-\omega}, \tag{1.6}$$

where $d_\omega := \operatorname{res}(\widetilde{\zeta}_A, \omega)$ for each $\omega \in \mathcal{D}$.

All of the above results (including the fractal tube formulas (1.2) and (1.6)) extend to relative fractal drums (RFDs) in $\mathbb{R}^N$ (with $\zeta_A, \widetilde{\zeta}_A$ and $\mathcal{D} = \mathcal{D}_A$ replaced by $\zeta_{A,\Omega}, \widetilde{\zeta}_{A,\Omega}$ and $\mathcal{D} = \mathcal{D}_{A,\Omega}$, respectively), which are very useful tools in their own right and enable us, in particular, to compute (by using appropriate decompositions and symmetry considerations) the fractal zeta functions and complex dimensions of many fractal compact subsets of $\mathbb{R}^N$.

At this stage, it is helpful to point out that many key results of the theory of complex dimensions of fractal strings [**Lap-vF4**] (briefly discussed in §2), including the fractal tube formulas (of which (2.10) is a typical example), can be recovered by specializing the higher-dimensional theory of complex dimensions of RFDs in $\mathbb{R}^N$ to the $N = 1$ case and by viewing fractal strings as RFDs in $\mathbb{R}$. In the process, a simple functional equation connecting the so-called geometric zeta function of a fractal string (described in the beginning of §2) and the distance zeta function of the associated RFD plays a key role; see §3.2.2 and §3.5.1.

Intuitively, a fractal, viewed as a geometric object, is like a musical instrument tuned to play certain notes with frequencies (respectively, amplitudes) essentially

---

[2]Throughout this paper, we use the following short-hand notation: given $\alpha \in \mathbb{R}$, we let

$$\{Re(s) \geq \alpha\} := \{s \in \mathbb{C} : Re(s) \geq \alpha\} \tag{1.4}$$

denote the closed right half-plane with abscissa α; and analogously for the vertical line $\{Re(s) = \alpha\}$ or the open right half-plane $\{Re(s) > \alpha\}$ with abscissa α, say. (If $\alpha = \pm\infty$, we adopt the obvious conventions $\{Re(s) \geq +\infty\} = \emptyset$ and $\{Re(s) \geq -\infty\} = \mathbb{C}$, for example.)

[3]For $Re(s)$ sufficiently large (in fact, precisely for $Re(s) > \overline{D}$, provided $\overline{D} < N$), $\widetilde{\zeta}_A$ is given by the Lebesgue (and hence, absolutely convergent) integral

$$\widetilde{\zeta}_A(s) := \int_0^\delta V(\varepsilon)\varepsilon^{s-N} \frac{d\varepsilon}{\varepsilon}, \tag{1.5}$$

for some arbitrary but fixed $\delta > 0$, the value of which is unimportant from the point of view of the definition (and the values) of the complex dimensions.

equal to the real parts (respectively, the imaginary parts) of the underlying complex dimensions. Alternatively, one can think of a "geometric wave" propagating through the fractal and with the aforementioned frequencies and amplitudes. This "physical" intuition is corroborated, for example, by the fractal tube formulas expressed via the distance (respectively, tube) zeta function, as in (1.2) (respectively, (1.6)).

As was mentioned just above, the theory of complex dimensions of fractal strings can be viewed essentially as the one-dimensional special case of the general theory of complex dimensions (of RFDs in $\mathbb{R}^N$) developed in [**LapRaZu1**]. Conversely, fractal string theory has provided the author and his collaborators with a broad and rich collection of examples with which to test various conjectures and formulate various definitions as well as elaborate tools that could eventually be used in more complicated higher-dimensional situations. Also, several of the key steps towards the proof of the higher-dimensional fractal tube formulas (such as in (1.2) and (1.6)) rely, in part, on techniques developed for dealing with the case of (generalized) fractal strings [**Lap-vF2**, **Lap-vF3**, **Lap-vF4**].

In addition, "fractality" is characterized (or rather, *defined*) in our general theory by the presence of *nonreal* complex dimensions.[4] This extends to any dimension $N \geq 1$ the definition of fractality given earlier in [Lap-vF2–4], thanks to the fact that we now have to our disposal a general definition of fractal zeta functions valid for arbitrary bounded (or, equivalently, compact) subsets of $\mathbb{R}^N$ (as well as, more generally, for all RFDs in $\mathbb{R}^N$).

We will also discuss (in §3.5.2 when $N \geq 1$ is arbitrary, and in Theorem 2.2 when $N = 1$) a general Minkowski measurability criterion expressed in terms of complex dimensions. Namely, under certain mild conditions (which imply that the Minkowski dimension D exists and is a complex dimension), a bounded set A (or, more generally, an RFD (A, Ω)) in $\mathbb{R}^N$ is Minkowski measurable[5] if and only if its only complex dimension with real part D (i.e., its only *principal complex dimension*) is D itself and D is simple. In other words, the existence of nonreal complex dimensions (i.e., the "*critical fractality*" of A or of (A, Ω); see §3.6), along with the simplicity of D (as a pole of ζ_A or equivalently, of $\zeta_{A,\Omega}$), characterizes the Minkowski *non*measurability of A (or of (A, Ω)). As a simple illustration, the Cantor set, the Cantor string, the Sierpinski gasket and the Sierpinski carpet, along with lattice self-similar strings (and more generally, in higher dimensions, lattice self-similar sprays with "sufficiently nice" generators), are all Minkowski nonmeasurable but are Minkowski nondegenerate; see §3.5.3. On the other hand, a "generic" (i.e., nonlattice) self-similar Cantor-type set (or string) or a "generic" self-similar carpet is Minkowski measurable (because it does not have any nonreal complex dimensions other than D, which is simple).

Beside this introduction (i.e., §1), this paper is divided into three parts:

(i) §2, a brief account of the theory of complex dimensions for fractal strings ($N = 1$) [**Lap-vF4**] and its prehistory, including a discussion (in §2.6) of natural direct and inverse spectral problems for fractal strings along with their intimate

[4] It is also very useful to extend the notion of "complex dimensions" by allowing more general (nonremovable) singularities than poles of the associated fractal zeta functions; see [**LapRaZu1**] and [LapRaZu6–7, 10], along with §2.5, §3.5.2, §3.6 and §4.4.

[5] Intuitively, Minkowski measurability is some kind of "fractal regularity" of the underlying geometry; for a precise definition, see §3.2 when $N \geq 1$ is arbitrary (or §2.1 when $N = 1$).

connections with the Riemann zeta function [**LapPo2**] and the Riemann hypothesis [**LapMa2**].

(ii) §3, an introduction to the higher-dimensional theory of complex dimensions and the associated fractal zeta functions (namely, the distance and tube zeta functions), based on [**LapRaZu1**] (and aspects of [LapRaZu2–9]), with emphasis on several key examples of bounded sets and relative fractal drums in $\mathbb{R}^N$ (with $N = 2, N = 3$ or $N \geq 1$ arbitrary) illustrating the key concepts of fractal zeta functions and their poles or, more generally, nonremovable singularities (i.e., the complex dimensions), as well as the associated fractal tube formulas. As was alluded to earlier, the latter explicit formulas provide a concrete justification of the use of the phrase "complex fractal dimensions" and help explain why both intuitively and in actuality, the theory of complex dimensions is a theory of oscillations that are intrinsic to fractal geometries.

It is noteworthy that even though we will mostly stress the aforementioned geometric oscillations in §3 (and in much of §2), the broad definition of "fractality" proposed in §3.6 and expressed in terms of the presence of nonreal complex dimensions encompasses oscillations that are intrinsic to number theories (via Riemann-type explicit formulas expressed in terms of the poles and the zeros of attached L-functions, or equivalently, in terms of the poles of the logarithmic derivatives of those L-functions; see §2.3 for the original example), or to dynamical systems (e.g., via explicit formulas for the counting functions of primitive periodic orbits; see [**Lap-vF4**, Ch. 7] for a class of examples), as well as to the spectra of fractal drums [both "drums with fractal boundary" (as, e.g., in [Lap1–3] and parts of [**Lap-vF4**]) and "drums with fractal membrane" (as, e.g., in [**Lap3**], [**KiLap1**] and [**Lap7**]) and other classical or quantum physical systems (via detailed spectral asymptotics or, essentially equivalently, via explicit formulas for the associated frequency or eigenvalue counting functions).

Much remains to be done in all of these directions for a variety of specific classes of dynamical systems and of fractal drums, for example. We point out, however, that the deep analogy between many aspects of fractal geometry and number theory (see, e.g., [Lap-vF1–5], [**Lap7**], [**HerLap1**] and [**LapRaZu1**]) was a key motivation for the author to want to develop (since the mid-1990s) a theory of "fractal cohomology", itself an important motivation for many aspects of the work described in §4.

(iii) §4, the epilogue, a very brief account (compared to the size of the corresponding material to be described) of "quantized number theory", both in the "real case" (§4.2, based on [**HerLap1**], [HerLap2–5] and [**Lap8**]) and in the "complex case" (§4.3, based principally on [CobLap1–2] and on aspects of [**Lap10**]) and the associated fractal cohomology (§4.1 and, especially, §4.4, as expanded upon in [**Lap10**]), with applications to several reformulations of the Riemann hypothesis (§4.2) expressed in terms of the "quasi-invertibility" [**HerLap1**] or the invertibility [**Lap8**] of so-called "spectral operators", in particular, as well as to the representations (§4.3) of various arithmetic (or number-theoretic) L-functions and other meromorphic functions (including the completed Riemann zeta function and the Weil zeta functions attached to varieties over finite fields [Weil–6, Gro1–4, Den1–6], and, e.g., [**Mani**], [**Kah**], [Tha1–2]) via (graded or supersymmetric) regularized (typically infinite dimensional) determinants of suitable unbounded linear operators (the so-called "generalized Polya–Hilbert operators") restricted to their eigenspaces (which are the proposed "fractal cohomology spaces").

These developments open-up a vast and very rich new domain of research, extending in a variety of directions and located at the intersection of many fields of mathematics, including fractal geometry, number theory and arithmetic geometry, mathematical physics, dynamical systems, harmonic analysis and spectral theory, complex analysis and geometry, geometric measure theory, as well as algebraic geometry and topology, to name a few.

We hope that the reader will be stimulated by the reading of this expository article (and eventually, of its much expanded sequel, the author's book in preparation [**Lap10**]) to explore the various ramifications and consequences of the theory, many of which are yet to be discovered. In other words, instead of offering here a complete and closed theory, we prefer to (and, in fact, must) offer here (especially, in §4) only glimpses of a possible future unifying and "universal" theory, resting on the contributions and conjectures or dreams of many past and contemporary mathematicians and physicists.

2. Fractal Strings and Their Complex Dimensions

A (*bounded*) *fractal string* can be viewed either as a bounded open set $\Omega \subseteq \mathbb{R}$ or else as a nonincreasing sequence of lengths (or positive numbers) $\mathcal{L} = (\ell_j)_{j=1}^{\infty}$ such that $\ell_j \downarrow 0$. (The latter condition is not needed if the sequence (ℓ_j) is finite.)

Let us briefly explain the connection between these two points of view. If Ω is a bounded open subset of $\mathbb{R}$, we can write $\Omega = \cup_{j\geq 1} I_j$ as an at most countable disjoint union of bounded open intervals I_j, of length $\ell_j > 0$. These intervals are nothing but the connected components of Ω. Since $|\Omega|_1 = \sum_{j\geq 1} \ell_j < \infty$ (i.e., Ω has finite total length), without loss of generality, we may assume (possibly after having reshuffled the intervals I_j, that $\ell_1 \geq \ell_2 \geq \cdots$ (counting multiplicities) and (provided the sequence $(\ell_j)_{j\geq 1}$ is infinite) $\ell_j \downarrow 0$.

Slightly more generally, in the definition of a *bounded fractal string*, one can assume that instead of being bounded, the open set $\Omega \subseteq \mathbb{R}$ has finite volume (i.e., length): $|\Omega|_1 < \infty$.

Unbounded fractal strings (i.e., strings $\mathcal{L} = (\ell_j)_{j\geq 1}$ such that $\sum_{j\geq 1} \ell_j = +\infty$) also play an important role in the theory (see, e.g., [**Lap-vF4**, Ch. 3 and parts of Chs. 9–11 along with §13.1]) but from now on, unless explicitly mentioned otherwise, we will assume that all of the (geometric) fractal strings under consideration are bounded. As a result, we will often drop the adjective "bounded" when referring to fractal strings.

From a physical point of view, the 'lengths' ℓ_j can also be thought of as being the underlying *scales* of the system. This is especially useful in the case of unbounded fractal strings but should also be kept in mind in the geometric situation of fractal strings.

A *geometric realization* of a (bounded) fractal string $\mathcal{L} = (\ell_j)_{j\geq 1}$ is any bounded open set Ω in $\mathbb{R}$ (or, more generally, any open set Ω in $\mathbb{R}$ of finite length) with associated length sequence $\mathcal{L}$.

The *geometric zeta function* $\zeta_{\mathcal{L}}$ of a fractal string $\mathcal{L} = (\ell_j)_{j\geq 1}$ is defined by

$$\zeta_{\mathcal{L}}(s) = \sum_{j\geq 1} \ell_j^s, \tag{2.1}$$

for all $s \in \mathbb{C}$ with $Re(s)$ sufficiently large. (Here and henceforth, we let $\ell_j^s := (\ell_j)^s$, for each $j \geq 1$.)

A simple example of a fractal string is the *Cantor string*, denoted by Ω_{CS} (or $\mathcal{L}_{CS}$) and defined by $\Omega_{CS} = [0,1]\backslash C$, the complement of the (classic ternary) Cantor set C in the unit interval. (Observe that the boundary of the Cantor string is the Cantor set itself: $\partial\mathcal{L}_{CS} := \partial\Omega_{CS} = C$.) Then, Ω_{CS} consists of the disjoint union of the deleted (open) intervals, in the usual construction of the Cantor set:

$$\Omega_{CS} = (0,1/3) \cup (1/9,2/9) \cup (7/9,8/9) \cup \cdots . \tag{2.2}$$

Hence, the associated sequence of lengths $\mathcal{L}_{CS}$ is given by

$$1/3, 1/9, 1/9, 1/27, 1/27, 1/27, 1/27, \cdots ; \tag{2.3}$$

equivalently, $\mathcal{L}_{CS}$ consists of the lengths $1/3^n$ counted with multiplicity 2^{n-1}, for $n = 1, 2, 3, \cdots$. It follows that ζ_{CS} can be computed by simply evaluating the following geometric series:

$$\begin{aligned} \zeta_{CS}(s) &= \sum_{n=1}^{\infty} 2^{n-1}(3^{-n})^s = 3^{-s}\sum_{n=0}^{\infty}(2 \cdot 3^{-s})^n \\ &= \frac{3^{-s}}{1 - 2 \cdot 3^{-s}} = \frac{1}{3^s - 2}. \end{aligned}$$

This calculation is valid for $Re(s) > \log_3 2$ (i.e., $|2 \cdot 3^{-s}| < 1$) but upon analytic continuation, we see that ζ_{CS} admits a (necessarily unique) meromorphic continuation to all of $\mathbb{C}$ (still denoted by ζ_{CS}, as usual) and that

$$\zeta_{CS}(s) = \frac{1}{3^s - 2}, \quad \text{for all } s \in \mathbb{C}. \tag{2.4}$$

The *complex dimensions* of the Cantor string are the poles of ζ_{CS}; that is, here, the complex solutions of the equation $3^s - 2 = 0$. Thus, the set $\mathcal{D}_{CS}$ of complex dimensions of $\mathcal{L}_{CS}$ is given by a single (discrete) vertical line,

$$\mathcal{D}_{CS} = \{D + in\mathbf{p} : n \in \mathbb{Z}\}, \tag{2.5}$$

where $D := D_{CS} = \log_3 2$ is the Minkowski dimension of the Cantor string (or of the Cantor set) and $\mathbf{p} := 2\pi/\log 3$ is its oscillatory period. (The definition of D is recalled in (2.7).)[6]

For an arbitrary fractal string $\mathcal{L}$, the *complex dimensions* of $\mathcal{L}$ (relative to a given domain $U \subseteq \mathbb{C}$ to which $\zeta_{\mathcal{L}}$ admits a meromorphic continuation), also called the *visible complex dimensions* of $\mathcal{L}$, are simply the poles of $\zeta_{\mathcal{L}}$ which lie in U. Thus, for the Cantor string, $\mathcal{D}_{CS} = \mathcal{D}_{CS}(\mathbb{C})$ is given by (2.5).

Recall that the *abscissa of convergence* $\alpha = \alpha_{\mathcal{L}}$ of the Dirichlet series defining $\zeta_{\mathcal{L}}$ in (2.1) is given by

$$\alpha := \inf\left\{\beta \in \mathbb{R} : \sum\nolimits_{j\geq 1} \ell_j^\beta < \infty\right\}; \tag{2.6}$$

so that α is the unique real number such that $\sum_{j\geq 1} \ell_j^s$ converges absolutely for $Re(s) > \alpha$ but diverges for $Re(s) < \alpha$.

[6]In the case of the Cantor string (or set), the Minkowski dimension exists and hence, there is not need to talk about (upper) Minkowski dimension; see §3.2 for the precise definitions.

THEOREM 2.1 (Abscissa of convergence and Minkowski dimension; [**Lap2, Lap3**], [**Lap-vF4**, Thm. 1.10]). *Let $\mathcal{L}$ be an arbitrary bounded fractal string $\mathcal{L}$ having infinitely many lengths. (When $\mathcal{L}$ has finitely many lengths, it is immediate to check that $\zeta_{\mathcal{L}}$ is entire and hence, $\alpha = -\infty$ while $D = 0$.) Then $\alpha = D$, the (upper) Minkowski dimension of $\mathcal{L}$ (i.e., of $\partial\Omega$, where the bounded open set Ω is any geometric realization of $\mathcal{L}$); see, respectively, (2.6) and (2.7) for the definition of α and D. In other words, the abscissa of convergence of $\zeta_{\mathcal{L}}$ and the Minkowski dimension of $\mathcal{L}$ coincide.*[7]

More precisely, here, the (*upper*) *Minkowski dimension* $D = D_{\mathcal{L}}$ of $\mathcal{L}$ is the nonnegative real number given by[8]

$$D := \inf\{\beta \geq 0 : V(\varepsilon) = O(\varepsilon^{1-\beta}) \quad \text{as } \varepsilon \to 0^+\}, \tag{2.7}$$

where

$$V(\varepsilon) = V_{\mathcal{L}}(x) := |\{x \in \Omega : d(x, \partial\Omega) < \varepsilon\}|_1 \tag{2.8}$$

is the volume (or length) of the ε-neighborhood of the boundary $\partial\Omega$ (relative to Ω) and $d(x, \partial\Omega)$ denotes the distance (in $\mathbb{R}$) from x to $\partial\Omega$.[9]

It follows at once from Theorem 2.1, along with the definition of α and D respectively given in (2.6) and (2.7), that for a fractal string, we have $0 \leq D \leq 1$.

For example, for the Cantor string, the computation leading to (2.4) shows that $\alpha = \log_3 2$ and it is well known that $D = \log_3 2$, in agreement with Theorem 2.1.

It is clear that the set $\mathcal{D}_{\mathcal{L}}$ of complex dimensions of a fractal string forms a discrete (and hence, at most countable) subset of $\mathbb{C}$ and (in light of Theorem 2.1, since $\zeta_{\mathcal{L}}$ is holomorphic for $Re(s) > D$)

$$\mathcal{D}_{\mathcal{L}} \subseteq \{Re(s) \leq D\},$$

where we use the short-hand notation

$$\{Re(s) \leq D\} := \{s \in \mathbb{C} : Re(s) \leq D\},$$

here and henceforth. (Similarly, for example, the notation $\{Re(s) = D\}$ stands for the vertical line $\{s \in \mathbb{C} : Re(s) = D\}$.)

The set of *principal complex dimensions* of $\mathcal{L}$, denoted by $\dim_{PC} \mathcal{L}$, is the set of complex dimensions with maximal real part:

$$\dim_{PC} \mathcal{L} = \{\omega \in \mathcal{D}_{\mathcal{L}} : Re(\omega) = D\}. \tag{2.9}$$

This set (or rather, multiset) plays an important role in the general theory of complex fractal dimensions. The same is true for its counterpart in the higher-dimensional theory, to be discussed in §3.

---

[7] In [**Lap2, Lap3**], the proof of this equality relied on a result obtained in [**BesTa**]. Then, several direct proofs were given in [Lap-vF2–4]. See, especially, [**Lap-vF4**, Thm. 1.10 and Thm. 13.111]; see also [**LapLu-vF2**]) and most recently, in [**LapRaZu1**, §2.1.4, esp. Prop. 2.1.59 and Cor. 2.1.61], via the higher-dimensional theory of complex dimensions (to be discussed in §3).

[8] The Minkowski dimension is also called the Minkowski–Bouligand dimension [**Bou**], the box dimension or the capacity dimension in the literature on fractal geometry; see, e.g., [**Man**], [**Fa1**], [**MartVuo**], [**Mat**], [Tri1–3], [Lap1–3], [**Lap-vF4**], [**LapRaZu1**] and [**LapRaRo**].

[9] In the present section (i.e., §2), for the simplicity of exposition, we will mostly ignore the distinction between upper Minkowski dimension and Minkowski dimension of $\mathcal{L}$. By contrast, in §3, we will denote, respectively, by $\overline{D}$ and D these two dimensions (when the latter exists); see §3.2 for the precise definitions. Note that in the terminology of §3, the notion introduced in (2.7) is that of upper Minkowski dimension of the bounded fractal string $\mathcal{L} = (\ell_j)_{j\geq 1}$, viewed as the relative fractal drum (or RFD) $(\partial\Omega, \Omega)$ in $\mathbb{R}$, where Ω is any geometric realization of $\mathcal{L}$.

For the Cantor string, in light of (2.5), we clearly have $\mathcal{D}_{\mathcal{L}} = \dim_{PC} \mathcal{L}$ but in general, $\dim_{PC} \mathcal{L}$ is often a strict subset of $\mathcal{D}_{\mathcal{L}} = \mathcal{D}_{\mathcal{L}}(U)$. (We implicitly assume here and in (2.9) that the connected open set U is a neighborhood of the vertical line $\{Re(s) = D\}$, or equivalently, of the closed half-plane $\{Re(s) \geq D\}$; observe, however, that the set $\dim_{PC} \mathcal{L}$ itself is independent of such a choice of U.)

We note for later use that since $\zeta_{\mathcal{L}}$ is a Dirichlet series with positive coefficients, $\zeta_{\mathcal{L}}(s) \to +\infty$ as $s \to D^+, s \in \mathbb{R}$ (or, more generally, as $s \in \mathbb{C}$ tends to D from the right within a sector of half-angle $< \pi/2$ and symmetric with respect to the real axis); see, e.g., [**Ser**] or [**Lap-vF4**, §1.2]. It follows that for a fractal string (with infinitely many lengths), the half-plane $\{Re(s) > D\}$ of absolute convergence of $\zeta_{\mathcal{L}}$ always coincides with the half-plane of holomorphic continuation of $\zeta_{\mathcal{L}}$, i.e., the maximal open right half-plane to which $\zeta_{\mathcal{L}}$ can be holomorphically continued. (See [**Lap-vF4**].) Hence, in the terminology and with the notation of [**LapRaZu1**] (to be introduced in §3.3), we have that $D = D_{hol}(\zeta_{\mathcal{L}})$, the abscissa of holomorphic continuation of $\zeta_{\mathcal{L}}$.

Observe that since D is always a singularity of $\zeta_{\mathcal{L}}$, then, provided $\zeta_{\mathcal{L}}$ can be meromorphically continued to a neighborhood of D, D must necessarily be a pole of $\zeta_{\mathcal{L}}$ (i.e., a complex dimension of $\mathcal{L}$).

2.1. Fractal tube formulas. Given a fractal string $\mathcal{L}$, under suitable hypotheses,[10] we can express its *tube function* $V(\varepsilon) = V_{\mathcal{L}}(\varepsilon)$ (or rather $\varepsilon \mapsto V(\varepsilon)$), as given by (2.8), in terms of its complex dimensions and the associated residues, as follows:

$$V(\varepsilon) = \sum_{\omega \in \mathcal{D}_{\mathcal{L}}} c_\omega \frac{(2\varepsilon)^{1-\omega}}{\omega(1-\omega)} + R(\varepsilon), \tag{2.10}$$

where $c_\omega := \operatorname{res}(\zeta_{\mathcal{L}}, \omega)$ is the residue of $\zeta_{\mathcal{L}}$ at $\omega \in \mathcal{D}_{\mathcal{L}}$ and $R(\varepsilon)$ is an error term which can be explicitly estimated.[11] If $R(\varepsilon) \equiv 0$ (which occurs, for example, for any self-similar string if we choose $U := \mathbb{C}$), the corresponding *fractal tube formula* (2.10) is said to be *exact*.[12]

In [**Lap-vF4**, Ch. 8], the interested reader can find the precise statement and hypotheses of the fractal tube formula. In fact, depending, in particular, on the growth assumptions made on the geometric zeta function $\zeta_{\mathcal{L}}$, there are a variety of fractal tube formulas, with or without error term (the latter ones being called *exact*), as well as interpreted *pointwise* or *distributionally*; see [**Lap-vF4**, §8.1].

Furthermore, in the important special case of self-similar strings (of which the Cantor string is an example), even more precise (pointwise) fractal tube formulas (exact or else with an *error term*, depending on the goal being pursued) are obtained in [**Lap-vF4**, §8.4].

---

[10]Namely, we assume that $\mathcal{L}$ is *languid* in a suitable connected open neighborhood U of $\{Re(s) \geq D\}$; i.e., roughly speaking, $\zeta_{\mathcal{L}}$ can be meromorphically continued to U and satisfies a suitable polynomial growth condition for a screen S bounding U (in the sense of [**Lap-vF4**, §5.3]).

[11]In this discussion, for clarity, we assume implicitly that all of the complex dimensions are simple (i.e., are simple poles of $\zeta_{\mathcal{L}}$). In the general case, (2.10) should be replaced by

$$V(\varepsilon) = \sum_{\omega \in \mathcal{D}_{\mathcal{L}}} \operatorname{res}\left(\frac{(2\varepsilon)^{1-s}}{s(1-s)}\, \zeta_{\mathcal{L}}(s), \omega\right) + R(\varepsilon). \tag{2.11}$$

[12]More generally, we obtain an exact tube formula whenever $\mathcal{L}$ (i.e., $\zeta_{\mathcal{L}}$) is *strongly languid* (which implies that $U := \mathbb{C}$), in the sense of [**Lap-vF4**, §5.3].

For the example of the Cantor string (which is a self-similar string because its boundary, the ternary Cantor set, is itself a self-similar set in $\mathbb{R}$), we have the following *exact fractal tube formula*, valid *pointwise* for all $\varepsilon \in (0, 1/2)$:

$$V_{CS}(\varepsilon) = \frac{1}{2\log 3} \sum_{n\in\mathbb{Z}} \frac{(2\varepsilon)^{1-D-in\mathbf{p}}}{(D+in\mathbf{p})(1-D-in\mathbf{p})} - 2\varepsilon, \tag{2.12}$$

with $D := \log_3 2$ and $\mathbf{p} := 2\pi/\log 3$.

Observe that we can rewrite (2.12) in the following form:

$$V_{CS}(\varepsilon) = \varepsilon^{1-D}\, G(\log_3 \varepsilon^{-1}) - 2\varepsilon, \tag{2.13}$$

where G is a nonconstant, positive 1-periodic function on $\mathbb{R}$ which is bounded away from zero and infinity. In fact,

$$0 < \mathcal{M}_* = \min_{u\in\mathbb{R}} G(u) \text{ and } \mathcal{M}^* = \max_{u\in\mathbb{R}} G(u) < \infty,$$

where $\mathcal{M}_*$ and $\mathcal{M}^*$ denote, respectively, the *lower* and *upper Minkowski content* of $\mathcal{L}$, defined by[13]

$$\mathcal{M}_* := \liminf_{\varepsilon\to 0^+} \varepsilon^{-(1-D)} V_{CS}(\varepsilon) \tag{2.14}$$

and

$$\mathcal{M}^* := \limsup_{\varepsilon\to 0^+} \varepsilon^{-(1-D)} V_{CS}(\varepsilon). \tag{2.15}$$

(Clearly, we have that $0 \leq \mathcal{M}_* \leq \mathcal{M}^* \leq \infty$.) For the Cantor string,

$$\mathcal{M}_* = 2^{1-D} D^{-D} \approx 2.4950 \text{ and } \mathcal{M}^* = 2^{2-D} \approx 2.5830$$

Hence, $\mathcal{M}_* < \mathcal{M}^*$ and thus, the limit of $V(\varepsilon)/\varepsilon^{1-D}$ as $\varepsilon \to 0^+$ does not exist; i.e., the Cantor string (and hence, also the Cantor set) is *not* Minkowski measurable.[14]

Recall that a fractal string $\mathcal{L}$ (or its boundary $\partial\Omega$) is said to be *Minkowski measurable* if the above limit exists in $(0, +\infty)$ and then,

$$\mathcal{M} := \lim_{\varepsilon\to 0^+} \varepsilon^{-(1-D)}\, V(\varepsilon) \tag{2.16}$$

is called the *Minkowski content* of $\mathcal{L}$ (or of $\partial\Omega$). In other words, $\mathcal{L}$ is Minkowski measurable if $\mathcal{M}_* = \mathcal{M}^*$, and this common value, denoted by $\mathcal{M}$, lies in $(0, +\infty)$.

There is another way to show that $\mathcal{L}$ (in the present case, $\mathcal{L} = \mathcal{L}_{CS}$, the Cantor string) is not Minkowski measurable. This can be seen by using the principal complex dimensions of $\mathcal{L}$, as defined by (2.9); in other words, the complex dimensions with maximal real part D. Indeed, the following useful Minkowski measurability criterion was obtained in [**Lap-vF1–4**].

THEOREM 2.2 (Minkowski measurability and complex dimensions; [**Lap-vF4**, Thm. 8.15]). *Under suitable hypotheses,*[15] *the following statements are equivalent:*

(i) *$\mathcal{L}$ is Minkowski measurable (with Minkowski dimension $D \in (0, 1)$).*

---

[13]An entirely analogous definition of $\mathcal{M}_*$ and $\mathcal{M}^*$ can be given for any fractal string $\mathcal{L}$; simply replace $V_{CS}(\varepsilon)$ by $V(\varepsilon) = V_{\mathcal{L}}(\varepsilon)$ in (2.14) and (2.15), respectively.

[14]This fact was first established in [LapPo1–2], by using a direct computation and Theorem 2.3, and then extended in [Lap-vF2–4] to a whole class of examples (including lattice self-similar strings and generalized Cantor strings; see [**Lap-vF4**, §8.4.2 and §10.1]). Another, more conceptual, proof was given in [**Lap-vF4**, Ch. 8] by using the existence of nonreal principal complex dimensions of the Cantor string; see Theorem 2.2 and the comments following it.

[15]In essence, we assume that $\mathcal{L}$ is languid (in the sense of footnote 10) for a screen S passing between the vertical line $\{Re(s) = D\}$ and all of the complex dimensions of $\mathcal{L}$ with real part $< D$.

(ii) *The only principal complex dimension of* $\mathcal{L}$ *is* D *itself, and it is simple.*

Observe that for the Cantor string $\mathcal{L}_{CS}$, there are infinitely many complex conjugate nonreal complex dimensions with real part D. Furthermore, $D = \log_3 2$ (like each of the complex dimensions of $\mathcal{L}_{CS}$ in (2.5)) is simple; i.e., it is a simple pole of ζ_{CS}. Therefore, this yields another proof of the fact that $\mathcal{L}_{CS}$ (or, equivalently, the Cantor set C) is not Minkowski measurable.

There is another, very useful, *characterization of Minkowski measurability*, obtained in [**LapPo2**] and announced in [**LapPo1**].

THEOREM 2.3 (Minkowski measurability and fractal strings; [**LapPo2**]). *Let* $\mathcal{L} = (\ell_j)_{j=1}^{\infty}$ *be an arbitrary fractal string* (*of Minkowski dimension* $D \in (0,1)$). *Then, the following statements are equivalent:*

(i) $\mathcal{L}$ *is Minkowski measurable.*

(ii) $\ell_j \sim Lj^{-1/D}$ as $j \to \infty$, *for some constant* $L \in (0, +\infty)$.[16]

In this case, the Minkowski content $\mathcal{M}$ *of* $\mathcal{L}$ *is given by*

$$\mathcal{M} = \frac{2^{1-D}}{1-D} L^D. \tag{2.17}$$

REMARK 2.4. (*a*) The proof of Theorem 2.3 given in [**LapPo2**] is analytical. Later, a different approach to a part of that proof was taken by Kenneth Falconer in [**Fa2**], based on a suitable dynamical system, and more recently, by Jan Rataj and Steffen Winter in [**RatWi**], based on aspects of geometric measure theory.

(*b*) If $\zeta_{\mathcal{L}}$ has a meromorphic continuation to a neighborhood of D and either condition (i) or (ii) of Theorem 2.3 is satisfied (or certainly, if the hypotheses and either condition (*i*) or (*ii*) of Theorem 2.2 hold), then

$$\mathcal{M} = \frac{2^{1-D}}{D(1-D)} \operatorname{res}(\zeta_{\mathcal{L}}, D). \tag{2.18}$$

(*c*) Even though the Cantor string $\mathcal{L}_{CS}$ is not Minkowski measurable, it is the case that its *average Minkowski content*, $\mathcal{M}_{av}$, defined as a suitable Cesaro average of $V_{CS}(\varepsilon)\varepsilon^{-(1-D)}$ (see the $N = 1$ case of footnote 65), exists and can be explicitly computed in terms of $\operatorname{res}(\zeta_{\mathcal{L}}, D)$; see [**Lap-vF4**, §8.4.3]. The same is true for any lattice self-similar string; see [**Lap-vF4**, Thm. 8.23].[17] More specifically, a lattice self-similar string is not Minkowski measurable but its average Minkowski content, $\mathcal{M}_{av}$, exists in $(0, +\infty)$ and is also given by the right-hand side of (2.18); see [**Lap-vF4**, Thm. 8.30].

(d) More generally, a self-similar string is Minkowski measurable if and only if it is nonlattice. In this case, its Minkowski content, $\mathcal{M}$, is given by either (2.17) or (2.18); see [**Lap-vF4**, Thms. 8.23 and 8.36]. Further, we have $\mathcal{M}_{av} = \mathcal{M}$, since there is no need to take any averaging anymore.

2.2. Other examples of fractal explicit formulas. Let $\mathcal{L} = (\ell_j)_{j \geq 1}$ be a fractal string. Then, it is a vibrating object and its (normalized frequency) spectrum consists of the numbers $f_{j,n} = n \cdot \ell_j^{-1}$, where $n, j \in \mathbb{N} = \{1, 2, \cdots\}$.

[16]Here, $\ell_j \sim m_j$ as $j \to \infty$ means that $\ell_j / m_j \to 1$ as $j \to \infty$.

[17]The precise definition of (bounded) self-similar strings is given in [**Lap-vF4**, Ch. 2]. Here, we simply recall that a self-similar string is said to be *lattice* if its distinct scaling ratios generate a (multiplicative) group of rank 1. It is said to be *nonlattice*, otherwise. The detailed structure of the complex dimensions of self-similar strings is discussed in [**Lap-vF4**, Chs. 2 and 3].

One can think of $\mathcal{L}$ as being composed of infinitely many ordinary Sturm–Liouville strings, with lengths ℓ_j, vibrating independently of one another and with their endpoints fixed (i.e., corresponding to homogeneous Dirichlet boundary conditions for the one-dimensional Laplacian $-d^2/dx^2$ on the open set $\Omega \subseteq \mathbb{R}$).

One of the major themes of fractal string theory is the study of the interplay between the geometry and the spectra of fractal strings.

Let $N_{\mathcal{L}}$ be the *geometric counting function* of $\mathcal{L}$, given by (here, $\#A$ denotes the cardinality of a finite set A)

$$N_{\mathcal{L}}(x) = \#\{j \geq 1 : \ell_j^{-1} \leq x\}, \text{ for } x > 0. \tag{2.19}$$

Similarly, let N_ν denote the (*frequency* or) *spectral counting function* of $\mathcal{L}$:

$$N_\nu(x) = \#\{f : f \text{ is a frequency of } \mathcal{L}, \text{ with } f \leq x\}, \text{ for } x > 0. \tag{2.20}$$

Then, $N_{\mathcal{L}}$ and N_ν are connected via the following identity, for all $x > 0$:[18]

$$N_\nu(x) = \sum_{j=1}^{\infty} N_{\mathcal{L}}\left(\frac{x}{j}\right). \tag{2.21}$$

Essentially equivalently, the geometric and spectral zeta functions $\zeta_{\mathcal{L}}$ and ζ_ν of $\mathcal{L}$ are connected by the following key identity (first observed in [**Lap2**], [**Lap3**]):[19]

$$\zeta_\nu(s) = \zeta(s) \cdot \zeta_{\mathcal{L}}(s), \tag{2.22}$$

where ζ denotes the classic Riemann zeta function, initially defined by $\zeta(s) := \sum_{n=1}^{\infty} n^{-s}$ for $Re(s) > 1$ and then meromorphically continued to all of $\mathbb{C}$ (see, e.g., [**Edw, Pat, Tit**]).

Note that in order to apprehend the principal complex dimensions of $\mathcal{L}$ and their effect on the spectrum of $\mathcal{L}$, one must work in the *closed critical strip* $\{0 \leq Re(s) \leq 1\}$ of ζ or, if one excludes the extreme cases when $D = 0$ or $D = 1$, in its open counterpart, $\{0 < Re(s) < 1\}$, henceforth referred to as the *critical strip*.

Now, let us give a few examples of *fractal explicit formulas*, analogous to the fractal tube formulas discussed in §2.1 above. In the spirit of this overview, we will not strive here for either mathematical precision or for the most general statements but instead refer to [**Lap-vF4**, Chs. 5 and 6] for all of the details and a much broader perspective.

Assume, for clarity, that all of the complex dimensions of $\mathcal{L} = (\ell_j)_{j=1}^{\infty}$ are simple. Then, under appropriate hypotheses, we obtain the following pointwise or distributional explicit formulas with error terms:[20]

$$N_{\mathcal{L}}(x) = \sum_{\omega \in \mathcal{D}_{\mathcal{L}}} c_w \frac{x^\omega}{\omega} + R_{\mathcal{L}}(x) \tag{2.23}$$

and

$$N_\nu(x) = \zeta_{\mathcal{L}}(1)x + \sum_{\omega \in \mathcal{D}_{\mathcal{L}}} c_\omega\, \zeta(\omega) \frac{x^\omega}{\omega} + R_\nu(x), \tag{2.24}$$

---

[18]Note that for each fixed $x > 0$, the sum in (2.21) contains only finitely many nonzero terms. However, as $x \to +\infty$, the number of these terms tends to $+\infty$.

[19]Here, $\zeta_\nu(s)$ is given for $Re(s) > 1$ by $\zeta_\nu(s) := \sum_f f^{-s}$, where f ranges through all the (normalized) frequencies of $\mathcal{L}$, and is then meromorphically continued wherever possible.

[20]Under somewhat stronger assumptions, we obtain exact formulas; namely, either $R_{\mathcal{L}}(x) \equiv 0$ or (more rarely) $R_\nu(x) \equiv 0$.

where, as before, $c_\omega := \text{res}(\zeta_{\mathcal{L}}, \omega)$ for every $\omega \in \mathcal{D}_{\mathcal{L}}$ and $R_{\mathcal{L}}$ and R_ν are error terms which can be suitably estimated (either pointwise or distributionally); see [**Lap-vF4**, §6.2]. (Note that $\zeta_{\mathcal{L}}(1) = |\Omega|_1 = \sum_{j=1}^{\infty} \ell_j$, the *total length* of the fractal string $\mathcal{L}$.)

Analogous formulas, now necessarily interpreted distributionally rather than pointwise, can be obtained for the positive measures η and ν, respectively defined as $dN_{\mathcal{L}}/dx$ and dN_ν/dx (the distributional derivatives of $N_{\mathcal{L}}$ and N_ν) and referred to as the *geometric* and *spectral densities of states*; see [**Lap-vF4**, §6.3.1]. Alternatively,

$$\eta([0,x]) := \frac{N_{\mathcal{L}}([0,x]) + N_{\mathcal{L}}([0,x))}{2}, \text{ for all } x > 0,$$

and similarly for ν and N_ν.

REMARK 2.5. (*Fractal string theory and its ramifications.*) Fractal string theory and the associated theory of complex dimensions has been developed in many directions and applied to a variety of fields, including harmonic analysis, fractal geometry, number theory and arithmetic geometry, complex analysis, spectral geometry, geometric measure theory, probability theory, nonarchimedean analysis, operator algebras and noncommutative geometry, as well as dynamical systems and mathematical physics.

In particular, in [**Lap-vF4**, Ch. 13], are discussed a variety of extensions or applications of fractal string theory in diverse settings (prior to the development of the higher-dimensional theory of complex dimensions and of fractal zeta functions in [LapRaZu1–10], to be partly surveyed in §3), including fractal tube formulas for fractal sprays (especially, self-similar sprays and tilings), in [**Lap-vF4**, §12.1] (based on [LapPe2–3, LapPeWi1–2], [**Pe, PeWi**]), complex dimensions and fractal tube formulas for p-adic fractal (and self-similar) strings, in [**Lap-vF4**, §12.2] (based on [LapLu1–3, LapLu-vF1–2]), multifractal zeta functions and strings, in [**Lap-vF4**, §12.3] (based on [**LapRo, LapLevyRo, ElLapMcRo**]), random fractal strings and zeta functions, in [**Lap-vF4**, §12.4] (based on [**HamLap**]), as well as fractal membranes (or 'quantized fractal strings') and their associated moduli space, in [**Lap-vF4**, §12.5] (based on [**Lap7,LapNes**]). See also [**Lap-vF4**, §12.2.1] for a brief description of a first direct attempt at a higher-dimensional theory, in [**LapPe1**], where a fractal tube formula was obtained for the Koch snowflake curve via a direct computation.

In addition to the aforementioned articles, we refer the interested reader to the research books [Lap-vF2–4], [**Lap7**], [**LapRaZu1**], [**HerLap1**], [**Lap10**], as well as [**Lap-vF6**], [**CaLapPe-vF**] and [**LapRaRo**], along with the research papers and survey articles [**Cae**], [CobLap1–2], [**CranMH**], [**DemDenKoÜ**], [**DemKoÖÜ**], [**DeniKoÖÜ**], [**DeniKoÖRÜ**], [**deSLapRRo**], [**DubSep**], [Es1–2], [EsLi1–2], [**Fa2**], [**Fr**], [**FreiKom**], [**Gat**], [**Ger**], [GerScm1–2], [**HeLap**], [HerLap2–5], [**KeKom**], [**Kom**], [**KomPeWi**], [**KoRati**], [LalLap1–2], [Lap1–6], [Lap8–9], [LapMa1–2], [LapPo1–3], [LapRaZu2–10], [**LapRoZu**], [**Lap-vF1**], [**Lap-vF5**], [**Lap-vF7**], [**LapWat**], [**LevyMen**], [**LiRadz**], [**MorSep**], [MorSepVi1–2], [Ol1–2], [Ra1–2], [**RatWi**], [Tep1–2], [vF1–2], [**Wat**] and [Zu1–2], for various aspects of fractal string theory and its applications.

2.3. Analogy with Riemann's explicit formula. One of the most beautiful formulas in mathematics, in the author's opinion, is Riemann's explicit formula.

The latter connects the *prime number counting function*

$$\Pi_{\mathcal{P}}(x) := \#\{p \in \mathcal{P} : p \leq x\}, \text{ for } x > 0 \tag{2.25}$$

(where $\mathcal{P}$ denotes the set of prime numbers) and the zeros of the Riemann zeta function. Because it is simpler to state (as well as to justify), although a precise proof was provided only about forty years after the publication of Riemann's celebrated 1858 paper [**Rie**], we will state a modern form of this formula. Namely, consider the *weighted prime powers counting function*

$$\varphi(x) := \sum_{p^n \leq x} \frac{1}{n}, \text{ for } x > 0, \tag{2.26}$$

where the sum ranges over all prime powers p^n (counted with a weight $1/n$, for every $n \in \mathbb{N}$). Then

$$\varphi(x) = Li(x) - \sum_{\rho} Li(x^{\rho}) + \int_x^{+\infty} \frac{1}{t^2 - 1} \frac{dt}{t \log t} - \log 2, \tag{2.27}$$

where $Li(x) := \int_0^x \frac{dt}{\log t}$ is the *integral logarithm* and the infinite sum in (2.27) is taken over all of the critical zeros ρ of ζ (in $\{0 < Re(s) < 1\}$), while the negative of the integral in (2.27) corresponds to the same sum but now taken over the trivial zeros $-2, -4, -6, \cdots$ of ζ. Furthermore, the leading term, $Li(x)$, corresponds to the (simple) pole of $\zeta = \zeta(s)$ at $s = 1$.

It follows from this formula (combined with an appropriate analysis, for instance based on a Tauberian theorem) that[21]

$$\Pi_{\mathcal{P}}(x) \sim Li(x) \quad \text{as } x \to +\infty, \tag{2.28}$$

or equivalently, that

$$\Pi_{\mathcal{P}}(x) \sim \frac{x}{\log x} \quad \text{as } x \to +\infty, \tag{2.29}$$

which is the statement of the famous Prime Number Theorem (PNT), as conjectured independently by Gauss (1792) and Legendre (1797) and proved independently by Hadamard ([**Had2**]) and de la Vallée Poussin ([**dV1**]) in the same year (1896), but a century later. Note, however, that it also took about forty years after the publication of [**Rie**] in 1858 in order to prove PNT, in the form of (2.28), about the same amount of time it took to rigorously justify Riemann's original explicit formula; see van Mangoldt's work [vM1–2], along with Ingham's book [**Ing**].[22] The latter formula is obtained by first proving (2.27) and then by using Möbius inversion [**Edw**, **Ove**] as follows (with φ now given by the explicit formula (2.27)):

$$\Pi_{\mathcal{P}}(x) = \sum_{n=1}^{\infty} \frac{\mu(n)}{n} \varphi\left(x^{1/n}\right), \tag{2.30}$$

[21]The expression $f(x) \sim g(x)$ as $x \to +\infty$ means that $f(x)/g(x) \to 1$ as $x \to +\infty$.

[22]There are more direct (but less insightful) ways to prove PNT. They also typically require to know that $\zeta(s)$ does not have any zero on the vertical line $\{Re(s) = 1\}$ (Hadamard, [**Had1**], 1893). However, in order to obtain a version of (2.28) with error term (PNT with error term), the Riemann–von Mangoldt explicit formula (2.27) (or one of its counterparts) is the most reliable tool (combined, for example, with an appropriate Tauberian theorem); see [**dV2**] along, e.g., with [**Edw**] (for a detailed history and analysis of Riemann's paper, [**Rie**]) and, especially, [**Ing**, **Ivi**, **KarVo**, **Pat**, **Tit**].

where μ denotes the *Möbius function* defined on $\mathbb{N}$ by $\mu(n) = (-1)^k$ if $n \geq 2$ is a product of k distinct primes, $\mu(1) = 1$, and $\mu(n) = 0$ otherwise. Riemann's original explicit formula is then deduced by substituting into (2.30) the expression of φ given by (2.27).

The analogy between Riemann's explicit formula (in any of its various disguises) and the fractal explicit formulas discussed in §2.1 and §2.2 is now apparent. The (critical) zeros of ζ correspond to the (nonreal) complex dimensions of $\mathcal{L}$, while the prime counting function $\Pi_{\mathcal{P}}$ in (2.25) (or the weighted prime powers counting function φ in (2.26)) corresponds to the geometric or spectral counting function $N_{\mathcal{L}}$ or N_ν (e.g., in (2.23) or (2.24), respectively), or else (in a more sophisticated but also more geometric form) to the tube function (or distribution) $V(\varepsilon)$ in (2.10).

In particular, the *oscillations* (in the counting function of the primes) associated with the (critical) zeros of ζ in the infinite sum appearing in (2.27) or in (2.30) correspond to the *geometric oscillations* (in the geometric counting function $N_{\mathcal{L}}$ in (2.23) or in the tube function $V(\varepsilon)$ in (2.10)) or to the *spectral oscillations* (in the frequency counting function N_ν in (2.24)). We will further discuss these oscillations in §2.4.

At this stage, it is natural for the reader to be troubled by the presence of zeros in (2.27) or (2.30), as opposed to just poles (or "complex dimensions") in (2.23), (2.24) and (2.10). However, this apparent discrepancy is quickly resolved by noting that the (necessarily simple) poles of (minus) the logarithmic derivative $-\zeta'(s)/\zeta(s)$ of $\zeta(s)$ correspond precisely to the zeros of $\zeta(s)$ and to its only pole (at $s = 1$, which accounts for the leading term $Li(x)$ in (2.27)). In addition, the residue of $-\zeta'(s)/\zeta(s)$ at a pole $s = \omega$ is a nonzero integer whose sign tells us whether it corresponds to a zero or a pole (here, $s = 1$) of $\zeta(s)$, and whose absolute value is the multiplicity of the zero or pole. As a simple exercise, the reader may wish to verify this statement and determine which sign of the residue corresponds to a zero or a pole.

In closing this subsection, we point out that the (pointwise or distributional) explicit formulas obtained in [**Lap-vF4**, Ch. 5], with or without error term, and used throughout [**Lap-vF4**, esp., in Chs. 6–11], extend Riemann's explicit formula (and its known number-theoretic counterparts) to a fractal, geometric, spectral, or dynamical setting in which the corresponding zeta functions do not necessarily have an Euler product or satisfy a functional equation. Furthermore, the general framework within which these explicit formulas are developed is sufficiently broad and flexible in order for the resulting formulas to be applied to a variety of situations (including arithmetic ones) and to help unify, in the process, aspects of fractal geometry and number theory, both technically and conceptually.[23]

2.4. The meaning of complex dimensions. In light of the fractal tube formulas and the other fractal explicit formulas discussed in §2.2 and §2.3, the following intuition of the notion of complex dimensions can easily be justified, mathematically. The *real parts* of the complex dimensions correspond to the *amplitudes* of '*geometric waves*' (propagating through the 'space of scales'), while the *imaginary parts* of the complex dimensions correspond to the *frequencies* of those waves.

---

[23]The interested reader can find in [**Lap-vF4**, §5.1.1 and §5.6] many references about number-theoretic and analytic explicit formulas in a variety of contexts, including [Wei4–5, Den1–3, DenSchr, Har1–3].

An analogous interpretation can be given in the spectral setting and in the dynamical setting. A common thread to these interpretations is provided by the (generalized) explicit formulas of [**Lap-vF4**] mentioned at the end of §2.3. Associated key words are *oscillations*, vibrations, and wave-like phenomena, which could also be applied to the number-theoretic setting corresponding to Riemann's explicit formula for the prime number counting function and discussed in §2.3.

The author has conjectured since the early 1990s that (possibly generalized or even virtual) fractal geometries and arithmetic geometries pertained to the same mathematical realm. Consequently, there should exist a fractal-like geometry whose complex dimensions are the Riemann zeros (the critical zeros of $\zeta = \zeta(s)$); see, especially, the author's book [**Lap7**]. There, in particular, an extension (and 'quantization') of the notion of fractal string is introduced and coined 'fractal membrane'. It turns out to be a noncommutative space, in the sense of [**Con1**]. The associated moduli space of fractal membranes (which can be thought of physically as a quantization of the moduli space of fractal strings) plays a fundamental role in [**Lap7**] in order to provide a conjectural explanation of why the Riemann hypothesis should be true, both for the classic Riemann zeta function and for all number-theoretic zeta functions (or L-functions, [ParsSh1–2], [**Sarn**], [**Lap-vF4**, App. A], [**Lap7**, Apps. B, C & E]) occurring in arithmetic geometry. It is expressed in terms of a (still conjectural) noncommutative dynamical system on the moduli space of fractal membranes, as well as of its counterparts on the associated moduli spaces of zeta functions (or 'partition functions') and of divisors (i.e., zeros and poles) on the Riemann sphere (which is the natural realm of the Riemann zeros and more generally, of the complex fractal dimensions). See, especially, [**Lap7**, Ch. 5].

2.5. Fractality, complex dimensions and irreality. Since, as we have seen, the imaginary parts of the complex dimensions give rise to oscillations in the intrinsic geometry (or in the spectra) of fractal strings, it is natural to wonder whether one could not define the elusive notion of fractality in terms of complex dimensions.

In [Lap-vF1–4] (as well as later, in higher dimensions, in [**LapRaZu1**] to be discussed in §3 below), an object is said to be '*fractal*' if it has at least one *nonreal* complex dimension[24] and hence, in other words, according to the explicit formulas discussed in §§2.1–2.4 (when $N = 1$) and in §3.5 (when $N \geq 1$ is arbitrary, where N is the dimension of the embedding space), if it has *intrinsic geometric, spectral, dynamical or arithmetic oscillations.*[25]

In the case of a fractal string, the complex dimensions are the poles of the associated geometric zeta functions, whereas (anticipating on the discussion of [**LapRaZu1**] given in §3), in higher dimensions (i.e., for bounded subsets of $\mathbb{R}^N$ or, more generally, for relative fractal drums in $\mathbb{R}^N$, for any $N \geq 1$), the complex dimensions are the poles of the associated fractal zeta functions. Furthermore, as was alluded to near the end of §2.3, in the arithmetic setting, the role played by the

[24]Since nonreal complex dimensions come in complex conjugate pairs, a fractal-like object must have at least two complex conjugate nonreal complex dimensions. In fact, in the geometric setting, it typically has infinitely many nonreal complex conjugate pairs of them.

[25]For various examples for which the source of the oscillations is of a dynamical (respectively, spectral) nature, see [**Lap-vF4**, Ch. 7 and §12.5.3] (respectively, [**Lap-vF4**, Chs. 6 and 9–11]), while for the case when it is of a geometric (respectively, arithmetic) nature, see [**Lap-vF4**, Chs. 6 and 9–13] (respectively, [**Lap-vF4**, Chs. 9 and 11]).

complex dimensions in fractal geometry is now essentially played by the Riemann zeros, or by their more general number-theoretic analogs (e.g., the critical zeros of automorphic L-functions or of zeta functions of varieties over finite fields).

REMARK 2.6. (*Reality principle.*) Geometrically, the fact that the nonreal complex dimensions come in complex conjugate pairs is significant. This is what enables us, for instance, to obtain a real-valued (and even positive) expression for the tube function $V(\varepsilon)$ in the fractal tube formulas of §2.1.[26] For example, the fractal tube formula for the Cantor string in (2.12) can be written as follows (with $D := \log_3 2$ and $\mathbf{p} := 2\pi/\log 3$):

(2.31)

$$V_{CS}(\varepsilon) = \frac{2^{-D}\varepsilon^{1-D}}{D(1-D)\log 3} + (2\varepsilon)^{1-D}\left(\sum_{n=1}^{\infty} Re\left(\frac{(2\varepsilon)^{-in\mathbf{p}}}{(D+in\mathbf{p})(D-in\mathbf{p})}\right)\right) - 2\varepsilon,$$

which in turn can be further expressed in terms of real-valued functions involving sine and cosine, by using Euler's identity

$$(2\varepsilon)^{-in\mathbf{p}} = \cos(n\mathbf{p}\log 2\varepsilon) - i\sin(n\mathbf{p}\log 2\varepsilon), \text{ for all } n \in \mathbb{Z}. \tag{2.32}$$

It is obvious that the Cantor string $\mathcal{L}_{CS}$ (and hence, also the Cantor set) is fractal according to the above definition. Indeed, in light of (2.5), it has infinitely many complex conjugate pairs of nonreal (principal) complex dimensions, $D \pm in\mathbf{p}$, with $n \in \mathbb{N}$ (as well as with D and $\mathbf{p}$ as in Remark 2.6). This is also apparent in the fractal tube formula (2.12) (or its "real" form (2.31) in Remark 2.6), as well as in the following pointwise explicit formulas for N_{CS} and $N_{\nu,CS}$, the geometric and spectral counting functions of $\mathcal{L}_{CS}$, respectively:[27]

$$N_{CS}(x) = \frac{1}{2\log 3}\sum_{n\in\mathbb{Z}} \frac{x^{D+in\mathbf{p}}}{D+in\mathbf{p}} - 1 \tag{2.33}$$

and

$$N_{\nu,CS}(x) = x + \frac{1}{2\log 3}\sum_{n\in\mathbb{Z}} \zeta(D+in\mathbf{p})\frac{x^{D+in\mathbf{p}}}{D+in\mathbf{p}} + O(1). \tag{2.34}$$

It is shown in [**Lap-vF4**, Ch. 11] that the Riemann zeta function $\zeta = \zeta(s)$ and many other arithmetic zeta functions (as well as other Dirichlet series) cannot have infinite vertical arithmetic progressions of critical zeros.[28] It follows that the spectral oscillations in (2.34), just like the geometric oscillations in (2.33), subsist. In fact, this result is established by reasoning by contradiction and using the counterparts of the explicit formulas (2.33) and (2.34), as well as by proving that (virtual) generalized Cantor strings always have both geometric and spectral oscillations of a suitable kind (namely, of leading order x^D as $x \to +\infty$); see [**Lap-vF4**, Ch. 10]. The sought for contradiction is then reached by making the nonreal complex

[26] An analogous comment can be made about the geometric and spectral counting functions of §2.2 or even the prime numbers counting functions of §2.3 (provided the "complex dimensions" are interpreted there as the zeros and the pole(s) of $\zeta = \zeta(s)$ or of its counterpart; that is, as the poles of (minus) the logarithmic derivative of the (arithmetic) zeta function under consideration.

[27] See [**Lap-vF4**, Eqns. (1.31) and (6.57)], where in the latter equations, we have slightly adapted the formula because our Cantor string (unlike in [**Lap-vF4**, Ch. 2]) has total length 1 (rather than 3). Note that (2.33) is an exact explicit formula whereas (2.34) has an error term.

[28] Unknown to the authors of [Lap-vF1–2] at the time, C. R. Putnam [Put1–2] had established a similar result in the case of $\zeta = \zeta(s)$, by completely different methods which only extended to a few arithmetic zeta functions.

dimensions $D + in\mathbf{p}$ (with $n \in \mathbb{Z}\backslash\{0\}$) of the (virtual, generalized) Cantor string coincide with the presumed zeros of ζ in infinite arithmetic progression along the vertical line $\{Re(s) = D\}$. (Here, $D \in (0,1)$ and the period $\mathbf{p} > 0$ can be chosen to be arbitrary.)[29] Then, in light of the counterpart of (2.34) in this context, we deduce that N_ν does not have any oscillations of leading order x^D, in contradiction to what is claimed just above. Observe the analogy with the method of proof (outlined in the latter part of §2.6.2 below) of a key result of [**LapMa2**] connecting the Riemann hypothesis and inverse spectral problems for fractal strings.

REMARK 2.7. (*Multiplicative oscillations.*) In order to better understand the nature of the (multiplicative) oscillations intrinsic to 'fractality', as expressed by fractal explicit formulas and the presence of nonreal complex dimensions (necessarily in complex conjugate pairs), it is helpful to first consider the following simple situation. Think, for instance, that each term of the form x^z (with $x > 0$ and $z \in \mathbb{C}$ such that $z = d + i\tau$, where $d \in \mathbb{R}$ and $\tau > 0$, say) arising in a given fractal (or arithmetic) explicit formula can be written as follows:

$$x^z = x^d x^{i\tau} = x^d(\cos(\tau \log x) + i \sin(\tau \log x)). \tag{2.35}$$

Now, clearly, the real part d of z governs the amplitude of the oscillations while the imaginary τ of z governs the frequency of the oscillations. Observe that physically, the term x^z can be viewed as a multiplicative analog of a plane wave (or a standing wave). The different terms of the form x^z (or ε^{-z}) occurring in the infinite sum ranging over all of the visible complex dimensions and appearing in a given fractal explicit formula (or fractal tube formula), such as (2.23), (2.24), (2.33), (2.34) (or (2.10), (2.12), (2.31)) provide a whole spectrum of amplitudes and frequencies associated with the corresponding superposition of 'standing waves'. The fact that these waves arise in *scale space* (rather than in ordinary frequency or momentum space, in physicists' terminology), explains why the corresponding oscillations are viewed multiplicatively (rather than additively) here.

This is very analogous to what happens for Fourier series. Observe, however, that unlike for Fourier series, the frequencies are no longer, in general, integer multiples of a given fundamental frequency.[30] In addition, the varying amplitudes of the oscillations do not have a counterpart for ordinary Fourier series. If one replaces the classic theory of Fourier series by Harald Bohr's less familiar but more general theory of almost periodic functions [**Boh**] (usually associated with purely imaginary rather than with arbitrary complex numbers z), one is getting closer to improving one's understanding of the situation. Nevertheless, the fact that typically, the complex dimensions form a countably infinite and discrete subset of $\mathbb{C}$ (rather than of $\mathbb{R}$) adds a lot of complexity and richness to the corresponding generalized 'almost periodic' functions (or distributions, [**Katzn**]).

[29] This is why the corresponding Cantor strings are not geometric, in general, but instead generalized (and virtual) fractal strings, in the sense of [**Lap-vF4**, Chs. 4 and 10–11].

[30] In that sense, the Cantor string and more generally, all lattice self-similar strings, are rather exceptional. In contrast, the complex dimensions of (bounded or unbounded) nonlattice self-similar strings have a much richer quasiperiodic structure; see [**Lap-vF4**, Ch. 3].

EXAMPLE 2.8. (*The a-string.*) For a simple example of a string that is not fractal, in the above sense, consider the a-string, where $a > 0$ is arbitrary.[31] Thus,

$$\Omega = \Omega_a := [0,1]\backslash\{j^{-a} : j \in \mathbb{N}\} = \bigcup_{j=1}^{\infty}((j+1)^{-a}, j^{-a}) \tag{2.36}$$

and hence,

$$\partial\Omega = \partial\Omega_a = \{j^{-a} : j \in \mathbb{N}\} \cup \{0\}, \tag{2.37}$$

while

$$\mathcal{L} = \mathcal{L}_a = (j^{-a} - (j+1)^{-a})_{j=1}^{\infty}. \tag{2.38}$$

It is shown in [**Lap-vF4**, Thm. 6.21] that the geometric zeta function $\zeta_{\mathcal{L}_a}$ of $\mathcal{L}_a$ admits a meromorphic continuation to all of $\mathbb{C}$ and that the complex dimensions of $\mathcal{L}_a$ are all simple with

$$\mathcal{D}_{\mathcal{L}_a} = \{D, -D, -2D, -3D, \cdots\}, \tag{2.39}$$

where $D = 1/(a+1)$ is the Minkowski dimension of $\mathcal{L}_a$ (or equivalently, of $\partial\Omega_a$).[32], Note that $D \in (0,1)$ whereas $H = 0$ for any $a > 0$, where H is the Hausdorff dimension of the compact set $\partial\Omega_a \subseteq \mathbb{R}$. This illustrates the fact that the Minkowski dimension, and not the Hausdorff dimension, is the proper notion of real dimension pertaining to the theory of complex fractal dimensions.

Since the a-string $\mathcal{L}_a$ (or $\partial\Omega_a$) does not have any nonreal complex dimension, it is *not* fractal, in the above sense. This conclusion is entirely compatible with our intuition according to which $\mathcal{L}_a$ (or the associated compact set $\partial\Omega_a \subseteq \mathbb{R}$ in (2.37)) does not have much complexity.

EXAMPLE 2.9. (*Self-similar strings.*) Next, we discuss a more geometrically interesting family of fractal strings, namely, *self-similar fractal strings* (as introduced and studied in detail in [Lap-vF1–4], see [**Lap-vF4**, Chs. 2 and 3]). It is shown in the just mentioned references that self-similar strings have infinitely many nonreal complex dimensions. In fact, their only real dimension is the Minkowski dimension D (and it is simple); as a result, D is also the maximal real part of the complex dimensions of such strings. Thus, as expected (since their boundaries are nontrivial self-similar sets in $\mathbb{R}$), self-similar strings are always fractal.

Now, as we may recall from our earlier discussion, there are two kinds of self-similar strings, *lattice strings* and *nonlattice* strings. For both types of (i.e., for all) self-similar strings, the geometric zeta function admits a meromorphic continuation to all of $\mathbb{C}$. In the *lattice case* (i.e., when $G = r^{\mathbb{Z}}$, for some $r \in (0,1)$, where G is the multiplicative group generated by the distinct scaling ratios and the 'gaps' of the self-similar string), the complex dimensions are periodically distributed (with the same vertical period $\mathbf{p} := 2\pi/\log(r^{-1}) > 0$, called the *oscillatory period* of the

[31] The a-string, then viewed in [**Lap1**] as a one-dimensional fractal drum, was the first example of fractal string (before that notion was formalized in [**LapPo2**]) and was used in [**Lap1**, Exple. 5.1 and Exple. 5.1'] to show that the remainder estimates obtained in [**Lap1**] for the spectral asymptotics of fractal drums are sharp, in general (and in every possible dimension).

[32] It is possible that some of the numbers $-nD$, with $n \in \mathbb{N}$, do not truly appear in (2.39), depending on the value of a. However, D is always a complex dimension and "typically", we have an equality (rather than an inclusion) in (2.39).

given lattice string) along finitely many vertical lines.[33] Furthermore, on each of these vertical lines, the multiplicity of the complex dimensions is the same, while the right most vertical line is $\{Re(s) = D\}$, on which lie the principal complex dimensions (which are necessarily all simple).

By contrast, in the *nonlattice case* (i.e., when the rank of the group G is strictly greater than 1), the complex dimensions are no longer periodically distributed. In fact, typically, on a given vertical line $\{Re(s) = \alpha\}$, with $\alpha \in \mathbb{R}$, there is either zero, or one complex dimension, necessarily D itself (only if $\alpha = D$), or else two complex conjugate nonreal complex dimensions $\omega, \overline{\omega}$ (with $Re(\omega) = \alpha < D$). (Note that there can be at most countably vertical lines containing at least one complex dimension.) Furthermore, on the right most vertical line, the only complex dimension is D itself, and it is simple. (Hence, according to a version of Theorem 2.2 obtained in [**Lap-vF4**, §8.4], a nonlattice string is always Minkowski measurable, in contrast to a lattice string, which is never so.)

Moreover, it is shown in [**Lap-vF4**, §3.4] by using Diophantine approximation that any nonlattice string $\mathcal{L}$ can be approximated by a sequence of lattice strings $\{\mathcal{L}_n\}_{n=1}^{\infty}$ with oscillatory periods $\mathbf{p}_n$ increasing exponentially fast to $+\infty$, as $n \to \infty$. Hence, the complex dimensions of $\mathcal{L}$ are themselves approximated by the complex dimensions of $\mathcal{L}_n$. As a result, the complex dimensions of a nonlattice string exhibit a *quasiperiodic structure.* (See *ibid* for more precision and for many examples of quasiperiodic patterns of complex dimensions of nonlattice strings.)

Finally, we refine (as in [**LapRaZu1**]) the notion of fractality by saying that a geometric object is '*fractal in dimension d*', where $d \in \mathbb{R}$, if it has a nonreal complex dimension with real part d. Hence, 'fractality' in our sense is equivalent to fractality in dimension d, for some $d \in \mathbb{R}$. (Clearly, $d \leq D$.) Also, fractality in dimension D amounts to the existence of nonreal principal complex dimensions. According to [**Lap-vF4**, Thms. 2.16 and 3.6], this is always the case for lattice strings (such as the Cantor string) but is never the case for nonlattice strings.

In light of the above discussion, a lattice string is fractal in dimension d for at least one value but at most finitely many values of d, whereas by contrast, nonlattice strings are fractal in dimension d for infinitely (but countably) many values of d. In fact, in the generic nonlattice case, a significantly stronger statement is true;[34] namely, the countable set of such d's is dense in a single compact interval of the form $[d_{\min}, D]$, with $-\infty < d_{\min} < D$. This latter density result was obtained by the authors of [**MorSepVi1**] who thereby proved a conjecture made in the generic nonlattice case in [**Lap-vF5**] (see also [Lap-vF3–4]).[35]

In §3.6, we will further discuss and broaden the notion of fractality (and the related notion of hyperfractality), but now by focusing on higher-dimensional examples, such as the devil's staircase (i.e., the Cantor graph), rather than on fractal strings.

---

[33]The number of vertical lines (counted according to multiplicity) is equal to the degree of the polynomial obtained after making the change of variable $w := r^s$ in the denominator of the expression for $\zeta_{\mathcal{L}}(s) = (\sum_{k=1}^{K} g_k^s)/(1 - \sum_{j=1}^{J} r_j^s)$, where $(r_j)_{j=1}^{J}$ are the (not necessarily distinct) scaling ratios and $(g_k)_{k=1}^{K}$ are the gaps of $\mathcal{L}$, with $J \geq 2$.

[34]A nonlattice string is said to be *generic* if the group G generated by its M distinct scaling ratios is of rank M and $M \geq 2$. It is *nongeneric*, otherwise.

[35]In general, in the nongeneric nonlattice case, the set of such d's should be dense in at most finitely (but at least one, ending at D) compact and pairwise disjoint intervals; this is noted, by means of examples, independently in [**Lap-vF4**] and in [**DubSep**].

REMARK 2.10. The lattice/nonlattice dichotomy arose in probabilistic renewal theory [**Fel**], where it is usually referred to as the arithmetic/nonarithmetic dichotomy. In fractal geometry, it was used in [Lall1–3] in connection with self-similar sets and generalizations thereof, and then, e.g., in [Lap2–7], [**KiLap1**], [**Gat**], [**LeviVa**], [Lap-vF1–7], [**Fr**], [Sab1–3], [**HamLap**], [LapPe2–3], [LapPeWi1–2], [LapLa-vF2–3], [LapLu-vF1–2], [**KeKom**], [**Kom**], [**MorSep**], [MorSepVi1–2], [**DubSep**], [LapRaZu1–9] and [**Lap10**]. The notion of *generic* nonlattice self-similar string (or, more generally, spray or even set) was introduced in [Lap-vF3–5].

2.6. Inverse spectral problems and the Riemann hypothesis. In this subsection, we briefly discuss the intimate connection between direct (respectively, inverse) spectral problems for fractal strings and the Riemann zeta function (respectively, the Riemann hypothesis). For a recent survey of this subject, we refer the interested reader to [**Lap9**].

2.6.1. *Direct spectral problems for fractal strings.* Let $\mathcal{L} = (\ell_j)_{j=1}^{\infty}$ (or any of its geometric realizations $\Omega \subseteq \mathbb{R}$) be a fractal string of dimension $D \in (0,1)$,[36] and let N_ν denote the associated spectral counting function of $\mathcal{L}$, as defined in §2.2. It turns out that the leading spectral asymptotics of $\mathcal{L}$ are given by the so-called *Weyl term*, W (named after the well-known N-dimensional result in [Weyl1–2]).[37] Namely,

$$N_\nu(x) \sim W(x) \quad \text{as } x \to +\infty, \tag{2.40}$$

where $W = W(x)$ is the *Weyl term* given by

$$W(x) := |\Omega|_1 x, \ \text{ for } x > 0, \tag{2.41}$$

with $|\Omega|_1 = \sum_{j=1}^{\infty} \ell_j$ being the total length of the string $\mathcal{L}$ (or the 'volume' of Ω).

This is a very special (one-dimensional) case of the spectral asymptotics with error term obtained in [**Lap1**] for fractal drums in $\mathbb{R}^N$, with $N \geq 1$ arbitrary. (In fact, $N_\nu(x) = W(x) + R(x)$, with $R(x) = O(x^D)$ if $\mathcal{M}^* < \infty$ and $R(x) = O(x^{D+\varepsilon})$ for any $\varepsilon > 0$, otherwise, where $D \in (N-1, N)$ is the (upper) Minkowski dimension of Ω, relative to $\partial\Omega$ or in the terminology of §3.2, of the RFD $(\partial\Omega, \Omega)$. Here, the Weyl term $W = W(x)$ is proportional to $|\Omega|_N x^N$, where $|\Omega|_N$ is the N-dimensional volume of Ω.) It also follows, for example, from a result of [**LapPo2**].

The following theorem (joint with C. Pomerance), obtained in [**LapPo2**] (and first announced in [**LapPo1**]), resolved in the affirmative the one-dimensional case of the modified Weyl–Berry (MWB) conjecture stated in [**Lap1**] for fractal drums.[38]

We refer to [**Lap1, Lap3, LapPo2, Lap9**] and [**Lap-vF4**, §12.5] for physical and mathematical motivations, as well as for many further references (including [**BirSol**], [**BroCar**], [**CouHil**], [**FlLap**], [**Ger**], [GerScm1–2], [**Gi**], [Ho1–3], [Ivr1–3], [**LeviVa**], [Mel1–2], [Met1–2], [**Ph**], [ReSi1–3], [See1–3], [Lap2–3] and [**Lap-vF4**, App. B], along with those cited in footnote 38), concerning the original conjectures of Weyl [Weyl1–2] for 'smooth drums', Berry [Berr1–2] for 'fractal drums' and their later modifications and extensions in [Lap1–3], in particular.

---

[36] Recall that for a fractal string, we always have that $D \in [0,1]$.

[37] Here, $\mathcal{L}$ is viewed as the RFD $(\partial\Omega, \Omega)$, in the terminology of §3.2, with (upper) Minkowski dimension $D \in (0,1)$.

[38] In higher dimensions, the situation is not as clear cut and the MWB conjecture itself needs to be further modified; see, e.g., [**FlVa, LapPo3, MolVai**].

THEOREM 2.11 ([**LapPo2**]). *Let $\mathcal{L}$ be a fractal string which is Minkowski measurable and of Minkowski dimension $D \in (0,1)$. Then, N_ν, the spectral counting function of $\mathcal{L}$, admits a* monotonic *asymptotic second term, proportional to x^D. More specifically,*

$$N_\nu(x) = W(x) - c_D \mathcal{M} x^D + o(x^D) \quad \text{as} \to +\infty, \tag{2.42}$$

where $W(x) = |\Omega|_1 x$ is the Weyl term given by (2.41) and $\mathcal{M}$ is the Minkowski content of $\mathcal{L}$. Furthermore, the constant c_D is positive, depends only on D and is given explicitly by

$$c_D := (1-D)2^{-(1-D)}(-\zeta(D)), \tag{2.43}$$

where $\zeta = \zeta(s)$ denotes the Riemann zeta function.

Theorem 2.11 relies, in particular, on (the easier direction of) the characterization of Minkowski measurability, given in [**LapPo2**] and recalled in Theorem 2.3. (Namely, $\mathcal{L}$ is Minkowski measurable iff $N_{\mathcal{L}}(x) \sim Q\,x^D$, for some constant $Q \in (0,+\infty)$, where $N_{\mathcal{L}}$ is the geometric counting function of $\mathcal{L}$.) It also makes use of the identity (2.21) connecting $N_\nu(x)$ and $N_{\mathcal{L}}(x)$ for any $x > 0$, as well as of a direct computation involving the *analytic continuation* of $\zeta(s)$ to the open right half-plane $\{Re(s) > 0\}$ and hence, to the (open) *critical strip* $\{0 < Re(s) < 1\}$ since $D \in (0,1)$).

REMARK 2.12. (*a*) (*Drums with fractal boundary.*) The aforementioned spectral error estimates obtained in [**Lap1**] are valid for Laplacians (or, suitably adapted, for more general elliptic operators with variable, nonsmooth coefficients and of order $2m$, with $m \in \mathbb{N}$) on bounded open sets $\Omega \subseteq \mathbb{R}^N$ (where $N \geq 1$ is arbitrary) with (possibly) fractal boundary ("drums with fractal boundary", in the sense of [**Lap3**]) and with Dirichlet or Neumann boundary conditions. For the Dirichlet problem, Ω is allowed to have finite volume (i.e., $|\Omega|_N < \infty$) rather than to be bounded. Furthermore, for the Neumann problem, one must assume, for example, that Ω satisfies the *extension property* for the Sobolev space $H^m = W^{m,2}$ [**Br**, **Maz**], where $m = 1$ in the present case of the Laplacian; see, e.g., [**Maz**], [**Jon**], [HajKosTu1–2], [**Lap1**], [**Vel-San**]. For instance, for a simply connected planar domain, Ω is an extension domain if and only if it is a quasidisk (i.e., $\partial\Omega$ is a quasicircle, [**Pomm**]); that is, Ω is the homeomorphic quasiconformal image of the open unit disk in $\mathbb{C}$.[39] Quasidisk and quasicircles (as well as John domains) are of frequent use in harmonic analysis, partial differential equations, complex dynamics and conformal dynamics; see, e.g., [**Be**], [**Maz**], [**Pomm**] and [**BedKS**].

(*b*) (*Drums with fractal membrane.*) An analog of the leading term (Weyl's asymptotic formula) and the associated error term in [**Lap1**] discussed in part (*a*) was obtained in [**KiLap1**] for Laplacians *on* fractals, rather than on bounded open sets of $\mathbb{R}^N$ with fractal boundary; that is, for "drums with fractal membrane" (in the sense of [**Lap3**]) rather than for "drums with fractal boundary" (as, e.g., in [**BirSol**], [**BroCar**], [Lap1–4], [LapPo1–3], [LapMa1–2], [**Ger**], [GerScm1–2], [**FlLap**], [**FlVa**], [**EdmHar**], [**Dav**], [**LeviVa**], [**HeLap**], [**MolVai**], [**vB-Gi**], [**HamLap**], [**Lap9**], [**LapPa**] and [**LapNRG**]). Examples of physics and mathematics references on Laplacians on fractals include the books [**Ki**] and [**Str**] along

[39] Another characterization of a quasidisk is that it is a planar domain which is both a John domain (see [**John**] and, e.g., [**CarlJonYoc**], [**McMul**], [**DieRuzSchu**], [**Dur-LopGar**], [**AcoDur-LopGar**] or [LopGar1–2]) and linearly connected; see, e.g., [**ChuOsgPomm**].

with the papers [**Ram**], [**RamTo**], [**Shi**], [**FukShi**], [KiLap1–2], [**Barl**], [Ham1–2], [Sab1–3], [Tep1–2], [**DerGrVo**], [Lap5–6], [**ChrIvLap**], [**CipGIS**], [**LapSar**] and [LalLap1–2], as well as the many relevant references therein.

2.6.2. *Inverse spectral problems for fractal strings.* Now that we have explicitly and fully solved the above direct spectral problem for fractal strings, it is natural to consider its converse, which is called an inverse spectral problem, (ISP). In fact, we have a one-parameter family of such problems, $(\mathrm{ISP})_D$, parametrized by the (Minkowski) dimension D. Recall that $D \in (0,1)$ is arbitrary, so that the parameter D sweeps out the entire '*critical interval*' $(0,1)$ for the Riemann zeta function $\zeta = \zeta(s)$.

$(\mathrm{ISP})_D$ *Let $\mathcal{L}$ be a fractal string of Minkowski dimension $D \in (0,1)$ such that its associated spectral counting function N_ν admits a* monotonic *asymptotic second term, proportional to x^D. Namely, with the Weyl term W given by* (2.41), *assume that*

$$N_\nu(x) = W(x) - \mathcal{C}x^D + o(x^D) \quad \text{as } x \to +\infty, \tag{2.44}$$

for some nonzero constant $\mathcal{C}$ (depending only on $\mathcal{L}$). Does it then follow that $\mathcal{L}$ is Minkowski measurable?

The above question à la Marc Kac, [**Kac**], could be coined *Can one hear the shape of a fractal string (of dimension $D \in (0,1)$)?* Note, however, that this question (or equivalently, the corresponding inverse spectral problem $(ISP)_D$), is of a very different nature from the original one, raised in [**Kac**].

REMARK 2.13. Equation (2.44) alone with $\mathcal{C} \neq 0$ and $D \in (0,1)$ implies that $\mathcal{L}$ has Minkowski dimension D. Furthermore, if $(\mathrm{ISP})_D$ has an affirmative answer, then it follows from Theorem 2.11 that $\mathcal{C} > 0$ and $\mathcal{C} = c_D\mathcal{M}$, where $c_D > 0$ is the constant (depending only on D) given by (2.43) and $\mathcal{M}$ is the Minkowski content of $\mathcal{L}$.

We can finally give the precise statement of the main result (Theorem 2.14 and Corollary 2.15) connecting the family of inverse spectral problems $(\mathrm{ISP})_D$, with $D \in (0,1)$, and the Riemann hypothesis (RH); see [LapMa1–2], joint with H. Maier.

THEOREM 2.14 (Critical zeros of ζ and inverse spectral problems; [**LapMa2**]). *Fix $D \in (0,1)$. Then, the inverse spectral problem $(ISP)_D$ has an affirmative answer if and only if $\zeta = \zeta(s)$ does not have any zeros on the vertical line $\{Re(s) = D\}$; i.e., if and only if the* 'partial RH' *(abbreviated $(RH)_D$) holds for this value of D. In short (and with the obvious notation), we have that*

$$(RH)_D \Leftrightarrow (ISP)_D, \text{ for any } D \in (0,1). \tag{2.45}$$

COROLLARY 2.15 (Spectral reformulation of RH; [**LapMa2**]). *The Riemann hypothesis is true if and only the inverse spectral problem $(ISP)_D$ has an affirmative answer for all $D \in (0,1)$, except in the 'midfractal case' when $D = 1/2$.*

PROOF. (*Proof of Corollary* 2.15.) Since it is known that ζ has a nonreal zero (and even infinitely many zeros, by a theorem of G. H. Hardy; see, e.g., [**Edw, Tit**]) on the critical line $\{Re(s) = 1/2\}$, this is a consequence of Theorem 2.14. □

REMARK 2.16. Observe that we could reformulate Corollary 2.15 by stating that RH is equivalent to $(\mathrm{ISP})_D$ having an affirmative answer for all $D \in (0, 1/2)$ (or equivalently, for all $D \in (1/2, 1)$). This follows from the well-known functional equation for ζ connecting $\zeta(s)$ and $\zeta(1-s)$, for all $s \in \mathbb{C}$; see, e.g., [**Edw**] or [**Tit**].

PROOF. (*Sketch of the proof of Theorem* 2.14.) The proof of one implication in Theorem 2.14 relies on the Wiener–Ikehara Tauberian theorem (see, e.g., [**Pos**]) or one of its later improvements (see, e.g., [**PitWie**], [**Pit**] and [**Kor**]).

The converse (i.e., the reverse implication $(ISP)_D \Rightarrow (RH)_D$ in (2.45)) is proved by contraposition. That is, we assume that $(RH)_D$ fails and we therefore want to show that the inverse spectral problem $(ISP)_D$ does not have an affirmative answer. Hence, suppose that there exists $\omega \in \mathbb{C}$, with $\omega = D + i\tau$ $(D \in (0,1), \tau > 0)$, such that $\zeta(\omega) = 0$. Then, clearly, $\overline{\omega} = D - i\tau$ also satisfies $\zeta(\overline{\omega}) = 0$.

Next, we use the intuition of the notion of complex dimension[40] in order to construct a fractal string $\mathcal{L} = (\ell_j)_{j=1}^{\infty}$ which has oscillations of leading order x^D in its *geometry* but which (because $0 = \zeta(\omega) = \overline{\zeta(\omega)} = \zeta(\overline{\omega})$) does *not* have oscillations of order D in its *spectrum* (so that $N_\nu(x)$ has a *monotonic* asymptotic second term proportional to x^D).

More specifically, we have that (for some positive constant β sufficiently small and with $[y]$ denoting the integer part of $y \in \mathbb{R}$)

$$N_{\mathcal{L}}(x) = [V(x)], \text{ for any } x > 0, \tag{2.46}$$

where

$$\begin{aligned} V(x) &:= x^D + \beta(x^\omega + x^{\overline{\omega}}) \\ &= x^D(1 + 2\beta\cos(\tau \log x)), \end{aligned} \tag{2.47}$$

for any $x > 0$. Note that it suffices to choose a positive number $\beta < 1/2$ so that $V(x) > 0$ for all $x > 0$ and $\beta < D/2(D+\tau)$ so that $V'(x) > 0$ for all $x > 0$.[41] Hence, we can simply choose $\beta \in (0, D/2(D+\tau))$. Since V is now strictly increasing from $(0, +\infty)$ to itself, we can uniquely define $\ell_j > 0$ so that $V(\ell_j^{-1}) = j$; i.e., in light of (2.46), $N_{\mathcal{L}}(\ell_j^{-1}) = j$, for each integer $j \geq 1$. This defines the sought for (bounded) fractal string $\mathcal{L} = (\ell_j)_{j=1}^{\infty}$;[42] one can then choose any geometric realization Ω of $\mathcal{L}$ as a subset of $\mathbb{R}$ with finite length.

Since by construction,

$$N_{\mathcal{L}}(x) \sim x^D(1 + 2\beta\cos(\tau \log x)) \quad \text{as } x \to +\infty, \tag{2.48}$$

we deduce from (the difficult part of) Theorem 2.3 (the characterization of Minkowski measurability) that the fractal string $\mathcal{L}$ is *not* Minkowski measurable.

Indeed, because $x^{-D}N_{\mathcal{L}}(x)$ oscillates asymptotically between the positive constants $1 - 2\beta$ and $1 + 2\beta$ (in light of (2.48)), $x^{-D}N_{\mathcal{L}}(x)$ cannot have a limit as

---

[40] At the time, in the early 1990s, the rigorous definition of complex dimension did not yet exist, even for fractal strings. This only came later, in the mid-1990s; see [**Lap-vF1**, **Lap-vF2**].

[41] Indeed, an elementary computation shows that

$$V'(x) = x^{D-1}(D + 2\beta D\cos(\omega \log x) - 2\beta\tau \sin(\omega \log x)), \text{ for all } x > 0.$$

[42] Note that $\mathcal{L}$ is bounded because, in light of (2.48), $N_{\mathcal{L}}(x)$ is of the order of x^D as $x \to +\infty$ and hence, ℓ_j is of the order of $j^{-1/D}$ as $j \to \infty$. Since $D \in (0,1)$, it follows that $\sum_{j\geq 1} \ell_j < \infty$.

$x \to +\infty$; equivalently, $j^{1/D}\ell_j$ cannot have a limit as $j \to \infty$, which violates condition (ii) of Theorem 2.3 and therefore shows that $\mathcal{L}$ is *not* Minkowski measurable, as desired.[43]

Next, a direct (but relatively involved) computation in [**LapMa2**] (based, in particular, on the identity (2.22) and several key properties of $\zeta = \zeta(s)$ and its meromorphic continuation to the critical strip $\{0 < Re(s) < 1\}$) shows that there exist nonzero constants E_D and E_ω such that as $x \to +\infty$,

$$N_\nu(x) = W(x) + E_D\zeta(D)x^D + E_\omega\zeta(\omega)x^\omega + \overline{E}_\omega\zeta(\overline{\omega}) + o(x^D), \tag{2.49}$$

where $W(x) = |\Omega|_1 x$ is the Weyl term (as given by (2.41)).

Now, since $\zeta(\omega) = \zeta(\overline{\omega}) = 0$, we see that $N_\nu(x)$ has a *monotonic* asymptotic second term, proportional to x^D, as claimed:

$$N_\nu(x) = W(x) + E_D\zeta(D)x^D + o(x^D) \quad \text{as } x \to +\infty. \tag{2.50}$$

This shows that the fractal string $\mathcal{L}$ which we have constructed is not Minkowski measurable but that its spectral counting function has a monotonic asymptotic second term, of the order of x^D. Therefore, the inverse spectral problem $(\mathrm{ISP})_D$ does not have an affirmative answer for this value of $D \in (0,1)$. Thus the implication $(ISP)_D \Rightarrow (RH)_D$ is now established. This concludes the sketch of the proof of Theorem 2.14 (and hence also of Corollary 2.15). □

REMARK 2.17. (*a*) The proof of Theorem 2.14 given in [**LapMa2**] actually shows that one can replace $o(x^D)$ by $O(1)$ in (2.49), and hence also in (2.50).

(*b*) In the language of the mathematical theory of complex dimensions, the fractal string $\mathcal{L}$ constructed just above has three (simple) complex dimensions; namely, the Minkowski dimension D and the complex conjugate pair of nonreal (principal) complex dimensions $(\omega, \overline{\omega}) = (D + i\tau, D - i\tau)$. Hence,

$$\mathcal{D}_{\mathcal{L}} = \{D, \omega, \overline{\omega}\}. \tag{2.51}$$

Accordingly, $\mathcal{L}$ (or, equivalently, any of its geometric realizations Ω as an open subset of $\mathbb{R}$ with finite length) is fractal (in the sense of §2.5) and even '*critically fractal*' (since it is fractal in the maximal possible dimension, D).

(*c*) Theorem 2.14 (suitably interpreted), along with Corollary 2.15, has been extended to a large class of arithmetic zeta functions (and other Dirichlet series) in [**Lap-vF2, Lap-vF3, Lap-vF4**]; see [**Lap-vF4**, Ch. 9]. This broad generalization relies on the mathematical theory of complex dimensions developed in those references, and especially, on the fractal explicit formulas obtained therein (see [**Lap-vF4**, Chs. 5–6]), of which a few examples were provided in §2.2.

(*d*) As was alluded to earlier, the second part of the proof of Theorem 2.14 (namely, the proof of the implication $(ISP)_D \Rightarrow (RH)_D$) was motivated by the intuition of the notion of complex dimensions. At the same time, it served as a powerful incentive for developing the mathematical theory of complex dimensions. Along with the statements of Theorem 2.14 and Corollary 2.15, it also provided a *natural geometric interpretation of the* (closed) *critical strip* $\{0 \le Re(s) \le 1\}$. In particular, the *midfractal case* when $D = 1/2$ corresponds to the critical line $\{Re(s) = 1/2\}$, while the *least* (respectively, *most*) *fractal case* when $D = 0$ (respectively, $D = 1$) corresponds to the left (respectively, right) most vertical line in

[43]It follows from another theorem in [**LapPo2**] that $\mathcal{L}$ is Minkowski nondegenerate (i.e., $0 < \mathcal{M}_*(\le)\mathcal{M}^* < \infty$) and has Minkowski dimension D; so that $0 < \mathcal{M}_* < \mathcal{M}^* < \infty$, by combining these two results.

the closed critical strip, $\{Re(s) = 0\}$ (respectively, $\{Re(s) = 1\}$). This observation has played a key role in later work, including [**Lap-vF4**], [Lap7–9] and [**HerLap1**].

It is also in agreement with the author's conjecture according to which there should exist a natural fractal-like geometry whose complex dimensions coincide with the union of $\{1/2\}$ and the critical zeros of ζ. Consequently, it would have Minkowski dimension 1/2 and apart from 1/2, the critical zeros ρ of ζ would be its principal complex dimensions. It is possible that instead, this "geometry" would only be *critically subfractal* (in dimension $\frac{1}{2}$), with the critical zeros ρ of ζ being precisely its complex dimensions with real part 1/2 (assuming RH, for clarity). Namely, in that case, it would have Minkowski dimension 1 (corresponding to the only pole of ζ, which is simple), midfractal complex dimensions the critical zeros of ζ, and possibly one other (simple) complex dimension at 0 (provided it has the perfect symmetry of the completed Riemann zeta function ξ, as described in the next paragraph). Accordingly (and still assuming RH, for clarity), in either the former case or the latter case, this geometry would be fractal in dimension 1/2, but not fractal in any other dimension $d \in \mathbb{R}$, with $d \neq 1/2$ (since both 0 and 1 are real).

Hence (in the latter case), in algebraic geometric terms, the set $\mathcal{D}_\zeta$ of complex dimensions of this elusive fractal geometry would coincide with the divisor of the completed (or global) Riemann zeta function $\xi(s) := \pi^{-s/2}\Gamma(s/2)\zeta(s)$ (or, equivalently, with the poles of minus the logarithmic derivative, $-\xi'(s)/\xi(s)$), leaving apart the multiplicities, which can be explicitly recovered by considering the corresponding residues. Namely, since the zeros of ξ coincide with the critical zeros of ζ and ξ satisfies the celebrated functional equation $\xi(s) = \xi(1-s)$ for all $s \in \mathbb{C}$, from which it follows that ξ has a (simple) pole at $s = 0$ and $s = 1$, we would have

$$\mathcal{D}_\zeta = \{0, 1\} \cup \{\text{critical zeros of } \zeta\}, \tag{2.52}$$

where each complex dimension is counted with multiplicity one. Accordingly, the sought for fractal-like geometry would be "*self-dual*" (in the sense of [**Lap7**]), reflecting the perfect symmetry of the functional equation with respect to the critical line $\{Re(s) = 1/2\}$. (See also §4.4 and [**Lap10**].)

3. A Taste of the Higher-Dimensional Theory: Complex Dimensions and Relative Fractal Drums (RFDs)

In this section, which by necessity of concision, will be significantly shorter than would be warranted, we limit ourselves to giving a brief overview of some of the key definitions and results of the higher-dimensional theory of complex dimensions, with emphasis on several key examples illustrating them. [We refer to [**Lap10**, Ch. 3] for a more extensive overview, and to the book [**LapRaZu1**] along with the accompanying papers [LapRaZu2–10] (including the two survey articles [LapRaZu8–9]) for a much more detailed account of the theory, with complete proofs.] First, however, we begin by providing a short history of one aspect of the subject.

3.1. Brief history. For a long time, the theory of complex dimensions was restricted to the one-dimensional case (corresponding to fractal strings and arbitrary compact subsets of $\mathbb{R}$) or in the higher-dimensional case, to very special (although useful) geometries; namely, to fractal sprays [**Lap3**], [**LapPo3**], obtained as countable disjoint unions of scaled copies of one or finitely many generators, and particularly, to self-similar sprays.

An approximate tube formula was first obtained in [Lap-vF2–4] for the devil's staircase (i.e., the graph of the well-known Cantor function), from which the complex dimensions of the Cantor graph could be deduced, by analogy with the fractal tube formula for fractal strings (discussed in §2.1). Then, for the important example of the snowflake curve (or, equivalently, of the Koch curve), an exact fractal tube formula was obtained by the author and Erin Pearse in [**LapPe1**] via a direct computation, based in part on symmetry considerations.[44] (See [**Lap-vF4**, §12.2.1] for an exposition.) Again, the complex dimensions of the Koch curve could be deduced by analogy with the case of fractal strings. However, no analog of the geometric zeta functions of fractal strings was used in the process, and therefore, the complex dimensions of the Koch curve (like those of the Cantor graph at that stage) could not yet be precisely defined.

Another important step was carried out by the author and Erin Pearse in [LapPe2–3] and significantly more generally, in joint work of those two authors and Steffen Winter in [**LapPeWi1**], where fractal tube formulas were obtained for a large class of fractal sprays [**LapPo3**], and especially, of self-similar sprays or self-similar tilings (as in [Lap2–3], [**Pe**], [LapPe1–2], [**PeWi**] and [LapPeWi1–2]).[45] This time, the resulting fractal tube formulas made use of certain "tubular zeta functions", but those zeta functions were of a rather ad hoc nature and could not be extended to more general types of fractal geometries. Also, of course, fractal sprays are rather special cases of fractals.[46]

Therefore, there still remained to find appropriate fractal-type zeta functions which enabled one, in particular, to both define the complex dimensions of arbitrary bounded (or, equivalently, compact) subsets of $\mathbb{R}^N$ (for any $N \geq 1$) and to obtain fractal tube formulas valid in this general higher-dimensional setting.

Finally, in 2009, this significant hurdle was overcome when the author introduced a fractal zeta function, now called the distance zeta function, which could be used to achieve the aforementioned goals. This was only the beginning of what turned out to be a very fruitful and creative period, extending from 2009 through 2017, during which large parts of the higher-dimensional theory of complex dimensions and associated fractal tube formulas were developed by the author, Goran Radunović and Darko Žubrinić in the series of papers [LapRaZu2–9] as well as in the nearly 700-page research book [**LapRaZu1**].

Our goal here is not to give a detailed account of the theory. Instead, we simply wish to highlight in the rest of this section a few definitions, results and useful examples. We refer to [LapRaZu2–10], [**Lap10**, Ch. 3] and, especially, to the research monograph [**LapRaZu1**], for a more detailed account, as well as for precise statements (and complete proofs) of the main results.

[44] In that context, an interesting open problem remains to explicitly determine the Fourier coefficients of a nonlinear (and periodic) analog of the Cantor function arising naturally in the computation leading to the corresponding fractal tube formula.

[45] See [**Lap-vF4**, §13.1] for an exposition of part of these results.

[46] In one dimension, however, fractal sprays reduce to fractal strings, while the latter enables us to deal with the general case of arbitrary bounded (or, equivalently, compact) subsets of the real line $\mathbb{R}$.

3.2. Fractal zeta functions and relative fractal drums (FZFs and RFDs). The theory of complex dimensions of [LapRaZu1–10] is developed for arbitrary bounded subsets A of $\mathbb{R}^N$ and, more generally, for arbitrary relative fractal drums (RFDs) (A,Ω) in $\mathbb{R}^N$, for any integer $N \geq 1$, as we now briefly explain.

Given A a (nonempty) subset of $\mathbb{R}^N$ and Ω a possibly unbounded open subset of $\mathbb{R}^N$ with finite volume and such that $\Omega \subseteq A_{\delta_1}$, for some $\delta_1 > 0$, the pair (A,Ω) is called a *relative fractal drum* (or *RFD*, in short) in $\mathbb{R}^N$.

Two important special cases of RFDs are (i) when $N = 1, \Omega$ is a bounded open subset of $\mathbb{R}^N$, and $A := \partial\Omega$, and (ii) when $N \geq 1$ is arbitrary, A is any bounded subset of $\mathbb{R}^N$, and $\Omega := A_{\delta_1}$, for some fixed (but arbitrary) $\delta_1 > 0$. Case (i) just above corresponds to the (ordinary or bounded) fractal strings discussed in §2 (but now viewed as the RFDs $(\partial\Omega,\Omega)$), while case (ii) corresponds to arbitrary bounded subsets A of $\mathbb{R}^N$ (for any integer $N \geq 1$); see also §3.2.2 or §3.2.1, respectively.

It is a simple matter to extend the definition of the Minkowski dimension and of the Minkowski content to RFDs, as we now explain. (In the sequel, given a measurable subset B of $\mathbb{R}^N$, we let $|B|_N = |B|$ denote the N-dimensional volume or Lebesgue measure of B.)

Given an RFD (A,Ω) in $\mathbb{R}^N$, we define its *tube function*

$$\varepsilon \mapsto V(\varepsilon) = V_{A,\Omega}(\varepsilon) := |A_\varepsilon \cap \Omega|, \text{ for } \varepsilon > 0, \tag{3.1}$$

and then,[47] its *upper Minkowski dimension*

$$\begin{aligned} \overline{D} &= \overline{\dim}_B(A,\Omega) \\ &= \inf\left\{\beta \in \mathbb{R} : V(\varepsilon) = O(\varepsilon^{N-\beta}) \quad \text{as } \varepsilon \to 0^+\right\} \\ &= \inf\left\{\beta \in \mathbb{R} : \limsup_{\varepsilon\to 0^+} \frac{V(\varepsilon)}{\varepsilon^{N-\beta}} < \infty\right\}. \end{aligned} \tag{3.2}$$

We define similarly $\underline{D} = \underline{\dim}_B(A,\Omega)$, the *lower Minkowski dimension* of (A,Ω), by simply replacing the upper limit by a lower limit on the right-hand side of the last equality of (3.2). In general, we have $\underline{D} \leq \overline{D}$ but if $\underline{D} = \overline{D}$ (which is the case for most classic fractals), we denote by D this common value and call it the *Minkowski dimension* of (A,Ω); the latter is then said to exist.

The *upper* (respectively, *lower*) *Minkowski content* of (A,Ω) is then defined (still with $V(\varepsilon) = V_{A,\Omega}(\varepsilon)$) by

$$\mathcal{M}^* = \limsup_{\varepsilon\to 0^+} \frac{V_{A,\Omega}(\varepsilon)}{\varepsilon^{N-D}} \tag{3.3}$$

and

$$\mathcal{M}_* = \liminf_{\varepsilon\to 0^+} \frac{V_{A,\Omega}(\varepsilon)}{\varepsilon^{N-D}}. \tag{3.4}$$

We clearly have $0 \leq \mathcal{M}_* \leq \mathcal{M}^* \leq \infty$. If $\mathcal{M}_* > 0$ and $\mathcal{M}^* < \infty$, we say that (A,Ω) is *Minkowski nondegenerate.* If, in addition, $\mathcal{M}^* = \mathcal{M}_*$ (i.e., if the limit in (3.3) and (3.4) exists and is in $(0,+\infty)$), then we denote by $\mathcal{M}$, and call the *Minkowski content* of (A,Ω), this common value and say that the RFD (A,Ω) is *Minkowski measurable.* It is easy to check that if (A,Ω) is Minkowski nondegenerate (and, in particular, if it is Minkowski measurable), then its Minkowski dimension D exists.

---

[47] If A is a bounded subset of $\mathbb{R}^N$ (viewed as an RFD, see §3.2.1), we then simply write $V(\varepsilon) = V_A(\varepsilon)$. Note that we then have $V_A(\varepsilon) = |A_\varepsilon|$, the volume of the ε-neighborhood of A.

We note that the upper Minkowski dimension of an RFD can be negative or even take the value $-\infty$. For instance, let $A := \{(0,0)\}$ and for $\alpha > 1$, let

$$\Omega := \{(x,y) \in \mathbb{R}^2 : 0 < y < x^\alpha, x \in (0,1)\}. \tag{3.5}$$

Then the RFD (A,Ω) in $\mathbb{R}^2$ has (relative) Minkowski dimension $D = 1 - \alpha < 0$, which takes arbitrary large negative values as $\alpha \to +\infty$. Furthermore, (A,Ω) is Minkowski measurable with (relative) Minkowski content $\mathcal{M} = 1/(1+\alpha)$. (See [**LapRaZu1**, Prop. 4.1.35].)

Further, another RFD (A,Ω) in $\mathbb{R}^2$, defined by $A := \{(0,0)\}$ and

$$\Omega := \{(x,y) \in \mathbb{R}^2 : 0 < y < e^{-1/x}, 0 < x < 1\}, \tag{3.6}$$

is such that its Minkowski dimension D exists (as in the previous example) but is no longer finite; namely, $D = -\infty$. (See [**LapRaZu1**, Cor. 4.1.38].)

For a fractal string or for an arbitrary bounded subset A of $\mathbb{R}^N$, however, we always have that $\underline{D} \geq 0$ and hence also, that $\overline{D} \geq 0$ (so that when the Minkowski dimension D exists, then $D \geq 0$).[48]

RFDs extend the notion of a bounded set in $\mathbb{R}^N$ (see §3.2.1 below), of a fractal string (when $N = 1$, see §3.2.2) and more generally, of a fractal drum $(\partial\Omega, \Omega)$, with Ω of finite volume in $\mathbb{R}^N$ (used, e.g., in [Lap1–3] in the case of Dirichlet boundary conditions). They are also very useful tools in the computation of the fractal zeta functions and complex dimensions of fractals (especially when some kind of self-similarity is present), by means of scaling and symmetry considerations; see [**LapRaZu1**] and [**LapRaZu4**].

REMARK 3.1. (*Relative box dimension.*) It is natural to wonder if there exists a notion of box dimension which is also valid for RFDs and thus, for which the corresponding values can be negative. It turns out to be the case, as is explained in a work in preparation by the authors of [**LapRaZu1**]. This (upper) "*relative box dimension*" not only exists but also coincides with the (upper) relative Minkowski dimension of the given RFD (A,Ω) in $\mathbb{R}^N$, as in the usual case of bounded subsets A of $\mathbb{R}^N$ (see, e.g., [**Fa1**]).[49] However, its definition now requires a *fractional counting* of the boxes involved (corresponding essentially to the relative volume of those boxes).

Furthermore, for RFDs of the form $(\partial\Omega, \Omega)$, with Ω a John domain in $\mathbb{R}^N$ $(N \geq 1)$,[50] one can show that as in the usual case of bounded subsets of $\mathbb{R}^N$, the relative box dimension of the RFD exists and not only coincides with the relative Minkowski dimension, $\dim_B(\partial\Omega, \Omega)$, but is also nonnegative.

We note that this notion of (possibly negative) relative box dimension (and that of relative Minkowski dimension, with which it coincides) could perhaps be used to make sense of the heuristic and elusive notion of "negative dimension" and "degree of emptiness" used (or sought for) in [**ManFra**]; see also [**Tri4**], in the context of the Hausdorff (rather than of the Minkowski) dimension.

---

[48] To see why this is true in the latter case, simply note that $A \cap A_\delta = A$ for every $\delta > 0$.

[49] It then follows that in the case of a bounded open set with an external cusp p (such as in (3.5) and (3.6), respectively, and the text surrounding them), the relative box dimension of the RFD $(\{p\}, \Omega)$ is negative or even equal to $-\infty$.

[50] See footnote 39 and the references therein, including [**John**], [**CarlJonYoc**], [**McMul**] and [**ChuOsgPomm**], for example.

We now define as follows the *distance zeta function* of an RFD (A,Ω) in $\mathbb{R}^N$:

$$\zeta_{A,\Omega}(s) = \int_{\Omega} d(x,A)^{s-N}\,dx, \tag{3.7}$$

for all $s \in \mathbb{C}$ with $Re(s)$ sufficiently large.[51]

Another very useful fractal zeta function is the closely related *tube zeta function* of the RFD (A,Ω),

$$\widetilde{\zeta}_{A,\Omega}(s) = \int_0^{\delta} V_{A,\Omega}(\varepsilon)\varepsilon^{s-N}\frac{d\varepsilon}{\varepsilon}, \tag{3.8}$$

for all $s \in \mathbb{C}$ with $Re(s)$ sufficiently large[52] and for $\delta > 0$ fixed but arbitrary.[53]

One can show that we have the following *functional equation*:

$$\zeta_{A,A_\delta\cap\Omega}(s) = \delta^{s-N}|A_\delta \cap \Omega| + (N-s)\,\widetilde{\zeta}_{A,\Omega}(s), \tag{3.9}$$

from which one deduces that $\zeta_{A,\Omega}$ and $\widetilde{\zeta}_{A,\Omega}$ contain essentially the same information, provided $\overline{D} < N$ and $\delta > 0$.[54] Indeed, under this very mild condition, $\zeta_{A,\Omega}$ has a meromorphic extension in a given domain $U \subseteq \mathbb{C}$ if and only if $\widetilde{\zeta}_{A,\Omega}$ does, and in this case, $\zeta_{A,\Omega}$ and $\widetilde{\zeta}_{A,\Omega}$ have the same poles (i.e., the same visible complex dimensions) in U, with the same multiplicities. Furthermore, still in this case, if $\omega \in U$ is a simple pole of $\widetilde{\zeta}_{A,\Omega}$, then it is also a simple pole of $\zeta_{A,A_\delta\cap\Omega}$ and

$$\operatorname{res}(\widetilde{\zeta}_{A,\Omega},\omega) = \frac{1}{N-\omega}\operatorname{res}(\zeta_{A,A_\delta\cap\Omega},\omega). \tag{3.10}$$

We stress that since according to the definition of an RFD, $\Omega \subseteq A_{\delta_1}$ for some $\delta_1 > 0$, we have $A_\delta \cap \Omega = \Omega$ for any $\delta \geq \delta_1$. Hence, under that condition, we can replace $\zeta_{A,A_\delta\cap\Omega}$ by $\zeta_{A,\Omega}$ in both (3.9) and (3.10), as well as in the above statements. The latter equations (3.9) and (3.10) then become (for every $\delta \geq \delta_1$), respectively,

$$\zeta_{A,\Omega}(s) = \delta^{s-N}|\Omega| + (N-s)\widetilde{\zeta}_{A,\Omega}(s) \tag{3.11}$$

and

$$\operatorname{res}(\widetilde{\zeta}_{A,\Omega},\omega) = \frac{1}{N-\omega}\operatorname{res}(\zeta_{A,\Omega},\omega). \tag{3.12}$$

In the sequel, we will assume, most of the time implicitly, that $\delta \geq \delta_1$ (and $\overline{D} < N$); so that (3.9) can be written in the simpler form of the functional equation (3.11).

Given a domain $U \subseteq \mathbb{C}$ containing the "critical line" $\{Re(s) = \overline{D}\}$ and assuming that $\zeta_{A,\Omega}$ has a (necessarily unique) meromorphic continuation to U, the poles of $\zeta_{A,\Omega}$ are called the *visible complex dimensions* of the RFD (A,Ω). If $U = \mathbb{C}$ (or if there is no ambiguity as to the choice of U), we simply call them the *complex dimensions* of (A,Ω).

---

[51] More specifically, according to property (b) of §3.3, this means that (3.7) holds for all $s \in \mathbb{C}$ with $Re(s) > \overline{D}$, where $\overline{D} := \overline{\dim}_B(A,\Omega)$, and that $\overline{D}$ is best possible.

[52] This can be interpreted exactly as in footnote 51 just above, provided $\overline{D} < N$.

[53] It turns out that the poles of $\widetilde{\zeta}_{A,\Omega}$ (i.e., the complex dimensions of (A,Ω)) are independent of the choice of $\delta > 0$ since changing the value of δ in (3.8) amounts to adding an entire function to $\widetilde{\zeta}_{A,\Omega}$.

[54] Note that it follows at once from (3.2) that we always have that $\overline{D} \leq N$, since $|\Omega| < \infty$ according to the definition of an RFD.

The set (really, multiset) of (visible) *complex dimensions* of (A,Ω) is denoted by $\mathcal{D}(\zeta_{A,\Omega})$ (or $\mathcal{D}(\zeta_{A,\Omega};U)$ if we want to specify U) and when $U=\mathbb{C}$, we also use the notation $\dim_{\mathbb{C}}(A,\Omega):=\mathcal{D}(\zeta_{A,\Omega};\mathbb{C})$.

We adopt a similar notation when $\zeta_{A,\Omega}$ is replaced by $\widetilde{\zeta}_{A,\Omega}$. In fact, as we alluded to earlier, we will always assume that $\overline{D}<N$; so that $\mathcal{D}(\widetilde{\zeta}_{A,\Omega};U)=\mathcal{D}(\zeta_{A,\Omega};U)$, in which case we also denote by $\mathcal{D}_{A,\Omega}(=\mathcal{D}_{A,\Omega}(U))$ the set (or really, multiset) of (visible) complex dimensions of (A,Ω).

3.2.1. *The special case of bounded sets in* $\mathbb{R}^N$. If A is a bounded subset of $\mathbb{R}^N$, then for any fixed $\delta>0$ and for all $s\in\mathbb{C}$ with $Re(s)$ large enough (really, for all $s\in\mathbb{C}$ with $Re(s)>\overline{D}$),

$$\zeta_A(s):=\zeta_{A,A_\delta}(s)=\int_{A_\delta} d(x,A)^{s-N}dx \tag{3.13}$$

and

$$\widetilde{\zeta}_A(s):=\zeta_{A,A_\delta}(s)=\int_0^\delta V_A(\varepsilon)\varepsilon^{s-N}\frac{d\varepsilon}{\varepsilon}, \tag{3.14}$$

where $V_A(\varepsilon):=|A_\varepsilon|$ (as mentioned earlier). Then, (3.11) reduces to the following simpler functional equation, valid for any $\delta>0$:

$$\zeta_A(s)=\delta^{s-N}|A_\delta|+(N-s)\,\widetilde{\zeta}_A(s). \tag{3.15}$$

Furthermore, (3.12) becomes (provided $\overline{D}<N$ and assuming that $\omega\in U$ is a simple pole of $\widetilde{\zeta}_A$ and hence also, of ζ_A):

$$\operatorname{res}(\widetilde{\zeta}_A,\omega)=\frac{1}{N-\omega}\operatorname{res}(\zeta_A,\omega). \tag{3.16}$$

Moreover, still provided $\overline{D}<N$, ζ_A has a meromorphic continuation to a given domain $U\subseteq\mathbb{C}$ if and only if $\widetilde{\zeta}_A$ does. In this case, ζ_A and $\widetilde{\zeta}_A$ have the same poles in U, called the *visible complex dimensions* of A and denoted by $\mathcal{D}_A(U)=\mathcal{D}(\zeta_A;U)=\mathcal{D}(\widetilde{\zeta}_A;U)$ or, when $U=\mathbb{C}$ or when no ambiguity may arise, by $\mathcal{D}_A=\dim_{\mathbb{C}}A=\mathcal{D}(\zeta_A)=\mathcal{D}(\widetilde{\zeta}_A)$. Indeed, the complex dimensions of the RFD (A,A_δ) (i.e., the complex dimensions of A) can be defined indifferently as the poles of ζ_A or of $\widetilde{\zeta}_A$. In addition, they are independent of the choice of $\delta>0$ because changing the value of δ in (3.13) or (3.14) amounts to adding an entire function to the original distance zeta function ζ_A or tube zeta function $\widetilde{\zeta}_A$, respectively.

Finally, we note that since $d(\cdot,A)=d(\cdot,\overline{A})$, we have that $A_\delta=(\overline{A})_\delta$ and hence, $V_A=V_{\overline{A}}$; so that $\zeta_A=\zeta_{\overline{A}}$ and $\widetilde{\zeta}_A=\widetilde{\zeta}_{\overline{A}}$. As a result, by simply replacing A by its closure $\overline{A}$ in $\mathbb{R}^N$, we may as well assume without loss of generality that $A\subseteq\mathbb{R}^N$ is compact instead of just being bounded.

3.2.2. *The special case of fractal strings.* Let $N=1$ and Ω be a bounded open subset of $\mathbb{R}$ (or more generally, an open subset of finite length in $\mathbb{R}$). Then, we view the RFD $(\partial\Omega,\Omega)$ in $\mathbb{R}$ as a geometric realization of the fractal string $\mathcal{L}=(\ell_j)_{j=1}^\infty$, the sequence of lengths of the connected components (i.e., open intervals) of Ω, as in §2.

Remarkably, the distance zeta function $\zeta_{\partial\Omega,\Omega}$ of the RFD $(\partial\Omega,\Omega)$ and the geometric zeta function $\zeta_{\mathcal{L}}$ of the fractal string $\mathcal{L}=(\ell_j)_{j\geq1}$ (see §2, Equation (2.1)) are related via the following very simple functional equation:

$$\zeta_{\partial\Omega,\Omega}(s)=2^{1-s}\frac{\zeta_{\mathcal{L}}(s)}{s}, \tag{3.17}$$

valid initially within the open half-plane $\{Re(s) > \overline{D}\}$, where

$$\overline{D} = \overline{\dim}_B(\partial\Omega, \Omega) = \overline{\dim}_B \mathcal{L} = D(\zeta_{\partial\Omega,\Omega}) = D(\zeta_{\mathcal{L}}), \tag{3.18}$$

and then (upon analytic continuation), within any domain $U \subseteq \mathbb{C}$ containing the closed half-plane $\{Re(s) \geq \overline{D}\}$ to which $\zeta_{\mathcal{L}}$ (and hence also, $\zeta_{\partial\Omega,\Omega}$) can be meromorphically continued.

One therefore deduces from (3.17) that the visible poles of $\zeta_{\partial\Omega,\Omega}$ in U (i.e., the visible complex dimensions of the RFD $(\partial\Omega, \Omega)$) are the same as those of $\zeta_{\mathcal{L}}$, except for the fact that (provided $0 \in U$) 0 is always a visible pole of $\zeta_{\partial\Omega,\Omega}$ (or more precisely, has multiplicity $m + 1$, where m is possibly equal to zero and is defined as the multiplicity of 0 as a pole of $\zeta_{\mathcal{L}}$). Furthermore, $\zeta_{\mathcal{L}}$ has a meromorphic continuation to U if and only if $\zeta_{\partial\Omega,\Omega}$ does, and in this case, the corresponding residues (at a simple pole $\omega \in U$, $\omega \neq 0$, of either $\zeta_{\mathcal{L}}$ or $\zeta_{\partial\Omega,\Omega}$) are related by the following identity:[55]

$$\operatorname{res}(\zeta_{\partial\Omega,\Omega}, \omega) = \frac{2^{1-\omega}}{\omega} \operatorname{res}(\zeta_{\mathcal{L}}, \omega). \tag{3.19}$$

If one starts instead with a given (bounded) fractal string $\mathcal{L} = (\ell_j)_{j\geq 1}$ such that $\sum_{j\geq 1} \ell_j < \infty$, then all the above facts and identities are independent of the choice of the geometric realization Ω of $\mathcal{L}$ as a bounded open subset of $\mathbb{R}$ and hence, of the choice of the RFD $(\partial\Omega, \Omega)$ associated with $\mathcal{L}$.

The above simple observations enable one, in particular, to view the theory of complex dimensions and the associated fractal tube formulas developed in [**LapRaZu1**] and [LapRaZu4,6] as containing (as a very special case) the corresponding theory of fractal strings (as developed in [**Lap-vF4**]). It also enables us to significantly simplify the statement of the fractal tube formulas for fractal strings[56] in the case when 0 happens to be a visible pole of $\zeta_{\mathcal{L}}$. Suffices to say here that, as a result, $\zeta_{\partial\Omega,\Omega}$ and therefore $\mathcal{D}_{\partial\Omega,\Omega} = \mathcal{D}(\zeta_{\partial\Omega,\Omega})$ should be considered as the fractal zeta function and the complex dimensions of a fractal string $\mathcal{L}$ or Ω (rather than $\zeta_{\mathcal{L}}$ and $\mathcal{D}_{\mathcal{L}} = \mathcal{D}(\zeta_{\mathcal{L}})$, respectively).

It follows from the functional equation (3.17) and the above discussion that $\mathcal{D}_{\mathcal{L}} = \mathcal{D}(\zeta_{\mathcal{L}})$ and $\mathcal{D}_{\partial\Omega,\Omega} = \mathcal{D}(\zeta_{\partial\Omega,\Omega})$, respectively the set of complex dimensions of the fractal string $\mathcal{L}$ (in the sense of [**Lap-vF4**] and §2 above) and the set of complex dimensions of the associated RFD $(\partial\Omega, \Omega)$ (in the sense of [**LapRaZu1**] and of the present §3.2), for any geometric realization $\Omega \subseteq \mathbb{R}$ of $\mathcal{L}$, are connected via the following relation (between multisets):

$$\mathcal{D}_{\partial\Omega,\Omega} = \mathcal{D}_{\mathcal{L}} \cup \{0\}, \tag{3.20}$$

in the sense that if 0 is a pole of $\zeta_{\mathcal{L}}$ (i.e., a complex dimension of $\mathcal{L}$, in the sense of §2) of multiplicity $m \geq 0$, then it is also a pole of $\zeta_{\partial\Omega,\Omega}$ (i.e., a complex dimension of the RFD $(\partial\Omega, \Omega)$, in the sense of §3.2) of multiplicity $m + 1$. All the other (i.e., nonzero) complex dimensions of the fractal string $\mathcal{L}$ and of the RFD $(\partial\Omega, \Omega)$ coincide, and have the same multiplicities.

[55]For the simplicity of exposition, we assume throughout this discussion in §3.2.1 that 0 is not a zero of $\zeta_{\mathcal{L}} : \zeta_{\mathcal{L}}(0) \neq 0$. Otherwise, we would have to make use of the notion of a divisor (in essence, the multiset of zeros and poles) of a meromorphic function, to be discussed and used later in §3 and §4.

[56]These formulas were briefly discussed in §2.1 and will be revisited in the relevant part of §3.5.3.

3.3. A few key properties of fractal zeta functions. In this section, we discuss several important properties of the fractal zeta functions (that is, of the distance and tube zeta functions) of relative fractal drums (RFDs) and, in particular, of bounded sets in $\mathbb{R}^N$. Throughout, $N \geq 1$ is a fixed (but arbitrary) positive integer. We will stress the case of distance zeta functions since (except for a few small differences) fractal tube zeta functions have entirely analogous properties.

3.3.1. *Abscissa of convergence and holomorphicity.* Let (A, Ω) be an RFD in $\mathbb{R}^N$ and consider its distance and tube zeta functions, $\zeta_{A,\Omega}$ and $\widetilde{\zeta}_{A,\Omega}$. In order to simplify the exposition, we still assume that $\overline{D} < N$.[57]

(a) (*Abscissa of convergence*). It is shown in [**LapRaZu1**] (see also [LapRaZu2,4]) that the *abscissa of convergence* $\sigma = D(\zeta_{A,\Omega}) = D(\widetilde{\zeta}_{A,\Omega})$ of $\zeta_{A,\Omega}$ and $\widetilde{\zeta}_{A,\Omega}$ is well defined and coincides with the upper Minkowski dimension $\overline{D} = \overline{\dim}_B(A, \Omega)$ of (A, Ω). Namely, the largest right half-plane $\{Re(s) > \alpha\}$ (with $\alpha \in (-\infty, N])$ in which the Lebesgue integral defining $\zeta_{A,\Omega}(s)$ (respectively, $\widetilde{\zeta}_{A,\Omega}(s)$) in (3.7) (respectively, in (3.8)) converges is $\{Re(s) > \overline{D}\}$, called the *half-plane of* (*absolute*) *convergence* of $\zeta_{A,\Omega}$ (respectively, of $\widetilde{\zeta}_{A,\Omega}$); for example,

$$\sigma := \inf\left\{\alpha \in \mathbb{R} : \int_\Omega d(x, A)^{\alpha - N} dx < \infty\right\} \\ = \overline{D} := \overline{\dim}_B(A, \Omega). \tag{3.21}$$

A part of the proof of this key fact relies on a suitable extension of a result obtained in [**HarvPol**] for a completely different purpose (the study of the singularities of the solutions of certain linear partial differential equations).

(b) (*Holomorphicity*). It is also shown in the aforementioned references that $\zeta_{A,\Omega}$ (and hence, in light of the functional equation (3.11), also $\widetilde{\zeta}_{A,\Omega}$) is holomorphic in the open right half-plane $\{Re(s) > \overline{D}\}$, where $\overline{D} := \overline{\dim}_B(A, \Omega)$, as before.

In addition, under mild assumptions (namely, if D, the Minkowski dimension of (A, Ω), exists and $\mathcal{M}_* > 0$), then this right half-plane $\{Re(s) > \overline{D}\}$ is also optimal from the point of view of holomorphicity because one can show that $\zeta_{A,\Omega}(s) \to +\infty$, as $s \to \overline{D}$, with $s \in \mathbb{R}$ and $s > \overline{D}$. Therefore, the half-plane of (absolute) convergence and the *half-plane of holomorphic continuation* coincide and are equal to $\{Re(s) > D\}$, in this case. One then says that D, the (relative) Minkowski dimension of (A, Ω), also coincides with $D_{hol}(\zeta_{A,\Omega}) = D_{hol}(\widetilde{\zeta}_{A,\Omega})$, the common *abscissa of holomorphic continuation* of $\zeta_{A,\Omega}$ and of $\widetilde{\zeta}_{A,\Omega}$.[58]

Finally, one can always (complex) differentiate $\zeta_{A,\Omega}$ or $\widetilde{\zeta}_{A,\Omega}$ under the integral sign and as many times as one wants. For example, provided $Re(s) > \overline{D}$,

$$\zeta'_{A,\Omega}(s) = \int_\Omega d(x, A)^{s-N} \log(d(x, A)) dx. \tag{3.23}$$

Naturally, all of the above properties (in part (a) or (b)) hold without change for ζ_A and $\widetilde{\zeta}_A$, the distance and tube zeta functions of bounded sets A in $\mathbb{R}^N$

---

[57] We also implicitly assume throughout §3.3 (and beyond) that $\delta \geq \delta_1$, where $\Omega \subseteq A_{\delta_1}$.

[58] In general, however, we only have that

$$D_{hol}(\zeta_{A,\Omega}) = D_{hol}(\widetilde{\zeta}_{A,\Omega}) \leq \overline{D}. \tag{3.22}$$

(initially defined for $Re(s) > \overline{D}$ by (3.13) and (3.14), respectively) by considering the associated RFDs (A, A_δ), for a fixed but arbitrary $\delta > 0$.

It is not known whether we can always have an equality in (3.22) or, equivalently, if $D_{hol}(\zeta_{A,\Omega}) = \overline{D}$. By contrast, for fractal strings, the analog of this property always holds;[59] see [**Lap-vF4**] or [**LapRaZu1**, §2.1.4]. Note that the property stated in part (a) above is the exact counterpart of Theorem 2.1.

Since $\zeta_{A,\Omega}$ and $\widetilde{\zeta}_{A,\Omega}$ admit a holomorphic continuation (necessarily unique) to $\{Re(s) > \overline{D}\}$), they cannot have any pole there. Consequently, the set of complex dimensions of (A, Ω) must be contained in $\{Re(s) \leq \overline{D}\}$. More specifically, if the domain $U \subseteq \mathbb{C}$ contains the vertical line $\{Re(s) = \overline{D}\}$ (or, equivalently, contains the closed half-plane $\{Re(s) \geq \overline{D}\}$),[60] then $\zeta_{A,\Omega}$ (and hence also, $\widetilde{\zeta}_{A,\Omega}$) can be meromorphically continued to a domain $U \subseteq \mathbb{C}$, and

$$\mathcal{D}(\zeta_{A,\Omega}; U) = \mathcal{D}(\widetilde{\zeta}_{A,\Omega}; U) \subseteq \{Re(s) \leq \overline{D}\}. \tag{3.24}$$

In addition, if $\overline{D} \in U$, which is certainly the case if U also contains the vertical line $\{Re(s) = \overline{D}\}$, then

$$\overline{D} := \overline{\dim}_B(A, \Omega) = \max\{Re(\omega) : \omega \in \mathcal{D}(\zeta_{A,\Omega}; U\}. \tag{3.25}$$

In other words, *the (upper) Minkowski dimension of (A, Ω) is equal to the maximal real part of the (visible) complex dimensions of (A, Ω).*

In the literature on this topic (see [LapRaZu1–10]), the vertical line $\{Re(s) = \overline{D}\}$ is often called the "*critical line*" (by analogy with the terminology adopted for the Riemann zeta function, but clearly, with a different meaning). Assume that $\zeta_{A,\Omega}$ (or, equivalently, $\widetilde{\zeta}_{A,\Omega}$) can be meromorphically extended (necessarily uniquely) to a domain U containing the critical line $\{Re(s) = \overline{D}\}$, then

$$\begin{aligned} \dim_{PC}(A, \Omega) &= \mathcal{D}_{PC}(A, \Omega) = \mathcal{D}_{PC}(\zeta_{A,\Omega})(= \mathcal{D}_{PC}(\widetilde{\zeta}_{A,\Omega})) \\ &:= \{\omega \in \mathcal{D}(\zeta_{A,\Omega}; U) : Re(\omega) = \overline{D}\} \end{aligned} \tag{3.26}$$

is called the set (really, multiset) of *principal complex dimensions* of the RFD (A, Ω). Clearly, $\mathcal{D}_{PC}(A, \Omega) = \mathcal{D}_{PC}(\zeta_{A,\Omega}) = \mathcal{D}_{PC}(\widetilde{\zeta}_{A,\Omega})$ is independent of the choice of the domain U satisfying the above assumption. (An entirely analogous notation and terminology is used for a bounded set A in $\mathbb{R}^N$; namely, $\dim_{PC}(A) = \mathcal{D}_{PC}(A) = \mathcal{D}_{PC}(\zeta_A) = \mathcal{D}_{PC}(\widetilde{\zeta}_A)$ denotes the set of *principal complex dimensions* of A.)

3.3.2. *Meromorphic continuation and Minkowski content.* In this subsection, we state a few results concerning the existence of a meromorphic continuation in a suitable region of the distance or tube zeta function of an RFD (A, Ω), along with related results concerning the (upper, lower) Minkowski content of (A, Ω). We will consider both the Minkowski measurable case and the (log-periodic) Minkowski nonmeasurable case, for which the residue evaluated at D (the Minkowski dimension of (A, Ω)) of the given fractal zeta function is directly connected to the Minkowski content or, respectively, the average Minkowski content of (A, Ω). For complete proofs of those results and of related results about the existence of a meromorphic

---

[59] More specifically, provided the (bounded) fractal string $\mathcal{L} = (\ell_j)_{j\geq 1}$ is infinite (i.e., has an infinite number of lengths), then $D_{hol}(\zeta_{\mathcal{L}}) = D(\zeta_{\mathcal{L}}) = \overline{D}$, the (upper) Minkowski dimension of $\mathcal{L}$.

[60] *Caution*: There exist RFDs for which such domains U do not exist; in fact, in [**LapRaZu1**, §4.6], one constructs RFDs in $\mathbb{R}^N$ (and also compact sets in $\mathbb{R}^N$, along with fractal strings) such that *every* point of the vertical line $\{Re(s) = \overline{D}\}$ is a (nonremovable) singularity of $\zeta_{A,\Omega}$ and of $\widetilde{\zeta}_{A,\Omega}$.

extension for various classes of RFDs (or of bounded sets) in $\mathbb{R}^N$, we refer to §2.2, §2.3, §3.5, §3.6 and §4.5 of [**LapRaZu1**], as well as to [LapRaZu3–4].

We shall state the results for RFDs in $\mathbb{R}^N$; they, of course, specialize to the case of bounded sets in $\mathbb{R}^N$, as explained in §3.2.1. (The proofs are the same in the special case of bounded sets as in the general situation of RFDs.) Throughout, as in §3.2, (A,Ω) is an RFD such that $\overline{D} < N$, even though this is only needed to easily state simultaneously the results both for the distance and the tube zeta functions.[61]

(a) (*Minkowski content and residue*). We begin by stating a simple result according to which, if $\zeta_{A,\Omega}$ (and hence also, $\tilde{\zeta}_{A,\Omega}$, in light of (3.11)) has a meromorphic extension to a connected open neighborhood of D (where the Minkowski dimension $D = D_{(A,\Omega)}$ of (A,Ω) is assumed to exist), and if $\mathcal{M}^* < \infty$, then[62]

$$(N-D)\mathcal{M}_* \leq \operatorname{res}(\zeta_{A,\Omega}, D) \leq (N-D)\mathcal{M}^*, \tag{3.27}$$

$s = D$ is a simple pole of $\zeta_{A,\Omega}$ (and thus also of $\tilde{\zeta}_{A,\Omega}$) and[63]

$$\mathcal{M}_* \leq \operatorname{res}(\tilde{\zeta}_{A,\Omega}, D) \leq \mathcal{M}^*. \tag{3.28}$$

In particular, if (A,Ω) is assumed to be Minkowski measurable, then

$$\operatorname{res}(\zeta_{A,\Omega}, D) = (N-D)\mathcal{M} \tag{3.29}$$

and

$$\operatorname{res}(\tilde{\zeta}_{A,\Omega}, D) = \mathcal{M}, \tag{3.30}$$

the Minkowski content of (A,Ω).

EXAMPLE 3.2. (*Cantor sets.*) As an illustration of the above result (as well as of the result stated in part (*c*) below), we consider the generalized Cantor set $A = C^{(a)} \subseteq [0,1]$, defined much like the usual ternary Cantor $C = C^{(1/3)}$, where the parameter a lies in $(0,1/2)$. Then, D exists and A is Minkowski nondegenerate but is not Minkowski measurable. More precisely, $D = \dim_B A = \log_{a^{-1}} 2$. Furthermore,

$$\mathcal{M}_* = \frac{1}{D}\Big(\frac{2D}{1-D}\Big)^{1-D}, \quad \mathcal{M}^* = 2(1-a)\Big(\frac{1}{2}-a\Big)^{D-1} \tag{3.31}$$

and

$$\operatorname{res}(\tilde{\zeta}_A, D) = \frac{2}{\log 2}\Big(\frac{1}{2}-a\Big)^{D}. \tag{3.32}$$

Moreover, we have strict inequalities in (3.28) and (3.27) for this example:

$$0 < \mathcal{M}_* < \operatorname{res}(\tilde{\zeta}_{A,D}) < \mathcal{M}^* < \infty, \tag{3.33}$$

and analogously for $\operatorname{res}(\zeta_A, D)$. Also, since $D < 1$, we have that

$$\operatorname{res}(\zeta_A, D) = (1-D)\operatorname{res}(\tilde{\zeta}_A, D), \tag{3.34}$$

[61] We also suppose, as before, that $\delta > 0$ is such that $\delta \geq \delta_1$, where $\Omega \subseteq A_{\delta_1}$, which can always be assumed without loss of generality.

[62] Clearly, if $\mathcal{M}_* = 0$ or $\mathcal{M}^* = +\infty$, then the corresponding inequality in (3.27) or in (3.28) is trivial. Hence, we may as well assume that (A,Ω) is Minkowski nondegenerate in order to obtain the full strength of the result.

[63] The value of the residue $\operatorname{res}(\tilde{\zeta}_{A,\Omega}, D)$ is independent of the choice of $\delta > 0$ in the definition (3.8) of $\tilde{\zeta}_{A,\Omega}$, as well as of the choice of δ_1 implicit in (3.7) and (3.8). And likewise for the values of the residues $\operatorname{res}(\zeta_A, D)$ and $\operatorname{res}(\tilde{\zeta}_A, D)$ in the definition of ζ_A and $\tilde{\zeta}_A$ in (3.13) and (3.14), in the counterpart of this result for a bounded subset A of $\mathbb{R}^N$; recall from (3.13) that $\zeta_A := \zeta_{A,A_\delta}$.

with $\operatorname{res}(\widetilde{\zeta}_A, D)$ given by (3.32).

We could discuss analogously the examples of the Sierpinski gasket and the Sierpinski carpet. In particular, we would find that the inequalities in (3.27) and (3.28) are also strict in those two cases, because the classic Sierpinski gasket and carpet are both Minkowski nondegenerate and not Minkowski measurable, as can be checked via a direct computation.

EXAMPLE 3.3. (*a-string.*) Another simple illustration of the above result stated in part (*a*) (as well as of the result stated in part (*b*) below) is the a-string discussed in §2 (see Example 2.8 in §2.5). Recall that for any $a > 0$,

$$\Omega = \Omega_a = \bigcup_{j=1}^{\infty} ((j+1)^{-a}, j^{-a}) \tag{3.35}$$

and so (with $\partial\Omega_a := \partial(\Omega_a)$)

$$\partial\Omega_a = \{j^{-a} : j \geq 1\} \cup \{0\} \tag{3.36}$$

and $\mathcal{L}_a = (\ell_j)_{j\geq 1}$, with $\ell_j := j^{-a} - (j+1)^{-a}$ for each $j \geq 1$. Furthermore, the RFD $(\partial\Omega_a, \Omega_a)$ (or, equivalently, the fractal string $\mathcal{L}_a$) is Minkowski measurable with Minkowski content

$$\mathcal{M} = \frac{2^{1-D} a^D}{1-D}, \tag{3.37}$$

where

$$D = \dim_B(\partial\Omega_a, \Omega_a) = D_{\mathcal{L}_a} = \frac{1}{a+1}. \tag{3.38}$$

Moreover, it follows from a direct computation, left as an exercise for the interested reader, that

$$\operatorname{res}(\zeta_{\mathcal{L}_a}, D) = D a^D, \tag{3.39}$$

which since (in light of (3.19))

$$\operatorname{res}(\zeta_{\partial\Omega_a, \Omega_a}, D) = \frac{2^{1-D}}{D} \operatorname{res}(\zeta_{\mathcal{L}_a}, D), \tag{3.40}$$

is in agreement with the exact counterpart for RFDs of (3.34); namely,

$$\operatorname{res}(\zeta_{\partial\Omega_a, \Omega_a}) = (1-D) \operatorname{res}(\widetilde{\zeta}_{\partial\Omega_a, \Omega_a}, D). \tag{3.41}$$

Indeed, by combining (3.37)–(3.40), we obtain that

$$\begin{aligned}(1-D)\mathcal{M} = 2^{1-D} a^D &= \frac{2^{1-D}}{D} (D a^D) \\ &= \frac{2^{1-D}}{D} \operatorname{res}(\zeta_{\mathcal{L}_a}, D) = \operatorname{res}(\zeta_{\partial\Omega_a, \Omega_a}, D),\end{aligned}$$

as desired.

(***b***) (*Existence of a meromorphic extension: Minkowski measurable case*). Let (A, Ω) be an RFD in $\mathbb{R}^N$ such that there exists $\alpha > 0$, $\mathcal{M} \in (0, +\infty)$ and $D \geq 0$ such that

$$V_{A,\Omega}(\varepsilon) := |A_\varepsilon \cap \Omega| = \varepsilon^{N-D}(\mathcal{M} + O(\varepsilon^\alpha)) \quad \text{as } \varepsilon \to 0^+. \tag{3.42}$$

Then, $\dim_B(A,\Omega)$ exists and $\dim_B(A,\Omega) = D$. Furthermore, (A,Ω) is Minkowski measurable with Minkowski content equal to $\mathcal{M}$. Moreover, the distance zeta function $\zeta_{A,\Omega}$ has for abscissa of convergence D and possesses a (necessarily unique) meromorphic continuation (still denoted by $\zeta_{A,\Omega}$, as usual) to (at least) the open right half-plane $\{Re(s) > D-\alpha\}$; that is, $D_{mer}(\zeta_{A,\Omega}) \leq D-\alpha$, where $D_{mer}(\zeta_{A,\Omega})$ is the *abscissa of meromorphic continuation* of $\zeta_{A,\Omega}$ (defined much as the abscissa of holomorphic continuation, except for the adjective "holomorphic" replaced by "meromorphic"). The only pole of $\zeta_{A,\Omega}$ (i.e., the only visible complex dimension of (A,Ω)) in this half-plane is $s = D$; it is simple and $\text{res}(\zeta_{A,\Omega}, D) = (N-D)\mathcal{M}$.

Clearly, the same result holds for $\widetilde{\zeta}_{A,\Omega}$, the tube (instead of the distance) zeta function of (A,Ω), except for the fact that in that case, $\text{res}(\widetilde{\zeta}_{A,\Omega}, D) = \mathcal{M}$.

EXERCISE 3.4. (*i*) Show that the hypotheses of part (*b*) are satisfied for the a-string of Example 3.3, viewed as the RFD $(\partial\Omega_a, \Omega_a)$.

(*ii*) Prove via a direct computation that $\zeta_{\mathcal{L}_a}$ has a meromorphic continuation to all of $\mathbb{C}$ with (simple) poles at $s = D = 1/(a+1)$ and (at possibly a subset of) $\{-D, -2D, \cdots, -nD, \cdots : n \geq 1\}$. (For a complete answer, see [**Lap-vF4**, Thm. 6.21 and its proof].) Deduce that

$$\mathcal{D}_{\mathcal{L}_a} = \mathcal{D}(\zeta_{\partial\Omega_a,\Omega_a}) \subseteq \{D, 0, -D, -2D, \cdots, -nD, \cdots : n \geq 1\},$$

where all the complex dimensions are simple and D and 0 are always complex dimensions of the fractal string $\mathcal{L}_a$ or equivalently, of the RFD $(\partial\Omega_a, \Omega_a)$.

(c) (*Existence of a meromorphic extension: Minkowski nonmeasurable case*). Let (A,Ω) be an RFD in $\mathbb{R}^N$ such that there exists $\alpha > 0$, $D \in \mathbb{R}$ and a nonconstant periodic function G with minimal period $T > 0$, satisfying

$$V_{A,\Omega}(\varepsilon) := |A_\varepsilon \cap \Omega| = \varepsilon^{N-D}(G(\log \varepsilon^{-1}) + O(\varepsilon^\alpha)) \quad \text{as } \varepsilon \to 0^+. \tag{3.43}$$

Then, $\dim_B(A,\Omega)$ exists and $\dim_B(A,\Omega) = D$. Furthermore, G is continuous and A is Minkowski nondegenerate with lower and upper Minkowski contents respectively given by

$$\mathcal{M}_* = \min G \text{ and } \mathcal{M}^* = \max G. \tag{3.44}$$

Moreover, the tube zeta function $\widetilde{\zeta}_{A,\Omega}$ also has for abscissa of convergence $D(\widetilde{\zeta}_{A,\Omega}) = D$ and possesses a (necessarily unique) meromorphic extension (still denoted by $\widetilde{\zeta}_{A,\Omega}$) to (at least) the open right half-plane $\{Re(s) > D-\alpha\}$; that is, $D_{mer}(\widetilde{\zeta}_{A,\Omega}) \leq D-\alpha$.[64]

In addition, the set of all the poles of $\widetilde{\zeta}_{A,\Omega}$ (i.e., the set of visible complex dimensions of the RFD (A,Ω)) in this half-plane is given by

(3.45)

$$\mathcal{D}_{PC}(A,\Omega) = \dim_{PC}(\widetilde{\zeta}_{A,\Omega}) = \mathcal{D}(\widetilde{\zeta}_{A,\Omega}) = \left\{ s_k := D + i\frac{2\pi}{T}k : \widehat{G}\Big(\frac{k}{T}\Big) \neq 0, k \in \mathbb{Z} \right\},$$

where

$$\widehat{G}_0(t) := \int_0^T e^{-2\pi i t\tau} G(\tau) d\tau, \text{ for all } t \in \mathbb{R}. \tag{3.46}$$

[64]We first state the results for $\widetilde{\zeta}_{A,\Omega}$ because they are more elegantly written in this situation; we will then mention the few changes needed for the corresponding statements about $\zeta_{A,\Omega}$.

We have, for every $k \in \mathbb{Z}$,

$$\operatorname{res}(\widetilde{\zeta}_{A,\Omega}, s_k) = \frac{1}{T}\widehat{G}_0\Big(\frac{k}{T}\Big), \tag{3.47}$$

and hence,

$$|\operatorname{res}(\widetilde{\zeta}_{A,\Omega}, s_k)| \leq \frac{1}{T}\int_0^T G(\tau)d\tau. \tag{3.48}$$

Also,

$$\lim_{k\to\pm\infty} \operatorname{res}(\widetilde{\zeta}_{A,\Omega}, s_k) = 0. \tag{3.49}$$

Finally, the residue of $\widetilde{\zeta}_{A,\Omega}$ at $s = D$ coincides both with the mean value of G and with $\widetilde{\mathcal{M}}$, the *average Minkowski content* of (A, Ω):[65]

$$\operatorname{res}(\widetilde{\zeta}_{A,\Omega}, D) = \frac{1}{T}\int_0^T G(\tau)d\tau = \widetilde{\mathcal{M}}. \tag{3.51}$$

In particular, the RFD (A, Ω) is Minkowski nondegenerate but is not Minkowski measurable and, in fact, we have that

$$\mathcal{M}_* < \operatorname{res}(\widetilde{\zeta}_{A,\Omega}, D) < \mathcal{M}^* < \infty. \tag{3.52}$$

For the distance (instead of the tube) zeta function $\zeta_{A,\Omega}$, entirely analogous results hold, except for the fact that in light of (3.16),

$$\operatorname{res}(\zeta_{A,\Omega}, s_k) = (N - s_k)\frac{1}{T}\widehat{G}_0\Big(\frac{k}{T}\Big), \text{ for all } k \in \mathbb{Z}, \tag{3.53}$$

and hence,

$$\operatorname{res}(\zeta_{A,\Omega}, s_k) - o(|k|) \quad \text{as } |k| \to \infty. \tag{3.54}$$

EXERCISE 3.5. For the generalized Cantor sets $A = C^{(a)}$ of Example 3.2 above, verify that both the hypotheses and the conclusions (of the counterpart for bounded sets) of part (*c*) just above are satisfied for an arbitrary $a > 0$. Deduce that for any $a \in (0, 1/2)$, both ζ_A and $\widetilde{\zeta}_A$ have a meromorphic extension to all of $\mathbb{C}$ and

$$\mathcal{D}_A = \mathcal{D}(\zeta_A) = \mathcal{D}(\widetilde{\zeta}_A) = \{D + ik\mathbf{p} : k \in \mathbb{Z}\},$$

where $D := \log 2/\log(1/a)$ and $\mathbf{p} := 2\pi/\log(1/a)$ are respectively the Minkowski dimension and the oscillatory period of $A = C^{(a)}$. Also, calculate the average Minkowski content $\widetilde{\mathcal{M}}$ of A both via a direct computation and by using one of the main results of part (*c*).

The next "exercise" is significantly more difficult than the previous one.[66]

---

[65]Here, $\widetilde{\mathcal{M}}$ is defined as the limit of a suitable Cesaro logarithmic average of $V_{A,\Omega}(t)/t^{N-D}$, much as in [**Lap-vF4**, Thm. 8.30] where $N = 1$. More specifically,

$$\widetilde{\mathcal{M}} := \lim_{\tau\to+\infty} \frac{1}{\log\tau}\int_{1/\tau}^1 \frac{V_{A,\Omega}(t)}{t^{N-D}}\frac{dt}{t}, \tag{3.50}$$

where the indicated limit is assumed to exist in $(0, +\infty)$.

[66]In fact, it essentially corresponds to Problem 6.2.35 in [**LapRaZu1**] and would also help solve one part of the much broader Problem 6.2.36 in [**LapRaZu1**].

EXERCISE 3.6. Let A be a self-similar set satisfying the open set condition. Find geometric conditions on A so that in the lattice case (respectively, in the nonlattice case), the hypotheses and hence also the conclusions of the main result (of the counterpart for bounded sets) of part (c) (respectively, part (b)) of this subsection (i.e., §3.3.2) are satisfied.[67]

A variety of significantly more complicated behaviors for the asymptotics (as $\varepsilon \to 0^+$) of the tube function $\varepsilon \mapsto |A_\varepsilon \cap \Omega| = V_{A,\Omega}(\varepsilon)$ of an RFD are considered in [**LapRaZu1**]. These include, most notably, transcendentally n-quasiperiodic behavior, for any $n \in \mathbb{N} \cup \{\infty\}$; see [**LapRaZu1**, §3.1 and §4.6]. Instead of giving the precise (but somewhat involved) definitions and results here, we limit ourselves for now to the following simple example. (Further information will be provided later in the paper, especially in §3.6.)

EXAMPLE 3.7. Suppose that the tube function of the bounded set A satisfies (3.43), where the function G is no longer assumed to be periodic (of period T) but is given instead by $G = G_1 + G_2$, where the nonconstant functions G_1 and G_2 are periodic of (minimal) periods T_1 and T_2, respectively, with T_1/T_2 irrational. Then, ζ_A has a meromorphic continuation to (at least) $\{Re(s) > D - \alpha\}$ and the set of principal complex dimensions of A consists of simple (nonremovable) singularities of ζ_A (and of $\widetilde{\zeta}_A$) and is given by

$$\mathcal{D}_{PC}(A) = \dim_{PC} A = \bigcup_{j=1}^{2} \left(D + i\frac{2\pi}{T_j}\mathbb{Z}\right) = D + i\Big(\bigcup_{j=1}^{2} \frac{2\pi}{T_j}\mathbb{Z}\Big), \tag{3.55}$$

where $D = D(\zeta_A) = D(\widetilde{\zeta}_A) = \dim_B A$. We note that since T_1 and T_2 are incommensurable, the imaginary parts of the principal complex dimensions of A have a rather different structure than in part (c) above. In particular, they are no longer in arithmetic progression.

Finally, we point out that the assumptions of this example are realized by the compact subset of $\mathbb{R}$ obtained by taking the disjoint union of two distinct and suitably chosen (two-parameter) generalized Cantor sets; see [**LapRaZu1**, Thm. 3.1.12] for the details and [**LapRaZu1**, Thm. 3.1.15] for the generalization to n such Cantor sets, corresponding to the case when A is (transcendentally) 2-quasiperiodic or more generally, n-quasiperiodic, respectively. The important (and highly nontrivial) extension to the case when $n = \infty$ is dealt with in [**LapRaZu1**, §4.6].

3.3.3. *Scaling property and invariance under isometries.* We first state the scaling invariance property of the distance zeta function $\zeta_{A,\Omega}$ of an RFD (A,Ω).[68] (We leave it as an exercise for the interested reader to state its counterpart for the tube zeta function $\widetilde{\zeta}_{A,\Omega}$; alternatively, see [**LapRaZu1**, §4.1.3].)

For any $\lambda > 0$, we have $D(\zeta_{\lambda A,\lambda\Omega}) = D(\zeta_{A,\Omega}) = \overline{D} := \overline{\dim}_B(A,\Omega)$ and

$$\zeta_{\lambda A,\lambda\Omega}(s) = \lambda^s \zeta_{A,\Omega}(s), \tag{3.56}$$

[67]Recall that a self-similar set is said to be lattice if the multiplicative group generated by its distinct scaling ratios is of rank 1, and is called nonlattice otherwise.

[68]The properties stated for RFDs in §3.3.3 and in fact, in all of §3.3.1–§3.3.4, have natural counterparts for bounded sets A in $\mathbb{R}^N$. In particular, recall from (3.21) and the discussion surrounding it that $D(\zeta_{A,A_\delta})$ denotes the abscissa of convergence of the RFD (A, A_δ).

for all $s \in \mathbb{C}$ with $Re(s) > \overline{\dim}_B A$ or, more generally, for all $s \in U$, where U is any domain containing the closed right half-plane $\{Re(s) \geq \overline{D}\}$ to which one, and hence both, of these fractal zeta functions has a meromorphic continuation.

Furthermore, if $\omega \in \mathbb{C}$ is a simple pole of the meromorphic extension of $\zeta_{A,\Omega}$ to some open connected neighborhood of the critical line $\{Re(s) = \overline{D}\}$ (or, equivalently, of $\{Re(s) \geq \overline{D}\}$), then

$$\operatorname{res}(\zeta_{\lambda A,\lambda\Omega}, \omega) = \lambda^{\omega} \operatorname{res}(\zeta_{A,\Omega}, \omega). \tag{3.57}$$

It is noteworthy that the scaling property of the residues of $\zeta_{A,\Omega}$, as stated in (3.57), is very analogous to the scaling property of Hausdorff measure; the latter, however, is restricted to a single exponent, namely, the Hausdorff dimension, whereas (3.57) is valid for any (visible) complex dimension of (A, Ω). We do not wish to elaborate on this point here but simply mention that under appropriate hypotheses, a suitable version of these residues should give rise to a family of complex measures, defined by the maps $\Omega \mapsto \operatorname{res}(\zeta_{A,\Omega}, \omega)$ and indexed by the (visible) complex dimensions ω of the bounded set A and where Ω is allowed to run through the Borel (and not necessarily open) subsets of $\mathbb{R}^N$. For more information, see [**LapRaZu1**, App. B].

Next, we simply mention that the distance and tube zeta functions of an RFD (A, Ω) are clearly invariant under the group of displacements of $\mathbb{R}^N$ (that is, under the group of isometries of the affine space $\mathbb{R}^N$, generated by the rotations and translations). More specifically, if f is such a displacement of $\mathbb{R}^N$, then

$$\zeta_{f(A),f(\Omega)} = \zeta_{A,\Omega} \tag{3.58}$$

(and analogously for $\widetilde{\zeta}_{A,\Omega}$, as well as for the corresponding upper, lower Minkowski dimensions and contents, and the visible complex dimensions, in particular).

The scaling and invariance properties stated in the present subsection (i.e., §3.3.3), along with other "covariance" properties of the fractal zeta functions, are very useful in the concrete computation of the distance and tube zeta functions of a variety of examples (including many of those discussed in §3.4), as well as of the corresponding complex dimensions. They also play an important role in the direct computation of fractal tube formulas for many concrete examples (including several of those discussed in §3.5.3). (See [**LapRaZu1**, esp., Chs. 3–5].)

3.3.4. *Invariance of the complex dimensions under embedding into higher-dimensional spaces.* Let (A, Ω) be an RFD in $\mathbb{R}^N$ and let $M \geq 1$ be an arbitrary integer. Denote by $(A, \Omega)_M$ the natural embedding of (A, Ω) into $\mathbb{R}^{N+M}$, where $(A, \Omega)_M := (A_M, \Omega_M)$, with

$$A_M := A \times \{0\} \times \cdots \times \{0\} \subseteq \mathbb{R}^{N+M} \tag{3.59}$$

and[69]

$$\Omega_M := \Omega \times (-1, 1)^M \subseteq \mathbb{R}^{N+M}. \tag{3.60}$$

Then, it is shown in [**LapRaZu1**, §4.7.2] that given any connected open neighborhood U of the critical line $\{Re(s) = \overline{D}\}$, where (as usual)

$$\overline{D} := \overline{\dim}_B A = D(\widetilde{\zeta}_A) = D(\zeta_A), \tag{3.61}$$

with $\overline{D} < N$, the tube zeta function $\widetilde{\zeta}_{A,\Omega}$ has a (necessarily unique) meromorphic extension to U if and only if $\widetilde{\zeta}_{(A,\Omega)_M}$ does, and in that case, the visible complex

[69]Here, $\{0\} \times \cdots \times \{0\}$ is the M-fold Cartesian product of $\{0\}$ by itself, viewed as a subset of $\mathbb{R}^M$.

dimensions of the RFD (A, Ω) in $\mathbb{R}^N$ and of the RFD $(A, \Omega)_M$ in $\mathbb{R}^{N+M}$ coincide (and similarly for the distance zeta functions $\zeta_{A,\Omega}$ and $\zeta_{(A,\Omega)_M}$):

$$\begin{aligned} \mathcal{D}_{A,\Omega}(U) &= \mathcal{D}(\widetilde{\zeta}_{A,\Omega}; U) = \mathcal{D}(\widetilde{\zeta}_{(A,\Omega)_M}; U) \\ &= \mathcal{D}(\zeta_{A,\Omega}; U) = \mathcal{D}(\zeta_{(A,\Omega)_M}; U) = \mathcal{D}_{(A,\Omega)_M}(U), \end{aligned} \tag{3.62}$$

as equalities between multisets. Moreover,

$$\begin{aligned} \mathcal{D}_{PC}(A, \Omega) &= \dim_{PC}(A, \Omega) = \mathcal{D}_{PC}(\widetilde{\zeta}_{A,\Omega}) = \mathcal{D}_{PC}(\widetilde{\zeta}_{(A,\Omega)_M}) \\ &= \mathcal{D}_{PC}(\zeta_{A,\Omega}) = \mathcal{D}_{PC}(\zeta_{(A,\Omega)_M}) = \dim_{PC}(\zeta_{(A,\Omega)_M}) = \mathcal{D}_{PC}(A, \Omega)_M \end{aligned} \tag{3.63}$$

and

$$\begin{aligned} \overline{D} := \overline{\dim}_B(A, \Omega) &= D(\widetilde{\zeta}_{A,\Omega}) = D(\widetilde{\zeta}_{(A,\Omega)_M}) \\ &= D(\zeta_{A,\Omega}) = D(\zeta_{(A,\Omega)_M}) = \overline{\dim}_B(A, \Omega)_M. \end{aligned} \tag{3.64}$$

Consequently, neither the values nor the multiplicities of the (visible) complex dimensions (and, in particular, of the principal complex dimensions) of the RFD (A, Ω) depend on the dimension of the ambient space.[70]

Moreover, we point out that since $\overline{\dim}_B(A, \Omega) < N$ (and hence also, we have $\overline{\dim}_B(A, \Omega)_M < N + M$), the exact same results hold for the tube zeta functions $\widetilde{\zeta}_{A,\Omega}$ and $\widetilde{\zeta}_{(A,\Omega)_M}$ replaced, respectively, by the distance zeta functions $\zeta_{A,\Omega}$ and $\zeta_{(A,\Omega)_M}$.

Finally, we note that as usual, the exact analog of the results stated in this subsection (i.e., §3.3.4) hold for the special case of bounded subsets A (instead of RFDs (A, Ω)) in $\mathbb{R}^N$. It suffices to replace the RFDs (A, Ω) and $(A, \Omega)_M$ by the bounded sets A and A_M (as given by (3.59)) in $\mathbb{R}^N$ and $\mathbb{R}^{N+M}$, respectively, in all of the corresponding statements; see [**LapRaZu1**, §4.7.1] for the details.

3.4. Examples of fractal zeta functions and complex dimensions. In this subsection, we give a variety of examples of bounded sets and of RFDs for which the associated distance zeta function (or, equivalently, in light of the functional equation (3.11) or (3.15), the tube zeta function) can be calculated explicitly and the corresponding poles (i.e., the complex dimensions) can be determined. We will limit ourselves to the simplest examples and omit the computation involved, often based in part on symmetry and scaling considerations, but refer instead to [**LapRaZu1**] or to [LapRaZu2–9] for the details and many further examples.

3.4.1. *The Sierpinski gasket.* Let $A \subseteq \mathbb{R}^2$ be the classic Sierpinski gasket. It is a self-similar set in $\mathbb{R}^2$ with three equal scaling ratios $r_1 = r_2 = r_3 = 1/2$ and hence, with Minkowski dimension $D = \dim_B A = \log 3/\log 2 = \log_2 3$ (coinciding with the similarity dimension of A). As is well-known, A is the unique (nonempty) compact subset of $\mathbb{R}^2$ satisfying the fixed point equation

$$A = \bigcup_{j=1}^{3} S_j(A), \tag{3.65}$$

[70]It is significantly simpler to check that the values of $\underline{\dim}_B(A, \Omega)$, and also the statements according to which (A, Ω) is Minkowski nondegenerate or is Minkowski measurable, are independent of the dimension of the ambient space; see [**LapRaZu1**, §4.7.2]. In the Minkowski measurable case, the corresponding Minkowski content can be suitably normalized (much as in [**Fed2**]) so as to also be independent of the embedding dimension; see [**Res**]. (And similarly for the normalized values of $\mathcal{M}_*$ and $\mathcal{M}^*$.)

where S_1, S_2, S_3 are contractive similarity transformations of $\mathbb{R}^2$ (with scaling ratios r_1, r_2, r_3, as above) defined in a simple way and with respective fixed points v_1, v_2, v_3, the vertices of the unit equilateral triangle, which is the generator of A.

Then, one can show (see [**LapRaZu1**, §5.5.3]) that for $\delta > 1/4\sqrt{3}$ (so that A_δ be connected),[71] ζ_A has a meromorphic extension to all of $\mathbb{C}$ given by

$$\zeta_A(s) = \frac{6(\sqrt{3})^{1-s}2^{-s}}{s(s-1)(2^s-3)} + 2\pi\frac{\delta^s}{s} + 3\frac{\delta^{s-1}}{s-1}, \tag{3.66}$$

for every $s \in \mathbb{C}$. Consequently, the set of principal complex dimensions of the Sierpinski gasket A is given by

$$\begin{aligned} \dim_{PC} A = \mathcal{D}_{PC}(\zeta_A) &= \left\{ \log_2 3 + i\frac{2\pi}{\log 2}k : k \in \mathbb{Z}\right\} \\ &= \log_2 3 + i\frac{2\pi}{\log 2}\mathbb{Z} \end{aligned} \tag{3.67}$$

and the set of all complex dimensions of A is given by

$$\begin{aligned} \mathcal{D}_A := \mathcal{D}(\zeta_A) = \{0\} \cup \left(\log_2 3 + i\frac{2\pi}{\log 2}\mathbb{Z}\right) &= \{0\} \cup \dim_{PC} A \\ = \{0\} \cup \left\{ s_k := \log_2 3 + i\frac{2\pi}{\log 2}k : k \in \mathbb{Z}\right\}. \end{aligned} \tag{3.68}$$

Each complex dimension 0 or $s_k := \log_2 3 + i(2\pi/\log 2)k$ ($k \in \mathbb{Z}$) is simple (i.e., is a simple pole of ζ_A) and the corresponding residue is given respectively by[72]

$$\operatorname{res}(\zeta_A, 0) = 3\sqrt{3} + 2\pi, \ \operatorname{res}(\zeta_A, 1) = 0 \tag{3.69}$$

and for each $k \in \mathbb{Z}$,

$$\operatorname{res}(\zeta_A, s_k) = \frac{6(\sqrt{3})^{1-s_k}}{(\log 2)4^{s_k}s_k(s_k-1)}. \tag{3.70}$$

Finally, note that in (3.67) and (3.68), $D := \log_2 3$ is the Minkowski dimension of A and $\mathbf{p} := 2\pi/\log 2$ is the oscillatory period of A. Also, the expression obtained for $\dim_{PC} A$ in (3.67) is compatible with the results of part (*c*) of §3.3.2; see (3.45).

3.4.2. *The Sierpinski carpet.* Let $A \subseteq \mathbb{R}^2$ be the classic Sierpinski carpet (with generator the unit square and 8 equal scaling ratios $r_1 = \cdots = r_8 = 1/3$). As is well known, A is the unique (nonempty) compact subset of $\mathbb{R}^2$ such that $A = \cup_{j=1}^8 S_j(A)$, where $S_1, \cdots, S_8$ are suitable contractive similarities of $\mathbb{R}^2$. Clearly, $D := \dim_B A$ exists and $D = \log_3 8$, the similarity dimension of the self-similar set A. Then, much as in the case of the Sierpinski gasket A from §3.4.1 just above, it can be shown (see [**LapRaZu1**, Prop. 3.21]) that ζ_A has a meromorphic extension to all

[71] Recall from §3.2 that the poles of $\zeta_{A,\Omega}$ (and of $\widetilde{\zeta}_{A,\Omega}$) are independent of the choice of $\delta > 0$; i.e., the set of complex dimensions of an RFD (A, Ω), $\mathcal{D}_{A,\Omega} = \mathcal{D}(\zeta_{A,\Omega}) = \mathcal{D}(\widetilde{\zeta}_{A,\Omega})$, is independent of the choice of $\delta > 0$. It is also true, in particular, for a bounded set A instead of an RFD (A, Ω) (by considering the RFD (A, A_{δ_2}), for any fixed $\delta_2 > 0$). This comment, being valid for any RFD (A, Ω) (and, in particular, bounded set) in $\mathbb{R}^N$, will no longer be repeated in this section (i.e., §3.4).

[72] A priori, in light of (3.66), $s = 1$ should be a pole of ζ_A. However, a direct computation shows that $\operatorname{res}(\zeta_A, 1) = -3 + 3 = 0$, as indicated in (3.69). Hence, 1 is *not* a complex dimension of the Sierpinski gasket. In fact, the corresponding fractal tube formula will *not* contain a term corresponding to $s = 1$; see the relevant parts of §3.5.3, along with [**LapRaZu1**, §5.5.3], especially, the last equation before Example 5.5.13 in *loc. cit.*

of $\mathbb{C}$, and that for every $\delta > 1/6$ (so that the δ-neighborhood A_δ of A be connected) and every $s \in \mathbb{C}$,

$$\zeta_A(s) = \frac{8}{2^s s(s-1)(3^s-8)} + 2\pi\frac{\delta^s}{s} + 4\frac{\delta^{s-1}}{s-1}. \tag{3.71}$$

It follows that

$$\dim_{PC} A = \mathcal{D}_{PC}(\zeta_A) = \left\{ \log_3 8 + i\frac{2\pi}{\log 3}k : k \in \mathbb{Z} \right\} = \log_3 8 + i\frac{2\pi}{\log 3}\mathbb{Z} \tag{3.72}$$

and the set of all complex dimensions of the Sierpinski carpet is given by

$$\begin{aligned} \mathcal{D}_A = \mathcal{D}(\zeta_A) &= \{0,1\} \cup \left(\log_3 8 + i\frac{2\pi}{\log 3}\mathbb{Z} \right) \\ &= \{0,1\} \cup \dim_{PC} A = \{0,1\} \cup \left\{ s_k := \log_3 8 + i\frac{2\pi}{\log 3}k : k \in \mathbb{Z} \right\}. \end{aligned} \tag{3.73}$$

Furthermore, the complex dimensions of A are all simple and the residues at $0, 1$ and s_k $(k \in \mathbb{Z})$ are given, respectively, by

$$\operatorname{res}(\zeta_A, 0) = 2\pi + \frac{8}{7}, \quad \operatorname{res}(\zeta_A, 1) = \frac{16}{5} \tag{3.74}$$

and

$$\operatorname{res}(\zeta_A, s_k) = \frac{2^{-s_k}}{(\log 3)s_k(s_k-1)}, \quad \text{for all } k \in \mathbb{Z}. \tag{3.75}$$

Again, in (3.72) and (3.73), $D := \log_3 8$ and $\mathbf{p} := 2\pi/\log 3$ are, respectively, the Minkowski dimension and the oscillatory period of A, in agreement with the results stated in part (*c*) of §3.3.2.

3.4.3. *The 3-d Sierpinski carpet.* We refer to [**LapRaZu1**, Exple. 5.5.13] for the precise definition of this version of the three-dimensional Sierpinski carpet A and for the corresponding results. It is shown there that for any $\delta > 1/6$ (so that A_δ be connected), ζ_A has a meromorphic extension to all of $\mathbb{C}$ given by

$$\zeta_A(s) = \frac{48 \,.\, 2^{-s}}{s(s-1)(s-2)(3^s-26)} + 4\pi\frac{\delta^s}{s} + 6\pi\frac{\delta^{s-1}}{s-1} + 6\frac{\delta^{s-2}}{s-2}, \tag{3.76}$$

for every $s \in \mathbb{C}$. Therefore,

$$\dim_{PC}(\zeta_A) = \mathcal{D}_{PC}(\zeta_A) = \log_3 26 + i\frac{2\pi}{\log 3}\mathbb{Z} \tag{3.77}$$

and

$$\mathcal{D}(\zeta_A) = \{0,1,2\} \cup \dim_{PC} A = \{0,1,2\} \cup \{s_k := D + ik\mathbf{p} : k \in \mathbb{Z}\}, \tag{3.78}$$

where $D := D(\zeta_A) = \log_3 26$ and $\mathbf{p} := 2\pi/\log 3$ are, respectively, the Minkowski dimension (as well as the similarity dimension) and the oscillatory period of A. Each complex dimension in (3.77) and (3.78) is simple and

$$\operatorname{res}(\zeta_A, j) = 4\pi - \frac{24}{25}, \; 6\pi + \frac{24}{23}, \; \frac{96}{17} \quad \text{for } j = 0, 1, 2, \tag{3.79}$$

respectively; also, for every $k \in \mathbb{Z}$,

$$\operatorname{res}(\zeta_A, s_k) = \frac{24}{13.2^{s_k} s_k(s_k-1)(s_k-2)\log 3}. \tag{3.80}$$

3.4.4. *The N-dimensional relative Sierpinski gasket.* Let (A_N, Ω_N) denote the *N-dimensional relative Sierpinski gasket,* also called the (*inhomogeneous*) *N-gasket* RFD, in short, and as introduced in [**LapRaZu1**, Exple. 4.2.26] (as well as in [**LapRaZu4**]). We refer the interested reader to *loc. cit.* for a detailed description of its geometric construction and for the corresponding figures. (See also Remark 3.8 below for a synopsis of the construction.) We simply mention here that for each fixed integer $N \geq 2, (A_N, \Omega_N)$ is an RFD in $\mathbb{R}^N$ which can also be viewed as a self-similar spray (or RFD) in $\mathbb{R}^N$ (in the refined sense of [**LapRaZu1**, §4.2.1 and §5.5.6] rather than in the original sense of [**Lap3**], [**LapPo3**] and [**LapPe2**] or [**LapPeWi1**]), with $N+1$ equal scaling ratios $r_1 = \cdots = r_{N+1} = 1/2$ and with a single generator RFD $(\partial\Omega_{N,0}, \Omega_{N,0})$, where the bounded open set $\Omega_{N,0}$ in $\mathbb{R}^N$ (called the N-plex) is described in Remark 3.8. Furthermore, unlike in our previous examples in §3.4.1–§3.4.3, A_N is *not* a self-similar set (in the usual sense of the term) but is instead an *inhomogeneous* self-similar set, in the sense of [**BarnDemk**] and (with a different terminology) of [**Hat**]. More specifically, it is the unique (nonempty) compact subset A of $\mathbb{R}^N$ satisfying the *inhomogeneous* fixed point equation

$$A = \bigcup_{j=1}^{N+1} S_j(A) \cup B, \tag{3.81}$$

where the maps S_j (for $j = 1, \cdots, N+1$) are $N+1$ contractive similarity transformations[73] of $\mathbb{R}^N$ (the same ones as those defining the N-dimensional analog of the usual self-similar gasket, which is an homogeneous or a classic self-similar set, satisfying the counterpart of (3.81) with $B := \emptyset$, the empty set) and B is a certain *nonempty* compact subset of $\mathbb{R}^N$; in fact, B can be chosen to be equal to $\partial\Omega_{N,0}$, the boundary of the N-plex $\Omega_{N,0}$ described in Remark 3.8.

REMARK 3.8. (*Construction of the generator $\Omega_{N,0}$ and of the inhomogeneous N-gasket RFD (A_N, Ω_N).*) Here, the generator $\Omega_{N,0}$ and the compact set A_N can be constructed as follows. Let $V_N = \{P_1, \cdots, P_{N+1}\}$ be a set of $N+1$ points in $\mathbb{R}^N$ such that $|P_j - P_k| = 1$, for any $j \neq k$. (Such a set can be constructed inductively.) Let Ω_N be the (necessarily closed) convex hull of V_N. Clearly, Ω_N is an N-simplex. Then, $\Omega_{N,0}$, called the N-plex, is the bounded open subset of $\mathbb{R}^N$ obtained by taking the interior of the convex hull of the set of midpoints of all of the $\frac{(N+1)N}{2} = \binom{N}{2}$ edges of the N-simplex Ω_N. (Note that for $N = 2, \Omega_{N,0}$ is the first deleted open triangle in the construction of the Sierpinski gasket, while for $N = 3$, it is an octahedron; see [**LapRaZu1**, Fig. 4.7] for an illustration.)

Now, the set $\overline{\Omega_N}\backslash\Omega_{N,0}$ is the union of $N+1$ congruent and compact N-simplices with disjoint interiors and having all their sides (edges) of length $1/2$. This is the first step in the construction of A_N. We proceed analogously with each of the aforementioned N-simplices. We then repeat the construction, ad infinitum. The compact subset A_N of $\mathbb{R}^N$ obtained in this manner is called the *inhomogeneous N-gasket.* For $N = 2$, it coincides with the classic Sierpinski gasket (studied in

[73]These similitudes $(S_j)_{j=1}^{N+1}$ have for respective fixed points $(P_j)_{j=1}^{N+1}$, the points chosen at the beginning of Remark 3.8.

§3.4.1), but when $N \geq 3$, it does not coincide with the usual N-dimensional Sierpinski gasket (studied, e.g., in [**KiLap1**]). In fact, still for $N \geq 3$, it is no longer self-similar (in the classic sense) but is instead an inhomogeneous self-similar set satisfying (3.81), with $B := \partial\Omega_{N,0}$, the boundary of the N-plex. (See [**LapRaZu1**, Fig. 4.8] for an illustration of the case when $N = 3$.)

Finally, the *relative* (or *inhomogeneous*) *N-gasket RFD* is given by (A_N, Ω_N), where A_N is the above inhomogeneous N-gasket and Ω_N is the above N-simplex.

Then (see [**LapRaZu1**, Exple. 4.2.26]), for the inhomogeneous N-gasket RFD (A_N, Ω_N), the distance zeta function ζ_{A_N,Ω_N} has a meromorphic extension to all of $\mathbb{C}$, given for every $s \in \mathbb{C}$ by

$$\zeta_{A_N,\Omega_N}(s) = \frac{g_N(s)}{s(s-1)\cdots(s-(N-1))(1-(N+1)2^{-s})}, \tag{3.82}$$

for some nowhere vanishing entire function g_N. (For example, when $N = 3$, we have $g_3(s) := 8(\sqrt{3})^{3-s}(2\sqrt{2})^{-s}$ and if $N = 2$, $g_2(s) := 6(\sqrt{3})^{1-s}2^{-s}$, still for all $s \in \mathbb{C}$; see, respectively, [**LapRaZu1**, Eq. (4.2.89) and Prop. 4.2.25].)

In order to explain the form of ζ_{A_N,Ω_N} given in (3.82), we recall that (A_N, Ω_N) is a self-similar RFD with generator the RFD $(\partial\Omega_{N,0}, \Omega_{N,0})$ and with equal scaling ratios $r_j \equiv 1/2$, for $j = 1, \cdots, N+1$. Thus, according to the results of [**LapRaZu1**, §4.2.1 and §5.5.6] about self-similar sprays (and recalled in §3.4.10 below),

$$\zeta_{A_N,\Omega_N}(s) = \zeta_{\mathfrak{s}}(s) \cdot \zeta_{\partial\Omega_{N,0},\Omega_{N,0}}(s), \tag{3.83}$$

where the *scaling zeta function* $\zeta_{\mathfrak{s}}$ (here, the geometric zeta function of the underlying unbounded self-similar string with equal scaling ratios $r_j \equiv 1/2$ for $j = 1, \cdots, N+1$) is given by

$$\zeta_{\mathfrak{s}}(s) = \frac{1}{1-(N+1)2^{-s}} \tag{3.84}$$

for all $s \in \mathbb{C}$, and where via a direct computation,[74] one can show that $\zeta_{\partial\Omega_{N,0},\Omega_{N,0}}$ is given for all $s \in \mathbb{C}$ by

$$\zeta_{\partial\Omega_{N,0},\Omega_{N,0}}(s) = \frac{g_N(s)}{s(s-1)\cdots(s-(N-1))}, \tag{3.85}$$

with g_N as above. Now combining (3.83)–(3.85), we obtain (3.82), as desired.

Next, since g_N is nowhere vanishing and is entire, we deduce from (3.82) that

$$\mathcal{D}_{A,\Omega} := \mathcal{D}(\zeta_{A,\Omega}) = \{0, 1, \cdots, N-1\} \cup \mathcal{D}(\zeta_{\mathfrak{s}}), \tag{3.86}$$

where (in light of (3.84))

$$\mathcal{D}(\zeta_{\mathfrak{s}}) = \log_2(N+1) + i\frac{2\pi}{\log 2}\mathbb{Z}. \tag{3.87}$$

Note that, by (3.85),

$$\mathcal{D}(\zeta_{\partial\Omega_{N,0},\Omega_{N,0}}) = \{0, 1, \cdots, N-1\}. \tag{3.88}$$

Therefore, for every $N \geq 2$, the set of complex dimensions of the inhomogeneous N-gasket is given by

$$\mathcal{D}(A_N, \Omega_N) := \mathcal{D}(\zeta_{A_N,\Omega_N}) = \{0, 1, \cdots, N-1\} \cup \Big(\log_2(N+1) + i\frac{2\pi}{\log 2}\mathbb{Z}\Big). \tag{3.89}$$

---

[74] See *loc. cit.* for the case when $N = 2$ or when $N = 3$.

Except when $\log_2(N+1) = j$, for some $j \in \{0, 1, \cdots, N-1\}$ (i.e., $N = 2^j - 1$, for some $j \in \{2, \cdots, N-1\}$, since $N \geq 2$ here), all of the complex dimensions of the RFD (A_N, Ω_N) in (3.86) are simple.[75]

It is instructive (although easy) to determine the dimensions $\dim_B(A_N, \Omega_N)$ as well as $\dim_{PC}(A_N, \Omega_N)$. In light of (3.83), we have that

$$(3.90) \quad \overline{\dim}_B(A_N, \Omega_N) = \max(D(\zeta_{\partial\Omega_{N,0},\Omega_{N,0}}),\ D(\zeta_{\mathfrak{s}})) = \max(N-1, \log_2(N+1)).$$

Observe that in (3.90), $\dim_B(\partial\Omega_{N,0}, \Omega_{N,0}) = N-1$ is the Minkowski dimension (which exists) of the generating RFD $(\partial\Omega_{N,0}, \Omega_{N,0})$ and $\sigma_N := N-1$ is the similarity dimension of the self-similar spray or RFD (A_N, Ω_N). Also, since one can show that $\dim_B(A_N, \Omega_N)$ exists, we conclude that

$$(3.91) \qquad D := \dim_B(A_N, \Omega_N)(= D(\zeta_{A_N,\Omega_N})) = \max(N-1, \log_2(N+1)) = \begin{cases} \log_2 3, & \text{for } N = 2, \\ N-1, & \text{for } N \geq 3. \end{cases}$$

Furthermore, we deduce from (3.89) and (3.91) that the set of principal complex dimensions of (A_N, Ω_N) is given by

$$(3.92) \qquad \dim_{PC}(A_N, \Omega_N) = \begin{cases} \log_2 3 + i\frac{2\pi}{\log 2}\mathbb{Z}, & \text{for } N = 2, \\ 2 + i\frac{2\pi}{\log 2}\mathbb{Z}, & \text{for } N = 3, \\ \{N-1\}, & \text{for } N \geq 4. \end{cases}$$

Observe that for $N = 2$, we have that $\sigma_2 = \log_2 3 > \dim_B(\partial\Omega_{2,0}, \Omega_{2,0}) = 1$, and hence, $\dim(A_2, \Omega_2) = \sigma_2$, the similarity dimension of the Sierpinski gasket, in agreement with a well-known result about classic or homogeneous self-similar sets (satisfying the open set condition); see, e.g., [**Hut**] or [**Fa1**, Ch. 9]. By contrast, when $N \geq 4$, we have the reverse inequality;[76] namely,

$$(3.93) \qquad \sigma_N = \log_2(N+1) \leq \dim_B(\partial\Omega_{N,0}, \Omega_{N,0}) = N-1.$$

Therefore, $\dim_B(A_N, \Omega_N) = \dim_B(\partial\Omega_{N,0}, \Omega_{N,0}) = N-1$, in this case.[77]

Finally, if $N = 3$, we have $\sigma_3 = \dim_B(\partial\Omega_{3,0}, \Omega_{3,0}) = 2$. This coincidence between the geometry of the generator $(\partial\Omega_{3,0}, \Omega_{3,0})$ and the scaling of the self-similar spray (A_3, Ω_3) explains why $D = 2$ is a complex dimension of multiplicity two if $N = 3$. In some sense, one can say that *there is a resonance between the underlying geometry and the underlying scaling of the relative 3-gasket* RFD (A_3, Ω_3).

The above facts have interesting geometric consequences, as is explained in detail in [**LapRaZu1**, §5.5.6], by using either the fractal tube formulas of [**LapRaZu1**, §5.1–§5.3] or the Minkowski measurability criteria of [**LapRaZu1**, §5.4] (both to be briefly discussed in §3.5). Firstly, if $N = 2$, the RFD (A_2, Ω_2) is not Minkowski measurable because in light of (3.92), it has nonreal principal complex dimensions; however, (A_2, Ω_2) is Minkowski nondegenerate. Secondly, if $N \geq 4$ (and since then,

[75]In view of the aforementioned results of *loc. cit.*, the residues of ζ_{A_N,Ω_N} at each complex dimension $\omega_j = j$ (for $j \in \{0, 1, \cdots, N-1\}$ and $s_k := \log_2(N+1) + i(2\pi/\log 2)k$ $(k \in \mathbb{Z})$ can be explicitly computed when $N = 2$ and when $N = 3$; see [**LapRaZu1**, Exple. 4.2.24 and Eq. (4.2.88)].

[76]In fact, this inequality (3.93) is always strict; indeed, it is easy to check by induction on N that we never have $N = 2^{N-1} - 1$, for some integer $N \geq 4$.

[77]There is no contradiction because, as we recall from our earlier discussion, (A_N, Ω_N) is an inhomogeneous (but unless $N = 2$) is *not* a standard (or homogeneous) self-similar set.

$2^{N-1} \neq N-1$, so that the dimension $D = N-1$ of (A_N, Ω_N) is simple), the RFD is not Minkowski measurable but is still Minkowski nondegenerate.

Lastly, if $N = 3$, (A_3, Ω_3) is not Minkowski measurable (since its Minkowski dimension $D = 2$ has multiplicity two); further, it is also Minkowski *degenerate*, which suggests that the usual power law is no longer appropriate to measure the "fractality" of (A_3, Ω_3).[78] However, one can use a suitably generalized Minkowski content (as in [**HeLap**] and [**LapRaZu1**, §6.1.1.2]), involving the choice of the gauge function $h(t) := \log(t^{-1})$ for all $t \in (0,1)$, so that the RFD (A_3, Ω_3) be not only Minkowski nondegenerate but also *Minkowski measurable, relative to h* (by contrast with the cases when $N = 2$ and $N = 3$); see [**LapRaZu1**, Thm. 5.4.27].

EXERCISE 3.9. Verify that when $N = 3$, we have

$$g_3(s) = 8(\sqrt{3})^{3-s}(2\sqrt{2})^{-s}, \tag{3.94}$$

for all $s \in \mathbb{C}$ in (3.82) and (3.85).

EXERCISE 3.10. Calculate the fractal zeta functions and the complex dimensions of the N-carpet RFD (A, Ω) (the N-dimensional relative Sierpinski carpet), which extends to $\mathbb{R}^N$ both the Sierpinski carpet ($N = 2$; see §3.4.2) and 3-carpet ($N = 3$; see §3.4.3).

Note that this example is significantly simpler than that of the N-gasket RFD studied in the present subsection (i.e., §3.4.4); indeed, unlike for the relative N-gasket, which is an inhomogeneous self-similar set, the compact set A is an homogeneous (i.e., classical) self-similar set in $\mathbb{R}^N$. In fact, A coincides with the standard N-Sierpinski carpet, while $\Omega = (0,1)^N$.

We refer the interested reader to [**LapRaZu1**, Exple. 4.2.31] for the complete answers and the corresponding computation. We simply mention here that

$$\dim_{PC}(A, \Omega) = \log_3(3^N - 1) + i\frac{2\pi}{\log 3}\mathbb{Z}, \tag{3.95}$$

where, as before, $D := \log_3(3^N - 1)$ is the Minkowski dimension of (A, Ω) and $\mathbf{p} := 2\pi/\log 3$ is the oscillatory period of (A, Ω), while

$$\mathcal{D}_{A,\Omega} = \mathcal{D}(\zeta_{A,\Omega}) = \{0, 1, \cdots, N-1\} \cup \dim_{PC}(A, \Omega). \tag{3.96}$$

Furthermore, in either (3.95) or (3.96), each complex dimension is simple.

3.4.5. *The $\frac{1}{2}$-square and $\frac{1}{3}$-square fractals.* We discuss here in parallel two related relative fractal drums, namely, the $\frac{1}{2}$-square and $\frac{1}{3}$-square fractals, which exhibit somewhat different properties.

(a) (*The $\frac{1}{2}$-square fractal*). Starting with the closed unit square $[0,1]^2 \subseteq \mathbb{R}^2$, we remove the two open squares of side length $\frac{1}{2}$, denoted by G_1 and G_2, along the main diagonal. Next, we repeat this step with the two remaining closed squares of side length $1/2$; and so on, ad infinitum. The *$\frac{1}{2}$-square fractal A* is the compact set that is left at the end of the process. (See also [**LapRaZu1**, Fig. 4.10] for an illustration.)

The $\frac{1}{2}$-square fractal is a *nonhomogeneous* self-similar fractal (as was the case of the set A in the construction of the relative Sierpinski N-gasket in §3.4.4); more

[78] All of these facts are established in [**LapRaZu1**, §5.5.6]; see, especially, part (*c*) of Remark 5.5.26 of *loc. cit.*.

specifically, it is the unique nonempty compact subset A of $\mathbb{R}^2$ satisfying the following inhomogeneous fixed point equation:

$$A = \bigcup_{j=1}^{2} S_j(A) \cup B, \tag{3.97}$$

where the nonempty compact set $B \subseteq \mathbb{R}^2$ is the union of the left and upper sides of the closed square $\overline{G}_1$ and of the right and lower sides of the closed square $\overline{G}_2$. Here, the contractive similitudes of $\mathbb{R}^2$ involved, namely, the maps S_1 and S_2, have respective fixed points at the lower left vertex and the upper right vertex of the unit square, and have scaling ratios $r_1 = r_2 = 1/2$.[79] (See [**LapRaZu1**, Fig. 4.11] for an illustration.)

Let $\Omega := (0,1)^2$ and consider the RFD (A, Ω); by construction, it is a self-similar spray (or RFD) with generator $G = G_1 \cup G_2$ and scaling ratios $r_1 = r_2 = 1/2$.

It is shown in [**LapRaZu1**, Exple. 4.2.33] that for every $\delta > 1/4$, $\zeta_{A,\Omega}$ and hence also, ζ_A, in light of (3.99) below, have a meromorphic continuation to all of $\mathbb{C}$ given for every $s \in \mathbb{C}$ respectively by

$$\zeta_{A,\Omega}(s) = \frac{\zeta_{\partial G, G}(s)}{1 - 2 \cdot 2^{-s}} = \frac{2^{-(s+1)}}{s(s-1)(2^{s-1}-1)} \tag{3.98}$$

and

$$\begin{aligned} \zeta_A(s) &= \zeta_{A,\Omega}(s) + \zeta_{[0,1]^2}(s) \\ &= \frac{2^{-(s+1)}}{s(s-1)(2^{s-1}-1)} + 4\frac{\delta^{s-1}}{s-1} + 2\pi\frac{\delta^s}{s}. \end{aligned} \tag{3.99}$$

It follows that

$$D := D(\zeta_A) = \dim_B A = D(\zeta_{A,\Omega}) = \dim_B(A, \Omega) = 1 \tag{3.100}$$

and

$$\dim_{PC} A = \dim_{PC}(A, \Omega) = 1 + \frac{2\pi}{\log 2}\mathbb{Z}, \tag{3.101}$$

as well as

$$\mathcal{D}_A = \mathcal{D}(\zeta_A) = \mathcal{D}_{A,\Omega} = \mathcal{D}(\zeta_{A,\Omega}) = \{0,1\} \cup \left(1 + i\frac{2\pi}{\log 2}\mathbb{Z}\right), \tag{3.102}$$

where these are equalities between multisets.

All of the complex dimensions in (3.101) and (3.102) (namely, 0 and $s_k := 1 + i\,(2\pi/\log 2)\,k$, for $k \in \mathbb{Z}$) are simple, except for the dimension $D = 1$ which is double.

Furthermore, for all $k \in \mathbb{Z}\backslash\{0\}$, we have

$$\operatorname{res}(\zeta_A, 0) = 1 + 2\pi \quad \text{and} \quad \operatorname{res}(\zeta_A, s_k) = \frac{4^{-i\mathbf{p}k}}{4s_k(s_k-1)}, \tag{3.103}$$

where $\mathbf{p} := 2\pi/\log 2$ is the oscillatory period of the self-similar spray (A, Ω).

(b) (*The $\frac{1}{3}$-square fractal*). As in part (*a*), we begin with the unit square $[0,1]^2$; we then divide it into nine congruent smaller squares. We further delete seven of those smaller squares; that is, we only keep the lower and upper right squares. We

[79] It is noteworthy that the homogeneous self-similar set E which is the unique nonempty compact subset of $\mathbb{R}^2$ satisfying the homogeneous fixed point equation associated with (3.97) (namely, $E = \cup_{j=1}^2 S_j(E)$), is the main diagonal of the unit square $[0,1]^2$.

then repeat the process, ad infinitum. What is left at the end of the process is denoted by A and called the $\frac{1}{3}$-*square fractal.*

If $\Omega := (0,1)^2$, then we consider the RFD (A,Ω) in $\mathbb{R}^2$. Note that (as in part (a)), A is an *inhomogeneous* self-similar fractal; more specifically, it is the unique (nonempty) compact subset of $\mathbb{R}^2$ satisfying the following inhomogeneous fixed point equation:

$$A = \bigcup_{j=1}^{2} S_j(A) \cup B, \tag{3.104}$$

where $B \subseteq \mathbb{R}^2$ is the nonempty compact set defined by $B := \partial G$ and G (called the generator of the self-similar spray (A,Ω)) is a suitable open convex polygon. Furthermore, S_1 and S_2 are contractive similitudes of $\mathbb{R}^2$, with respective fixed points located at the lower left vertex and the upper right vertex of the unit square.

The RFD (A,Ω) is a self-similar spray (or RFD) with generator G and scaling ratios $r_1 = r_2 = r_3 = 1/3$.

Much as in part (a), it is shown in [**LapRaZu1**, Exple. 4.2.34] that $\zeta_{A,\Omega}$ (and thus also ζ_A) admits a meromorphic continuation to all of $\mathbb{C}$ given for every $s \in \mathbb{C}$ (and for all sufficiently large positive δ) by

$$\zeta_{A,\Omega}(s) = \frac{\zeta_{\partial G,G}(s)}{1 - 2\cdot 3^{-s}} = \frac{2}{s(3^s-2)}\Big(\frac{6}{s-1} + \Psi(s)\Big), \tag{3.105}$$

where Ψ is a suitable entire function (which is explicitly known), and

$$\begin{aligned} \zeta_A(s) &= \zeta_{A,\Omega}(s) + \zeta_{[0,1]^2}(s) \\ &= \frac{2}{s(3^s-2)}\Big(\frac{6}{s-1} + \Psi(s)\Big) + 4\frac{\delta^{s-1}}{s-1} + 2\pi\frac{\delta^s}{s}. \end{aligned} \tag{3.106}$$

As a result,

$$D = D_A = D(\zeta_A) = D_{A,\Omega} = D(\zeta_{A,\Omega}) = 1 \tag{3.107}$$

and (see (3.112) below for a more precise statement)

$$\dim_{PC} A = \dim_{PC}(A,\Omega) \subseteq \{1\} \cup \Big(\log_3 2 + i\frac{2\pi}{\log 3}\mathbb{Z}\Big), \tag{3.108}$$

as well as

$$\{0,1\} \cup F \subseteq \mathcal{D}_A = \mathcal{D}(\zeta_A) = \mathcal{D}_{A,\Omega} = \mathcal{D}(\zeta_{A,\Omega}) \subseteq \{0,1\} \cup \Big(\log_3 2 + i\frac{2\pi}{\log 3}\mathbb{Z}\Big), \tag{3.109}$$

each complex dimension in (3.108) and (3.109) being simple. Here, F is a subset of $\log_3 2 + i(2\pi/\log 3)\mathbb{Z}$ containing $\log_3 2$ and at least finitely many (but more than two) nonreal principal complex dimensions.[80] We conjecture that the set F is in fact (countably) infinite and furthermore, that the inclusions in (3.109) should actually be equalities and $\dim_{PC} A = \dim_{PC}(A,\Omega) = \{1\}$.

It is noteworthy that $\log_3 2$ is the dimension of the homogeneous self-similar set E associated with (3.104) (i.e., $E = \cup_{j=1}^2 S_j(E)$, with $E \subseteq \mathbb{R}^2$ nonempty and compact); indeed, E is just the ternary Cantor set located along the main diagonal of the unit square $[0,1]^2$.

[80] At this stage, the inclusion appearing on the left of (3.109) is only verified numerically. The difficulty here is due to the presence of the entire function Ψ.

Finally, a simple computation yields that

$$\operatorname{res}(\zeta_A, 0) = 12 + \pi, \ \operatorname{res}(\zeta_{A,1}) = 16 \tag{3.110}$$

and (with $\mathbf{p} := 2\pi/\log 3$ and $s_k := \log_3 2 + i(2\pi/\log 3)k$, for each $k \in \mathbb{Z}$)

$$\operatorname{res}(\zeta_A, s_k) = \frac{3^{-i\mathbf{p}k}}{(\log 3)s_k}\Big(\frac{6}{s_k - 1} + \Psi(s_k)\Big). \tag{3.111}$$

Therefore, in light of (3.110), we can now specify the statement made in (3.108) by affirming that

$$\dim_{PC} A = \dim_{PC}(A, \Omega) = \{1\}. \tag{3.112}$$

3.4.6. *The $(N-1)$-sphere and its associated RFD.* In this subsection, we study the complex dimensions of the $(N-1)$-sphere

$$A := S^{N-1} = \{x \in \mathbb{R}^N : |x| = 1\}, \tag{3.113}$$

where $|\cdot|$ denotes the Euclidean norm in $\mathbb{R}^N$, and of the associated RFD (A, Ω), relative to the open unit ball Ω of $\mathbb{R}^N$, called the *$(N-1)$-sphere RFD.* We shall see that the answer obtained in the latter case is very natural. The difference between the answers in the former case (the $(N-1)$-sphere) and the latter case (the $(N-1)$-sphere RFD) is simply due to the fact that in the former case, we consider two-sided ε-neighborhoods of $A = S^{N-1}$ whereas in the latter case, we deal with one-sided (or "inner") ε-neighborhoods of $A = S^{N-1}$.

(a) (*The $(N-1)$-sphere*). Let A be the $(N-1)$-dimensional sphere, as given by (3.113). Then, in [**LapRaZu1**, Exple. 2.2.21], the tube zeta function ζ_A of A is shown to have a meromorphic extension to all of $\mathbb{C}$ given for every $s \in \mathbb{C}$ and for any fixed $\delta \in (0,1)$ by

$$\widetilde{\zeta}_A(s) = \Theta_N \sum_{k=0}^{N} (1 - (-1)^k) \binom{N}{k} \frac{\delta^{s-N+k}}{s - (N-k)}, \tag{3.114}$$

where Θ_N denotes the volume of the unit ball in $\mathbb{R}^N$ and the numbers $\binom{N}{k}$ are the usual binomial coefficients.[81] Therefore, independently of the value of $\delta > 0$,

$$D := D(\widetilde{\zeta}_A) = D(\zeta_A) = \dim_B A = N - 1, \tag{3.115}$$

$$\dim_{PC} A = \{N - 1\} \tag{3.116}$$

and

$$\mathcal{D}_A = \mathcal{D}(\widetilde{\zeta}_A) = \mathcal{D}(\zeta_A) = \Big\{N-1, N-3, \cdots, N - \Big(2\Big[\frac{N-1}{2}\Big] + 1\Big)\Big\}, \tag{3.117}$$

each complex dimension in (3.116) and (3.117) being simple. Note that for $N \geq 1$ odd (respectively, even), the last number in this set is equal to 0 (respectively, 1).

Finally, for every $d \in \mathcal{D}_A$,

$$\operatorname{res}(\widetilde{\zeta}_A, d) = 2\Theta_N \binom{N}{d}. \tag{3.118}$$

[81] In light of (3.15), we deduce at once the value of the distance zeta function $\zeta_A(s)$ for $s \in \mathbb{C}$.

In particular, for $d := D = N-1$, a direct computation (based on the definition of the Minkowski content given in §3.2 above) yields

$$\mathcal{M} = \mathcal{M}(A) = 2N\Theta_N = \operatorname{res}(\widetilde{\zeta}_A, D), \tag{3.119}$$

in agreement with a result stated in part (a) of §3.3.2.[82] Note that, clearly, A is Minkowski measurable with Minkowski content $\mathcal{M}$.

EXERCISE 3.11. Show directly that $A = S^{N-1}$ is Minkowski measurable, with Minkowski content $\mathcal{M}$ given by the second equality of (3.119).

(*b*) (*The* $(N-1)$*-sphere RFD*). Consider the $(N-1)$-sphere RFD (A,Ω), where $A := S^{N-1}$ is the unit sphere of $\mathbb{R}^N$ (as in part (a) just above) and Ω is the open unit ball in $\mathbb{R}^N$; so that $A = \partial\Omega$ and hence, $(A,\Omega) = (\partial\Omega, \Omega)$. Clearly, for any $N \geq 1$, the $(N-1)$-sphere RFD (or *relative* $(N-1)$*-sphere*) is an RFD in $\mathbb{R}^N$.

It is shown in [**LapRaZu1**, Exple. 4.1.19] that the distance zeta function $\zeta_{A,\Omega}$ of (A,Ω) admits a (necessarily unique) meromorphic extension to all of $\mathbb{C}$ given for every $s \in \mathbb{C}$ and for any fixed $\delta \in (0,1)$ by the following expression:[83]

$$\zeta_{A,\Omega}(s) = N\Theta_N \sum_{j=0}^{N-1} \binom{N-1}{j} \frac{(-1)^{N-j-1}}{s-j}. \tag{3.120}$$

Therefore, one deduces at once that

$$D = D(\zeta_{A,\Omega}) = D(\widetilde{\zeta}_{A,\Omega}) = \dim_B(A,\Omega) = N-1, \tag{3.121}$$

$$\dim_{PC} = \{N-1\} \tag{3.122}$$

and

$$\mathcal{D}_{A,\Omega} = \mathcal{D}(\zeta_{A,\Omega}) = \mathcal{D}(\widetilde{\zeta}_{A,\Omega}) = \{0, 1, \cdots N-1\}. \tag{3.123}$$

Observe the contrast between the result obtained for $\mathcal{D}_A$ and $\mathcal{D}_{A,\Omega}$ in (3.117) and (3.123), respectively, as was alluded to in the introduction to this subsection (i.e., §3.4.6). In particular, the set of complex dimensions $\mathcal{D}_{A,\Omega} = \{0, 1, \cdots, N-1\}$ of the relative $(N-1)$-sphere obtained in (3.123) is exactly the one we would have expected, a priori.

Finally, for every $j \in \{0, 1, \cdots, N-1\}$,

$$\operatorname{res}(\zeta_{A,\Omega}, j) = (-1)^{N-j-1} N\Theta_N \binom{N-1}{j}. \tag{3.124}$$

In particular, for $j := D = N-1$, we see that the RFD (A,Ω) is Minkowski measurable with (relative) Minkowski content given by

$$\mathcal{M} = \mathcal{M}(A,\Omega) = (N-D)\operatorname{res}(\zeta_{A,\Omega}, D) = N\Theta_N; \tag{3.125}$$

note that here, $N - D = 1$. (Compare with (3.29).)

[82] If the (absolute) $(N-1)$-sphere A had radius R instead of radius 1, then for any $\delta \in (0,R]$, one should simply substitute $\Theta_N R^d$ and $\Theta_N R^{N-1}$ for Θ_N in (3.118) and (3.119), respectively. (See also part (iii) of Exercise 3.12.)

[83] Unlike in part (a) of the present subsection, it is easier to compute directly $\zeta_{A,\Omega}$ rather than $\widetilde{\zeta}_{A,\Omega}$. Of course, in light of (3.11), one can then deduce $\widetilde{\zeta}_{A,\Omega}$ from (3.120).

EXERCISE 3.12. (*i*) Show via a direct computation that $\zeta_{A,\Omega}$ is given by (3.120).

(*ii*) Address the same question as in (*i*) for $\widetilde{\zeta}_A$ (in part (*a*) of this subsection) in order to recover the expression stated in (3.114).

(*iii*) How are the expressions of $\widetilde{\zeta}_A$ in (3.114) and of $\zeta_{A,\Omega}$ in (3.120) modified if A and Ω are, respectively, the $(N-1)$-sphere and the (open) N-ball of radius R (instead of radius 1)?

3.4.7. *The Cantor grill.* Let $A := C \times [0,1]$ be the Cartesian product of the ternary Cantor set by the unit interval. Henceforth, $A \subseteq \mathbb{R}^2$ is referred to as the *Cantor grill.* (See [**LapRaZu1**, Fig. 2.15] for an illustration.) Then, according to [**LapRaZu1**, Exple. 2.2.34 in §2.2.3],

$$D = D_C = \dim_B A = 1 + \log_3 2, \tag{3.126}$$

the set of principal complex dimensions of A is given by

$$\dim_{PC} A = (1 + \log_3 2) + i\frac{2\pi}{\log 3}\mathbb{Z}, \tag{3.127}$$

while the set of all complex dimensions (in $\mathbb{C}$) of A is given by

$$\begin{aligned} \mathcal{D}_A = \mathcal{D}(\zeta_A) &= \{0,1\} \cup \bigcup_{m=0}^{1} \left((m + \log_3 2) + i\frac{2\pi}{\log 3}\mathbb{Z}\right) \\ &= \{0,1\} \cup (D_C + i\mathbf{p}\mathbb{Z}) \cup ((1 + D_C) + i\mathbf{p}\mathbb{Z}), \end{aligned} \tag{3.128}$$

where $D_C = \log_3 2 = \dim_B C$ is the Minkowski dimension of the Cantor set C and $\mathbf{p} := 2\pi/\log 3$ is the oscillatory period of C. Each complex dimension in (3.127) and (3.128) is simple.

EXERCISE 3.13. (*i*) (*Higher-dimensional Cantor grills*) Generalize the results of the present subsection to the higher-dimensional Cantor grill $A := C \times [0,1]^d$, where d is an arbitrary positive integer. In particular, show that (with $\mathbf{p} := 2\pi/\log 3$ and $D_C = \log_3 2$, as above)

$$D_A = \dim_B A = d + \dim_B C = d + D_C \tag{3.129}$$

$$\dim_{PC} A = D_A + i\mathbf{p}\mathbb{Z}, \tag{3.130}$$

and

$$\mathcal{D}_A = \mathcal{D}(\zeta_A) = \{0, 1, \cdots, d\} \cup \bigcup_{m=0}^{d} ((m + D_C) + i\mathbf{p}\mathbb{Z}). \tag{3.131}$$

[*Hint*: In order to establish (3.129) and (3.130), compare $\zeta_A(s)$ and $\zeta_A(s-d)$ and show that the difference of these two functions is holomorphic in a suitable half-plane, namely,

$$\{Re(s) > D_C + (d-1)\} \supseteq \{Re(s) \geq D_A\}.$$

The proof of (3.131) is significantly more complicated; if needed, see [**LapRaZu1**, Exple. 2.2.34 and §4.7.1].]

(*ii*) (*Fractal combs*). Let K be any compact subset of $\mathbb{R}$ and let $A := K \times [0,1]^d$, with $d \in \mathbb{N}$. Extend the results of part (*i*) to this more general situation.

EXERCISE 3.14. (*Two different Cantor grill RFDs.*) Let $A := C$, the ternary Cantor set, and $\Omega_1 := (0,1)^2$ while $\Omega_2 := (-1,0) \times (0,1)$. Then, show that the complex dimensions of the RFDs (A, Ω_1) and (A, Ω_2) are very different. More specifically, as is observed in [**LapRaZu1**, §1.1], it turns out that

$$\mathcal{D}_{A,\Omega_1} = \mathcal{D}_{C\times[0,1]} = \{0,1\} \cup (D_C + i\mathbf{p}\mathbb{Z}) \cup ((1+D_C) + i\mathbf{p}\mathbb{Z}), \tag{3.132}$$

as in (3.128), where $D_C = \log_3 2$ and $\mathbf{p} = 2\pi/\log 3$ are, respectively, the Minkowski dimension and the oscillatory period of C. By contrast,

$$\mathcal{D}_{A,\Omega_2} = \{0,1\} \cup \mathcal{D}_C = \{0,1\} \cup (D_C + i\mathbf{p}\mathbb{Z}). \tag{3.133}$$

(Further, all the complex dimensions in either (3.132) or (3.133) are simple.) Thus, the RFD (A, Ω_2) no longer "sees" the principal complex dimensions of the Cantor grill $C \times [0,1]$ in (3.127) (or of the RFD (A, Ω_1), according to (3.132)).

In addition to establishing (3.132) and (3.133), provide an intuitive explanation for the striking difference between these two results.

3.4.8. *The Cantor dust.* Let $A := C \times C$, where as before, C is the ternary Cantor set. Henceforth, A, the Cartesian product of C by itself, is referred to as the *Cantor dust.* (See [**LapRaZu1**, Fig. 1.2] for a depiction of A.) The associated *Cantor dust RFD* (A, Ω) is defined by $\Omega := (0,1)^2$ and $A := C \times C$, as above.

Then, it is shown in [**LapRaZu1**, Exple. 4.7.15] that $\zeta_{A,\Omega}$ has (for all $\delta > 0$ large enough) a meromorphic continuation to all of $\mathbb{C}$ given for every $s \in \mathbb{C}$ by

$$\zeta_{A,\Omega}(s) = \frac{8}{s(3^s-4)} \left(\frac{J(s)}{6^s} + \frac{\Gamma(\frac{1-s}{2})}{\Gamma(\frac{2-s}{2})} \frac{\sqrt{\pi}}{6^s s(3^s-2)} + K(s) \right), \tag{3.134}$$

where $\Gamma = \Gamma(s)$ denotes the classic gamma function (which, as we recall, does not have any zeros anywhere in $\mathbb{C}$ but has simple poles at $0, -1, -2, -3, \cdots$. Here, $J(s) := \int_0^{\pi/4} (\cos\theta)^{-s} d\theta$ is an entire function and $K = K(s)$ is a meromorphic function in all of $\mathbb{C}$ with (simple) poles at $1, 3, 5, \cdots$.

It follows from (3.134) and the aforementioned properties of Γ, J and K that

$$D_{A,\Omega} = D(\zeta_{A,\Omega}) = \dim_B(A,\Omega) = \log_3 4 = D_A = \dim_B A, \tag{3.135}$$

as expected since $\log_3 4 = \log_3 2 + \log_3 2 = 2 \dim_B C$, and that the set of complex dimensions (in $\mathbb{C}$) of the Cantor dust RFD consists of simple poles of $\zeta_{A,\Omega}$ and is given by

$$\mathcal{D}_{A,\Omega} = \mathcal{D}(\zeta_{A,\Omega}) = \{0\} \cup \left(\log_3 2 + i\frac{2\pi}{\log 3}\mathbb{Z}\right) \cup \left(\log_3 4 + i\frac{2\pi}{\log 3}\mathbb{Z}\right). \tag{3.136}$$

More specifically, due to possible zero-pole cancellations, $\mathcal{D}_{A,\Omega}$ is a subset of the set given on the right-hand side of (3.136) and contains $D_C = \log_3 2, D_A = D_{C\times C} = \log_3 4$, as well as at least two nonreal complex conjugate principal complex dimensions. It is conjectured in *loc. cit.* that, in fact, we have a true equality in (3.136) and hence, in particular, that the set of principal complex dimensions of the RFD (A, Ω) (as well as of the compact set $A \subseteq \mathbb{R}^2$) is given by

$$\dim_{PC}(A,\Omega) = \dim_{PC} A = \log_3 4 + i\frac{2\pi}{\log 3}\mathbb{Z} \tag{3.137}$$

or is, at least, an infinite subset of the 'periodic set' $\log_3 4 + i(2\pi/\log 3)\mathbb{Z}$. Furthermore, the author conjectures entirely similar statements about the set of subcritical complex dimensions; namely, the set of complex dimensions of (A, Ω) (and of A)

should be equal to (or, at least, coincide with an infinite subset of) the periodic set $\log_3 2 + i(2\pi/\log 3)\mathbb{Z}$.

We refer to Conjecture 3.16 below for a much more general and precise statement about the complex dimensions (interpreted there in an extended sense) of Cartesian products.

EXERCISE 3.15. (*i*) Deduce from the above results about the Cantor dust RFD (A,Ω) analogous results concerning the Cantor dust itself, $A := C \times C$.

(*ii*) Generalize the above results about the Cantor dust RFD and the Cantor dust to $A = C^d := C \times \cdots \times C$, the d-fold Cartesian product of C with itself (where $d \geq 1$ is arbitrary) and to the associated RFD (A,Ω), with A as just above and with $\Omega := (0,1)^d$. We expect that $\dim_{PC} A = \dim_{PC}(A,\Omega)$ and, similarly, $\mathcal{D}_A = \mathcal{D}_{A,\Omega}$.

(*iii*) Finally, replace the ternary Cantor set C by other Cantor-type sets and by more general compact subsets of $\mathbb{R}$ (including lattice and nonlattice self-similar sets).

The following conjecture of the author was motivated, in part, by the results from [**LapRaZu1**] (and [LapRaZu2–6]) stated about the complex dimensions of the Cantor grill (in §3.4.7) and of the Cantor dust (in §3.4.8), along with the results of *loc. cit.* briefly discussed in §3.3.4 about the invariance of the complex dimensions under embeddings into higher-dimensional Euclidean spaces. More specifically, the author was first led to stating this conjecture (in December 2016), on the basis of his joint work on quantized number theory and fractal cohomology ([CobLap1–2], [**Lap10**]), and especially, due to the requirement that 'fractal cohomology' should satisfy an appropriate analog of the Künneth formula for the cohomology of Cartesian products. (See, especially, [**Lap10**, esp., Chs. 4–6] for more details and motivations; see also §4.1 and §4.4 in the epilogue for a brief discussion of the general context.)

For simplicity, we state the conjecture for compact sets rather than for general RFDs, but clearly, an analogous conjecture can be made about RFDs. We also implicitly assume that the corresponding fractal zeta functions are meromorphic in all of $\mathbb{C}$,[84] but we can also state the conjecture relative to a common window (in the sense of [**Lap-vF4, LapRaZu1**]) or more generally, a domain of $\mathbb{C}$ to which the fractal zeta functions can be meromorphically continued.[85] Finally, the mathematical notion of *divisor* (denoted by $\mathfrak{D}(f)$) of a meromorphic function f is well known and will be recalled in Definition 3.18 below (following [**Lap-vF4**, §3.4]). For now, we simply mention that $\mathfrak{D}(f)$ is the multiset of zeros and poles of f; that is, the graded set of zeros and poles of f, counted according to their multiplicities (with the zeros counted positively and the poles counted negatively).

CONJECTURE 3.16 (Complex dimensions of Cartesian products, [**Lap10**]). *Let A_1 and A_2 be two bounded (or, equivalently, compact) subsets of $\mathbb{R}^{N_1}$ and $\mathbb{R}^{N_2}$, respectively. For $j = 1, 2$, let $\mathfrak{D}(A_j)$ denote the divisor of $\zeta_{A_j}, \mathfrak{D}(A_j) := \mathfrak{D}(\zeta_{A_j})$, and $\mathcal{D}(A_j) = \mathcal{D}(\zeta_{A_j})$ denote the (multi)set of complex dimensions of A_j.*[86]

---

[84]See, in particular, Remark 3.17 for a more general situation.

[85]If f and g are two meromorphic functions defined on the same domain U of $\mathbb{C}$, we simply let $\mathfrak{D}(f) = \mathfrak{D}(f;U)$ and $\mathfrak{D}(g) = \mathfrak{D}(g;U)$ in the statement of Conjecture 3.16, and with the notation of Definition 3.18.

[86]For simplicity, we assume implicitly that $\overline{\dim}_B A_j < N_j$, for $j = 1, 2$. We also use ζ_{A_j} in order to define both $\mathfrak{D}_{A_j}$ and $\mathcal{D}_{A_j}$ for $j = 1, 2$; namely, for $j = 1, 2$, $\mathfrak{D}_{A_j} := \mathfrak{D}(\zeta_{A_j})$ and

Then, we have the identity

$$\mathfrak{D}(A_1 \times A_2) = \mathfrak{D}(A_1) + \mathfrak{D}(A_2),^{87} \tag{3.138}$$

Also, we have the inclusion

$$\mathcal{D}(A_1 \times A_2) \subseteq \mathcal{D}(A_1) + \mathcal{D}(A_2), \tag{3.140}$$

the Minkowski sum of $\mathcal{D}(A_1)$ *and* $\mathcal{D}(A_2)$. *Furthermore, typically (or "generically", in a vague sense), we have an equality in* (3.140), *because we do not have zero-pole cancellations in such cases*:

$$\mathcal{D}(A_1 \times A_2) = \mathcal{D}(A_1) + \mathcal{D}(A_2). \tag{3.141}$$

REMARK 3.17. If $\zeta_{A_1 \times A_2}$ is not necessarily meromorphic in all of $\mathbb{C}$ (or in the given domain $U \subseteq \mathbb{C}$ under consideration), but ζ_{A_1} and ζ_{A_2} still are, then, under appropriate hypotheses, we expect (based in part on cohomological and spectral considerations; see §4.3 and §4.4, along with [**Lap10**]) that Conjecture 3.16 can be suitably modified and extended by substituting for the ordinary divisor $\zeta_{A_1 \times A_2}$ a "generalized divisor" (still denoted by $\boldsymbol{\mathcal{D}}(\zeta_{A_1 \times A_2})$ or, in short, $\boldsymbol{\mathcal{D}}(A_1 \times A_2)$) which takes into account the (nonremovable) singularities (and not just the poles) of $\zeta_{A_1 \times A_2}$; in that case, we must also replace the Minkowski sum by its closure in the right-hand side of the counterpart of (3.138):

$$\boldsymbol{\mathcal{D}}(A_1 \times A_2) = cl(\boldsymbol{\mathcal{D}}(A_1) + \boldsymbol{\mathcal{D}}(A_2)). \tag{3.142}$$

In particular, the counterpart of (3.140) becomes

$$\mathcal{D}(A_1 \times A_2) \subseteq cl(\mathcal{D}(A_1) + \mathcal{D}(A_2)). \tag{3.143}$$

Next, as promised, we recall the definition of the divisor of a meromorphic function (see, e.g., [**Lap-vF4**, Def. 3.11]). It goes back at least to Riemann in the related context of Riemann surfaces and has counterparts and various generalizations in many fields, including arithmetic and algebraic geometry, as well as in algebraic combinatorics. The notion of divisor is ideally suited to making precise sense of the possible cancellations between the zeros and poles of a meromorphic function. Hence, its key use in the statement of Conjecture 3.16 above.

DEFINITION 3.18. (*Divisor of a meromorphic function*). Let f be a meromorphic function on a given domain $U \subseteq \mathbb{C}$. Then, the *divisor* of f, denoted by $\mathfrak{D} = \mathfrak{D}(f)$, is defined as the formal sum[88]

$$\mathfrak{D}(f) = \mathfrak{D}(f; U) := \sum_{\omega \in U} ord(f; \omega) \cdot \omega, \tag{3.144}$$

$\mathcal{D}_{A_j} := \mathcal{D}(\zeta_{A_j})(= \mathcal{D}(\tilde{\zeta}_{A_j})$, in this case). [Note that in spite of the functional equation (3.15) (or, more generally, (3.11), in the case of RFDs), ζ_{A_j} and $\tilde{\zeta}_{A_j}$ have the same poles but not necessarily the same zeros.] One could make a similar conjecture without assuming that $\overline{\dim}_B A_j < N_j$ for $j = 1, 2$ and by using $\tilde{\zeta}_{A_j}$ instead of ζ_{A_j} in order to define both $\mathfrak{D}_{A_j}$ and $\mathcal{D}_{A_j}$, for $j = 1, 2$.

[87]Given two subsets (or submultisets) E_1 and E_2 of the same additive group, their Minkowski sum $E_1 + E_2$ is defined by

$$E_1 + E_2 = \{e_1 + e_2 : e_1 \in E_1, e_2 \in E_2\}, \tag{3.139}$$

viewed as a subset (or submultiset) of this same group.

[88]This is an at most countable sum since clearly, $ord(f, \omega) = 0$ whenever $\omega \in U$ is neither a pole nor a zero of f. Also, $\mathfrak{D}(f)$ can be viewing as lying in the free ablian group generated by the distinct zeros and poles of f.

where the order of f at $\omega \in U$ is defined as the integer $m \in \mathbb{Z}$ such that the function $|f(s)(s-\omega)^{-m}|$ is bounded away from 0 and ∞ in a neighborhood of ω. Therefore, if ω is a zero of (positive) multiplicity m, then $order(f,\omega) = m$, a positive integer, whereas if ω is a pole of f of (positive) multiplicity n, then $ord(f,\omega) = -n =: m$, a negative integer. Furthermore, $ord(f,\omega) = 0$ if ω is neither a zero nor a pole of f (in U).

Formally, the *Minkowski sum* of the divisors $\mathfrak{D}(f)$ and $\mathfrak{D}(g)$ of two meromorphic functions on the same domain U of $\mathbb{C}$ is denoted by $\mathfrak{D}(f) + \mathfrak{D}(g)$ and is given by

$$\mathfrak{D}(f) + \mathfrak{D}(g) = \sum_{\omega\in U} (ord(f,\omega) + ord(g,w)) \cdot \omega. \tag{3.145}$$

Note that on the right-hand side of (3.145), some of the coefficients given by $ord(f,\omega) + ord(g,\omega)$ may vanish, corresponding precisely to the aforementioned (exact) zero-pole cancellations.

REMARK 3.19. (*a*) Conjecture 3.16 above is consistent with all the known examples and results, including those described in §3.3.4, §3.4.7 and §3.4.8. (See, in particular, parts (*b*) and (*c*) of the present remark for the examples of the Cantor grill and the Cantor dust, respectively.) We warn the interested reader, however, that in the general case, this conjecture is likely to be quite difficult to prove although the coherence and consistency of the fractal cohomology theory partly developed in [**Lap10**] (and briefly discussed in §4.1 and §4.4) clearly requires that Conjecture 3.16 must be true.

(*b*) (*Cantor grill, revisited*) As a simple verification, for the example of the Cantor grill $C \times [0,1]$ discussed in the main text of §3.4.7 (and corresponding to the case when $d = 1$) or, more generally, of its higher-dimensional counterpart $A := C \times [0,1]^d$ considered in Exercise 3.13(*i*), we have

$$\mathcal{D}_{[0,1]^d} := \mathcal{D}(\widetilde{\zeta}_{[0,1]^d}) = \{0,1,\cdots,d\}, \tag{3.146}$$

as can be easily verified,[89] and

$$\mathcal{D}_C = \{0\} \cup \mathcal{D}_{CS} = \{0\} \cup (D_C + i\mathbf{p}\mathbb{Z}), \tag{3.147}$$

where $D_C = \log_3 2$ and $\mathbf{p} = 2\pi/\log 3$. (See (3.17) and (3.20) for a closely related fact.) Therefore, the Minkowski sum $\mathcal{D}_C + \mathcal{D}_{[0,1]^d}$ is given by

$$\{0,1,\cdots,d\} \cup \left(\bigcup_{m=0}^{d} ((m + D_C) + i\mathbf{p}\mathbb{Z}) \right), \tag{3.148}$$

which is precisely $\mathcal{D}_A = \mathcal{D}_{C\times[0,1]^d}$, as given by (3.131) (and, in particular, for the usual Cantor grill corresponding to the choice $d = 1$, as given by (3.128)). Of course, this is also in agreement with the result predicted by Conjecture 3.16 in (3.140), and even in (3.141) since we have

$$\mathcal{D}_C + \mathcal{D}_{[0,1]^d} = \mathcal{D}_{C\times[0,1]^d} \tag{3.149}$$

in the present situation, as expected in the "generic" case.

Observe that it is absolutely crucial here that $\mathcal{D}_C$ be given by (3.147), as obtained above via the higher-dimensional theory of fractal zeta functions, and not just by $\mathcal{D}(\zeta_{CS}) = D_C + i\mathbf{p}\mathbb{Z}$, as given by the theory of fractal strings and their associated geometric zeta functions. Recall that $\mathcal{D}_C = \mathcal{D}(\zeta_C) = \mathcal{D}(\widetilde{\zeta}_C) =$

[89] Note that we only use $\widetilde{\zeta}_{[0,1]^d}$ here because $D_{[0,1]^d} = \dim_B[0,1]^d = d$.

$\{0\} \cup \mathcal{D}(\zeta_{CS})$, where $\mathcal{D}(\zeta_{CS})$ is given just above. (An entirely analogous comment can be made about the example of the Cantor dust dealt with in part (*c*).)

(*c*) (*The Cantor dust, revisited*). Let $A = C \times C$ be the Cantor dust, as in §3.4.8. Then, in light of (3.147) in part (*b*) just above, it is easy to check that $\mathcal{D}_C + \mathcal{D}_C$ is given by (with the same notation as in part (*b*) of this remark)

$$\mathcal{D}_C + \mathcal{D}_C = \{0\} \cup (D_C + i\mathbf{p}\mathbb{Z}) \cup (2D_C + i\mathbf{p}\mathbb{Z}), \tag{3.150}$$

with $2D_C = \log_3 4 = D_A$, in accord with the result stated in (3.136) and more precisely, immediately after (3.136), and also in agreement with the containment (3.140) of Conjecture 3.16, even though it has not yet been fully proved for this example. In fact, we expect once again that we have an equality in the present situation, as predicted for the "generic" case in (3.141) of Conjecture 3.16; namely, we expect that

$$\mathcal{D}_{C\times C} = \mathcal{D}_C + \mathcal{D}_C. \tag{3.151}$$

We leave it to the interested reader to calculate the d-fold Minkowski sum of $\mathcal{D}_C$ and to deduce from it (assuming that we are in the generic case of Conjecture 3.16) the expression for $\mathcal{D}_{C^d}$, where C^d is the d-fold Cartesian product of C by itself, as sought for in part (*ii*) of Exercise 3.15.

The following example (the Cantor graph RFD) will play an important role in §3.6 in order to illustrate the definition of fractality in terms of the existence of nonreal complex dimensions.

3.4.9. *The devil's staircase (or Cantor graph) RFD.* Let A denote the graph of the classic Cantor function, also called from now on the devil's staircase or the Cantor graph. It is well known that A is a self-affine (rather than a self-similar) set in $\mathbb{R}^2$; more specifically, it scales distances by the factors $1/2$ and $1/3$ along the horizontal and vertical directions, respectively. (See, for example, [**LapRaZu1**, Remark 1.2.1] for a more detailed description, along with [**Man**, Plate 83, p. 83] and [**LapRaZu1**, Figs. 1.5–1.7] for an illustration.)

Next, let Ω be the union of the triangles Δ_k and $\widetilde{\Delta}_k$ above and below the horizontal parts of the Cantor graph A, for $k = 1, 2, \cdots$. (For each $k \geq 1$, at the k-th stage of the construction of A and of the RFD (A, Ω) below, there are 2^{k-1} pairs of congruent triangles Δ_k and $\widetilde{\Delta}_k$.) Then, (A, Ω) is called the *Cantor graph RFD*; clearly, it is an RFD in $\mathbb{R}^2$. Intuitively, it can be thought of as the Cantor graph viewed from the perspective of the (non-Euclidean) ℓ^∞ metric of $\mathbb{R}^2$, given by $||(x, y)||_\infty = \max(|x|, |y|)$, for $x, y \in \mathbb{R}^2$. In spite of that, (A, Ω) captures the essence of the Cantor graph.

It is shown in [**LapRaZu1**, Exple. 5.5.14] that $\zeta_{A,\Omega}$ admits a (necessarily unique) meromorphic continuation to all of $\mathbb{C}$, given for every $s \in \mathbb{C}$ by

$$\zeta_{A,\Omega}(s) = \frac{2}{s(3^s - 2)(s-1)}. \tag{3.152}$$

It follows that the set of principal complex dimensions of the Cantor graph RFD (A, Ω) is given by

$$\dim_{PC}(A, \Omega) = \{1\}, \tag{3.153}$$

where

$$D_{A,\Omega} = \dim_B(A, \Omega) = 1, \tag{3.154}$$

and that the set of all complex dimensions of (A,Ω) is

$$\mathcal{D}_{A,\Omega} = \mathcal{D}(\zeta_{A,\Omega}) = \{0,1\} \cup \left(\log_3 2 + i\frac{2\pi}{\log 3}\mathbb{Z}\right), \tag{3.155}$$

with each complex dimension being simple. Apart from the complex dimension at $s=0$, this is in complete agreement with the values of the complex dimensions of the Cantor graph A predicted in [Lap-vF2–4] (see, especially, [**Lap-vF4**, §12.1.2]), on the basis of an approximate tube formula for A (but, at the time, without an appropriate definition of higher-dimensional fractal zeta functions).

Furthermore, with $s_k := \log_3 2 + i(2\pi/\log 3)$ for each $k \in \mathbb{Z}$, we have that

$$\operatorname{res}(\zeta_{A,\Omega}, s_k) = \frac{1}{(\log 3)(s_k - 1)s_k}. \tag{3.156}$$

Moreover,

$$\operatorname{res}(\zeta_{A,\Omega}, 0) = \operatorname{res}(\zeta_{A,\Omega}, 1) = 2. \tag{3.157}$$

Let us next briefly consider the *Euclidean Cantor graph RFD* $(A, A_{1/3})$, where $A_{1/3}$ is the $(1/3)$-neighborhood (with respect to the Euclidean metric of $\mathbb{R}^2$) of the Cantor graph $\mathbb{R}^2$. Then, it is also shown in [**LapRaZu1**, §5.5.4] that

$$\mathcal{D}_{A,A_{1/3}} = \mathcal{D}(\zeta_{A,A_{1/3}}) \subseteq \mathcal{D}_{(A,\Omega)} = \{0,1\} \cup \left(\log_3 2 + i\frac{2\pi}{\log 3}\mathbb{Z}\right) \tag{3.158}$$

and that[90]

$$\dim_{PC}(A, A_{1/3}) = \dim_{PC}(A,\Omega) = \{1\}, \tag{3.161}$$

where

$$\dim_B(A, A_{1/3}) = \dim_B(A,\Omega) = D(\zeta_{A,\Omega}) = D(\zeta_{A,A_{1/3}}) = 1. \tag{3.162}$$

It can be checked numerically (see *loc. cit.*) that $\mathcal{D}_{A,A_{1/3}}$ contains several pairs of *nonreal* complex conjugate complex dimensions with real part $D_{CS} = D_C = \log_3 2$. In fact, we expect that a lot more is true, as is stated in the following conjecture. (Compare with the inclusion obtained in (3.158).)

CONJECTURE 3.20. ([LapRaZu4, 7], [**LapRaZu1**, §5.5.4]). *We not only have that $\mathcal{D}_{A,A_{1/3}} \subseteq \mathcal{D}_{A,\Omega}$ (as stated in* (3.158)*) but also that*

$$\mathcal{D}_A = \mathcal{D}_{A,A_{1/3}} = \mathcal{D}_{A,\Omega}, \tag{3.163}$$

where (as in (3.155) *and with the notation of footnote* 90)

$$\mathcal{D}_{A,\Omega} = \{0,1\} \cup (D_{CS} + i\mathbf{p}\mathbb{Z}), \tag{3.164}$$

with $D_{CS} = \log_3 2$ and $\mathbf{p} = (2\pi/\log 3)$.

[90] Observe that we can rewrite (3.158) as follows:

$$\mathcal{D}_{A,\Omega} = \mathcal{D}_{CS} \cup \{1\}, \tag{3.159}$$

where the set of complex dimensions of the Cantor string CS (and of the set C) is given by

$$\mathcal{D}_{CS} = \mathcal{D}_C = \{0\} \cup (D_{CS} + i\mathbf{p}\mathbb{Z}). \tag{3.160}$$

Here, $D_{CS} = D_C = \log_3 2$ is the Minkowski dimension of the Cantor string (and set) and $\mathbf{p} = (2\pi/\log 3)$ is its oscillatory period.

3.4.10. *Self-similar sprays.* The notion of a self-similar (or, more generally, fractal) spray was introduced in [**LapPo3**], formalizing a number of examples discussed in [**Lap1, Lap3**]. It was then used extensively, in particular, in [LapPe1–2, LapPeWi1–2] where were established fractal tube formulas extending (and using) those obtained for fractal strings in [Lap-vF1–4]. (See also, e.g., [**DemDenKoÜ**], [**DemKoÖÜ**], [**DeniKoÖÜ**] and [**KoRati**].) The notion of a self-similar spray or RFD introduced in [**LapRaZu1**] (and in [LapRaZu4,6]) is slightly different from the one used in those references; in particular, its greater flexibility is well suited to a variety of examples discussed in *loc. cit.* as well as in the present section (e.g., the relative N-gasket in §3.4.4). Since the precise definition is a bit technical, we simply mention that, roughly speaking, a *self-similar spray* (or RFD) is an RFD (A,Ω) in $\mathbb{R}^N$ which, up to displacements, is obtained from a single *generator* (also called the *base RFD*) $(\partial G, G)$),[91] itself an RFD in $\mathbb{R}^N$, via a *scaling sequence* $\mathcal{L} = (\ell_j)$, which is a possibly unbounded self-similar fractal string (in the sense of [**Lap-vF4**, Ch. 3]) with (not necessarily distinct) scaling ratios $r_1, \cdots, r_J$. Here, $J \geq 2$ and $0 < r_1, \cdots, r_J < 1$; furthermore, $\mathcal{L}$ is comprised of all the finite products of elements of the *ratio list* $\{r_1 \cdots, r_J\}$.

It follows that (A,Ω) is the disjoint union of scaled copies of the generating RFD $(\partial\Omega,\Omega)$, scaled by the self-similar string $\mathcal{L}$. In order for Ω to have finite total volume, we assume that G is open, $|G| < \infty, \dim_B(\partial G, G) < N$ and $\sum_{j=1}^J r_j^N < 1$.

Let (A,Ω) be an arbitrary self-similar spray (or RFD) in $\mathbb{R}^N$. It is shown in [**LapRaZu1**, Thm. 4.2.17] that $\zeta_{A,\Omega}$ admits a meromorphic continuation to all of $\mathbb{C}$ given for every $s \in \mathbb{C}$ by

$$\zeta_{A,\Omega}(s) = \frac{\zeta_{\partial G,G}(s)}{1 - \sum_{j=1}^J r_j^s} \tag{3.165}$$

or equivalently, by the following *factorization formula:*[92]

$$\zeta_{A,\Omega}(s) = \zeta_{\mathfrak{s}}(s) \cdot \zeta_{(\partial G,G)}(s), \tag{3.166}$$

where $\zeta_{\mathfrak{s}}(s) := \zeta_{\mathcal{L}}(s)$, the geometric zeta function of the possibly unbounded self-similar string $\mathcal{L}$, is called the *scaling zeta function* of the fractal spray (A,Ω) and is given (also for every $s \in \mathbb{C}$) by

$$\zeta_{\mathfrak{s}}(s) = \frac{1}{1 - \sum_{j=1}^J r_j^s}. \tag{3.167}$$

It then follows from (3.165)–(3.167) that (with, as before, the notation $\mathcal{D}_{A,\Omega} = \mathcal{D}(\zeta_{A,\Omega}), \mathcal{D}_{(\partial G,G)} = \mathcal{D}(\zeta_{(\partial G,G)})$ and $\mathcal{D}_{\mathfrak{s}} = \mathcal{D}(\zeta_{\mathfrak{s}}) = \mathcal{D}(\zeta_{\mathcal{L}})$)

$$\mathcal{D}(\zeta_{A,\Omega}) \subseteq \mathcal{D}(\zeta_{\mathfrak{s}}) \cup \mathcal{D}(\zeta_{\partial G,G}), \tag{3.168}$$

where the containment comes from the fact that there could, in general, be zero-pole cancellations in the expression on the right-hand side of (3.166) (or, equivalently,

[91] For the simplicity of exposition, we consider here the case of a single generator. The case of multiple generators (i.e., finitely many generators) is an immediate consequence of the present case of a single generator; see part (a) of Remark 3.21.

[92] A priori, the equivalent identities (3.165) and (3.166) are valid for $Re(s) > \overline{D}$, with $\overline{D} = D_{A,\Omega} = D(\zeta_{A,\Omega})$, as given by (3.170). However, upon meromorphic continuation, it remains valid in any domain U to which $\zeta_{\partial G,G}$ can be meromorphically extended. For example, we can take $U = \mathbb{C}$ if G is sufficiently "nice" (e.g., monophase or even pluriphase, in the sense of [**LapPe2, LapPeWi1**], and in particular, a nontrivial polytope [**KoRati**].

of (3.165)). Note that we also have the following equality:

$$\mathfrak{D}(\zeta_{A,\Omega}) = \mathfrak{D}(\zeta_{\mathfrak{s}}) \cup \mathfrak{D}(\zeta_{\partial G,G}), \tag{3.169}$$

where (as in (3.144)) $\mathfrak{D}(f)$ denotes the divisor of the meromorphic function f.

We also deduce from (3.166) that (still with the notation $D_{A,\Omega} = D(\zeta_{A,\Omega})$ and $D_{\partial G,G} = D(\zeta_{\partial G,G})$)

$$D_{A,\Omega} = \max(\sigma_0, D_{\partial G,G}) \tag{3.170}$$

or, equivalently,

$$\overline{\dim}_B(A,\Omega) = \max(\sigma_0, \overline{\dim}_B(\partial G, G)), \tag{3.171}$$

where $\sigma_0 = D(\zeta_{\mathfrak{s}})$ is the *similarity dimension* of the self-similar spray (A,Ω) defined as the unique *real* solution of the Moran equation [**Mora**]:[93]

$$\sum_{j=1}^{J} r_j^{\sigma_0} = 1. \tag{3.172}$$

In addition, in light of (3.167), $\mathcal{D}_{\mathfrak{s}} = \mathcal{D}(\zeta_{\mathfrak{s}})$ is given as the *set of complex solutions of the complexified Moran equation*, also called the *set of scaling complex dimensions of* (A,Ω):

$$\mathcal{D}_{\mathfrak{s}} = \left\{ s \in \mathbb{C} : \sum_{j=1}^{J} r_j^s = 1 \right\}, \tag{3.173}$$

so that this latter expression for $\mathcal{D}_{\mathfrak{s}}$ can be substituted in (3.168):

$$\mathcal{D}_{A,\Omega} \subseteq \mathcal{D}_{\partial G,G} \cup \left\{ s \in \mathbb{C} : \sum_{j=1}^{J} r_j^s = 1 \right\}. \tag{3.174}$$

If there are no zero-pole cancellations in (3.165) (or, equivalently, in (3.166)), which (for "nice" generators) is the case "generically" (in a vague sense), then we have an actual equality in (3.174). (See [**LapRaZu1**, Thm. 4.2.19].)

If the generator G (a bounded open subset of $\mathbb{R}^N$) is sufficiently nice (e.g., according to the main result of [**KoRati**] proving a conjecture in [LapPe1–2], if G is a nontrivial polytope, a very frequent situation for the classic self-similar fractals), then it is shown in [**LapRaZu1**, §5.5.6] that

$$\mathcal{D}_{\partial G,G} \subseteq \{0, 1, \cdots, N-1\}. \tag{3.175}$$

It therefore follows from (3.169) and (3.174) that

$$\mathcal{D}_{A,\Omega} \subseteq \mathcal{D}_{\partial G,G} \cup \mathcal{D}_{\mathfrak{s}} \subseteq \{0, 1, \cdots N-1\} \cup \left\{ s \in \mathbb{C} : \sum_{j=1}^{J} r_j^s = 1 \right\}, \tag{3.176}$$

with equalities in the "generic" case. In such a situation (much as in [LapPe1–2), LapPeWi1–2]), $\mathcal{D}_{\partial G,G} \subseteq \{0, 1, \cdots, N-1\}$ and $\mathcal{D}_{\mathfrak{s}} \subseteq \{s \in \mathbb{C} : \sum_{j=1}^{J} r_j^s = 1\}$ are respectively called the *integer dimensions* and the *scaling dimensions* of the self-similar spray (A,Ω).

[93] Clearly, in light of the definition (3.172) of σ_0, we have $\sigma_0 \in (0,N)$; indeed, by assumption, $J \geq 2 > 1$ (hence, $\sigma_0 > 0$) and $\overline{\dim}_B(\partial G, G) < N$ (hence, $\sigma_0 < N$). Observe that as a result, and in light of (3.170) and (3.171), we have that $\overline{\dim}_B(A,\Omega) < N$. Thus the complex dimensions of (A,Ω) can be defined indifferently via $\zeta_{A,\Omega}$ or $\widetilde{\zeta}_{A,\Omega}$.

Sketch of the proof of the factorization formulas (3.165)–(3.166). It is instructive to provide the proof of formula (3.165) (and hence, equivalently, of the factorization formula (3.166)), as given in the proof of [**LapRaZu1**, Thms. 4.2.17 and 4.2.19, p. 289].

First, we note that since the (possibly unbounded) self-similar string $\mathcal{L} = (\ell_j)_{j\geq 1}$ has for '*scales*' $\{\ell_j : j \geq 1\}$, the free monoid with generators $r_1, \cdots, r_J$, $\mathcal{L}$ satisfies the following *self-similar identity* (much as in [**Lap-vF4**, §4.4.1]):[94]

$$\mathcal{L} = \mathcal{L}_0 \sqcup \bigsqcup_{j=1}^{J} (r_j \mathcal{L}), \text{ where } \mathcal{L}_0 := \{1\}. \tag{3.177}$$

Further, a moment's reflection shows that we can deduce from (3.177) the following *self-similar identity* satisfied by the self-similar spray (A, Ω):

$$(A, \Omega) = (\partial G, G) \sqcup \bigsqcup_{j=1}^{J} (r_j(A, \Omega)), \tag{3.178}$$

where much as in (3.177), the symbol $\sqcup$ denotes the disjoint union of RFDs in $\mathbb{R}^N$. (See [**LapRaZu1**, Def. 4.1.43] for the precise definition of such a disjoint union of RFDs in $\mathbb{R}^N$, which is itself an RFD in $\mathbb{R}^N$.)

Now, by combining the scaling (and the invariance) property of the distance zeta function (see §3.3.3 and §3.3.4) according to which (for all $s \in \mathbb{C}$)

$$\zeta_{r_j(A,\Omega)}(s) = r_j^s\, \zeta_{A,\Omega}(s), \text{ for each } j = 1, \cdots, J, \tag{3.179}$$

where $r_j(A, \Omega) := (r_j A, r_j \Omega)$, along with the (finite) additivity of the distance zeta function under disjoint unions (see a special case of [**LapRaZu1**, Prop. 4.1.17]),[95] we obtain the following functional equation,

$$\zeta_{A,\Omega}(s) = \zeta_{\partial G,G}(s) + \sum_{j=1}^{J} r_j^s \zeta_{A,\Omega}(s), \tag{3.180}$$

which is equivalent to (3.165) (and to (3.166)), after an elementary factorization.

REMARK 3.21. (*a*) (*Multiple generators*). In the case of a self-similar spray (A, Ω) with multiple generators, $G_1, \cdots, G_Q$, we simply add up the results obtained for each of the generators. More specifically, in light of (3.165)–(3.167) applied to each of the generators, we then have that

$$\zeta_{A,\Omega}(s) = \frac{\sum_{q=1}^{Q} \zeta_{\partial G_q, G_q}(s)}{1 - \sum_{j=1}^{J} r_j^s}. \tag{3.181}$$

(*b*) (*Fractal sprays*). For (not necessarily self-similar) fractal sprays, several (but not all) of the above results are still valid. More specifically, if (A, Ω) is a fractal spray RFD in $\mathbb{R}^N$, with a single generator $(\partial G, G)$ scaled by the (not necessarily bounded or self-similar) fractal string $\mathcal{L} = (\ell_j)_{j\geq 1}$ and such that $|\Omega| < \infty$ and $\overline{\dim}_B(\partial G, G) < N$, then

$$\overline{D} = \overline{\dim}_B(A, \Omega) = \max(\overline{D}_{\mathfrak{s}}, \overline{\dim}_B(\partial G, G)), \tag{3.182}$$

[94] Here, in (3.177), the symbol $\sqcup$ denotes the disjoint union of fractal strings.

[95] This finite additivity property can be easily established. The significantly more delicate countable additivity property (precisely stated and established in [**LapRaZu1**, Prop. 4.1.17]) is not needed here but is frequently used throughout [**LapRaZu1**].

where $D_{\mathfrak{s}} := D(\zeta_{\mathcal{L}}) = D(\zeta_{\mathfrak{s}}), \overline{D} := D(\zeta_{A,\Omega})$ and $\overline{D}_{\partial G,G} := D(\zeta_{\partial G,G}) = \overline{\dim}_B(\partial G, G)$.

Then, for all $s \in \mathbb{C}$ such that $Re(s) > \overline{D}$ (and, hence, upon meromorphic continuation and for G nice enough,[96] for all $s \in \mathbb{C}$),

$$\zeta_{A,\Omega}(s) = \zeta_{\mathfrak{s}}(s) \cdot \zeta_{\partial G,G}(s), \tag{3.183}$$

where $\zeta_{\mathfrak{s}} := \zeta_{\mathcal{L}}$ denotes the meromorphic continuation of $\zeta_{\mathcal{L}}$:

$$\zeta_{\mathfrak{s}}(s) := \zeta_{\mathcal{L}}(s) \left(= \sum_{j \geq 1} \ell_j^s, \text{ for } Re(s) > \overline{D}_{\mathfrak{s}} \right). \tag{3.184}$$

It then follows from (3.183) that (with $\mathcal{D}_{\mathfrak{s}} := \mathcal{D}(\zeta_{\mathcal{L}}) = \mathcal{D}(\zeta_{\mathfrak{s}})$)

$$\mathcal{D}_{A,\Omega} \subseteq \mathcal{D}_{\partial G,G} \cup \mathcal{D}_{\mathfrak{s}}, \tag{3.185}$$

with equality instead of an inclusion, unless there are zero-poles cancellations. Furthermore, we always have the following identity between divisors:

$$\mathfrak{D}_{A,\Omega} = \mathfrak{D}_{\partial G,G} \cup \mathfrak{D}_{\mathfrak{s}}. \tag{3.186}$$

Naturally, a comment entirely similar to the one made in part (a) of this remark applies here if the fractal spray has multiple (i.e., finitely many) generators instead of a single generator.

In the next exercise, the interested reader is asked implicitly to use, in particular, the results of the present subsection in order to answer the various questions.

EXERCISE 3.22. (a) (*Sierpinski gasket*). Let A be the classic Sierpinski gasket. Then, calculate $\zeta_A, \zeta_{A,T}$ (where T is the equilateral triangle with side lengths 1), and deduce from this computation the two sets of complex dimensions $\mathcal{D}_A$ and $\mathcal{D}_{A,T}$, respectively. Compare your results with those stated in §3.4.1 (or with the $N = 2$ case of §3.4.4).

(b) (*Sierpinski carpet*). Let A be the classic Sierpinski carpet. Then, answer the same questions as in part (a) just above, except with the unit triangle T replaced by the open unit square $S := (0,1)^2$, and with §3.4.1 and §3.4.4 replaced by §3.4.2 and §3.4.3, respectively.

(c) Answer analogous questions for the relative (or inhomogeneous) N-gasket, as discussed in §3.4.4, as well as for the $\frac{1}{2}$-square and $\frac{1}{3}$-square fractals, as discussed in §3.4.5.

We refer the interested reader to [LapRaZu1–10] for many other examples of computations of fractal zeta functions and complex dimensions (as well as of fractal tube formulas, in regard to §3.5 just below). These examples include non self-similar fractals or RFDs such as fractal nests (see [**LapRaZu1**, §3.5 and Exples. 5.5.16 and 5.5.24 in §5.5.5]) as well as bounded and unbounded geometric chirps (see [**LapRaZu1**, §3.6, §4.4.1 and Exple. 5.5.19 in §5.5.5]). They also include examples of fractal strings and RFDs, as well as compact sets, with principal complex dimensions having arbitrarily prescribed (finite or infinite) multiplicities (see [**LapRaZu1**, Thms. 3.3.6 and 4.2.19]).

[96]This is the case, for example, if G is a nontrivial polytope [**KoRati**] or more generally, if G is either monophase or pluriphase (in the sense of [**LapPe2, LapPeWi1**]).

3.5. Fractal tube formulas and Minkowski measurability criteria: Theory and examples. The goal of this subsection is to briefly state (in §3.5.1) the higher-dimensional analog (obtained in [**LapRaZu6**] and [**LapRaZu1**, §§5.1–5.3]) of the fractal tube formulas obtained for fractal strings in [Lap-vF2–4] (see, especially, [**Lap-vF4**, Ch. 8]). In fact, the latter tube formulas (briefly discussed in §2.1) are now but a very special case of their higher-dimensional counterparts.[97]

We will also state (in §3.5.2) some of the Minkowski measurability criteria obtained in [**LapRaZu1**, §5.4.3 and §5.4.4].

In fact, the main goal of this section is to illustrate (in §3.5.3) the aforementioned results concerning fractal tube formulas and Minkowski measurability criteria (and established, in particular, in [**LapRaZu7**] and [**LapRaZu1**, §5.4]), in §3.5.1 and §3.5.2, respectively) via a variety of examples, many of which have been discussed from other points of view earlier in the paper, especially in §3.4.

3.5.1. *Fractal tube formulas for RFDs in $\mathbb{R}^N$, via distance and tube zeta functions.* Recall from §2.1, adapted to the present much more general situation of RFDs in $\mathbb{R}^N$, that a fractal tube formula enables us to express the tube function

$$\varepsilon \mapsto V(\varepsilon) = V_{A,\Omega}(\varepsilon) := |A_\varepsilon \cap \Omega|_N \tag{3.187}$$

of a RFD (A, Ω) in $\mathbb{R}^N$ in terms of the complex dimensions of (A, Ω) and the associated residues of the corresponding fractal zeta function (here, either the distance or the tube zeta function of (A, Ω)).

An important feature of the higher-dimensional theory of complex dimensions is that such fractal tube formulas can be established under great generality for RFDs in $\mathbb{R}^N$, via either distance zeta functions or tube zeta functions, as well as via other fractal zeta functions, such as the so-called shell zeta functions and the Mellin zeta functions, which are of interest in their own right but are also used in [**LapRaZu1**, Ch. 5] in key steps towards the proof of the fractal tube formulas and of related Minkowski measurability criteria. We will limit ourselves here to the case of distance and tube zeta functions.

Depending on the growth assumptions made about the fractal zeta functions under consideration,[98] we obtain fractal tube formulas with error term or without error term (i.e., *exact*), as well as interpreted either pointwise or distributionally. Therefore, just as in the one-dimensional case of fractal strings, but now in the much more general case of RFDs (and, in particular, of bounded sets) in $\mathbb{R}^N$, there is a lot of flexibility for obtaining and applying such tube formulas.

Let us now be a little bit more specific, while avoiding technicalities and cumbersome (although useful) definitions. We state all of the results for RFDs (A, Ω) in $\mathbb{R}^N$ but, as usual, the special case of bounded subsets A of $\mathbb{R}^N$ is obtained by simply considering the associated RFD (A, A_{δ_1}) for some fixed $\delta_1 > 0$; as before, which $\delta_1 > 0$ is chosen turns out to be unimportant.

---

[97]However, as is often the case in such situations, the results and techniques developed in [**Lap-vF4**, Chs. 5 and 8] in order to prove fractal tube formulas and other explicit formulas for (generalized) fractal strings are key to establishing the higher-dimensional (pointwise and distributional, exact or with error term) fractal tube formulas. Nevertheless, a significant amount of additonal work is required in order to prove those tube formulas for RFDs (or, in particular, bounded sets) in $\mathbb{R}^N$; see, especially, [**LapRaZu1**, §§5.1–5.3].

[98]These polynomial-type growth conditions are referred to (much as in [**Lap-vF4**, Chs. 5 and 8]) as *languidity conditions* in the case of tube formulas with error term and as *strong languidity conditions* in the case of exact tube formulas. These conditions are slightly different for distance and tube zeta functions; see [**LapRaZu1**, Ch. 5] for more details.

We begin by (loosely) stating the fractal tube formula via distance zeta functions. Let (A,Ω) be a RFD in $\mathbb{R}^N$ such that $\overline{D} = \overline{\dim}_B(A,\Omega) < N$. Then, if (A,Ω) is d-languid (which roughly means that the distance zeta function satisfies some mild polynomial growth conditions), we have the following fractal formula, expressed via $\zeta_{A,\Omega}$:[99]

$$(3.188) \qquad V_{A,\Omega}(\varepsilon) = \sum_{\omega \in \mathcal{D}_{A,\Omega}} c_\omega \frac{\varepsilon^{N-\omega}}{N-\omega} + R(\varepsilon),$$

where $\mathcal{D}_{A,\Omega} = \mathcal{D}_{A,\Omega}(U)$ is the set of visible complex dimensions of (A,Ω) (defined as the poles of $\zeta_{A,\Omega}$ belonging to U) and U is a suitable domain of $\mathbb{C}$ (to which $\zeta_{A,\Omega}$ can be meromorphically extended to a d-languid function). Furthermore, for each $\omega \in \mathcal{D}_{A,\Omega}$,

$$(3.189) \qquad c_\omega := \operatorname{res}(\zeta_A, \omega);$$

so that (3.188) can be rewritten equivalently in the following form (still in the case of simple complex dimensions):

$$(3.190) \qquad V_{A,\Omega}(\varepsilon) = \sum_{\omega \in \mathcal{D}_{A,\Omega}} \operatorname{res}(\zeta_{A,\Omega}, \omega) \frac{\varepsilon^{N-\omega}}{N-\omega} + R(\varepsilon).$$

Moreover, in (3.188) and (3.190), $R(\varepsilon)$ is an error tem (of lower order than the sum over the complex dimensions) which can be estimated explicitly.[100]

If, in addition, (A,Ω) (i.e., $\zeta_{A,\Omega}$) is strongly d-languid (a growth condition which requires that $U := \mathbb{C}$ and is stronger than d-languidity), then $R(\varepsilon) \equiv 0$ in (3.188) (and equivalently, in (3.190) or in (3.192)). Hence, we obtain an *exact* fractal tube formula in this case (i.e., a tube formula without error term):

$$(3.191) \qquad V_{A,\Omega}(\varepsilon) = \sum_{\omega \subset \mathcal{D}_{A,\Omega}} \operatorname{res}(\zeta_{A,\Omega}, \omega) \frac{\varepsilon^{N-\omega}}{N-\omega}.$$

Finally, we note that if the visible complex dimensions are not necessarily all simple, then (for example) the tube formula (3.190) takes the following, slightly more complicated, form (except for this modification, all the other statements and hypotheses are identical, otherwise):[101]

$$(3.192) \qquad V_{A,\Omega}(\varepsilon) = \sum_{\omega \in \mathcal{D}_{A,\Omega}} \operatorname{res}\left(\frac{\varepsilon^{N-s}}{N-s}\zeta_{A,\Omega}(s), \omega\right) + R(\varepsilon),$$

with $R(\varepsilon) \equiv 0$ in the case of an exact formula.

Entirely analogous fractal tube formulas are also obtained via the tube (instead of the distance) zeta function; that is, for $\widetilde{\zeta}_{A,\Omega}$ instead of $\zeta_{A,\Omega}$. The main difference is that in the counterparts of the fractal tube formulas (3.188) and (3.190)–(3.192), we must replace $\varepsilon^{N-\omega}/(N-\omega)$ by $\varepsilon^{N-\omega}$, while in the counterpart of (3.192), we must replace $\frac{\varepsilon^{\omega-s}}{N-s}\zeta_{A,\Omega}(s)$ by $\varepsilon^{N-s}\zeta_{A,\Omega}(s)$. The only other difference, truly minor

[99] For the simplicity of exposition, we assume at first that all of the (visible) complex dimensions (i.e., the visible poles of $\zeta_{A,\Omega}$) are simple; see (3.192) for the general case when some (or possibly all) of the complex dimensions are multiple.

[100] Naturally, in the case of a pointwise (respectively, distributional) tube formula, $R(\varepsilon)$ is a pointwise (respectively, distributional) error term.

[101] Here, for clarity, we write $\operatorname{res}(f(s),\omega)$ (instead of $\operatorname{res}(f,\omega)$) to denote the residue of a meromorphic function $f = f(s)$ at $s = \omega$.

this time, is that the d-languidity (respectively, strong d-languidity) condition must be replaced by the slightly different languidity (respectively, strong languidity) condition, in the case of a (pointwise or distributional) fractal tube formula with (respectively, without) error term.

For example, the (pointwise or distributional) fractal tube formula, expressed via the tube zeta function $\widetilde{\zeta}_{A,\Omega}$, takes the following form (in the case of simple complex dimensions):[102]

$$V_{A,\Omega}(\varepsilon) = \sum_{\omega \in \mathcal{D}_{A,\Omega}} \operatorname{res}(\widetilde{\zeta}_{A,\Omega}, \omega)\varepsilon^{N-\omega} + R(\varepsilon). \tag{3.193}$$

In addition, if (A, Ω) (i.e., $\widetilde{\zeta}_{A,\Omega}$) is strongly languid (which requires that $U := \mathbb{C}$), then we can let $R(\varepsilon) \equiv 0$ in (3.193); that is, we obtain an exact fractal tube formula in this case.

REMARK 3.23. (*a*) For the precise statements and the full proofs of the above (as well as of other pointwise and distributional) fractal tube formulas, we refer the interested reader to [**LapRaZu6**] or [**LapRaZu1**, §§5.1–5.3].

(*b*) Much as was discussed in §2.5 (see, especially, Remark 2.7), and as is apparent in (3.188), (3.190), (3.191) and (3.193), each (visible) simple complex dimension $\omega \in \mathcal{D}_{A,\Omega}$ gives rise to an *oscillatory term*, proportional to $\varepsilon^{N-\omega}$, with exponent the *fractal complex co-dimension* $N - \omega$ of ω.[103] (Naturally, for each individual complex dimension, these oscillations are multiplicatively periodic. In order to obtain the associated additive or ordinary oscillations, it suffices to let $x := \log(\varepsilon^{-1})$.) Furthermore, as was mentioned before in §2 in the case of fractal strings, as ω varies in $\mathcal{D}_{A,\Omega}$, the amplitudes (respectively, frequencies) of these oscillations are governed by the real (respectively, imaginary) parts of the complex dimensions. Observe that in the case of simple poles (as in (3.188), (3.190)–(3.191) and (3.193) in §3.5.1), the (typically) infinite sum over the complex dimensions of the RFD (A, Ω) can be thought of as (pointwise or distributional) natural generalizations of Fourier series or of almost periodic functions (or distributions). (Compare with [**Schw**, §VII, I] and [**Boh**], respectively.)

We close this subsection by placing the above results in a broader context and providing related references in geometric measure theory and convex geometry. Prior to that, let us recall that the first fractal tube formulas (expressed in terms of complex dimensions and geometric zeta functions) were obtained in [Lap-vF1–4] (see, especially, [**Lap-vF4**, Ch. 8]), in the case of fractal strings. In the case of fractal sprays (roughly, higher-dimensional analogs of fractal strings), fractal tube formulas were obtained in [LapPe2–3] and, in greater generality, in [**LapPeWi1**], but without a natural notion of associated fractal zeta function. Finally, the fractal tube formulas described in the present subsection (i.e., §3.5.1) were obtained in [**LapRaZu6**] and [**LapRaZu1**, Ch. 5]. They are expressed in terms of complex dimensions and natural fractal zeta functions and include the earlier fractal tube formulas for fractal strings from [Lap-vF1–4] and for fractal sprays [LapPe2–3, LapPeWi1]; see, in particular, Example 3.26 and Example 3.36.

---

[102] Compare with its counterpart, (3.190), expressed via the distance zeta function $\zeta_{A,\Omega}$.

[103] In the case when ω is a multiple complex dimension, this oscillatory term is modulated by the multiplication by a suitable polynomial in the variable $x := \log(\varepsilon^{-1})$ and of degree equal to the multiplicity minus one.

REMARK 3.24. (*Tube formulas and curvatures: Brief history, references, and beyond.*) It is well known that for a (nonempty) compact convex set A of $\mathbb{R}^N$, the tube function $V_A(\varepsilon) := |A_\varepsilon|_N$ is a polynomial of degree (at most) N in ε, the coefficients of which can be interpreted geometrically. This is known as Steiner's formula ([**Stein**], 1840). More specifically,

$$V_A(\varepsilon) = \sum_{\alpha=0}^{N} \gamma_\alpha \mu_\alpha(A) t^{N-\alpha}, \tag{3.194}$$

where for each $\alpha \in \{0, 1, \cdots, N\}$, γ_α is the volume of the unit ball in $\mathbb{R}^{N-\alpha}$ and the (normalized) coefficients $\mu_\alpha(A)$ have either a geometric, combinatorial or algebraic interpretation. (See, e.g., [**Schm**] and [**KlRot**].) For example, $\mu_N(A) = |A|_N$ is the volume of A, $\mu_{N-1}(A)$ is the surface area of $A, \cdots, \mu_1(A)$ is its mean width, while $\mu_0(A)$ is its Euler characteristic (equal to 1 in the present case, but equal to any integer in more general situations).

Furthermore, H. Weyl ([**Wey3**], 1939) has obtained an analog of Steiner's formula when A is a smooth compact submanifold of Euclidean space $\mathbb{R}^N$, thereby interpreting the coefficients $\mu_\alpha(A)$ as suitable curvatures, now called the *Weyl curvatures* of A (see, e.g., [**BergGos**] and [**Gra**]).

Moreover, H. Federer ([**Fed1**], 1969) has unified and extended both Steiner's formula and Weyl's tube formula by establishing their counterpart for sets of positive reach.[104] In the process, he introduced (real or signed) measures μ_α (for $\alpha \in \{0, 1, \cdots, N\}$), now called *Federer's curvature measures*, whose total mass $\mu_\alpha(A)$ coincide with the normalized coefficients of (3.194). He also obtained a localized version of the tube formula (3.194), expressed in terms of the values of the measures μ_α, for $\alpha \in \{0, 1, \cdots, N\}$, at suitable Borel subsets of $\mathbb{R}^N$ (or of $\mathbb{R}^N \times S^{N-1}$).

Much later, 'fractal curvatures' were introduced by S. Winter in [**Wi**], M. Zähle [Zä3–4], and S. Winter and M. Zähle [**WiZä**], for certain classes of self-similar deterministic or random fractals, but still for integer values of the index. The author has long conjectured that there should exist suitable notions of complex measures ([**Coh**], [**Fo**], [**Ru1**]) or distributions ([**Schw**], [**GelSh**]), called '*fractal curvature measures*', associated with each (visible) complex dimension $\omega \in \mathcal{D}_A$ and enabling us to interpret geometrically the (normalized) coefficients of the fractal tube formulas (see, for example, (3.188) and (3.193)). Also, a *local* fractal tube formula, yet to be precisely formulated and to be rigorously established, should extend Federer's local tube formula, and be expressed (in the case of simple poles) in terms of the residues of a suitably defined *local fractal zeta function* evaluated at each of the visible complex dimensions. (See [**LapRaZu1**, Pb. 6.2.3.8 and App. B].)[105]

We close this discussion of tube formulas by mentioning several relevant references (beside [**Stein**], [**Mink**], [**Wey3**] and [**Fed1**]), including the books [**Fed2**], [**KlRot**], [**BergGos**], [**Gra**], and [**Schn**], along with the papers [**HugLasWeil**],

[104] A (compact) subset A of $\mathbb{R}^N$ is said to be of *positive reach* if there exists $\eta \in (0, +\infty]$ such that each point of A_η, the η-neighborhood of A, has a unique metric projection onto A. Clearly, a convex set has infinite reach.

[105] See also the earlier work in [LapPe2–3] and [**LapPeWi1**] for the very special case of fractal sprays, as well as [**LapPe1**] which dealt with the example of the Koch snowflake curve (but without the use of any zeta functions).

[**KeKom**], [**Kom**], [LapPe1–3], [**LapPeWi1**], [LapLu-vF1–2], [**LapRaZu6**], [Ol1–2], [**Sta**], [**Wi**], [**WiZä**], [Za1–4], as well as [**Lap-vF4**, §13.1] and [**LapRaZu1**, Ch. 5], and the many relevant references therein.

3.5.2. *Minkowski measurability criteria for RFDs in* $\mathbb{R}^N$. In this subsection, we briefly discuss, in particular, a few of the Minkowski measurability criteria obtained in [LapRaZu5,7] or in [**LapRaZu1**, §5.4], to which we refer for further information and for closely related necessary or sufficient conditions for Minkowski measurablity. The criteria will be stated for RFDs in $\mathbb{R}^N$, but as usual, can also be applied to bounded subsets (which, as we know by now, are special cases of RFDs).

The main criterion (see [**LapRaZu1**, Thm. 5.4.20]) is expressed in terms of the distance zeta function:

Let (A, Ω) *be a RFD in* $\mathbb{R}^N$. *Assume that* $D := \dim_B A$ *exists and* $D < N$. *Then, under suitable hypotheses,*[106] *the following statements are equivalent*:

(*i*) *The RFD* (A, Ω) *is Minkowski measurable.*

(*ii*) *The Minkowski dimension* D *is the only principal complex dimension of* (A, Ω) (*i.e., the only pole of* $\zeta_{A,\Omega}$ *with real part equal to* D) *and it is simple.*[107]

The above Minkowski measurability criterion is the higher-dimensional counterpart of Theorem 2.2 in §2.1, the Minkowski measurability criterion for fractal strings obtained in [**Lap-vF4**, §8.3]. Its proof involves several key ingredients including, especially, the Wiener–Pitt Tauberian theorem [**PitWie**] (stated, e.g., in [**Pit**], [**Kor**] and [**LapRaZu1**, Thm. 5.4.1]), a version of the fractal tube formula (with error term) via $\zeta_{A,\Omega}$ discussed in §3.5.1 just above, as well as a uniqueness theorem for almost periodic functions (or rather, distributions [**Schw**, §VI.9.6]).

An analog for the tube (rather than the distance) zeta function $\widetilde{\zeta}_{A,\Omega}$ of the above Minkowski measurability criterion for RFDs is also obtained in [**LapRaZu1**, Thm. 5.4.25].

Moreover, a necessary (respectively, sufficient) condition for the Minkowski measurability of RFDs is obtained in [**LapRaZu1**, Thm. 5.4.15] (respectively, [**LapRaZu1**, Thm. 5.4.2]).

In addition, the case of RFDs for which the underlying scaling law is no longer a power law but is governed instead by a nontrivial gauge function $h = h(t)$ (with $t \in (0, 1)$, say)[108] is examined in [LapRaZu6–7]; see also [**LapRaZu4**]) and [**LapRaZu1**, §5.4.4] where both an h-Minkowski measurability criterion and an optimal h-fractal tube expansion are obtained, especially for gauge functions of the form $h(t) := (\log t^{-1})^{m-1}$ for some integer $m \geq 2$ (corresponding, e.g., under appropriate hypotheses, to $D = \dim_B(A, \Omega)$ being a multiple pole of order m.

In the definition of the (upper, lower) Minkowski content, relative to a given gauge function h, where for some $\varepsilon_0 > 0$, $h : (0, \varepsilon_0) \to (0, +\infty)$ is a function of slow growth satisfying suitable conditions near 0 (including the fact that $h(t) \to +\infty$ as $t \to 0^+$), one simply replaces ε^{N-D} by $\varepsilon^{N-D}h(\varepsilon)$. For example, assuming that

[106]Namely, (A, Ω) (i.e., $\zeta_{A,\Omega}$) is assumed to be d-languid (as in §3.5.1) for a screen S passing between the critical line $\{Re(s) = D\}$ and all of the complex dimensions of (A, Ω) with real part $< D$. (Roughly speaking, a screen S is a suitable curve bounding the region U, with $\overline{U} \supseteq \{Re(s) = D\}$, to which $\zeta_{A,\Omega}$ is meromorphically continued, and extending to infinity in the vertical direction. Also, S is required not to contain any pole of $\zeta_{A,\Omega}$; see [**Lap-vF4**, §5.3] or [**LapRaZu1**, §5.1.1].)

[107]Equivalently, the RFD (A, Ω) does *not* have any *nonreal* complex dimension, and D is simple.

[108]The standard case when the underlying scaling law is a power law corresponds to the trivial gauge function $h(t) \equiv 1$.

$D = \dim_B(A,\Omega)$ exists, the *upper h-Minkowski content* of (A,Ω) is given by

$$\mathcal{M}^*((A,\Omega),h) := \overline{\lim_{\varepsilon\to 0^+}} \frac{V_{A,\Omega}(\varepsilon)}{\varepsilon^{N-D}h(\varepsilon)}, \tag{3.195}$$

and similarly for the *lower h-Minkowski content*, $\mathcal{M}_*((A,\Omega),h)$, and the *h-Minkowski content*, $\mathcal{M}(A,\Omega,h)$. (See [**HeLap**] and [**LapRaZu1**, Eq. (4.5.10), p. 352, and §6.1.1.2, pp. 544–545].) Examples of allowable gauge functions considered in [**HeLap**], [**LapRaZu1**], [**LapRaZu7**] and [**LapRaZu10**] include $\log^k(t^{-1})$, for all $t\in(0,1)$, where $k\in\mathbb{N}$ is arbitrary and $\log^k$ denotes the k-th iterated logarithm. The reciprocals $1/h(t)$ of allowable gauge functions are also considered in the above definitions and references.

Although we do not wish to go into the technical details here, we mention that in [**LapRaZu10**], connections between generalized Minkowski contents with logarithm-type gauge functions, Minkowski measurability criteria, and appropriate Riemann surfaces (see, e.g., [**Ebe, Schl**]), are explored. These types of generalized Minkowski contents arise naturally in certain geometric situations and in the study of certain dynamical systems.

REMARK 3.25. (*Extension to Ahlfors metric spaces.*) It is noteworthy that essentially the entire (higher-dimensional) theory of complex dimensions, fractal zeta functions and fractal tube formulas (including the Minkowski measurability criteria) from [**LapRaZu1**] and the accompanying series of papers can be extended without change to a large class of metric measure spaces (see, e.g., [**DaMcCS**]) called Ahlfors spaces, doubling spaces or else, spaces of homogeneous type, and of frequent use in harmonic analysis, nonsmooth analysis, fractal geometry and the theory of dynamical systems.[109] This extension is carried out in [**Wat**] and [**LapWat**] and the corresponding theory is illustrated by several examples of computation of complex dimensions and concrete fractal tube formulas, both in the setting of Ahlfors spaces and a little beyond.

3.5.3. *Examples.* In the present subsection, we illustrate by means of a variety of examples some of the results about fractal tube formulas and Minkowski measurability criteria discussed in §3.5.1 and §3.5.2, respectively. In the process, in order to avoid unnecessary repetitions, we often refer to the corresponding examples in §3.4.

EXAMPLE 3.26. (*Fractal strings*). We briefly explain here how to recover the fractal tube formulas for fractal strings from [**Lap-vF4**]) discussed in §2.1 above.

Let $\mathcal{L} = (\ell_j)_{j\geq 1}$ be a (bounded) fractal string and let $\Omega\subseteq\mathbb{R}$ be any geometric realization of $\mathcal{L}$ as an open set with finite length (i.e., $|\Omega|_1<\infty$). Furthermore, let $(\partial\Omega,\Omega)$ be the associated RFD in $\mathbb{R}$. Then, as we have seen in §3.2.2, the *distance zeta function* $\zeta_{\partial\Omega,\Omega}$ of the RFD $(\partial\Omega,\Omega)$ and the *geometric zeta function* $\zeta_{\mathcal{L}}$ of the fractal string $\mathcal{L}$ are connected via the following functional equation:

$$\zeta_{\partial\Omega,\Omega}(s) = \frac{2^{1-s}}{s}\zeta_{\mathcal{L}}(s), \tag{3.196}$$

[109] A metric space $(X,\mathbf{d})$ of finite diameter and equipped with a positive Borel measure μ (i.e., a *metric measure space*) is said to be an *Ahlfors space of Ahlfors dimension* α if $\mu(B_r(x))$ is comparable to r^α, where $B_r(x)$ is the closed ball (with respect to the metric $\mathbf{d}$) of center x and radius r, and with $x\in X$ arbitrary. (The implicit constants must, of course, be independent of $x\in X$ and of $r>0$ sufficiently small.) In this case, α coincides with the Hausdorff and Minkowski dimensions of X.

for all $s \in U$, where U is any domain of $\mathbb{C}$ to which $\zeta_{\mathcal{L}}$ (or, equivalently, $\zeta_{\partial\Omega,\Omega}$) can be meromorphically continued. As a result, provided $0 \in U$ (and assuming for simplicity that $\zeta_{\mathcal{L}}(0) \neq 0$), then

$$\mathcal{D}_{\partial\Omega,\Omega} = \mathcal{D}_{\mathcal{L}} \cup \{0\}, \tag{3.197}$$

where the union is taken between multisets, as usual. More specifically, if 0 is a pole of $\zeta_{\mathcal{L}}$ of multiplicity $m \geq 0$ (the case when $m = 0$ corresponding to 0 not being a pole of $\zeta_{\mathcal{L}}$), then it is a pole of $\zeta_{\partial\Omega,\Omega}$ of multiplicity $m+1$.

The identity (3.197) explains why the expressions for the fractal tube formulas in the case of simple poles to be discussed a little further on looked somewhat awkward in [**Lap-vF4**, §8.1] but are now significantly simplified, both conceptually and concretely. In particular, as is now clear, even in the present case when $N = 1$, the distance zeta function $\zeta_{\partial\Omega,\Omega}$ is the proper theoretical tool to define and understand the complex dimensions of fractal strings as well as to formulate the associated fractal tube formulas.[110]

In light of (3.196)–(3.197) and since $N = 1$, the fractal tube formula (3.192) yields its counterpart for fractal strings (for every $\delta_1 \geq \ell_1/2$):[111]

$$\begin{aligned} V_{\mathcal{L}}(\varepsilon) &= V_{\partial\Omega,\Omega}(\varepsilon) \\ &= \sum_{\omega\in\mathcal{D}_{\partial\Omega,\Omega}} \operatorname{res}\left(\frac{\varepsilon^{1-s}}{1-s}\zeta_{\partial\Omega,\Omega}(s),\omega\right) + R(\varepsilon) \\ &= \sum_{\omega\in\mathcal{D}_{\mathcal{L}}\cup\{0\}} \operatorname{res}\left(\frac{(2\varepsilon)^{1-s}}{s(1-s)}\zeta_{\mathcal{L}}(s),\omega\right) + R(\varepsilon) \\ &= \sum_{\omega\in\mathcal{D}_{\mathcal{L}}} \operatorname{res}\left(\frac{(2\varepsilon)^{1-s}}{s(1-s)}\zeta_{\mathcal{L}}(s),\omega\right) + \{2\varepsilon\zeta_{\mathcal{L}}(0)\}_{0\in U\setminus\mathcal{D}_{\mathcal{L}}} + R(\varepsilon), \end{aligned} \tag{3.198}$$

when $\mathcal{L}$ (i.e., $\zeta_{\mathcal{L}}$) is languid, and with $R(\varepsilon) \equiv 0$ when $\mathcal{L}$ (i.e., $\zeta_{\mathcal{L}}$) is strongly languid. Here, by definition, the term $\{2\varepsilon\zeta_{\mathcal{L}}(0)\}_{0\in U\setminus\mathcal{D}_{\mathcal{L}}}$ between braces in the last equality of (3.198) is included only if $0 \in U\setminus\mathcal{D}_{\mathcal{L}}$.

We note that the expression obtained in (3.198) is in complete agreement with the (pointwise or distributional) fractal tube formulas obtained for fractal strings in [**Lap-vF4**, Thm. 8.1 or Thm. 8.7]).[112]

Since, in view of (3.196),

$$\operatorname{res}(\zeta_{\partial\Omega,\Omega},\omega) = \frac{2^{1-\omega}}{\omega}\operatorname{res}(\zeta_{\mathcal{L}},\omega) \tag{3.199}$$

for every simple complex dimension $\omega \in U\setminus\{0\}$, we deduce from (3.198) the following (pointwise or distributional) tube formula in the special case when all of the

[110]This is so even though the geometric zeta function has been (and will continue to be) a very useful tool as well. We note that an entirely analogous comment can be made about (not necessarily self-similar) fractal sprays; recall from §3.4.10 that in the latter case, $\mathcal{L}$ may be unbounded and $\zeta_{\mathcal{L}}$ is then called the *scaling zeta function* of the fractal spray (see [**LapPe2, LapPeWi1**] and [**Lap-vF4**, §13.1]).

[111]Note that in light of (2.8) and (3.187), we have that $V_{\mathcal{L}}(\varepsilon) = V_{\partial\Omega,\Omega}(\varepsilon)$. Also, we let $\mathcal{D}_{\mathcal{L}} = \mathcal{D}_{\mathcal{L}}(U)$ and $\mathcal{D}_{\partial\Omega,\Omega} = \mathcal{D}_{\partial\Omega,\Omega}(U)$.

[112]In §2.1, for the simplicity of exposition, we gave a less precise statement of the formula; compare with formula (2.11).

visible complex dimensions are simple (and $\zeta_{\mathcal{L}}$ is languid):

$$
\begin{aligned}
V_{\mathcal{L}}(\varepsilon) &= V_{\partial\Omega,\Omega}(\varepsilon) \\
&= \sum_{\omega\in\mathcal{D}_{\mathcal{L}}(W)\setminus\{0\}} \operatorname{res}(\zeta_{\mathcal{L}},\omega)\frac{(2\varepsilon)^{1-\omega}}{\omega(1-\omega)} \\
&\quad + \{2\varepsilon(1-\log(2\varepsilon))\operatorname{res}(\zeta_{\mathcal{L}},0) + 2\varepsilon\zeta_{\mathcal{L}}(0)\}_{0\in U} + R(\varepsilon),
\end{aligned}
\tag{3.200}
$$

where, by definition, the term between braces in the last equality of (3.200) is included only if $0 \in U$. Also, as before, if $\mathcal{L}$ (i.e., $\zeta_{\mathcal{L}}$) is strongly languid (which implies that $U = \mathbb{C}$), then we obtain an exact tube formula; that is, we can let $R(\varepsilon) \equiv 0$ in (3.200).

We point out that (3.200) is in agreement with the results stated in [**Lap-vF4**, Cor. 8.3 or Cor. 8.10] in the distributional or pointwise case, respectively.[113]

EXAMPLE 3.27. (*The Sierpinski gasket*). We briefly revisit the first example from §3.4, studied in §3.4.1, in which $A \subseteq \mathbb{R}^2$ is the classic Sierpinski gasket and, by (3.68), all of the complex dimensions $s = 0$ and $s_k := \log_2 3 + i(2\pi/\log 2)k$ (for any $k \in \mathbb{Z}$) are simple with associated residues given by (3.69) and (3.70), respectively. Then, in light of (3.191) and since $N = 2$ here, we obtain the following pointwise, exact fractal tube formula (the hypotheses of which are shown to be satisfied in [**LapRaZu1**, Expl. 5.5.12]), valid for all $\varepsilon \in (0, 1/2\sqrt{3})$:

$$
\begin{aligned}
V_A(\varepsilon) &= \sum_{k\in\mathbb{Z}} \operatorname{res}(\zeta_A, s_k)\frac{\varepsilon^{2-s_k}}{2-s_k} + \operatorname{res}(\zeta_A, 0)\frac{\varepsilon^{2-0}}{2} \\
&= \sum_{k\in\mathbb{Z}} \frac{6(\sqrt{3})^{1-s_k}}{4^{s_k}(\log 2)s_k(s_k-1)}\frac{\varepsilon^{2-s_k}}{2-s_k} + (3\sqrt{3}+2\pi)\frac{\varepsilon^2}{2}.
\end{aligned}
\tag{3.201}
$$

Since $s_k = D + ik\mathbf{p}$, for all $k \in \mathbb{Z}$, where $D = \log_2 3$ and $\mathbf{p} = (2\pi/\log 2)$ are, respectively, the Minkowski dimension and the oscillatory period of A, we can clearly rewrite (3.201) in the following form:

$$
V_A(\varepsilon) = \varepsilon^{2-D}G(\log_2 \varepsilon^{-1}) + \left(\frac{3\sqrt{3}+2\pi}{2}\right)\varepsilon^2, \tag{3.202}
$$

where G is a continuous, nonconstant 1-periodic function, which is bounded away from zero and infinity, and is given by the following absolutely convergent (and hence, pointwise convergent) Fourier series expansion, for all $x \in \mathbb{R}$:

$$
G(x) := \frac{6\sqrt{3}}{\log 2}\sum_{k\in\mathbb{Z}} \frac{(4\sqrt{3})^{-s_k}}{(2-s_k)(s_k-1)s_k} e^{i2\pi kx}. \tag{3.203}
$$

It is apparent from (3.203) that since G is nonconstant, it follows that the function $\varepsilon^{-(2-D)}V_A(\varepsilon)$ is oscillatory, and hence does not have a limit as $\varepsilon \to 0^+$. Therefore, the Sierpinski gasket is *not* Minkowski measurable, in agreement with the Minkowski measurability criterion stated in §3.5.2, the hypotheses of which are easily verified since $D = \log_2 3$ is simple and A has infinitely many nonreal principal complex dimensions (here, $s_k = D + ik\mathbf{p}$, with $k \in \mathbb{Z}\setminus\{0\}$).

---

[113] Again, we note that in §2.1, for the simplicity of exposition, the corresponding formula stated in (2.10) was not as precise as in (3.200).

EXAMPLE 3.28. (*The Sierpinski carpet*). Let $A \subseteq \mathbb{R}^2$ be the classic Sierpinski carpet studied in §3.4.2. Then, in light of (3.71)–(3.73), (3.191) yields the following pointwise, exact fractal tube formula:

$$V_A(\varepsilon) = \sum_{k\in\mathbb{Z}} \frac{2^{-s_k}}{(\log 3)(2-s_k)(s_k-1)s_k}\varepsilon^{2-s_k} + \frac{16}{5}\varepsilon^2 + \frac{1}{2}\Big(2\pi + \frac{8}{7}\Big)\varepsilon^2, \tag{3.204}$$

where $s_k := D + ik\mathbf{p}$, for each $k \in \mathbb{Z}$; here, $D := \log_3 8$ and $\mathbf{p} := (2\pi/\log 3)$ are, respectively, the Minkowski dimension and the oscillatory period of A.

EXERCISE 3.29. (i) Much as in (3.202) and (3.203) above, rewrite the leading term (i.e., the sum over all $k \in \mathbb{Z}$) in (3.204) in the form $\varepsilon^{2-D}G(\log_3 \varepsilon^{-1})$, where G is a continuous, nonconstant 1-periodic function, which is bounded away from zero and infinity.

(ii) Deduce from part (i) via a direct computation that the Sierpinski carpet is *not* Minkowski measurable, also in agreement with the Minkowski measurability criterion stated in §3.5.2 (and of which you should verify the hypotheses).

(iii) Finally, by means of a direct computation (based, e.g., on part (i)), show that the Sierpinski carpet is Minkowski nondegenerate and calculate its average Minkowski content $\widetilde{\mathcal{M}}$. Furthermore, verify the latter results by using (3.51), connecting $\widetilde{\mathcal{M}}$ and the residue of $\zeta_A(s)$ at $s = D$ in the non-Minkowski measurable case (and of which the hypotheses are satisfied, in light of part (i)).

EXAMPLE 3.30. (*The 3-d carpet*). Let $A \subseteq \mathbb{R}^2$ be the 3-d carpet studied in §3.4.3. Then, in light of (3.76)–(3.80), (3.191) yields the following pointwise, exact fractal tube formula (valid for all $\varepsilon \in (0, 1/2)$):

$$V_A(\varepsilon) = \frac{24}{13\log 3}\varepsilon^{3-D}G(\log_3 \varepsilon^{-1}) + \Big(6 - \frac{6}{17}\Big)\varepsilon + \Big(3\pi + \frac{12}{23}\Big)\varepsilon^2 + \Big(\frac{4\pi}{3} - \frac{8}{25}\Big)\varepsilon^3, \tag{3.205}$$

where $D = \log_3 26$ is the Minkowski dimension of A and G is a continuous, nonconstant 1-periodic function which is bounded away from zero and infinity, and is given by the following pointwise (absolutely convergent and hence) convergent Fourier series expansion:

$$G(x) = \frac{24}{3\log 3}\sum_{k\in\mathbb{Z}} \frac{2^{-s_k}}{(3-s_k)(s_k-1)(s_k-2)s_k} e^{i2\pi kx}, \text{ for all } x \in \mathbb{R}, \tag{3.206}$$

with $(s_k := D + ik\mathbf{p})_{k\in\mathbb{Z}}$ denoting the sequence of principal complex dimensions of A and $\mathbf{p} := 2\pi/\log 3$ denoting the oscillatory period of A.

EXERCISE 3.31. (a) (*3-d carpet, revisited*). For the Sierpinski 3-d carpet A in Example 3.30, answer the analog of questions (ii) and (iii) of Exercise 3.29.

(b) (*3-gasket RFD*). For the relative (or inhomogeneous) Sierpinski 3-gasket (A_3, Ω_3) studied in §3.4.4 (specialized to $N = 3$), use (3.191), along with the $N = 3$ case of (3.82) and (3.89) (with $g_3 = g_3(s)$ given by formula (3.94) of Exercise 3.9), in order to obtain a pointwise, exact fractal tube formula for (A_3, Ω_3). (Recall from the discussion following (3.92) that $D = \log_2 4 = 2$ is a complex dimension of (A_3, Ω_3) of multiplicity *two*, whereas the other complex dimensions are simple.)

Show via a direct computation that the RFD (A_3, Ω_3) is Minkowski degenerate and therefore not Minkowski measurable but that with respect to the gauge function $h(t) := \log(t^{-1})$, for all $t \in (0, 1)$, (A_3, Ω_3) is h-Minkowski measurable (and hence also h-Minkowski nondegenerate), as was stated towards the end of §3.4.4. In order to establish the latter facts, you may use the results from [**LapRaZu6**] and [**LapRaZu1**, §5.4.4] briefly discussed towards the end of §3.5.2.

EXAMPLE 3.32. (*The* $\frac{1}{2}$*-square and* $\frac{1}{3}$*-square fractals*). We revisit and complete here part (a) (the $\frac{1}{2}$-square fractal) and part (b) (the $\frac{1}{3}$-square fractal) of §3.4.5.

(a) (*The* $\frac{1}{2}$*-square fractal*). Recall from part (a) of §3.4.5 that all of the fractal complex dimensions of the $\frac{1}{2}$-square fractal A are simple, except for $s=1$, which is equal to $\dim_B A$. As a result, it follows from (3.191) in light of (3.98), (3.102) and (3.103) that the following pointwise, exact fractal tube formula holds (for all $\varepsilon \in (0,1/2)$):

$$V_A(\varepsilon) = \frac{1}{4\log 2}\varepsilon \log \varepsilon^{-1} + \varepsilon G(\log_2(4\varepsilon)^{-1}) + \frac{1+2\pi}{2}\varepsilon^2, \tag{3.207}$$

where G is a nonconstant, continuous 1-periodic function which is bounded away from zero and infinity and is given by the following convergent (because absolutely convergent) Fourier expansion:

$$G(x) := \frac{29\log 2 - 4}{8\log 2} + \frac{1}{4}\sum_{k\in\mathbb{Z}\setminus\{0\}} \frac{e^{2\pi i k x}}{(2-s_k)(s_k-1)s_k}, \quad \text{for all } x \in \mathbb{R}, \tag{3.208}$$

with $s_k := ik\mathbf{p}$, for all $k \in \mathbb{Z}\setminus\{0\}$, and $\mathbf{p} := (2\pi/\log 2)$, the oscillatory period of A.

Let us briefly explain how to obtain (3.207) and (3.208). In light of (3.192) (applied with $R(\varepsilon)\equiv 0$ because we are in the strongly d-languid case),

$$\begin{aligned} V_A(\varepsilon) &= \sum_{\omega\in\mathcal{D}(\zeta_A)} \operatorname{res}\left(\frac{\varepsilon^{2-s}}{2-s}\zeta_A(s),\omega\right) \\ &= \operatorname{res}\left(\frac{\varepsilon^{2-s}}{2-s}\zeta_A(s),1\right) + \sum_{\omega\in\mathcal{D}(\zeta_A)\setminus\{1\}} \operatorname{res}(\zeta_A(s),\omega)\frac{\varepsilon^{2-\omega}}{2-\omega}. \end{aligned} \tag{3.209}$$

In order to calculate the above residue at $s=1$ in the last equality of (3.209), one computes the Laurent series expansion of ζ_A around $s=1$ (which is a double pole of ζ_A), as follows:

$$\zeta_A(s) = \frac{d_{-2}}{(s-1)^2} + \frac{d_{-1}}{s-1} + O(1) \quad \text{as } s \to 1, \tag{3.210}$$

with

$$d_{-2} := \frac{1}{4\log 2} \quad \text{and} \quad d_{-1} := \frac{29\log 2 - 2}{8\log 2}. \tag{3.211}$$

One then deduces from combining (3.210) and (3.211) that

$$\begin{aligned} \operatorname{res}\left(\frac{\varepsilon^{2-s}}{2-s}\zeta_A(s),1\right) &= \varepsilon(d_{-1} - d_{-2} + d_{-2}\log\varepsilon^{-1}) \\ &= \frac{1}{4\log 2}\varepsilon\log\varepsilon^{-1} + \frac{29\log 2 - 4}{8\log 2}. \end{aligned} \tag{3.212}$$

Finally, in light of (3.212) and the expression of $\operatorname{res}(\zeta_A(s),s_k)$ for any $k \in \mathbb{Z}\setminus\{0\}$ given in (3.103), as well as of the value of $\operatorname{res}(\zeta_A(s),0)$ obtained in the just mentioned equation, we deduce the exact tube formula (3.207) from (3.209).

Next, recalling from (3.100) that $\dim_B A = 1$, we can easily deduce from the fractal tube formula (3.207) that A is Minkowski degenerate with Minkowski content $\mathcal{M} = +\infty$. In particular, A is not Minkowski measurable, in the usual sense.

Moreover, it follows from the h-Minkowski measurability criterion discussed at the end of §3.5.2, that for the choice of the gauge function $h(t) := \log t^{-1}$ (for all $t \in (0,1)$), A is h-Minkowski measurable with h-Minkowski content $\mathcal{M}(A,h)$ given by

$$\mathcal{M}(A,h) = \frac{1}{4\log 2}. \tag{3.213}$$

This concludes the discussion of the $\frac{1}{2}$-square fractal, for now.

(b) (*The $\frac{1}{3}$-square fractal*). Let us next briefly consider the $\frac{1}{3}$-square fractal $A \subseteq \mathbb{R}^2$ studied in part (*b*) of §3.4.5. Recall from that discussion (see, especially, (3.107), (3.109) and (3.112)) that $\dim_B A = 1, \dim_{PC} A = \{1\}$, and all of the complex dimensions of A are simple with

$$F \cup \{0,1\} \subseteq \mathcal{D}_A \subseteq \{0,1\} \cup \{s_k := \log_3 2 + i\mathbf{p}k : k \in \mathbb{Z}\}, \tag{3.214}$$

where F is a nonempty finite subset of $\log_3 2 + i\mathbf{p}\mathbb{Z}$ containing $\log_3 2$ as well as several nonreal complex dimensions (and conjectured to be infinite). Here and henceforth, $\mathbf{p} := 2\pi/\log 3$.

Then, in light of the exact fractal tube formula (in the case of simple poles) stated in (3.191), combined with (3.214) and (3.109)–(3.112), we obtain the following pointwise, exact fractal tube formula for A (valid for all $\varepsilon \in (0, 1/\sqrt{2})$):

$$V_A(\varepsilon) = 16\varepsilon + \varepsilon^{2-\log_3 2} G(\log_3(3\varepsilon)^{-1}) + \frac{12+\pi}{2}\varepsilon^2, \tag{3.215}$$

where G is a nonconstant, continuous 1-periodic function which is bounded away from zero infinity and is given by the following absolutely convergent (and hence, pointwise convergent) Fourier series, for all $x \in \mathbb{R}$:

$$G(x) := \frac{1}{\log 3} \sum_{k\in\mathbb{Z}} \frac{e^{2\pi i k x}}{(2-s_k)s_k} \Big(\frac{6}{s_k - 1} + \Psi(s_k) \Big), \tag{3.216}$$

where $\Psi = \Psi(s)$ is the entire function occurring in (3.105) and (3.111).

Finally, it follows from (3.215) that A is Minkowski measurable with Minkowski content (in the usual sense) given by $\mathcal{M} = 16$. This concludes for now our discussion of the $\frac{1}{3}$-square fractal.

EXERCISE 3.33. (*a*) (*Cantor grill*). Use the results of §3.4.7 to calculate the residues of ζ_A at the complex dimensions, and then to obtain a (pointwise, exact) fractal tube formula for the Cantor grill $A = C \times [0,1]$, where C is the ternary Cantor set.

(*b*) (*Cantor dust*). Answer an analogous question for the Cantor dust $A = C \times C$ studied in §3.4.8. Then, extend your result to $A = C^d$, where C^d is the Cartesian product of d copies of the Cantor set C, with $d \geq 2$.

[*Caution*: Question (*b*) is more difficult than question (*a*).]

EXAMPLE 3.34. (*The Cantor graph RFD*). Let (A,Ω) denote the Cantor graph RFD (in $\mathbb{R}^2$) described and studied in §3.4.9. Recall that the compact set $A \subseteq \mathbb{R}^2$ is the graph of the Cantor function (also called the devil's staircase). Then, in light of (3.154)–(3.155), $D_{A,\Omega} = D_A = 1$ and

$$\mathcal{D}_{A,\Omega} = \{0,1\} \cup \{s_k := \log_3 2 + ik\mathbf{p} : k \in \mathbb{Z}\}, \tag{3.217}$$

with $\mathbf{p} := 2\pi/\log 3$ and each complex dimension being simple. We therefore deduce from (3.191) the following pointwise, exact fractal tube formula, valid for every $\varepsilon \in (0,1)$ (where we use the values of the residues of ζ_A given in (3.156)–(3.157)):

$$\begin{aligned} V_{A,\Omega}(\varepsilon) &= 2\varepsilon + \varepsilon^{2-\log_3 2} G(\log_3 \varepsilon^{-1}) + \varepsilon^2 \\ &= 2\varepsilon^{2-D_{A,\Omega}} + \varepsilon^{2-D_{CS}} G(\log_3 \varepsilon^{-1}), \end{aligned} \tag{3.218}$$

where (as above) $D_{A,\Omega} = D_A = 1$, the dimension of the Cantor graph, and $D_{CS} = D_C = \log_3 2$, the dimension of the Cantor set (or of the Cantor string). Furthermore, in (3.218), G is a nonconstant, continuous 1-periodic function which is bounded away from zero and infinity, and is given by the following absolutely convergent (and hence, pointwise convergent) Fourier series expansion, for all $x \in \mathbb{R}$:

$$G(x) := \frac{1}{\log 3} \sum_{k \in \mathbb{Z}} \frac{e^{2\pi i k x}}{(2 - s_k)(s_k - 1)s_k}. \tag{3.219}$$

Finally, it easily follows from (3.218)–(3.219) that (A, Ω) is Minkowski measurable, with Minkowski content given by

$$\mathcal{M}_{A,\Omega} = \frac{\operatorname{res}(\zeta_{A,\Omega}, 1)}{2 - 1} = 2, \tag{3.220}$$

where we have used (3.157) in the second equality. Exactly the same property and identity as in (3.220) holds for the Cantor graph A instead of for (A, Ω), as the interested reader can verify; in particular, $\mathcal{M}_A = 2$. This concludes our discussion of the Cantor graph RFD for now. We will return to this example (and to the associated Cantor graph) in §3.6, when discussing the notion of fractality; see the text surrounding (3.239)–(3.245).

EXERCISE 3.35. Directly calculate the length of the Cantor graph (i.e., of the devil's staircase) A and compare your result with the value of the Minkowski content of A given above. Furthermore, give a heuristic, geometric argument that enables you to guess the length of A without any computation.

EXAMPLE 3.36. (*Self-similar sprays*). We briefly discuss the important class of self-similar sprays, studied in §3.4.10 above. Since this discussion could be quite lengthy, otherwise, we refer to [**LapRaZu1**, §5.5.6] for the details.

Let (A, Ω) be a self-similar spray with scaling ratios $r_1, \cdots, r_J$ (with $J \geq 2$) and (for simplicity, but without loss of generality) with a simple generator G (or rather, generating or base RFD $(\partial G, G)$), as in §3.4.10. Also as in §3.4.10, we assume that G is a (nonempty) bounded open subset of $\mathbb{R}^N$, with $D_{\partial G, G} = \dim_B(\partial G, G) < N$, and that $\sum_{j=1}^J r_j^N < \infty$, so that the fractal spray has finite total volume.

Recall that $\zeta_{A,\Omega}$ is then given by the key factorization formula (3.165) or (3.166), expressing $\zeta_{A,\Omega}$ in terms of the distance zeta function $\zeta_{\partial G, G}$ of the generating RFD and of the scaling zeta function $\zeta_{\mathfrak{s}}$ of the spray. Namely,

$$\zeta_{A,\Omega}(s) = \zeta_{\mathfrak{s}}(s) \cdot \zeta_{\partial G, G}(s) = \frac{\zeta_{\partial G, G}(s)}{1 - \sum_{j=1}^J r_j^s}. \tag{3.221}$$

Furthermore, in light of (3.170), (3.173)–(3.174) and (3.176),

$$D_{A,\Omega} = \max(\sigma_0, D_{\partial G, G}), \tag{3.222}$$

where $\sigma_0 = D(\zeta_{\mathfrak{s}}) \in (0, N)$ (the similarity dimension of the spray) is the unique real solution of the Moran equation [**Mora**]; i.e., $\sigma_0 \in \mathbb{R}$ and $\sum_{j=1}^J r_j^{\sigma_0} = 1$. Moreover, in light of (3.173)–(3.174) and (3.176), if we assume that the generator G is sufficiently "nice" [e.g., G is monophase, in the sense of [LapPe1–2, LapPeWi1–2][114] and, in particular, if ∂G is a nontrivial polytope (by a result in [**KoRati**])], we have that

$$\mathcal{D}_{A,\Omega} = \mathcal{D}_{\partial G,G} \cup \mathcal{D}_{\mathfrak{s}} \subseteq \{0,1,\cdots,N-1\} \cup \left\{ s \in \mathbb{C} : \sum_{j=1}^{J} r_j^s = 1 \right\}, \tag{3.223}$$

with frequent or "typical" equality in (3.223) and with $\mathcal{D}_{A,\Omega}$ always containing $D_{A,\Omega}$ and (by the results in [**Lap-vF4**, Thm. 3]) also containing infinitely many (scaling) complex dimensions with real part σ_0.

We then deduce from (3.192) the following exact, pointwise fractal tube formula:

$$V_{A,\Omega}(\varepsilon) = \sum_{\omega \in \mathcal{D}_{\mathfrak{s}} \cup \{0,1,\cdots,N-1\}} \operatorname{res}\left(\frac{\varepsilon^{N-s}\left(\sum_{\alpha=0}^{N} \kappa_\alpha \frac{g^{s-\alpha}}{s-\alpha}\right)}{(N-s)(1-\sum_{j=1}^{J} r_j^s)}, \omega \right). \tag{3.224}$$

Here, g is the inner radius of the generator G and the coefficients κ_α (some of which could vanish) are the coefficients of the polynomial expansion of $V_{\partial G,G}(\varepsilon)$.[115]

In the important special case when all of the scaling complex dimensions are simple and when σ_0 is not an integer, the fractal tube formula (3.224) takes the following simpler form:

$$V_{A,\Omega}(\varepsilon) = \sum_{\omega \in \mathcal{D}_{\mathfrak{s}} \cup \{0,1,\cdots,N-1\}} d_\omega \varepsilon^{N-\omega}, \tag{3.226}$$

where

$$d_\omega = \operatorname{res}(\zeta_{\mathfrak{s}}, \omega) \left(\sum_{\alpha=0}^{N} \frac{\kappa_\alpha g^{\omega-\alpha}}{\omega-\alpha} \right), \quad \text{if } \omega \in \mathcal{D}_{\mathfrak{s}}, \tag{3.227}$$

and

$$d_\omega = \zeta_{\mathfrak{s}}(\omega)\kappa_\omega, \quad \text{if } \omega \in \{0,1,\cdots,N-1\}. \tag{3.228}$$

Therefore,

$$V_{A,\Omega}(\varepsilon) = \sum_{\omega \in \mathcal{D}_{\mathfrak{s}}} \operatorname{res}(\zeta_{\mathfrak{s}}, \omega) \left(\sum_{\alpha=0}^{N} \frac{\kappa_\alpha g^{\omega-\alpha}}{\omega-\alpha} \right) + \sum_{\alpha=0}^{N-1} \zeta_{\mathfrak{s}}(\alpha)\kappa_\alpha \varepsilon^{N-\alpha}. \tag{3.229}$$

We note that if the generator G is pluriphase[116] instead of monophase, then we can easily extend the above results and at the same time recover (as well as significantly extend) the results of [**LapPe2**] and, especially, of [**LapPeWi1**]. Also, even if G is not pluriphase (and thus certainly not monophase), it is clear from (3.221) that under suitable polynomial-type growth assumptions on $\zeta_{\partial G,G}$, one can use the tube formulas (from [**LapRaZu6**] and [**LapRaZu1**, §§5.1–5.3]) recalled in §3.5.1 in order to obtain pointwise or distributional fractal tube formulas (with or

[114]Roughly, this means that $V_{\partial G,G}(\varepsilon)$ is polynomial for all ε sufficiently small.

[115]More specifically,

$$V_{\partial G,G}(\varepsilon) = \sum_{\alpha=0}^{N-1} \kappa_\alpha t^{N-\alpha}, \quad \text{for } 0 < \varepsilon < g. \tag{3.225}$$

[116]That is, roughly speaking, if $V_{\partial G,G}(\varepsilon)$ is a piecewise polynomial function; see [**LapPe2**, **LapPeWi1**].

without error term) for $(\partial G, G)$, expressed via $\zeta_{\partial G, G}$. This is so even if the fractal spray RFD (A, Ω) is not necessarily self-similar. (Indeed, in that general case, the first equality in (3.221) still holds.) We let the interested reader elaborate on the latter comments.

As was already pointed out in §3.4.10, in the present case of self-similar sprays (and unlike for ordinary self-similar sets A for which one always has $D_A = \sigma_0$), we must distinguish three cases here, all of which are realized (see, e.g., the example of the relative Sierpinski N-gasket discussed in §3.4.4 and at the end of §3.6.1):[117]

Case (i): $D_G < \sigma_0$. Then, by (3.222), $D_{A,\Omega} = \sigma_0$ and is simple. Hence,

$$\dim_{PC}(A, \Omega) = \mathcal{D}_{\mathfrak{s}} = \left\{ s \in \mathbb{C} : \sum_{j=1}^{J} r_j^s = 1 \right\}. \tag{3.230}$$

Consequently, in light of the structure of the scaling complex dimensions of self-similar sprays given in §3.4.10 (and based on the results of [**Lap-vF4**, Ch. 3]), we obtain the precise counterpart of the Minkowski measurablity criterion of self-similar strings from [**Lap-vF4**, §8.4] (see part (d) of Remark 2.4). Namely, *the self-similar spray RFD (A, Ω) is Minkowski measurable if and only if it is nonlattice; that is, if and only if it does not have any nonreal principal complex dimensions. Hence, if (A, Ω) is lattice, it is not Minkowski measurable, whereas if it is nonlattice, it is Minkowski measurable.* This result follows from the refined version of the Minkowski measurability criterion stated in §3.5.2, and more precisely, from the necessary (respectively, sufficient) condition for Minkowski measurability obtained in [**LapRaZu1**, §5.4] and in [**LapRaZu6**] (and briefly alluded to in §3.5.2).[118]

Case (ii): $D_G = \sigma_0$ (and hence, in light of (3.222), $D_{A,\Omega} = D_G = \sigma_0$ is an integer). Then, as was noted in §3.4.10, it follows from the factorization formula (3.221) that $D_{A,\Omega}$ is a complex dimension of (A, Ω) of multiplicity two. As a result, due to the Minkowski measurability criterion discussed in §3.5.2, (A, Ω) cannot be Minkowski measurable (in the usual sense), irrespective of whether (A, Ω) is lattice or nonlattice. However, if we use the gauge function $h(t) := \log t^{-1}$, for all $t \in (0, 1)$, then provided its hypotheses are satisfied, the h-Minkowski measurability criterion briefly discussed towards the end of §3.5.2 ([**LapRaZu1**, Thm. 5.4.32]), the RFD (A, Ω) is h-Minkowski measurable.

Case (iii): $D_G > \sigma_0$. Then, by (3.222), $D_{A,G} = D_G$. Since all of the scaling complex dimensions of (A, Ω) have real parts not exceeding σ_0 and hence, strictly less than $D_{A,\Omega}$, we deduce from (3.223) that $D_{A,\Omega} = D_G$ is the only principal complex dimension of (A, Ω) and (since D_G is a simple pole of $\zeta_{\partial G, G}$) that it is simple. Consequently, *in this case (i.e., in case (iii)), (A, Ω) is always Minkowski measurable*, whether or not the self-similar spray (A, Ω) is lattice or nonlattice.

However, this does not preclude (A, Ω) from having lower-order oscillations in its geometry (e.g., in its fractal tube formula (3.229)). This is indeed what happens (generically) for the relative Sierpinski N-gasket when $N \geq 4$.

---

[117]Henceforth, we use the above notation and write $D_G := D_{\partial G, G} = \dim_B(\partial G, G)$.

[118]*Caution*: The hypotheses of the Minkowski measurability criterion stated in §3.5.2 are not always satisfied in the nonlattice case (see [**Lap-vF4**, Exple. 5.32] for a counterexample). This is why we have to use a refined form (the aforementioned sufficient condition) in order to establish the Minkowski measurability of (A, Ω) in the nonlattice case. On the other hand, in the lattice case, the hypotheses of the above criterion are clearly satisfied and therefore we can conclude that (A, Ω) is not Minkowski measurable; alternatively, one can use the aforementioned sufficient condition (see [**LapRaZu1**, Thm. 5.4.15]) in order to reach the same conclusion.

EXERCISE 3.37. (*Relative N-gasket*). By using the trichotomy outlined in cases (i)–(iii) just above, obtain (as explicitly as possible) fractal tube formulas for the relative (or inhomogeneous) Sierpinski N-gasket studied in §3.4.4. Also, determine the Minkowski measurability (or, more generally and when necessary, the h-Minkowski measurability) of the relative N-gasket, depending on the value of N. When appropriate, calculate the corresponding Minkowski content $\mathcal{M}$ or the average Minkowski content $\widetilde{\mathcal{M}}$.

[*Hint*: Distinguish the three cases when $N = 2, N = 3$ and $N \geq 4$, respectively.]

REMARK 3.38. (*Ordinary self-similar sets.*) (a) It has long been conjectured by the author (see [**Lap3**, Conj. 3, p. 163]) that (classic or homogeneous) self-similar sets in $\mathbb{R}^N$ satisfying the open set condition (in the sense of [**Hut**]; see also [**Fa1**]) are Minkowski measurable if and only if they are nonlattice (and, equivalently, are not Minkowski measurable if and only if they are lattice). When $N = 1$ (i.e., for self-similar strings), the fact that nonlattice self-similar sets (i.e., strings) are Minkowski measurable was first proved by the author in [**Lap3**] and then, independently, by K. Falconer in [**Fa2**], in both cases by using the renewal theorem (first used in a related context by S. Lalley in [Lall1–3]). Then, this result was extended to higher dimensions (and to certain random fractals, as was also conjectured in [**Lap3**]) by D. Gatzouras in [**Gat**]. There remained to prove that the nonlattice condition was also necessary for obtaining the Minkowski measurability of a given self-similar set. This was first established when $N = 1$ (i.e., for self-similar strings) by the author and M. van Frankenhuijsen in [**Lap-vF2**] (see [**Lap-vF4**, §8.4]), where both the necessary and sufficient conditions were proved by using the theory of complex dimensions of fractal strings (combined with a suitable Tauberian theorem) and the associated fractal tube formulas; see [**Lap-vF4**, §8.4, Thms. 8.23 and 8.36]. Finally, in higher dimensions (i.e., when $N \geq 2$), the sufficient condition was recently established (independently of the above results from [**LapRaZu1**] and the accompanying papers about self-similar sprays or RFDs) by S. Kombrink, E. Pearse and S. Winter in [**KomPeWi**], also by using the renewal theorem but now combined with several nice new observations.

(b) We conjecture that under suitable hypotheses and still for (classic or homogeneous) self-similar sets satisfying the open set condition, the above characterization expressed in terms of complex dimensions is still valid; that is, the presence of a nonreal principal complex dimension is equivalent to the self-similar set being not Minkowski measurable (and hence, Minkowski measurability is equivalent to the Minkowski dimension D being the only principal complex dimension; see [**LapRaZu1**, Pb. 6.2.36] for more details. This problem still remains open, for now, and its resolution will require, in particular, suitably extending the factorization formula (3.221) to this setting or finding an appropriate substitute for it.

3.6. Fractality, hyperfractality and unreality, revisited. We pursue and complete here the discussion of fractality and unreality (as well as of the closely related topic of the meaning of complex dimensions) started in §2.4 and §2.5. In light of the higher-dimensional theory of complex dimensions developed in [**LapRaZu1**] (and in the accompanying series of papers, [LapRaZu2–10]) and, in particular, of the fractal tube formulas obtained in [**LapRaZu1**, Ch. 5] and [**LapRaZu6**] (as well as discussed in §3.5.1 and §3.5.3), the interpretation of (necessarily complex conjugate pairs of) *nonreal* complex dimensions as giving rise to (or detecting) the *intrinsic oscillations* of a given geometric object is exactly the same as in the one-dimensional case of fractal strings discussed in §2.4. Therefore, we refer the interested reader to §2.4 for the corresponding discussion, which can easily be adapted to higher dimensions and is illustrated by the many examples of complex dimensions and fractal tube formulas provided in §3.4 and §3.5.3, respectively.

We further mention that for clarity, we will focus here on geometric objects which are relative fractal drums or RFDs (and, in particular, bounded sets) in $\mathbb{R}^N$. However, since the entire theory of fractal zeta functions and complex dimensions (in particular, of the associated fractal tube formulas) extends (essentially without change) to suitable metric measure spaces (namely, Ahlfors-type spaces and beyond), as is shown in [**LapWat**, **Wat**], we could instead work within the much greater generality of RFDs (and, in particular, of bounded sets) in such metric measure spaces. (See Remark 3.25.) The main difference is that the embedding dimension N would have to be replaced by the Ahlfors dimension (or its appropriate analog) of the embedding metric measure space; see footnote 109.

Perhaps more importantly, we could also apply to various counting functions the very general (pointwise or distributional) explicit formulas from [**Lap-vF4**, Ch. 5] in order to detect the potential geometric, spectral, dynamical, or arithmetic oscillations that are intrinsic to fractal-like (physical or mathematical) objects. These counting functions could be geometric (such as the box-counting function), spectral (such as the eigenvalue or frequency counting function or else, the partition function or trace of the heat semigroup) or dynamical (such as the prime orbit counting function, counting the number of homology classes of primitive periodic orbits of the corresponding dynamical system[119]).

Moreover, in a similar spirit, we could use other types of fractal-like zeta functions (such as, for example, suitably weighted Ruelle's dynamical zeta functions; see, e.g., [Rue1–4, ParrPol1–2], [**Lag**], along with footnote 119) could be used to define "fractality" via the existence of nonreal complex dimensions. Finally, as will be further explained below, even the notion of "complex dimensions" itself can be relaxed, by considering (nonremovable) singularities that are not just poles of the associated fractal zeta functions.

Accordingly, the proposed notion of fractality can be potentially applied to a great variety of settings in mathematics, physics, cosmology, chemistry, biology, medicine, geology, computer science, engineering, economics, finance, and the arts, as well as can be illustrated via many different kinds of fractal-type explicit formulas. In mathematics or physics alone, the corresponding fields involved would include, for instance, harmonic analysis, partial differential equations, geometric measure theory, spectral theory, spectral geometry, probability theory, dynamical systems, combinatorics and graph theory, number theory and arithmetic geometry, algebraic geometry, operator algebras and noncommutative geometry, quantum groups, mathematical physics, along with, on the more physical side, condensed matter physics, astronomy and cosmology, quantum theory and its myriad of applications, quantum gravity, classical and quantum chaos, quantum computing and

[119]See [**Lap-vF4**, Ch. 7] for a simple but illuminating example. The author has long thought that the just referred work could be greatly extended to a variety of hyperbolic and other dynamical systems, for example, within the setting of the theory developed by Parry and Pollicott in [**ParrPol1**, **ParrPol2**], where Ruelle or dynamical zeta functions ([Rue1–4], [**Lag**]) were used. This potentially significant extension still remains to be achieved.

string theory. For the simplicity of exposition, however, and with one single exception,[120] we will limit our discussion to the geometric setting, that of RFDs (and in particular, of bounded sets) in an N-dimensional Euclidean space $\mathbb{R}^N$, as well as briefly illustrate it by means of the fractal tube formulas and the Minkowski measurability results described in §§3.5.1–3.5.3.

As in §2.5, we say that a geometric object (e.g., a fractal drum (A,Ω) in $\mathbb{R}^N$ or, in particular, a bounded subset A of $\mathbb{R}^N$) is *fractal* if it has at least one nonreal complex dimension (and hence, at least one pair of nonreal complex conjugate complex dimensions). For now, "complex dimensions" are interpreted as being the (visible) poles of the associated fractal zeta function $\zeta_{A,\Omega}$ or, equivalently, $\widetilde{\zeta}_{A,\Omega}$ of (A,Ω).[121] However, we will further broaden this notion later on in this subsection; see, especially, §3.6.2 and §3.6.3. This definition is formally identical to the one proposed in [**Lap-vF4**, Ch. 12] (and, prior to that, in [Lap-vF1–3]),[122] but it is worth pointing out that at the time, there was no suitable general definition of fractal zeta functions, and hence also of complex dimensions, that was compatible with the existence of fractal tube formulas for compact subsets of $\mathbb{R}^N$, with $N \geq 2$.

More specifically, given $d \in \mathbb{R}$, an RFD (A,Ω) (and, in particular, a bounded set A) in $\mathbb{R}^N$ is said to be *fractal in dimension* d if it has at least one (and hence, a pair of complex conjugate) complex dimension(s) of real part d.[123] Therefore, by definition, fractality is equivalent to fractality in some dimension $d \in \mathbb{R}$.

Moreover, (A,Ω) (or A) is said to be *critically fractal* if it is fractal in dimension D, the (upper) Minkowski dimension of (A,Ω) (or of A); in other words, if and only if it has at least one nonreal complex dimension. Otherwise (A,Ω) (or A) is said to be *subcritically fractal*; in that case, it is therefore fractal in some dimension $d < D$ but not in dimension D.

In addition, the RFD (A,Ω) is said to be *hyperfractal* if the associated fractal zeta function $\zeta_{A,\Omega}$ (or equivalently, $\widetilde{\zeta}_{A,\Omega}$, since $D < N$ here) cannot be meromorphically extended to a connected open neighborhood of a suitable curve or contour S in $\mathbb{C}$ extending in both vertical directions (i.e., S is a screen, in the sense of [**Lap-vF4**] or [**LapRaZu1**]), and it is said to be *critically hyperfractal* (or *strongly hyperfractal*, as in [**LapRaZu1**]) if $S = \{Re(s) = D\}$, and *maximally hyperfractal* if the critical line $\{Re(s) = D\}$ consists solely of (nonisolated and nonremovable) singularities of $\zeta_{A,\Omega}$ (or equivalently, of $\widetilde{\zeta}_{A,\Omega}$).[124] (Naturally, a similar terminology is used in the special case of a bounded subset A of $\mathbb{R}^N$.)

[120]This exception will have to do with the spectra of fractal drums; see §3.6.4.

[121]For simplicity, we assume from now on that $D := \overline{\dim}_B(A,\Omega) < N$, so that the distance and tube zeta functions $\zeta_{A,\Omega}$ and $\widetilde{\zeta}_{A,\Omega}$ have the same visible poles in a given domain U of $\mathbb{C}$ to which one (and hence, both) of these fractal zeta function has a (necessarily unique) meromorphic extension; in particular, $D(\zeta_{A,\Omega}) = D(\widetilde{\zeta}_{A,\Omega}) = D$. (See §3.3.1 and §3.3.2.) If one wants to deal with the case when $D = N$, one should then work with $\widetilde{\zeta}_{A,\Omega}$ alone.

[122]There is one small difference; namely, we no longer require the real part of the nonreal complex dimension to be positive (a condition that was included mainly for aesthetic reasons and is fulfilled, for example, by classic self-similar geometries). Indeed, for an RFD (A,Ω), even $D = \overline{\dim}_B(A,\Omega)$ can be negative (or equal to $-\infty$); see the two examples provided between (3.4) and (3.7) in §3.2.

[123]Then, clearly, we must have $D > -\infty$ and $d \leq D \leq N$, where $D := \overline{\dim}_B(A,\Omega)$ or $(D := \overline{\dim}_B A)$. Also, note that d itself need not be a (nonremovable) singularity, let alone a pole, of $\zeta_{A,\Omega}$ (or of $\widetilde{\zeta}_{A,\Omega}$).

[124]Precise definitions of the notion of a singularity (of a complex-valued function on a domain of $\mathbb{C}$) can be found in [**LapRaZu1**, §1.3.2] and the references therein; see also [**LapRaZu1**, §4.6.3].

Clearly, "maximally hyperfractal" implies "critically hyperfractal", which itself implies "hyperfractal". Furthermore, maximal hyperfractals are such that $\zeta_{A,\Omega}$ (or equivalently, $\widetilde{\zeta}_{A,\Omega}$) have nonreal (complex conjugate pairs of) singularities (which are not necessarily poles). Moreover, in some sense, they are among the most complicated fractals.

In [**LapRaZu1**, §4.6] (and [**LapRaZu3**]) are constructed maximally hyperfractal compact sets as well as RFDs in $\mathbb{R}^N$, for $N \geq 1$ arbitrary (and, in particular, when $N = 1$, maximally hyperfractal fractal strings), with any prescribed (upper) Minkowski dimension $D \in (0, N)$. In fact, the family of examples constructed in *loc. cit.* consists not only of maximally hyperfractal but also of transcendentally ∞-quasiperiodic sets or RFDs. Recall from [**LapRaZu1**, §4.6] that an RFD (A, Ω) (or a bounded set A) in $\mathbb{R}^N$, of (upper) Minkowski dimension D is said to be *transcendentally ∞-quasiperiodic* if its tube function $V = V(\varepsilon)$ (where $V := V_{A,\Omega}$ or $V := V_A$) satisfies

$$V(\varepsilon) = \varepsilon^{N-D}(G(\log \varepsilon^{-1}) + o(1)) \quad \text{as } \varepsilon \to 0^+, \tag{3.231}$$

where the function $G : \mathbb{R} \to \mathbb{R}$ is *transcendentally ∞-quasiperiodic*; i.e., where $G = G(t)$ is the restriction to the diagonal of a nonconstant function $H : \mathbb{R}^\infty \to \mathbb{R}$ which is separately periodic (of minimal period T_j) in each variable t_j, for $j = 1, 2, \cdots$:

$$G(t) = H(t, t, t, \cdots), \quad \text{for all } t \in \mathbb{R}, \tag{3.232}$$

where (for all $(t_1, t_2, \cdots) \in \mathbb{R}^\infty$)

$$H(t_1, t_2, \cdots, t_{j-1}, t_j + T_j, t_{j+1}, \cdots) = H(t_1, t_2, \cdots, t_{j-1}, t_j, t_{j+1}, \cdots). \tag{3.233}$$

In addition, the use of the adverb "transcendentally" in the above definition means that the resulting sequence of *quasiperiods* $(T_j)_{j \geq 1}$ is linearly independent over the field of algebraic numbers.[125]

The aforementioned construction of maximally hyperfractal and transcendentally ∞-quasiperiodic RFDs is rather complicated. Its first (and main) step consists in constructing fractal strings with the above properties. In essence, these highly complex fractal strings are obtained by taking the countable disjoint union of a suitable sequence extracted from a two parameter family of topological Cantor sets (or strings), and then to appropriately apply a key result from transcendental number theory (namely, Baker's theorem [**Bak**]; see also [**LapRaZu1**, Thm. 3.114]).[126] In some sense, this construction can be viewed as a fractal-geometric interpretation of Baker's theorem.

We close this discussion by providing several examples of critical and subcritical fractals.

First, we note that under suitable hypotheses (namely, the hypotheses of the Minkowski measurability criterion stated in §3.5.2), and supposing that the Minkow-

[125]Recall that the field of *algebraic numbers* can be viewed (up to isomorphism) as the algebraic closure $\overline{\mathbb{Q}}$ of $\mathbb{Q}$, the field of rational numbers. It is obtained by adjoining to $\mathbb{Q}$ the complex roots of all of the monic polynomials with coefficients in $\mathbb{Q}$. By reasoning by absurdum, one can easily check that it is a countable set.

[126]Recall that Baker's theorem states that given $n \in \mathbb{N}$ with $n \geq 2$, if $m_1, \cdots, m_n$ are positive algebraic numbers such that $\log m_1, \cdots, \log m_n$ are linearly independent over the rationals, then $1, \log m_1, \cdots, \log m_n$ are linearly independent over the field of algebraic numbers (i.e., algebraically independent). In particular, $\log m_1, \cdots, \log m_n$ are transcendental (i.e., not algebraic) numbers, and so are their pairwise quotients.

ski dimension $D = \dim_B(A, \Omega)$ exists and is simple,[127] the RFD (A, Ω) in $\mathbb{R}^N$ is subcritically fractal if and only if it is Minkowski measurable. Stated another way, and still under the hypotheses of the aforementioned criterion, (A, Ω) is critically fractal if and only if it is not Minkowski measurable. Naturally, the same statements hold for bounded subsets A of $\mathbb{R}^N$.

Moreover, self-similar strings are subcritically fractal if and only if they are nonlattice, and also if and only if they are Minkowski measurable. (See [**Lap-vF4**, §8.4] and part (d) of Remark 2.4.)[128] Equivalently, self-similar strings are critically fractal if and only if they are lattice, and also if and only if they are not Minkowski measurable (even though they are always Minkowski nondegenerate).

As was mentioned in Example 3.36, exactly the same statements (as just above for self-similar strings) hold for self-similar fractal spray RFDs (in $\mathbb{R}^N, N \geq 1$) with "nice" generators (e.g., monophase or even pluriphase generators, and in particular, with generators that are nontrivial polytopes), in case (i) of Example 3.36; i.e., when $D_G := \dim_B(\partial G, G) < \sigma_0$, the similarity dimension of (A, Ω), where $(\partial G, G)$ is the generator (or base) of the self-similar spray RFD (A, Ω).

In addition, irrespective of whether we are in case (i), (ii) or (iii) of Example 3.36, it follows from the results of [**Lap-vF4**, Ch. 3] combined with (3.223) and the text following it, that self-similar sprays (with nice generators) are always fractal, and more specifically, that lattice (respectively, nonlattice) sprays are fractal in dimension d for finitely (respectively, countably infinitely) many values of $d \in \mathbb{R}$. Also, such lattice sprays are critically fractal whereas nonlattice sprays are subcritically fractal. Furthermore, still in light of those results, but now also combined with the main theorem in [**MorSepVi1**] (proving and extending a conjecture in [**Lap-vF5**], see also [**Lap-vF3**]), it is known that the set of d's for which a given nonlattice spray is fractal in dimension d is dense in a single compact (nonempty) interval $[D_\ell, D]$ in the generic case, with $D_\ell < D$, while we conjecture that an analogous statement is true in the nongeneric case, but now with finitely many (but more than one) compact nonempty intervals instead of a single one.[129]

3.6.1. *The $\frac{1}{2}$-square fractal, the $\frac{1}{3}$-square fractal and the relative Sierpinski N-gasket.* The above results can be illustrated by the $\frac{1}{2}$-square fractal RFD, the $\frac{1}{3}$-square fractal RFD and (for suitable values of N) the relative N-gasket.

[127] If D is multiple (as a pole of $\zeta_{A,\Omega}$), then it follows from [**LapRaZu1**, Thm. 5.4.27] briefly discussed towards the end of §3.5.2 that under the hypotheses of that theorem, the RFD (A, Ω) is h-Minkowski measurable with respect to the gauge function $h(t) := (\log t^{-1})^{m-1}$, for all $t \in (0, 1)$, where $m \geq 2$ is the multiplicity of D.

[128] We caution the reader that this statement is not a direct consequence of the general Minkowski measurability criterion (in terms of nonreal principal complex dimensions) used above and stated in §3.5.2. Indeed, the hypotheses of the corresponding theorem are not satisfied by all nonlattice self-similar strings. However, as was briefly mentioned in Example 3.36 of [**LapRaZu1**, §5.4], the extension of this statement to self-similar sprays with nice generators is proved by using separately the necessary condition and the sufficient condition for Minkowski measurability obtained in *loc. cit.*

[129] We conjecture that under suitable hypotheses, analogous results hold for the scaling complex dimension of (classic or homogeneous) self-similar sets satisfying the open set condition (as in [**Hut**] and, e.g., [**Fa1**]); in particular, all such self-similar sets are fractal. A key open problem in this context is to first obtain a factorization formula of the type (3.221) (see also (3.165)), possibly up to the addition of a holomorphic function in an appropriate right half-plane $\{Re(s) > \beta\}$, for some $\beta < D$. (See also [**LapRaZu1**, Pb. 6.2.36 and Rem. 6.2.37].)

(a) More specifically, for the $\frac{1}{2}$-square fractal RFD (A, Ω), $D := \dim_B(A, \Omega) = 1$ is a complex dimension of multiplicity two; hence, (A, Ω) is not Minkowski measurable but is also Minkowski degenerate, with Minkowski content $\mathcal{M} = +\infty$. However, (A, Ω) is h-Minkowski measurable with respect to the gauge function $h(t) = \log t^{-1}$ (for all $t \in (0,1)$) and with h-Minkowski content $\mathcal{M}((A, \Omega), h) = 1/4 \log 2$. (See part ($a$) of §3.4.5 and part ($a$) of Example 3.32.) Further, by (3.100)–(3.102),

$$D := D_{A,\Omega} = 1, \tag{3.234}$$

$$\mathcal{D}_{A,\Omega} = \{0, 1\} \cup \left(1 + i\frac{2\pi}{\log 2}\mathbb{Z}\right) \tag{3.235}$$

and

$$\dim_{PC}(A, \Omega) = 1 + i\frac{2\pi}{\log 2}\mathbb{Z}. \tag{3.236}$$

Therefore, the $\frac{1}{2}$-square fractal (A, Ω) is critically fractal, and is only fractal in dimension $D = 1$, the Minkowski dimension of (A, Ω). According to *loc. cit.*, the same statements are true for the $\frac{1}{2}$-square fractal A (instead of for (A, Ω)).

(b) Next, let (A, Ω) be the $\frac{1}{3}$-square fractal, as in part (b) of §3.4.5 and in part (b) of Example 3.32. Then, in light of (3.109) and the discussion surrounding it, (A, Ω) is subcritically fractal in dimension $d := \log_3 2$ and is only fractal in that dimension, the Minkowski dimension of the ternary Cantor set. Furthermore, $D := D_{A,\Omega} = 1$ and 1 is simple (as well as the only principal complex dimension of (A, Ω)). Moreover, according to part (b) of Example 3.32, the $\frac{1}{3}$-square fractal RFD (A, Ω) is Minkowski measurable. In light of *loc. cit.*, the exact same results hold for the $\frac{1}{3}$-square fractal A itself.

(c) We now consider the relative N-gasket (A_N, Ω_N), studied for any $N \geq 2$ in §3.4.4. Recall that with the notation used above for self-similar sprays, $D_G = N-1$ and $\sigma_0 = \log_2(N+1)$, and in light of (3.89) and (3.91), that

$$\mathcal{D}_{A_N,\Omega_N} = \{0, 1, \cdots, N-1\} \cup \left(\log_2(N+1) + i\frac{2\pi}{\log 2}\mathbb{Z}\right) \tag{3.237}$$

and

$$D := \dim_B(A_N, \Omega_N) = \max(N-1, \log_2(N+1)) = \begin{cases} \log_2 3, & \text{if } N = 2, \\ N-1, & \text{if } N \geq 3. \end{cases} \tag{3.238}$$

Therefore, as we next explain, we recover the three cases (i), (ii) and (iii) discussed above for general self-similar spray RFDs:

Case (i) when $D_G < \sigma_0$ corresponds to the $N = 2$ case; then, (A_2, Ω_2) is not Minkowski measurable but is Minkowski nondegenerate. Also, it is critically fractal and fractal only in dimension $d = \log_2 3$, the dimension of the relative Sierpinski gasket (A_2, Ω_2). In light of the results of §3.2.1, exactly the same statements hold for the (classic) Sierpinski gasket A_2 itself.

Case (ii) when $D_G = \sigma_0$ (i.e., $N - 1 = \log_2(N+1)$) corresponds to $N = 3$. Then, (A_3, Ω_3) is not Minkowski measurable (because $D = 2$ is of multiplicity two) and is also Minkowski degenerate. However, it is h-Minkowski measurable with respect to the gauge function $h(t) := \log t^{-1}$ (for all $t \in (0,1)$). Furthermore, in light of (3.237) and (3.238), (A_3, Ω_3) is critically fractal (necessarily in dimension $d := D = \log_2 4 = 2$), and is fractal only in that dimension.

Finally, case (*iii*) when $D_G > \sigma_0$ corresponds to every value of $N \geq 4$. In this case, $D = D_G = N - 1, (A_N, \Omega_N)$ is subcritically fractal in dimension $d := \sigma_0 = \log_2(N+1)$, and is fractal only in that dimension. Also, it is Minkowski measurable but has geometric oscillations of lower order (corresponding to the nonreal complex dimensions of real part $\log_2(N+1)$). This concludes our discussion of the relative Sierpinski N-gasket.

3.6.2. *The devil's staircase and fractality.* We close this part of the discussion by considering the emblematic example of the Cantor graph RFD (A, Ω) studied in §3.4.9 and in Example 3.34. This example is closely related to the Cantor graph A, also called the "devil's staircase" in [**Man**]. It is important for a variety of reasons:

(*a*) First, (A, Ω) is not a self-similar spray RFD. Indeed, the devil's staircase A is not a self-similar set; instead, it is a self-affine set, which makes the corresponding computation significantly more complicated, if not impossible to carry out.

(*b*) Secondly, the devil's staircase is *not* fractal according to Mandelbrot's definition of fractality (to be recalled just below), even though everyone with an exercised eye (including Benoit Mandelbrot himself)[130] would agree that it must be "fractal".[131] At this point, it may be helpful to the reader to recall that in [**Man**] (and in later work), Mandelbrot called "fractal" any (bounded) subset A of Euclidean space $\mathbb{R}^N$ such that its topological dimension $\dim_T A$ and Hausdorff dimension $\dim_H A$ do not coincide:[132]

$$\dim_H A \neq \dim_T A, \tag{3.239}$$

or, equivalently, since it is always true that $\dim_T A \leq \dim_H A$,

$$\dim_H A > \dim_T A. \tag{3.240}$$

In the present case when A is the devil's staircase, this definition clearly fails (as Mandelbrot was well aware of). Indeed, since the Cantor graph is a rectifiable curve (i.e., a curve of finite length), we have that

$$\dim_H A = \dim_B A = \dim_T A = 1, \tag{3.241}$$

according to a well-known result in elementary geometric measure theory (see [**Fed1**]) and as can also be directly checked here.

(*c*) Lastly, we mention that the unambiguous and frustrating contradiction between the visual impression one gets when contemplating the devil's staircase and Mandelbrot's definition of fractality (which he only adopted reluctantly, in response to the many queries and criticisms he received after the publication of his earlier French book on fractals) led the present author to wonder how to resolve this paradox and more importantly, how to much better capture the intuition underlying the informal notion of "fractality". Combined with the author's early work on fractal drums [Lap1–4] and his joint work on fractal strings and the Riemann zeta

[130]See the very explicit and enlightening comments in [**Man**, p. 82] and [**Man**, Plate 83, p. 83]; see also the very convincing and beautiful figure in *loc. cit.* Let us quote from [**Man**, p. 82]: "*One would love to call the present curve a fractal, but to achieve this goal, we would have to define fractal less stringently on the basis of notions other than D* [the Hausdorff dimension] *alone.*"

[131]The exact same comment can be made verbatim about the Cantor graph RFD provided, in Mandelbrot's original definition, one substitutes the (relative) Minkowski dimension for the Hausdorff dimension.

[132]It is noteworthy that if A is nonempty, $\dim_T A$ is always a nonnegative integer. For example, for the ternary Cantor set, it is equal to 0 while for the classic Sierpinski gasket and for the Koch snowflake curve, it is equal to 1.

function as well as the Riemann hypothesis [LapPo1–3, LapMa1–3], this quest eventually led (first, for fractal strings, in [Lap-vF1–4] and then, in any dimension $N \geq 1$, in [**LapRaZu1**]) to the notion of complex fractal dimensions and to the present notion of fractality expressed in terms of the irreality (or the "unreality") of some of the underlying complex dimensions.

Now, let us return to the Cantor graph RFD (A, Ω) and its complex dimensions. In light of the discussion of that example provided in §3.4.9 and in Example 3.34,

$$D := D_{A,\Omega} = D_A = 1, \tag{3.242}$$

$$\{1\} \subseteq \mathcal{D}_A \subseteq \mathcal{D}_{A,\Omega} = \{0, 1\} \cup \Big(\log_3 2 + i\frac{2\pi}{\log 3}\mathbb{Z}\Big) \tag{3.243}$$

and hence,

$$\dim_{PC} A = \dim_{PC}(A, \Omega) = \{1\}, \tag{3.244}$$

where (in (3.243)) $D_{CS} = \log_3 2$ is the Minkowski dimension of the ternary Cantor set. In addition, it is conjectured (first in [Lap-vF2–4], via an approximate computation of the corresponding tube function, and then, with significantly more precise supporting arguments, in [LapRaZu1,4]) that

$$\log_3 2 + i\frac{2\pi}{\log 3}\mathbb{Z} \subseteq \mathcal{D}_A \tag{3.245}$$

or even that $\mathcal{D}_A = \mathcal{D}_{A,\Omega}$ in (3.243).

The Cantor graph RFD is Minkowski measurable; furthermore, it is fractal in dimension $d := D_{CS} = \log_3 2$ and is fractal only in that dimension. Consequently, it is critically subfractal. The presence of the nonreal complex (and subcritical) dimensions $s_k = \log_3 2 + i(2\pi/\log 3)k$ (with $k \in \mathbb{Z}\backslash\{0\}$) gives rise to logarithmically (or multiplicatively) periodic oscillations of order $D_{CS} = \log_3 2 < 1$.[133] (See the fractal tube formula (3.218) along with (3.219).) Therefore, it has a real effect on the geometry. According to the aforementioned conjecture (see (3.245) and the text following it), the devil's staircase (or Cantor graph) A itself should have exactly the same qualitative and quantitative properties as the Cantor graph RFD (A, Ω). In particular, it is *not* critically fractal but is critically fractal in dimension $d := D_{CS} = \log_3 2$.

The following open problem is technically very challenging but also conceptually important, in light of the above discussion.

OPEN PROBLEM 3.39. (*Complex dimensions of the devil's staircase.*) Prove the conjecture stated in (3.245) or even its stronger form stated just afterward.

As a first but important step towards that conjecture, try to show that there are at most finitely many exceptions to the inclusion appearing in (3.245) or at least, that infinitely many elements of the left-hand side of the inclusion in (3.245) belong to $\mathcal{D}_A$.

[133] More precisely, they correspond to the term $\varepsilon^{2-D_{CS}}G(\log_3 \varepsilon^{-1})$, which is exactly of order $\varepsilon^{2-D_{CS}}$ since the nonconstant periodic function G in (3.219) is bounded away from 0 and ∞.

3.6.3. *Extended notion of complex dimensions, scaling laws and Riemann surfaces.* We continue this discussion by broadening the notion of complex dimensions and correspondingly, of fractality. As was alluded to earlier, it is natural to call "complex dimensions" of an RFD (A, Ω) not only the poles but also other types of (nonremovable) singularities of the associated fractal zeta function $\zeta_{A,\Omega}$ (or, equivalently, $\widetilde{\zeta}_{A,\Omega}$, provided $D := \dim_B(A, \Omega) < N$). Typically, the (necessarily closed) set of singularities is obtained as the closure of a countable (or finite) set of "geometric singularities", which we propose to call the *kernel* or simply, the *set of "geometric singularities"* of (A, Ω).[134] For example, for each member (A, Ω), say, of the family of (transcendentally ∞-quasiperiodic) maximally hyperfractal RFDs or compact sets constructed in [LapRaZu1,3] and discussed earlier, the kernel consists of a countably infinite set of geometric singularities (which are not poles of $\zeta_{A,\Omega}$ or of $\widetilde{\zeta}_{A,\Omega}$) and is dense in the whole critical line $\{Re(s) = D\}$, where $D := \overline{\dim}_B(A, \Omega)$ can be prescribed a priori in the interval $(0, N)$. However, only the geometric singularities will contribute to the associated fractal tube formulas.[135]

Several other examples are provided in [**LapRaZu1**]. In one of those examples, the kernel consists of a countable set of essential singularities and is shown to contribute to the fractal tube formulas on the same footing as mere poles of $\zeta_{A,\Omega}$. The remaining singularities do not contribute to the main term in the fractal tube formula; they may, however, contribute to the error term (the 'noise').

It is then immediate to extend the above notion of fractality by saying that an object (say, an RFD (A, Ω) in $\mathbb{R}^N$) is *fractal* if it has at least one nonreal (geometric) singularity. The notions of fractality in dimension $d \in \mathbb{R}$, as well as of critical and subcritical fractality, are similarly extended.

In the author's opinion, a deep understanding of the scaling laws in mathematics and physics (among many other fields, including cosmology, computer science, economics, chemistry and biology), as well as of many aspects of dynamics and of fractal, spectral and arithmetic geometry, could be gained by pursuing this venue and extending the theory of complex dimensions by merging it with aspects of the theory of Riemann surfaces[136] and using the notion of h-Minkowski content and related notions for a variety of (admissible) gauge functions. The beginning of such a theory is provided in a joint work under completion, [**LapRaZu10**].

As is suggested in [**Lap10**], in the long term, a potentially far-reaching further extension of the theory of complex dimensions to multivariable fractal zeta functions and their analytic varieties of singularities, merged with aspects of the theory of complex manifolds and sheaf theory (see, e.g., [**Ebe**] and [**GunRos**]), should be even more fruitful in this context; see also the end of §4.4.2 (and of §4) below.

We close this subsection by discussing originally unexpected connections between hyperfractality and the spectra of fractal drums.

3.6.4. *Maximal hyperfractals and meromorphic extensions of spectral zeta functions of fractal drums.* The error estimates obtained in [**Lap1**] for the spectral

[134]Accordingly, the complement of the kernel in the set of all (nonremovable) singularities could be referred to as the *set of "analytic singularities"*. In practice, it can be ignored because it is not expected to contribute (in a significant way) to the fractal tube formula for (A, Ω), although this remains to be proved in general.

[135]Simpler examples of this type can be obtained by considering the Cartesian product of two different self-similar (and lattice) Cantor sets with incommensurable oscillatory periods.

[136]See, e.g., [**Ebe**], [**Schl**] and beyond, in the spirit of Riemann's original broader intuition of Riemann surface.

asymptotics of fractal drums (i.e., drums with fractal boundary) are, in general, best possible. (See the paragraph following (2.41), along with part (a) of Remark 2.12.) They also imply that the (normalized) spectral zeta function ζ_ν associated with a given fractal drum[137] can be meromorphically extended to the open right half-plane $\{Re(s) > D\}$, where D is the (upper) Minkowski dimension of the boundary of the fractal drum.[138]

In [**LapRaZu1**, §4.3.2] and [**LapRaZu8**], it is shown that the construction of (transcendentally ∞-quasiperiodic) maximal hyperfractal RFDs $(\partial\Omega, \Omega)$, where Ω is a bounded open set in $\mathbb{R}^N$ (and hence, has compact boundary $\partial\Omega$), carried out in [**LapRaZu1**, §4.6.1] and [LapRaZu2–4] (as was briefly discussed earlier in this subsection), implies that the above right half-plane $\{Re(s) > D\}$ is in general optimal (i.e., as large as possible among all open right half-planes to which ζ_ν can be meromorphically continued). This is so for any possible value of the (upper) Minkowski dimension of the RFD $(\partial\Omega, \Omega)$; namely, for any $N \geq 1$ and for every $D \in (0, N)$.

This last result establishes an interesting new connection between (maximal) hyperfractality and the vibrations of fractal drums. In particular, for each of the maximally hyperfractal drums or RFDs $(\partial\Omega, \Omega)$ constructed in [LapRaZu1,8], the half-plane of meromorphic convergence of their spectral zeta function ζ_ν coincides with the half-plane of (absolute) convergence of their distance zeta function $\zeta_{\partial\Omega,\Omega}$ (or equivalently, since $D < N$ here, of their tube zeta function $\widetilde{\zeta}_{\partial\Omega,\Omega}$).

4. Epilogue: From Complex Fractal Dimensions to Quantized Number Theory and Fractal Cohomology

In [Lap-vF2–4], [**Lap-vF7**] and [**Lap7**] was proposed a search for a 'fractal cohomology theory' that would be naturally associated with the theory of complex fractal dimensions (at the time, as developed for fractal strings and described in part in §2, but now also extended as in [**LapRaZu1**] and [LapRaZu2–10] to the much broader higher-dimensional setting, as described in §3), as well as help unify at a deeper level several important aspects of fractal geometry, number theory and arithmetic geometry. (See [**Lap-vF4**, §12.3 and §12.4], along with [**Lap7**, Chs. 4 and 5].)

In particular, from this perspective (see [**Lap-vF4**, §12.4]), an analogy was developed between lattice self-similar geometries and finite-dimensional (algebraic) varieties over finite fields, while nonlattices self-similar geometries could be seen as a limiting case corresponding to infinite dimensional varieties (seemingly over a "field of characteristic one"). Indeed, the zeros and the poles of the scaling zeta functions[139] of lattice self-similar strings are periodically distributed along finitely many vertical lines. The same is true of self-similar sprays (as well as, conjecturally,

[137]For the Dirichlet Laplacian with eigenvalue spectrum $(\lambda_j)_{j=1}^{\infty}$, where the eigenvalues are repeated according to their multiplicities, we let $\zeta_\nu(s) := \sum_{j=1}^{\infty} f_j^{-s}$, for all $s \in \mathbb{C}$ with $Re(s)$ sufficiently large, where for each $j \geq 1, f_j := \sqrt{\lambda_j}$ is the j-th frequency of the fractal drum (written in nonincreasing order and repeated according to multiplicity).

[138]That is, $D := \dim_B(\partial\Omega, \Omega)$ in the notation of relative fractal drums.

[139]By definition, the scaling zeta function of a self-similar string $\mathcal{L}$ coincides with the geometric zeta function of $\mathcal{L}$.

of self-similar sets) with "nice generators". (For the case of self-similar strings, see Example 2.9, and for the more general case of self-similar sprays, see §3.4.10.)[140]

4.1. Analogy between self-similar geometries and varieties over finite fields. Recall that for a (smooth, projective, finite-dimensional, algebraic) variety V over a finite field $\mathbb{F}_q$ (where $q = p^m$, with $m \in \mathbb{N}$ and the prime number p being the underlying prime characteristic), the corresponding zeta function ζ_V is periodic with complex period $i\mathbf{p}$ (i.e., $\zeta_V(s) = \zeta_V(s + i\mathbf{p})$, for all $s \in \mathbb{C}$), where $\mathbf{p} := 2\pi/\log q = 2\pi/m\log p$. Therefore, as $q = \#\mathbb{F}_q \to \infty$ (e.g., if we successively consider V over the finite field extensions $\mathbb{F}_{q^n}$, with $n \in \mathbb{N}$ increasing to ∞), then the 'oscillatory period' $\mathbf{p}$ tends to 0.

Similarly, for a lattice string (or spray), the corresponding scaling zeta function $\zeta_{\mathfrak{s}}$ is periodic with complex period $i\mathbf{p}$ (where $\mathbf{p} := 2\pi/\log r^{-1}$ is the oscillatory period and r in $(0, 1)$ is the multiplicative generator of the underlying scaling ratios), in the same sense as above. Furthermore, in the approximation of a nonlattice string (or spray) by a sequence of lattice strings (or sprays) with increasing oscillatory periods $\mathbf{p}_n$ (as described in detail in [**Lap-vF4**, Ch. 3, esp., §§3.4–3.5] and briefly discussed in Example 2.9), so that $\zeta_{\mathfrak{s}}(s + iu\mathbf{p}_n)$ and $\zeta_{\mathfrak{s}}(s)$ are close for $u \in \mathbb{C}$ with $|u|$ not too large), we have $\mathbf{p}_n \to \infty$. In this sense, nonlattice strings (or sprays) satisfy $\mathbf{p} = \infty$ (or $\mathbf{p} \to \infty$) and behave as though $r \to 1^+$ and hence, as though the "underlying prime characteristic were equal (or tending) to 1", as was mentioned above.[141]

Moreover, for a (finite-dimensional) variety V over $\mathbb{F}_q$ (really, over the algebraic closure $\overline{\mathbb{F}}_q$ of $\mathbb{F}_q$), the total number of vertical lines, along which the zeros and the poles of ζ_V (counted according to their multiplicities) are distributed is equal to $2d + 1$, where $d := \dim V$ is the dimension of the variety. In addition, the zeros (respectively, poles) correspond to the even (respectively, odd) cohomology spaces.[142]

The aforementioned (Weil-type or étale) cohomology spaces played a key role in the proof of the Weil conjectures [Weil1–3] (and, in particular, of the counterpart of the Riemann hypothesis) for curves over finite fields by A. Weil in *loc. cit.* and then, for higher-dimensional (but still finite-dimensional) varieties over finite fields, by P. Deligne in [Del1–2].[143]

In the process, a certain map (on the underlying variety V and induced by the self-map $x \mapsto x^q$ of $\mathbb{F}_q$), called the *Frobenius morphism*, also plays a central role, via

140We leave aside here the 'integer dimensions' since we use the scaling (rather than the fractal) zeta functions.

141In the case of a self-similar spray with multiple generators, the scaling zeta function has both zeros and poles; in fact, it is of the form $\zeta_{\mathfrak{s}}(s) = \sum_{k=1}^{K} g_k^s/(1 - \sum_{j=1}^{J} r_j^s)$, where the positive numbers g_k and scaling ratios r_j are not necessarily assumed to be distinct. By definition, the lattice case then corresponds to the situation when the group generated by the *distinct values* of the g_k's and r_j's is of rank 1 (with generator denoted by r and assumed to lie in $(0, 1)$); the associated oscillatory period is then $\mathbf{p} := 2\pi/\log r^{-1}$.

142In other words, the total cohomology space is naturally $\mathbb{Z}_2$-graded, with $\dot{0}$ corresponding to the zeros and $\dot{1}$ corresponding to the poles, for this choice of grading.

143For a brief introduction to curves (or, more generally, varieties) over finite fields, as well as to the associated zeta functions and Weil conjectures (including RH in this context), we point out, for example, [**Dieu1**, **Katz**, **Oort**, **ParsSh1**] and [**Lap7**, App. B], along with the many references therein (or since), including [**Art**], [**Has**], [**Schm**], [Weil1–3], [Gro1–4], [Del1–2], [Den1–6], [Har1–3], [CobLap1–2] and [**Lap10**]. Also, for relevant notions from algebraic geometry, we refer, e.g., to [**Hart**].

the linear endomorphism F it induces on the total cohomology space and called the *Frobenius operator*.[144] The (graded or alternating) '*characteristic polynomial*' of F is then shown to coincide with the zeta function of V (in the variable $z = q^{-s}$). Symbolically, and with $Z_V(z) := \zeta_V(q^{-s})$, the *Weil zeta function* of the variety V, we have (ignoring multiplication by an unessential nowhere vanishing entire function)

$$Z_V(z) = s\text{-det}(I - zF), \tag{4.1}$$

where "s-det" stands for the graded (or super, [**Del3**, **Wein**]) determinant of F over the (reduced) total cohomology space. It was later conjectured by C. Deninger in [Den1–6] (see, especially, [Den1–3]) that a similar procedure (but now involving, in general, possibly infinite dimensional cohomology spaces) could be used to deal with the Riemann zeta function $\zeta = \zeta(s)$ and other L-functions (in characteristic zero, for example, in the case of ζ, over the field $\mathbb{Q}$).

Since the early to mid-1990s, the author's intuition has been that there should exist a suitable notion of "fractal cohomology" such that the nonlattice case (for self-similar geometries) would be analogous to the situation expected to hold for ζ (the latter being a very special case, however, one typical of arithmetic geometries and for which RH would hold). In [**Lap7**], building on [Lap-vF1–3] (see also [**Lap-vF4**, §12.4]), it was conjectured that a (generalized) Polya–Hilbert operator [having for spectrum (the reciprocals of) the zeros and poles (i.e., the reciprocal of the divisor) of the underlying fractal or arithmetic zeta function] should exist in this context so that the analog of (4.1) would hold and could be rigorously established. Partially realizing this dream has required significantly building on another semi-heuristic proposal made in [Lap-vF3–4] (regarding a so-called "spectral operator"), then the development in [HerLap1–5] of quantized number theory (in the "real case"), followed by the development in [CobLap1–2] of quantized number theory (in the "complex case") and the construction of a corresponding (generalized) Polya–Hilbert operator, along with the foundations of an associated fractal cohomology, in [**Lap10**].

4.2. Quantized number theory: The real case. We begin by presenting (in a very concise form) aspects of the first version of quantized number theory (in the "real case"), as developed by Hafedh Herichi and the author in [HerLap1–5] and in [**Lap8**], based on a rigorous notion of the infinitesimal shift ∂ of the real line and of the corresponding spectral operator (as proposed heuristically in [Lap-vF3–4]; see, especially, [**Lap-vF4**, §6.3] and [**HerLap1**, Ch. 4]). In particular, the theory developed in [HerLap1–5] explains how to "quantize" the Riemann zeta function $\zeta = \zeta(s)$ in this context in order to view the spectral operator $\mathfrak{a}$ (which sends the geometry of fractal strings onto their spectrum) as follows (see [**HerLap1**, §7.2]):

$$\mathfrak{a} = \zeta(\partial), \tag{4.2}$$

where $\zeta(\partial)$ is interpreted in the sense of the functional calculus for the unbounded normal operator $\partial = \partial_c$ (with $\partial = d/dt$, the differentiation operator) acting on a suitable complex Hilbert space $\mathbb{H}_c := L^2(\mathbb{R}, e^{-2ct}dt)$, the weighted L^2-space with respect to the (positive) weight function $w_c(t) := e^{-2ct}$, defined for all $t \in \mathbb{R}$.

[144]By using the periodicity of ζ_V (i.e., by making the changing of variable $z := q^{-s}$), one formally obtains a reduced (total) cohomology space, which is a *finite-dimensional* vector space. In fact, in [Weil–3], [Del1–2] or [Gro1–4], only finite-dimensional vector spaces (over the underlying field) are considered.

(See [**HerLap1**, Ch. 5] for the precise definition of ∂, including of its domain of definition, as well as for the proof of the normality of ∂: $\partial^*\partial = \partial\partial^*$ since it is shown in *loc. cit.* that the adjoint ∂^* of ∂ is given by $\partial^* = 2c - \partial$.)

Here, the "dimensional parameter" $c \in \mathbb{R}$ is *fixed*, but enables us, in particular, to sweep out the entire critical strip $0 < Re(s) < 1$ (corresponding to the "critical interval" $0 < c < 1$) as well as the half-plane of absolute convergence for $\zeta = \zeta(s)$, namely, the open half-plane $Re(s) > 1$ corresponding to the half-line $c > 1$).

The (strongly continuous) semigroup of bounded linear operators of $\mathbb{H}_c$ generated by $\partial = \partial_c$ is given by $\{e^{-h\partial}\}_{h\geq 0}$ and acts as the semigroup of translations (or shifts) of the real line (see [**HerLap1**, §6.3]):[145]

$$(e^{-h\partial})(f)(u) = f(u-h), \quad \text{for all } f \in \mathbb{H}_c, \tag{4.3}$$

as well as for almost every $u \in \mathbb{R}$ and for every $h \geq 0$. In fact, formula (4.3) is also valid for $h \leq 0$ and hence, the *real infinitesimal shift* $\partial = \partial_c$ is the infinitesimal generator of the one-parameter group of operators $\{e^{-h\partial}\}_{h\in\mathbb{R}}$.

We note that the semigroup $\{e^{-h\partial}\}_{h\geq 0}$ is a contractive (respectively, expansive) semigroup for $c \geq 0$ (respectively, $c \leq 0$), while the group $\{e^{-h\partial}\}_{h\in\mathbb{R}}$ is a group of isometries if and only if $c = 0$. Finally, since the adjoint of ∂ is given by $\partial^* = 2c - \partial$, as is shown in [**HerLap1**, §5.3], the adjoint group $\{e^{-h\partial^*}\}_{h\in\mathbb{R}}$ coincides with $\{e^{-2ch}e^{h\partial}\}_{h\in\mathbb{R}}$.

It is shown in [**HerLap1**, §6.2] that the spectrum, $\sigma(\partial)$, of the real infinitesimal shift $\partial = \partial_c$ is given by L_c, the vertical line going through $c \in \mathbb{R}$:[146]

$$\sigma(\partial) = L_c := \{Re(s) = c\}. \tag{4.4}$$

Hence, ∂ is an unbounded (normal) operator, for any value of $c \in \mathbb{R}$. Further, in light of (4.2) and (4.4), one deduces from a suitable version of the spectral mapping theorem for unbounded normal operators given in [**HerLap1**, App. E] that the spectrum, $\sigma(\mathfrak{a})$, of the spectral operator $\mathfrak{a} = \zeta(\partial)$ coincides with the closure of the range of $\zeta = \zeta(s)$ along the vertical line L_c (see [**HerLap1**, §9.1]):[147]

$$\sigma(\mathfrak{a}) = c\ell(L_c) = c\ell(\{\zeta(s) : Re(s) = c\}). \tag{4.5}$$

It follows from (4.5) that, in light of the Bohr–Courant theorem [**BohCou**], itself now viewed as a mere consequence of the universality of ζ among all (suitable) analytic functions on the right critical strip $\{1/2 < Re(s) < 1\}$ (as established in [**Vor**]), that

$$\sigma(\mathfrak{a}) = \mathbb{C}, \quad \text{for every } c \in (1/2, 1). \tag{4.6}$$

[145]For the general theory of strongly continuous semigroups of bounded linear operators and its applications, from different points of view, we refer, e.g., to [**Go, HilPh, JoLap, JoLapNie, Kat**], [ReSi1–3] and [**Ru2, Sc, Yo**].

[146]For the spectral theory of unbounded operators in various settings, we refer, e.g., to [**DunSch, ReSi1, JoLap, Kat, Ru2, Sc, Yo**]. We note, however, that the references [**ReSi1**] and [**Sc**] are mostly limited to the spectral theory of unbounded self-adjoint (rather than normal) operators, as are many of the references focusing on the applications to quantum physics.

[147]If $c = 1$, the unique (and simple) pole of ζ, one must exclude $s = 1$ from the line L_c in interpreting (4.5). Alternatively, one can consider the extended spectrum of $\mathfrak{a}$, denoted by $\widetilde{\sigma}(\mathfrak{a})$ and defined by $\widetilde{\sigma}(\mathfrak{a}) := \sigma(\mathfrak{a})$ if $\mathfrak{a}$ is bounded and $\widetilde{\sigma}(\mathfrak{a}) := \sigma(\mathfrak{a}) \cup \{\infty\}$ otherwise, and view the meromorphic function ζ as a continuous function from $\mathbb{C}$ to the Riemann sphere $\widetilde{\mathbb{C}} := \mathbb{C} \cup \{\infty\}$. Then, the closure in (4.5) is interpreted in the compact space $\widetilde{\mathbb{C}}$ instead of in the unbounded complex plane $\mathbb{C}$.

In particular, for any $c \in (1/2, 1)$, the spectral operator is *not* invertible (in the usual sense of unbounded operators); see [**HerLap1**, §9.2].

Next, we examine what happens when $c > 1$ (which, as we recall and in light of the identity (4.5), corresponds to the half-plane of absolute convergence for $\zeta = \zeta(s) = \sum_{n=1}^{\infty} n^{-s}$, namely, the open half-plane $\{Re(s) > 1\}$); see [**HerLap1**, Ch. 7] for the full details.

For $c > 1$, the spectral operator $\mathfrak{a} = \zeta(\partial)$ can be represented by a norm convergent *quantized* (or operator-valued) *Dirichlet series* and a *quantized* (also norm convergent) *Euler product*:

$$\begin{aligned} \mathfrak{a} = \zeta(\partial) &= \sum_{n=1}^{\infty} n^{-\partial} \\ &= \prod_{p \in \mathcal{P}} (1 - p^{-\partial})^{-1}, \end{aligned} \tag{4.7}$$

where $\mathcal{P}$ denotes the set of prime numbers. It follows from (4.7) that for every $c > 1$, $\mathfrak{a}$ is invertible (in the strong sense of $\mathcal{B}(\mathbb{H}_c)$, the space of bounded linear operators on $\mathbb{H}_c$), and that its (bounded) inverse $\mathfrak{a}^{-1}$ is given by

$$\mathfrak{a}^{-1} = \sum_{n=1}^{\infty} \mu(n) n^{-\partial} = \prod_{p \in \mathcal{P}} (1 - p^{-\partial}), \tag{4.8}$$

where $\mu = \mu(n)$ is the Möbius function, defined by $\mu(1) = 1, \mu(n) = 0$ if the integer $n \geq 2$ is not square-free, and $\mu(n) = k$ if $n \geq 2$ is the product of k distinct primes. We stress that for $c > 1$, all of the infinite series and infinite products in (4.7) and (4.8) are convergent in the operator norm (i.e., in the Banach algebra $\mathcal{B}(\mathbb{H}_c)$).

Moreover, for $0 < c < 1$ (i.e., "within" the critical strip $0 < Re(s) < 1$), the spectral operator $\mathfrak{a} = \mathfrak{a}_c$ is represented (when applied to a state function f belonging to a suitable dense subspace of $\mathbb{H}_c$, which is also an operator core for $\partial = \partial_c$) by the same (but now weakly convergent) operator-valued Dirichlet series and Euler product as in (4.7). A similar comment applies to $\mathfrak{a}^{-1}$, which is then also a possibly unbounded operator when $0 < c < 1$.

Therefore, as was conjectured in §6.3.2 of [**Lap-vF3**] and of [**Lap-vF4**], but now in a very precise sense, the spectral operator $\mathfrak{a} = \zeta(\partial)$, which can be viewed as the *quantized Riemann zeta function*, has an operator-valued Euler product representation (as well as a convergent Dirichlet series, which was not conjectured to exist in [**Lap-vF3, Lap-vF4**]) that is convergent (in a suitable sense) even in the critical strip $0 < Re(s) < 1$ (i.e., even when the dimensional parameter c lies in the critical interval $(0, 1)$); see [**HerLap1**, §7.5].

One of the key results of [**HerLap1**] is to provide various characterizations of the "quasi-invertibility" of the spectral operator $\mathfrak{a} = \mathfrak{a}_c$.[148] More specifically, the Riemann hypothesis (RH) is shown to be equivalent to the fact that for $c \in (0, 1)$, the spectral operator $\mathfrak{a} = \mathfrak{a}_c$ is quasi-invertible if and only if $c \neq 1/2$, thereby providing an exact operator-theoretic counterpart of the reformulation of RH obtained in [**LapMa2**] (and briefly discussed in §2.6.2) in terms of inverse spectral problems for fractal strings. (See [**HerLap1**, Ch. 8, esp., §8.3].)

---

[148]The possibly unbounded, normal operator $\mathfrak{a} = \zeta(\partial)$ is said to be *quasi-invertible* if for every $T > 0$, the truncated spectral operator $\mathfrak{a}^{(T)} := \zeta(\partial^{(T)})$ is invertible, where $\partial^{(T)} := \varphi^{(T)}(\partial)$ (defined via the functional calculus for unbounded normal operators) is the *truncated infinitesimal shift* and the function $\varphi^{(T)}$ is chosen so that $\sigma(\mathfrak{a}^{(T)}) = [c - iT, c + iT]$; see [**HerLap1**, §6.4].

Another, seemingly very different, reformulation of the Riemann hypothesis, obtained by the author in [**Lap8**] (and also discussed in [**HerLap1**, §9.4]) is the following statement: The Riemann hypothesis holds true if and only if the spectral operator $\mathfrak{a} = \mathfrak{a}_c$ is invertible (in the usual sense of unbounded operators) for every $c \in (0, 1/2)$.[149] This result is referred to in [**Lap8**] as *an asymmetric criterion for RH.*

Many other results are obtained in [HerLap1–5] and [**Lap8**], concerning, in particular, a quantized analog of the functional equation for $\zeta = \zeta(s)$ (or for its completion $\xi = \xi(s)$, see [**HerLap1**, §7.6] where the global spectral operator $\mathcal{A} = \mathcal{A}_c := \xi(\partial_c)$ is studied) and of the universality of the Riemann zeta function [**KarVo, Lau, Steu**] (see [**HerLap1**, Ch. 10, esp., §10.2]), as well as about the form of the inverse of the spectral operator, when it exists (see [**Lap8**] and [**HerLap1**, §9.4]), and the nature of the mathematical phase transitions at $c = 1/2$ and at $c = 1$ concerning the shape of the spectrum, the quasi-invertibility, the invertibility and the boundedness of $\mathfrak{a} = \mathfrak{a}_c$ (see [**HerLap1**, §9.3]).

We note that the mathematical phase transitions occurring in [**HerLap1**] at $c = 1$ are analogous to (but different from) the one studied from an operator-algebraic point of view by J.-B. Bost and A. Connes in [BosCon1–2] and [**Con1**, §V.II] (because the latter also corresponds to the pole at $s = 1$ of $\zeta = \zeta(s)$), whereas the phase transitions occurring in [**HerLap1**] and in [**Lap8**] in the midfractal case when $c = 1/2$ are of a very different nature. They were expected to occur in the author's early conjecture (and open problem) formulated in [**Lap3**, Quest. 2.6, p. 14] about the existence of a notion of complex fractal dimensions that would enable us to interpret the Riemann hypothesis as a phase transition in the midfractal case, in the sense of Wilson's work [**Wils**] on critical phenomena in condensed matter physics and quantum field theory.

For a full exposition of these results, we refer the interested reader to the book [**HerLap1**]. We note that the above results could likely be extended to a large class of meromorphic functions (instead of just the Riemann zeta function), and especially of the arithmetic L-functions ([**Sarn**], [Murt1–3], [**Lap7**, App. C]) for which the counterpart of the Riemann hypothesis or of the universality theorem is expected to hold. In particular, conjecturally, the results concerning RH should have an appropriate counterpart for all the elements of the Selberg class ([**Sel1, Sarn**], [Murt1–3] and [**Lap7**, App. E]).

4.3. Quantized number theory: The complex case. We continue this epilogue by providing an extremely brief overview of the results of Tim Cobler and the author on "quantized number theory" (in the "complex case") obtained in [CobLap1–2] and pursued in the book in preparation [**Lap10**], in which the real infinitesimal shift $\partial = d/dt$ acting on $\mathbb{H}_c = L^2(\mathbb{R}, e^{-2ct}dt)$ (with $c \in \mathbb{R}$ and discussed in §4.2) is replaced by the complex infinitesimal shift $\partial = d/dz$, now acting on a suitable weighted Bergman space $\mathcal{H}$ of entire functions (as introduced originally and for completely different purposes by A. Atzmon and B. Brive in [**AtzBri**]). Namely, $\mathcal{H}$ consists of the entire functions belonging to the complex

[149] By using a result in [**GarSteu**], this is shown to be equivalent to the non-universality of $\zeta = \zeta(s)$ in the left critical strip $\{0 < Re(s) < 1/2\}$. On the other hand, in light of (4.5), the universality of the Riemann zeta function $\zeta = \zeta(s)$ in the right critical strip $\{1/2 < Re(s) < 1\}$ (due to S. M. Voronin in [**Vor**] and extended by B. Bagchi and A. Reich in [**Bag**] and [Rei1–2]) implies that $\mathfrak{a} = \mathfrak{a}_c$ is *never* invertible for any $c \in (1/2, 1)$.

weighted Hilbert space $L^2(\mathbb{C}, e^{-|z|^\alpha} dz)$, for some fixed, but otherwise unimportant, parameter $\alpha \in (0,1)$.[150]

For these values of α, the operator ∂ is bounded (but not normal) and its spectrum, $\sigma(\partial)$, is given by a single point, namely, the origin:[151]

$$\sigma(\partial) = \{0\}. \tag{4.9}$$

From this, it follows that in the notation introduced below (namely, with $\partial_\tau := \partial + \tau$), we have that $\sigma(\partial_\tau) = \{\tau\}$, for every $\tau \in \mathbb{C}$. We note that 0 is an eigenvalue of ∂ (in fact, its only eigenvalue) and that the associated eigenfunction is the (possibly suitably normalized) constant function 1, the 'vacuum state'. With 0 replaced by $\tau \in \mathbb{C}$, the same statement holds for the shifted differentiation operator ∂_τ.

The *complex infinitesimal shift* ∂ is the infinitesimal generator of the group of translations (or shifts) on the complex plane, $\{e^{-z\partial}\}_{z\in\mathbb{C}}$, acting on every $\psi \in \mathcal{H}$ as follows (for almost every $u \in \mathbb{C}$ and every $z \in \mathbb{C}$):

$$(e^{-z\partial})(\psi)(u) = \psi(u - z). \tag{4.10}$$

In [**CobLap1**] are rigorously constructed "*generalized Polya–Hilbert operators*" (GPOs, in short), in a sense close to that of [**Lap7**], and infinite dimensional regularized determinants of the restrictions to their eigenspaces which enable one, in particular, to recover (under appropriate hypotheses) the corresponding meromorphic functions (or zeta functions) as suitable (graded or alternating) "characteristic polynomials" of the restrictions of the GPOs to their total eigenspaces, in the spirit of §4.1 (but well beyond).

More specifically, by working with the family of translates $\{\partial_\tau\}_{\tau\in\mathbb{C}}$ of the complex infinitesimal shift $\partial = d/dz$ (i.e., $\partial_\tau := \partial + \tau$, for every $\tau \in \mathbb{C}$) and the associated Bergman spaces, one can then construct (via orthogonal direct sums) a kind of 'universal Polya–Hilbert operator' (GPO), $\mathbb{D}$, acting on a typically countably infinite direct sum $\mathcal{H}$ of Bergman spaces, whose spectrum (when it is discrete) consists only of (isolated) eigenvalues with finite multiplicities and coincides with any prescribed discrete subset of $\mathbb{C}$, for example, a (multi)set of complex dimensions or (the reciprocal of) the divisor [viewed as a $\mathbb{Z}_2$-graded (multi)set of zeros and poles] of a suitable meromorphic function, such as (an appropriate version of) the global Riemann zeta function.

We can then write the given meromorphic function $g = g(s)$ (assumed to be a quotient of two entire functions of finite orders) as a (typically infinite dimensional) regularized $\mathbb{Z}_2$-graded (or supersymmetric) determinant:

$$g(s) = s\text{-}\det(I - s\mathcal{F}), \tag{4.11}$$

where $\mathcal{F} = \mathcal{F}_g$ is the restriction of the GPO $\mathbb{D}$ to its total eigenspace; so that

$$\sigma(\mathbb{D}) = \sigma(\mathcal{F}) = \mathfrak{D}(g)^{-1}, \tag{4.12}$$

the reciprocal of the divisor of g, in the sense of Definition 3.18, and the (graded) spectra of $\mathbb{D}$ and $\mathcal{F}$ are discrete and consist only of eigenvalues, which are the reciprocals of the zeros and the poles of $g = g(s)$ repeated according to their multiplicities.

---

[150] For the theory of (not necessarily weighted) Bergman spaces, see, e.g., [**DureSchu**] and [**HedKonZhu**].

[151] If we were to allow the value $\alpha = 1$, then in (4.9), $\sigma(\partial)$, the spectrum of ∂, would be the closed unit disk (with center the origin) in $\mathbb{C}$ instead of being reduced to $\{0\}$.

The operator $\mathcal{F} = \mathcal{F}_g$ occurring in (4.11) and (4.12) and defined as the restriction of the GPO $\mathbb{D} = \mathbb{D}_g$ to its total eigenspace is called (in [**Lap10**]) the *generalized Frobenius operator* (GFO, in short) associated with g. Indeed, in the special case of varieties over finite fields $\mathbb{F}_q$ discussed in §4.1, it is clearly a version (in the s-variable) of the usual Frobenius operator F (defined by means of the z-variable, with $z := q^{-s}$) induced by the Frobenius morphism on the variety and acting on the underlying total cohomology space (which also coincides with the total eigenspace of F).

We stress that here and thereafter, $\mathbb{D} = \mathbb{D}_g$ is the GPO associated with the discrete (and $\mathbb{Z}_2$-graded) multiset $\mathfrak{D}(g)^{-1}$ consisting of 0 (if 0 is either a zero or a pole of g) and of the reciprocals of the (nonzero) zeros and poles of g. Under the aforementioned conditions on g (which are relaxed in [**CobLap2**]), this choice of the GPO $\mathbb{D}_g$ guarantees the compactness (and the normality) of the GFO $\mathcal{F}_g$ and even, that $\mathcal{F}_g$ belongs to some Schatten class (i.e., the sequence of characteristic eigenvalues of $\mathcal{F}_g$ belongs to $\ell^n = \ell^n(\mathbb{C})$, for some integer $n \geq 1$); so that the regularized determinant in (4.11) is well defined (and is of order n).

The regularized superdeterminant (or Berezinian, in the terminology of supersymmetric quantum field theory; see, e.g., [**Del3, Wein**]) arising in (4.11) above can be written as a quotient of regularized determinants which are natural generalizations of the well-known (infinite dimensional) Fredholm determinants of trace class compact operators [**Fred**]. For a detailed exposition and for the genesis of the theory of these generalized Fredholm determinants, we refer to [Sim1–3]; see also [CobLap1–2] and [**Lap10**]. We simply mention that it relies in part on the work in [**Fred**], [**Plem**], [**Poin5**], [**Smit**], [**Lids**], [**GohKre**] and [Sim1–3].

In the special case of the zeta function of a variety V over a finite field $\mathbb{F}_q$ and after having made the change of variable $z := q^{-s}$, we obtain a rational function $Z_V = Z_V(z)$ which can be written as an alternating product of (finite-dimensional) determinants, as in (4.1) in the setting of the Weil conjectures.

Furthermore, in the other very important special case when g is an appropriate version of the completed (or global) Riemann zeta function, for example, $g(s) = \xi(s) := \pi^{-s/2}\Gamma(s/2)\zeta(s)$ or $g(s) = \Xi(s) := (s-1)\xi(s)$, the corresponding regularized determinant is then truly infinite dimensional since $\zeta = \zeta(s)$ (and hence also $\xi = \xi(s)$ as well as $\Xi = \Xi(s)$) has infinitely many (critical) zeros . Moreover, the regularized determinant corresponding to the zeros is not a mere Fredholm operator [**Fred**] but involves a renormalization of second order (that is, in the hierarchy of such regularized determinants indexed by integers $n \geq 1$, as in [**Sim2**], it is of level $n = 2$).

These results provide, in particular, a partial but completely rigorous and quite general mathematical realization of what was sought for by a number of authors, including A. Weil [Weil–6], A Grothendieck [Gro1–4], P. Deligne [Del1–2], and especially, in the present context, C. Deninger [Den1–6]. Naturally, we should point out that they do not provide a full realization of those authors and their successors' rich set of ideas and conjectures, especially concerning the existence of suitable cohomology theories ([Weil–3], [Gro1–4], [Den1–6] and, e.g., [Tha1–2], along with the relevant references therein) and of an appropriate positivity condition ([Wei4–6], [Har1–3], [**Con2**]), and therefore cannot (for now) be used to try to prove the Riemann hypothesis, although they may constitute a significant first step, particularly once (and if) one can obtain an appropriate geometric and topological (i.e., cohomological) interpretation (which may, however, be very difficult and take a very long time to achieve; see §4.4).

We also mention that the search for suitable Polya–Hilbert operators has been the object of a number of works, both in physics (see, e.g., [Berr3–4] and [**BerrKea**]) and in mathematics (see, e.g., [Den1–5], [**Con2**] and [**Lap7**, Chs. 4 & 5]). The abstract (functional analytic) framework for GPOs provided in [CobLap1–2] and extended in [**Lap10**] is, in some ways, too general for certain purposes (for example, for proving RH, at least in our current state of knowledge) but presents the advantage of being completely categorical and widely applicable, well beyond the settings of fractal geometry and number theory which originally motivated it.

The proof of the identity (4.11) given in [**CobLap1**] makes use of Hadamard's factorization theorem for entire functions of finite order ([Had1–2], [**Conw**]).

As is shown in [**Lap10**], the above formalism (and, in particular, the identity (4.11)) can also be applied to all the elements of the Selberg class ([**Sel2, Sarn**], [Murt1–3] and [**Lap7**, App. E]) and hence, at last conjecturally, to all of the arithmetic zeta function or L-functions (more specifically, the automorphic L-functions) for which the extended Riemann hypothesis (ERH) is expected to hold (see, e.g., [**Sarn**], [Murt1–3, ParsSh1–2] and [**Lap7**, App. C]). In this case, one also needs to use regularized determinants of order two, as for the various versions of the completed Riemann zeta function discussed above.

4.4. Towards a fractal cohomology. Here, we briefly discuss the emerging theory of 'fractal cohomology', as expounded upon in [**Lap10**]. This theory builds on the work from [**CobLap1**] (and from [**CobLap2**]) described in §4.3, itself inspired in part by the theory from [**HerLap1**] described in §4.2, as well as on the many references provided in §4.1, including [Weil–3], [**Gro1**], [**Den3**] and [Lap-vF2–4].

The idea underlying the notion of a fractal cohomology is that to every complex dimension (now understood in the extended sense of a zero or a pole of the given fractal zeta function or arithmetic zeta function or else, more generally, of a suitable meromorphic function $g = g(s)$),[152] one can associate a finite-dimensional complex Hilbert space $H^{\pm}(\omega)$, with the plus (respectively, minus) sign corresponding to ω being a zero (respectively, a pole) of dimension (over $\mathbb{C}$) equal to the multiplicity of ω.

More generally, to (the reciprocal $\mathfrak{D}^{-1}(g)$ of) the divisor $\mathfrak{D}(g)$ of the meromorphic function g, we associate a $\mathbb{Z}_2$-graded (or supersymmetric) complex (and separable) Hilbert space[153]

$$H = H^+ \oplus_s H^-, \tag{4.13}$$

where

$$H^+ := \bigoplus_{\omega\in\mathcal{Z}^+} H^+(\omega) \quad \text{and} \quad H^- := \bigoplus_{\omega\in\mathcal{Z}^-} H^-(\omega), \tag{4.14}$$

[152] Note that the zeros and poles of the meromorphic function $g = g(s)$, can be viewed as the (necessarily simple) poles of (minus) the logarithmic derivative of g, the function $-g'/g$, with the associated residues being equal to the orders of the zeros or of the poles (and with corresponding signs identifying whether the poles of $-g'/g$ comes from a zero or a pole of g).

[153] Here and thereafter, for notational simplicity, we use the same symbol $\oplus_s$ to indicate the $\mathbb{Z}_2$-graded (or supersymmetric) direct sum of Hilbert spaces or of operators acting on them, as well as of (multi)sets, depending on the context.

with $\mathcal{Z}^+$ (respectively, $\mathcal{Z}^-$) denoting the set of the reciprocals of the distinct zeros (respectively) poles of g (with the convention according to which one does not have to take the reciprocal of 0, if this value occurs in either $\mathcal{Z}^+$ or $\mathcal{Z}^-$). The finite-dimensional (complex) Hilbert space $H^{\pm}(\omega)$ has dimension equal to the multiplicity of the corresponding pole or zero ω of $g = g(s)$.

As was explained in §4.3, H is the eigenspace of the *generalized Polya–Hilbert space* (GPO) $\mathbb{D} = \mathbb{D}_g = \mathbb{D}_{\mathcal{Z}}$, with $\mathcal{Z} := \mathcal{Z}^+ \oplus_s \mathcal{Z}^-$ and $\mathbb{D} = \mathbb{D}^+ \oplus_s \mathbb{D}^-$. Furthermore, $\mathcal{F} = \mathcal{F}^+ \oplus_s \mathcal{F}^-$, defined as the restriction of the GPO $\mathbb{D} = \mathbb{D}^+ \oplus_s \mathbb{D}^-$ to its (total) graded eigenspace $H = H^+ \oplus_s H^-$ (as given by (4.13) and (4.14)). The (possibly unbounded) operator $\mathcal{F}$ is called the *generalized Frobenius operator* (GFO) associated with g (or with $\mathcal{Z}$). Under suitable assumptions on g (specified in [CobLap1–2] and in [**Lap10**]),[154] the GFO $\mathcal{F}$ enables us to recover the meromorphic function g via the determinant formula (4.11) showing that $g = g(s)$ is the "characteristic polynomial" of $\mathcal{F}$ (viewed as a supersymmetric regularized determinant).

REMARK 4.1. By construction, the multiplicity of $\omega \in \mathcal{Z}^+$ (respectively, $\omega \in \mathcal{Z}^-$), with $\omega \neq 0$, as an eigenvalue of the GPO $\mathbb{D}^+$ (respectively, $\mathbb{D}^-$), or equivalently, of the GFO $\mathcal{F}^+$ (respectively, $\mathcal{F}^-$), coincides with the multiplicity of the corresponding zero (respectively, pole) ω^{-1} of the meromorphic function $g = g(s)$. Hence, the finite-dimensional complex Hilbert space $H(\omega)$ has dimension over $\mathbb{C}$ equal to the multiplicity of the corresponding zero (respectively, pole) ω^{-1} of $g = g(s)$.[155] Consequently, since finite-dimensional Hilbert spaces of the same dimension are isomorphic, the (total) cohomology space H in (4.13) could be equivalently defined by letting ω run through the divisor $\mathfrak{D}(g)$ of g (rather than through its reciprocal $\mathfrak{D}^{-1}(g)$, as indicated in (4.14)): symbolically,

$$H = \bigoplus_{\omega \in \mathfrak{D}(g)} H(\omega). \tag{4.15}$$

Accordingly, in (4.14), the "even" (respectively, "odd") cohomology space H^+ (respectively, H^-) can be defined by replacing $\mathcal{Z}^+$ (respectively, $\mathcal{Z}^-$) by the set of *distinct poles* (respectively, of *distinct zeros*) of $g = g(s)$: symbolically,

$$H^+ = \bigoplus_{\omega \in \{\text{zeros of } g\}} H^+(\omega) \quad \text{and} \quad H^- = \bigoplus_{\omega \in \{\text{poles of } g\}} H^-(\omega) . \tag{4.16}$$

When $\mathcal{Z}$ is infinite (that is, when g has at least infinitely many zeros or poles), then the *total cohomology space* $H = H_g = H_{\mathcal{Z}}$ is infinite dimensional, and vice versa. Furthermore, since $\mathcal{Z}$ is at most countable, the cohomology space H is a separable (complex) Hilbert space. This is the case, for example, when $g = g(s)$ is one of the completed Riemann zeta functions, $\xi = \xi(s)$ or $\Xi = \Xi(s)$. Therefore, the (total) cohomology space H_ξ or H_Ξ is *intrinsically infinite dimensional* and cannot be reduced to a finite-dimensional one.

On the other hand, in the case of the zeta function attached to a variety V over a finite field $\mathbb{F}_q$ (as in §4.1), the periodicity of the zeta function combined with the change of variable $z := q^{-s}$ yields a rational function $Z_V = Z_V(z)$ to which is associated a finite-dimensional cohomology space (corresponding to the zeros and

[154]More specifically, in light of [**CobLap1**], one assumes that the meromorphic function $g = f/h$ is the ratio of two entire functions f and h of finite orders (which is sufficient for all of the potential applications to number theory, as well as for many of the potential applications to fractal geometry), while according to [**CobLap2**], one should be able to remove the hypothesis that the entire functions f and h are of finite orders.

[155]Of course, the same is true if $\omega = 0$ provided we replace ω^{-1} by ω, in the latter statement.

the poles of the rational function, counted according to their multiplicities). *Hence, in this case, the total cohomology space is a priori infinite dimensional but can also naturally be reduced to a finite-dimensional one.*

We thus obtain a cohomology theory which seems to satisfy several (if not all) of the main properties expected by Grothendieck (in [Gro1–4]) and by Deninger (in [Den1–6]), for example. (See also, e.g., [Tha1–2] for an exposition of some of those ideas.) Let us further explain this statement, but without going into the details. The key is that this "fractal cohomology theory" satisfies the counterpart (in the present context) of the *Künneth formula* (for the product of two algebraic varieties or of two differentiable manifolds; see, e.g., [Dieu1,2] and [**MacL**]). This is, in fact, a key motivation for the author's conjecture (Conjecture 3.16) made in §3.4 about the set of complex dimensions (really, the divisor of the associated fractal zeta function) of the Cartesian product of two bounded sets and, more generally, of two RFDs (in $\mathbb{R}^N$). If correct, this conjecture would help provide a geometric interpretation of the Künneth formula of fractal cohomology theory in terms of Cartesian products (at least in the geometric setting of bounded sets and RFDs in $\mathbb{R}^N$).

A suitable version of Poincaré duality (or of an appropriate extension thereof) in the context of fractal cohomology still remains to be found; see, e.g., [Poin1–4], [Dieu1,2] and [**MacL**] for the classic notions of Poincaré duality. A natural extension of that notion could clearly be formulated, however, in terms of the fractal cohomology spaces associated with $g(s)$ and with $g(1-s)$.

It is noteworthy that Grothendieck's beautiful dream and elusive notion of a "motive" is in fact intimately connected with the search for a "universal cohomology theory" satisfying several axioms, including especially, the analog of Künneth's formula and of Poincaré duality. (See, e.g., [**Kah**], [**Tha1**] and [**Tha2**].)[156] The emerging fractal cohomology theory proposed in [**Lap10**] (and building, in particular, on [**Lap-vF4**], [**Lap7**] and [CobLap1,2]) is evolving in that spirit. It associates (in a functorial way) to a suitable meromorphic function g (or to its divisor $\mathfrak{D}(g)$) a cohomology space H_g, which is the eigenspace of the generalized Frobenius operator (GFO) $\mathcal{F} = \mathcal{F}_g$. In addition, the identity (4.11) which enables one (under appropriate hypotheses) to recover the meromorphic function from the action of $\mathcal{F}$ on H can be viewed as naturally providing an associated inverse functor in this context.

4.4.1. *Grading of the fractal cohomology by the real parts.* It is natural to wonder why, unlike in standard algebraic or differential toplogy (see, e.g., [**Dieu2**] and [**MacL**]), the fractal cohomology spaces are no longer (in our context) indexed by integers. In fact, a natural grading of the total fractal cohomology space $H = H^+ \oplus_s H^-$ is provided by the real parts (of the reciprocals) of the zeros and poles of the meromorphic function g.[157] Hence, in general, H is graded by real numbers and not just by (nonnegative) integers.[158]

More specifically, let $\mathcal{Z} = \mathcal{Z}^+ \oplus_s \mathcal{Z}^-$, where $\mathcal{Z}^+$ and $\mathcal{Z}^-$ denote, respectively, the reciprocals of the distinct zeros and poles of g (with the same convention concerning

[156] See also [**Cart**] for an interesting discussion of the evolution of the notion of a 'geometric space'.

[157] If 0 is either a zero or a pole of g, then H^+ or H^- is also graded by zero.

[158] In light of Remark 4.1, we could replace throughout this discussion the real parts of the reciprocals of the zeros and of the poles by the zeros and the poles themselves; see also Remark 4.2.

0 as usual). Furthermore, let $R^{\pm}$ denote the (at most countable) set of real parts of the elements of $\mathcal{Z}^{\pm}$. Moreover, for $\alpha \in R^{\pm}$, let

$$\mathcal{Z}^{\pm}_{\alpha} := \{\omega \in \mathcal{Z}^{\pm} : Re(\omega) = \alpha\} \tag{4.17}$$

and

$$H^{\pm}_{\alpha} := \underset{\omega \in \mathcal{Z}^{\pm}_{\alpha}}{\oplus} H(\omega), \tag{4.18}$$

where, as earlier, $H(\omega)$ denotes the eigenspace corresponding to the eigenvalue ω of $\mathcal{F}_g$ (and also of $\mathbb{D}_g$); so that the dimension (over $\mathbb{C}$) of the finite-dimensional Hilbert space $H^{\pm}(\omega)$ is equal to the multiplicity of ω as an eigenvalue of $\mathcal{F}_g$ (or, equivalently, if $\omega \neq 0$, as the reciprocal of a zero or a pole of g; if $\omega = 0$, then we can omit the word "reciprocal" here).

Then, we can now provide the following precise formulation of the statement according to which the (total) fractal cohomology space is graded by the real parts, of the elements of $\mathfrak{D}(g)^{-1}$:

$$H^{\pm} = \underset{\alpha \in R^{\pm}}{\oplus} H^{\pm}_{\alpha}, \tag{4.19}$$

with $R^{\pm}$ defined just before (4.17) and $H^{\pm}_{\alpha}$ defined by (4.18) for all $\alpha \in R^{\pm}$.

REMARK 4.2. For notational simplicity, we have worked above with the grading provided by the real parts of the elements of $\mathcal{Z}^{+}$ and $\mathcal{Z}^{-}$. However, in light of Remark 4.1, we could just as well replace $\mathcal{Z}^{+}$ and $\mathcal{Z}^{-}$ by the sets of distinct zeros and poles of $g = g(s)$, respectively. And similarly, for the definition of $R^{\pm}$, $\mathcal{Z}^{\pm}_{\alpha}$, $H^{\pm}_{\alpha}$, H and $H^{\pm}$ in (or around) (4.13), (4.14) and (4.17)–(4.19). Consequently, in that setting, the fractal cohomology spaces H, H^{+} and H^{-} would actually be indexed by the real parts of the elements of the divisor $\mathfrak{D}(g)$, the zero set and the pole set of the meromorphic function $g = g(s)$, respectively. Accordingly, the corresponding grading index sets R^{+} and R^{-} would instead consist of the real parts of the distinct zeros and of the distinct poles of $g = g(s)$, respectively.

We note that $H^{\pm}$ is graded by the (possibly infinite, [**BruDKPW**]) matroid ([**Oxl, Wel**]) $R^{\pm}$, of rank (possibly infinite and defined as in [**BruDKPW**]) equal to the number of vertical lines in $\mathcal{Z}^{\pm}$ (or, equivalently, to the cardinality of $R^{\pm}$, viewed either as the set of distinct zeros/poles of $g = g(s)$).

For example, for a d-dimensional variety V over a finite field $\mathbb{F}_q$, R^{-} (respectively, R^{+}) consists of the integers j (respectively, the half-integers $j/2$), where $j \in \{0, 1, \cdots, d\}$. This is the case relative to the original s-variable. On the other hand, if $z := q^{-s}$, then the corresponding sets R^{-} and R^{+} are $\{q^{-j} : j = 0, 1, \cdots, d\}$ and $\{q^{-j/2} : j = 0, 1, \cdots, d\}$, respectively. Thus, in light of the analog of the Riemann hypothesis proved by Deligne in [Del1–2] in the general case when $d \geq 1$ (and by Weil in [Weil–3] when $d = 1$), the determinant formula (4.11) yields (still in the z-variable)

$$Z_V(z) = \frac{Q_1(z) \cdots Q_{2d-1}(z)}{Q_0(z) \cdots Q_{2d}(z)}, \tag{4.20}$$

where for $k \in \{0, \cdots, d\}$, the polynomial Q_{2k} is the characteristic polynomial of the linear operator $\mathbf{F}_{2k} := \mathcal{F}^{-}_{q-k}$ (the restriction of the GFO $\mathcal{F}$ to $\mathbf{H}_{2k} := H^{-}_{q-k}$) and for $k \in \{1, \cdots, d\}$, the polynomial Q_{2k-1} is the characteristic polynomial of the linear operator $\mathbf{F}_{2k-1} := \mathcal{F}^+_{q^{-k/2}}$ (the restriction of the GFO $\mathcal{F}$ to $\mathbf{H}_{2k-1} := H^+_{q^{-k/2}}$). With this notation, (4.19) becomes

$$H^+ = \bigoplus_{k=1}^{d} \mathbf{H}_{2k-1} \quad \text{and} \quad H^- = \bigoplus_{k=0}^{d} \mathbf{H}_{2k}, \tag{4.21}$$

while the total cohomology space H is given by

$$H = H^+ \oplus_s H^- = \bigoplus_{j=0}^{2d} \mathbf{H}_j. \tag{4.22}$$

In the case of the completed Riemann zeta function $\xi(s)$, *and assuming the truth of the classic Riemann hypothesis*, we have that $R^+ := \{1/2\}$ and $R^- := \{0, 1\}$, while $\mathcal{Z}^- = \{0, 1\}$ and

$$\mathcal{Z}^+ := \{\text{reciprocals of the critical zeros of } \zeta(s)\}. \tag{4.23}$$

Hence, with the notation[159]

$$\mathbf{H}_0 := H_0^-, \quad \mathbf{H}_2 := H_1^-, \quad \text{and} \quad \mathbf{H}_1 := H^+_{1/2}, \tag{4.24}$$

while $\boldsymbol{\mathcal{F}}_j$ is defined as the restriction of the generalized Frobenius operator $\mathcal{F}$ to $\mathbf{H}_j$, for $j \in \{0, 1, 2\}$, then (4.11) becomes (neglecting an unimportant factor):[160]

$$\xi(s) = \frac{\det(I - s\boldsymbol{\mathcal{F}}_1)}{\det(I - s\boldsymbol{\mathcal{F}}_0)\det(I - s\boldsymbol{\mathcal{F}}_2)}, \tag{4.25}$$

with $\det(I - s\boldsymbol{\mathcal{F}}_0) = s$ and $\det(I - s\boldsymbol{\mathcal{F}}_2) = s - 1$ corresponding to the pole of $\xi = \xi(s)$ at $s = 0$ and at $s = 1$, respectively.

Note that with the above notation (which is inspired by the one used for varieties over finite fields as well as by what is expected for Weil-type cohomologies in the present context as well as for more general arithmetic L-functions),[161] we have (in light of (4.24))

$$H^+ = \mathbf{H}_1 \quad \text{and} \quad H^- = \mathbf{H}_0 \oplus \mathbf{H}_2. \tag{4.26}$$

The classic Riemann hypothesis (for ζ or, equivalently, for ξ) is equivalent to the fact that the total even cohomology space of $\xi = \xi(s)$ (namely, $H^+ = H^+_\xi = \mathbf{H}_1$) is '*monofractal*', i.e., with the notation of Remark 4.2, RH is equivalent to the fact that R^+ consists of a single point (of necessity, then, $R^+ = \{1/2\}$).[162]

Entirely analogous statements can be made about the zeta functions of number fields [ParsSh1–2] and of more general arithmetic L-functions, such as automorphic L-functions or conjecturally equivalently (see, e.g., [**Lap7**, App. C and App. E]), the members of the Selberg class.

In contrast, the total odd cohomology space $H^- = H^-_{\zeta_{\mathcal{L}}}$ (still relative to the underlying $\mathbb{Z}_2$-grading) of a generic nonlattice self-similar string $\mathcal{L}$ is '*multifractal*' in a very strong sense; namely, R^-, the set of reciprocals of the real parts of the

---

[159] We are now using the grading of the total cohomology space by the real parts of the zeros and the poles of $\xi = \xi(s)$, as in Remark 4.2.

[160] This factor is a nonvanishing entire function involving the number π and Euler's constant γ; see [**CobLap1**] and [**Lap10**] for its specific value.

[161] See, e.g., [Gro1–3], [**Dieu1**], [**Den3**], [**Con2**], [**Lap7**, esp., Ch. 4 & App. B], [**Kah**], [Tha1–2] and the relevant references therein.

[162] Observe that by the functional equation for ξ (i.e., $\xi(s) = \xi(1-s)$, for all $s \in \mathbb{C}$) and since ξ has zeros on the critical line $\{Re(s) = 1/2\}$, we must have $R^+ = \{1/2\}$ since otherwise, R^+ would contain at least three points, $1/2, \rho$ and $1/2 - \rho$, for some $\rho \in (0, 1/2)$. Also, conversely, if R^+ had more than one point, then RH would obviously be violated.

complex dimensions of $\mathcal{L}$ (in the usual sense of §2 and §3) is a countably infinite set which is dense in a compact interval $[D^{-1}, D_\ell^{-1}]$, where $-\infty < D_\ell \leq D$ and $D = D_{\mathcal{L}}$ is the Minkowski dimension of $\mathcal{L}$.[163]

4.4.2. *Open problems and perspectives: Geometric interpretation and beyond.* We close this section (and paper) by mentioning a few open problems and long-term research directions related to the nascent fractal cohomology theory and its possible geometric and topological interpretations:

(*i*) *Geometric interpretation.* It would be very useful and interesting to obtain a good geometric interpretation of the theory. It is natural to wonder what is the 'curve' (or 'variety') underlying fractal cohomology. By analogy with the case of curves (or, more generally, varieties) over finite fields, one may guess that the underlying supersymmetric 'fractal curve' is the divisor $\mathfrak{D}(g)$ of the given meromorphic function g (or perhaps, its inverse, $\mathfrak{D}(g)^{-1}$, as defined earlier).[164] Ignoring the distinction between $\mathfrak{D}(g)$ and its inverse, let us denote symbolically this 'curve' by $C = C_g$. Then, abusing geometric language, one can think of the original Hilbert space $\mathcal{H}_g$, the eigenspace of the GPO $\mathbb{D}_g$ (namely, the fractal cohomology space H_g) and even the GPO $\mathbb{D}_g$ and the GFO $\mathcal{F}_g$, as 'bundles' or 'sheaves' over C.

If we let the function g vary in a suitable 'moduli space' of meromorphic functions (or zeta functions),[165] we then obtain geometric, analytic and algebraic structures (such as $C_g, \mathcal{H}_g, \partial_g, \mathbb{D}_g, \mathcal{F}_g$ and H_g) which are naturally defined on that moduli space.

Pursuing the analogy with curves (or, more generally, varieties) over finite fields, one may wonder if the curve C should not be augmented and replaced by a new (and larger) 'curve' $\widetilde{C}$ whose (generic) points would coincide with the fixed points of the generalized Frobenius operator (GFO) $\mathcal{F}$ and of its iterates $\mathcal{F}^n$ (with $n \in \mathbb{N}$ arbitrary). In the present discussion, we allow the solutions (i.e., the fixed points of $\mathcal{F}^n$, for some $n \in \mathbb{N}$) to be distributional solutions (in a suitably defined weighted space of tempered distributions) and not just the Hilbert space solutions.[166] Then, a simple differential equation computation (see [**Lap10**]) shows that $\widetilde{C} = \widetilde{C}_g$ consists of the points of $C = C_g$ augmented by a copy of the group $\mu(\infty)$ of all (complex) roots of unity attached to (and acting on) each point of C.[167] In other words, $\widetilde{C}$ can be viewed as a 'principal bundle' over C, with structure group $\mu(\infty)$. Alternatively, $\widetilde{C}$ can be viewed as a 'sheaf' over C.

Amazingly and likely not coincidentally, the group $\mu(\infty)$ (and variants thereof) also plays a key role in the theory of motives over the elusive field of one element,

---

[163]We assume here implicitly that there are no (drastic) cancellations between the zeros and the poles of $\zeta_{\mathcal{L}}$. If we work instead with $H^- = H^-_{\partial\Omega,\Omega}$, where the RFD $(\partial\Omega, \Omega)$ is a geometric realization of $\mathcal{L}$, then $R\backslash\{0\}$ is of the above form.

[164]For a nonlattice self-similar string $\mathcal{L}$ (in the case of a single gap), according to the results obtained in [**Lap-vF4**, Ch. 3, esp., §3.4], the (multi)set of complex dimensions $\mathfrak{D}(g) = \mathcal{D}(\zeta_{\mathcal{L}})$ obeys a quasiperiodic pattern and is conjectured to form a '*generalized quasicrystal*' (in a sense to be yet fully specified); see [**Lap-vF4**, Problem 3.22]. More broadly, in [**Lap7**, esp., Ch. 5 and App. F], this type of generalized quasicrystal plays a key conceptual role.

[165]Compare with the moduli spaces of fractal membranes (and of the associated zeta functions or spectral 'partition functions') that play a central role in [**Lap7**, Ch. 5].

[166]If we only allowed Hilbert space solutions, then $\widetilde{C}$ would simply coincide with C.

[167]Hence, $\mu(\infty) = \cup_{n\geq 1}\mu(n)$, where for each $n \in \mathbb{N}$, $\mu(n)$ is the group of (complex) n-th roots of unity; clearly, here, $\mu(n)$ does not denote the value of the Möbius function at n.

$\mathbb{F}_1$ (which originated in Y. Manin's seminal article [**Mani**]); see, especially, [**Kah**], [**Tha1**], [**Tha2**] and the relevant references therein.[168]

In the aforementioned references, working over $\mathbb{F}_1$ typically leads to adding new points to the underlying curve or variety (or motif) that were formally *invisible*. The same is true in the present situation of 'fractal motives' and the associated fractal cohomology. Indeed, going from C to $\widetilde{C}$ amounts to adding previously invisible points. Analytically, as was alluded to above, this amounts to going outside the original Hilbert space $\mathcal{H}_g$ and finding all of the fixed points of the iterates $\mathcal{F}_g^n$ (for any $n \geq 1$) that lie in a suitable weighted space of tempered distributions containing $\mathcal{H}_g$; see [**Lap10**].

In this context, for every given $n \in \mathbb{N}$, the fixed points of $\mathcal{F}^n$ correspond to the 'points' of the curve $\widetilde{C}$ which lie in the n-th field extension of $\mathbb{F}_1$. Heuristically, they can be thought of as a 'principal bundle' over C with structure group $\mu(n)$, the group of n-th complex roots of unity. (Note the close analogy with the case of curves (or varieties) over finite fields discussed in §4.1.)

Therefore, at the most fundamental level, we should aim at interpreting the present theory of fractal cohomology (as developed in [CobLap1–2] and [**Lap10**]) as a universal cohomology theory of '*fractal motives*' over $\mathbb{F}_1$, the elusive field of one element. Finding a coherent and completely rigorous geometric and arithmetic interpretation of this type is clearly a long-term open problem.

REMARK 4.3. The 'curve' $C = C_g$ itself, or its augmentation (by invisible points, the distributional fixed points of $\mathcal{F}^n$, for any $n \in \mathbb{N}$) $\widetilde{C} = \widetilde{C}_g$, can be viewed as the '*sheaf of complex dimensions*' (here, visible zeros and poles of g) associated with 'the' analytic continuation of g (i.e., with 'the' meromorphic continuation of g on any given connected open set U of $\mathbb{C}$ on which the latter exists). Interestingly, the notion of a sheaf was introduced in the 1940s by J. Leray to solve certain problems in cohomology theory and then used to precisely deal with the a priori ill-defined notion of analytic continuation of a function of several complex variables, in connection with the Cousin problem. (See, e.g., [**GunRos**], [**Ebe**] and [**Dieu2**].) The previously mentioned extensions of the notion of complex dimensions (as non-removable singularities of suitable complex analytic functions on Riemann surfaces, or in higher dimensions, on complex analytic varieties or even, on analytic spaces) and of the associated theory (see §2.5 and §3.6, as well as part (*iii*) of the present subsection) would, in the long term, provide a natural and significantly broader framework within which to consider the (yet to be precisely defined) sheaves C and $\widetilde{C}$ in this extended context.

We next very briefly mention two other key research directions in this area:

(*ii*) *Fractal homology.* As is well known, standard cohomology theories in topology and in differential geometry, for example) are often the dual theories of suitable homology theories. (See, e.g., [**Dieu2**] and [**MacL**].) We wish to ask here what could be an appropriate '*fractal homology theory*' in the present context. More specifically, in the geometric context of bounded subsets of $\mathbb{R}^N$ (and, more generally, of RFDs in $\mathbb{R}^N$), can one construct a suitable fractal homology theory whose

[168]See also the recent work [**Har4**] in which $\mu(\infty)$ does not seem to play a role but whose proposed geometric and algebraic language and formalism for number theory and 'varieties' over $\mathbb{F}_1$ is quite appealing and should, in the long term, be connected with aspects of the present theory.

dual theory would be the corresponding fractal cohomology theory (associated with the divisors of fractal zeta functions, as described above).[169] Is there an appropriate generalization of Poincaré duality and of Künneth's formula in this context? What about the more general (but at present, less geometric) context of general (and suitable) meromorphic functions?

(iii) Singularities of functions of several complex variables. Finally, we mention that as was alluded to in Remark 4.3, it would also be interesting to extend the theory of complex fractal dimensions and the associated fractal cohomology theory to (suitable) meromorphic functions of several complex variables naturally defined on complex manifolds (or, more generally on complex analytic spaces), as well as to develop an associated theory of (generalized) sheaves and sheaf cohomology (see, e.g., [**GunRos**] and [**Ebe**] for the classic theory).

Just as in the case of a single complex variable, we need not restrict ourselves to meromorphic functions defined on domains of $\mathbb{C}^r$, with $r \geq 1$. One is therefore naturally led to consider (nonremovable) '*singularities*' of functions that go beyond the mere poles, as is already apparent in the present geometric theory of complex fractal dimensions; see, e.g., §2.5 and §3.6 above, along with [**LapRaZu10**] for a first model in the case when $r = 1$ and thus when suitable Riemann surfaces would be associated with the given singularities.

For more information on fractal cohomology theory and the many associated open problems, we refer the interested reader to (the latter part of) the author's forthcoming book, [**Lap10**].

Acknowledgements

This may be a fitting place to express my deep gratitude to Erin P. J. Pearse, John A. Rock and Tony Samuel, the organizers of the International Conference and Summer School (supported by the National Science Foundation) on *Fractal Geometry and Complex Dimensions* held on the occasion of my sixtieth birthday in the beautiful setting of the campus of the California Polytechnic State University at San Luis Obispo, as well as to the editors of this volume of *Contemporary Mathematics* (the proceedings volume of the above SLO meeting), the aforementioned organizers along with Robert G. Niemeyer, for their great patience in waiting (with some anxiety, I presume) for my contribution and for their helpful suggestions for shortening the original text (which has now become the foundation for the author's forthcoming book, [**Lap10**]).

I wish to thank all of my wonderful collaborators over the past twenty five to thirty years on many aspects of the theory of complex fractal dimensions and its various extensions or related topics, including (in rough chronological order) Jacqueline Fleckinger ([**FlLap**] and other papers), Carl Pomerance [LapPo1–3], Helmut Maier [LapMa1–2], Jun Kigami [KiLap1–2], Michael M. H. Pang [**LapPa**], Cheryl A. Griffith, John W. Neuberger and Robert J. Renka [**LapNRG**], Christina Q. He [**HeLap**], Machiel van Frankenhuijsen [Lap-vF1–7], Ben M. Hambly [**HamLap**], Erin P. J. Pearse [LapPe1–3], Hung (Tim) Lu ([LapLu1–3], [LapLu-vF1–2]), John A. Rock ([**LapRo**], [**LapRoZu**], [**LapRaRo**]), Erik Christensen and Cristina Ivan [**ChrIvLap**], Jacques Lévy Vehel [**LapLevyRo**], Steffen Winter [LapPeWi1–2], Nishu Lal [LalLap1–2], Hafedh Herichi [HerLap1–5], Rolando de Santiago and

[169] At least for certain self-similar geometries, there seems to be glimpses of such a theory, in which the underlying complex dimensions play a natural role.

Scott A. Roby [**deSLapRRo**], Robyn L. Miller and/or Robert Niemeyer (in papers on fractal billiards), Jonathan J. Sarhad [**LapSar**], Goran Radunović and Darko Žubrinić [LapRaZu1–10], Kate E. Ellis and Michael Mackenzie [**ElLapMcRo**], Tim Cobler [CobLap1–2] and Ryszard Nest [**LapNes**].

I also wish to thank my many diverse and talented Ph.D. students (soon twenty of them, this year, in 2018) and postdocs, a number of whom have been mentioned among my collaborators just above.

Furthermore, I am most grateful to the two very conscientious and thorough referees of this paper for their very helpful suggestions, questions and comments.

Moreover, I am very indebted to my personal Administrative Assistant, Ms. Yu-Tzu Tsai, without whose invaluable help this paper would never have been completed (almost) on time.

Finally, I would like to gratefully acknowledge the long-term support of the U.S. National Science Foundation (NSF) over the past thirty years, since the beginning of the theory of complex dimensions and throughout much of its prehistory, under the research grants DMS-8703138, DMS-8904389, DMS-9207098, DMS-9623002, DMS-0070497, DMS-0707524 and DMS-1107750.

Glossary

$a_k \sim b_k$ as $k \to \infty$, asymptotically equivalent sequences 154, 157

$\mathfrak{a} = \mathfrak{a}_c$, spectral operator 231

$\mathfrak{a}^{(T)} = \mathfrak{a}_c^{(T)}$, truncated spectral operator 233

A_ε, ε-neighborhood of $A \subseteq \mathbb{R}^N$ 145, 171

$\mathcal{A} = \mathcal{A}_c$, global spectral operator 234

$\mathcal{B}(\mathbb{H}_c)$, Banach algebra of bounded linear operators on the Hilbert space $\mathbb{H}_c$ 233

CS, Cantor string 150

$\mathbb{C}$, the field of complex numbers 152

$\widetilde{\mathbb{C}} := \mathbb{C} \cup \{\infty\}$, the Riemann sphere 232

$c\ell(A)$ or $\overline{A}$, closure of a subset A of $\mathbb{C}$ or of $\mathbb{R}^N$ 174, 232

$\underline{D}$, $\overline{D}$, lower, upper Minkowski dimension (of a RFD (A, Ω) in $\mathbb{R}^N$) .. 171

D, Minkowski dimension (of a fractal string $\mathcal{L}$ or of $A \subseteq \mathbb{R}^N$, or else of a RFD (A, Ω) in $\mathbb{R}^N$) 151, 171

$\mathcal{D} = \mathcal{D}_\mathcal{L}$, the set of complex dimensions of a fractal string $\mathcal{L}$ (or of its boundary) 151, 152

$\dim_\mathbb{C} \mathcal{L} = \mathcal{D}_\mathcal{L}(\mathbb{C})$, the set of all complex dimensions (in $\mathbb{C}$) of the fractal string $\mathcal{L}$ 151, 152

$\dim_{PC} \mathcal{L} = \mathcal{D}_{PC}(\mathcal{L})$, the set of principal complex dimensions of the fractal string $\mathcal{L}$ 151, 152

$\dim_\mathbb{C}(A, \Omega) = \mathcal{D}_{A,\Omega}(\mathbb{C})$, the set of all complex dimensions (in $\mathbb{C}$) of the RFD (A, Ω) in $\mathbb{R}^N$ 174

$\dim_{PC}(A, \Omega) = \mathcal{D}_{PC}(A, \Omega)$, the set of principal complex dimensions of the RFD (A, Ω) in $\mathbb{R}^N$ 177

$\mathcal{D}(\zeta_\mathcal{L}) = \mathcal{D}(\zeta_\mathcal{L}; U)$, the set of (visible) poles (in U) of the geometric zeta function $\zeta_\mathcal{L}$ of a fractal string $\mathcal{L}$ 151, 152

$\mathcal{D}_\mathcal{L} = \mathcal{D}_\mathcal{L}(U)$, the set of (visible) complex dimensions of a fractal string $\mathcal{L}$ ($\mathcal{D}_\mathcal{L} = \mathcal{D}(\zeta_\mathcal{L})$) 151, 152

References

[AcoDur-LopGar] G. Acosta, R. G. Durán, and F. López García, *Korn inequality and divergence operator: Counterexamples and optimality of weighted estimates*, Proc. Amer. Math. Soc. **141** (2013), no. 1, 217–232, DOI 10.1090/S0002-9939-2012-11408-X. MR2988724

[Art] E. Artin, Quadratische Körper im gebiet der höheren Kongruenzen, I and II. *Math. Zeitschrift* **19** (1924), 153–206 and 207–246.

[AtzBri] A. Atzmon and B. Brive, *Surjectivity and invariant subspaces of differential operators on weighted Bergman spaces of entire functions*, Bergman spaces and related topics in complex analysis, Contemp. Math., vol. 404, Amer. Math. Soc., Providence, RI, 2006, pp. 27–39, DOI 10.1090/conm/404/07632. MR2244002

[Bag] B. Bagchi, *The Statistical Behaviour and Universality Properties of the Riemann Zeta-Function and Other Allied Dirichlet Series*, Ph.D. Thesis, Indian Statistical Institute, Calcutta, India, 1981.

[Bak] A. Baker, *Transcendental number theory*, Cambridge University Press, London-New York, 1975. MR0422171

[Barl] M. T. Barlow, Diffusions on fractals, in: *Lectures on Probability Theory and Statistics* (P. Bernard, ed.), Ecole d'Eté de Probabilités de Saint-Flour XXV–1995, Lecture Notes in Mathematics, vol. 1690, Springer-Verlag, 1998, pp. 1–121.

[BarnDemk] M. F. Barnsley and S. Demko, *Iterated function systems and the global construction of fractals*, Proc. Roy. Soc. London Ser. A **399** (1985), no. 1817, 243–275. MR799111

[Be] A. F. Beardon, *Iteration of rational functions*, Graduate Texts in Mathematics, vol. 132, Springer-Verlag, New York, 1991. Complex analytic dynamical systems. MR1128089

[BedKS] T. Bedford, M. Keane, and C. Series (eds.), *Ergodic theory, symbolic dynamics, and hyperbolic spaces*, Oxford Science Publications, The Clarendon Press, Oxford University Press, New York, 1991. Papers from the Workshop on Hyperbolic Geometry and Ergodic Theory held in Trieste, April 17–28, 1989. MR1130170

[BergGos] M. Berger and B. Gostiaux, *Differential geometry: Manifolds, curves, and surfaces*, Graduate Texts in Mathematics, vol. 115, Springer-Verlag, New York, 1988. Translated from the French by Silvio Levy. MR917479

[Berr1] M. V. Berry, *Distribution of modes in fractal resonators*, Structural stability in physics (Proc. Internat. Symposia Appl. Catastrophe Theory and Topological Concepts in Phys., Inst. Inform. Sci., Univ. Tübingen, Tübingen, 1978), Springer Ser. Synergetics, vol. 4, Springer, Berlin, 1979, pp. 51–53, DOI 10.1007/978-3-642-67363-4_7. MR556688

[Berr2] M. V. Berry, *Some geometric aspects of wave motion: Wavefront dislocations, diffraction catastrophes, diffractals*, Geometry of the Laplace operator (Proc. Sympos. Pure Math., Univ. Hawaii, Honolulu, Hawaii, 1979), Proc. Sympos. Pure Math., XXXVI, Amer. Math. Soc., Providence, RI, 1980, pp. 13–28. MR573427

[Berr3] M. V. Berry, *Riemann's zeta function: A model for quantum chaos*, in: Lecture Notes in Physics, vol. 263, Springer-Verlag, Berlin, 1986, pp. 1–17.

[Berr4] M. V. Berry, *The Bakerian lecture, 1987. Quantum chaology*, Proc. Roy. Soc. London Ser. A **413** (1987), no. 1844, 183–198. MR909277

[BerrKea] M. V. Berry and J. P. Keating, *The Riemann zeros and eigenvalue asymptotics*, SIAM Rev. **41** (1999), no. 2, 236–266, DOI 10.1137/S0036144598347497. MR1684543

[BesTa] A. S. Besicovitch and S. J. Taylor, *On the complementary intervals of a linear closed set of zero Lebesgue measure*, J. London Math. Soc. **29** (1954), 449–459, DOI 10.1112/jlms/s1-29.4.449. MR0064849

[BirSol] M. Š. Birman and M. Z. Solomjak, *Spectral asymptotics of nonsmooth elliptic operators. I, II* (Russian), Trudy Moskov. Mat. Obšč. **27** (1972), 3–52; ibid. **28** (1973), 3–34. MR0364898

[Boh] H. Bohr, *Almost Periodic Functions*, Chelsea Publishing Company, New York, N.Y., 1947. MR0020163

[BohCou] H. Bohr and R. Courant, *Neue Anwendungen der Theorie der Diophantischen Approximationen auf die Riemannsche Zetafunktion* (German), J. Reine Angew. Math. **144** (1914), 249–274, DOI 10.1515/crll.1914.144.249. MR1580903

[BosCon1] J.-B. Bost and A. Connes, *Produits eulériens et facteurs de type* III (French, with English and French summaries), C. R. Acad. Sci. Paris Sér. I Math. **315** (1992), no. 3, 279–284. MR1179720

[BosCon2] J.-B. Bost and A. Connes, *Hecke algebras, type III factors and phase transitions with spontaneous symmetry breaking in number theory*, Selecta Math. (N.S.) **1** (1995), no. 3, 411–457, DOI 10.1007/BF01589495. MR1366621

[Bou] G. Bouligand, Ensembles impropres et nombre dimensionnel, *Bull. Sci. Math.* (2) **52** (1928), 320–344 and 361–376.

[Br] H. Brezis, *Functional analysis, Sobolev spaces and partial differential equations*, Universitext, Springer, New York, 2011. (English translation and revised and enlarged edition of H. Brezis, *Analyse fonctionnelle: Théorie et applications*, Masson, Paris, 1983.)

[BroCar] J. Brossard and R. Carmona, *Can one hear the dimension of a fractal?*, Comm. Math. Phys. **104** (1986), no. 1, 103–122. MR834484

[BruDKPW] H. Bruhn, R. Diestel, M. Kriesell, R. Pendavingh, and P. Wollan, *Axioms for infinite matroids*, Adv. Math. **239** (2013), 18–46, DOI 10.1016/j.aim.2013.01.011. MR3045140

[CaLapPe-vF] D. Carfi, M. L. Lapidus, E. P. J. Pearse and M. van Frankenhuijsen, *Fractal geometry and dynamical systems in pure and applied mathematics*, I & II, Contemporary Mathematics, vols. 600 and 601, Amer. Math. Soc., Providence,

RI, 2013. (I: *Fractals in pure mathematics*, vol. 600. II: *Fractals in applied mathematics*, vol. 601.)

[Cae] A. M. Caetano, *On the search for the asymptotic behaviour of the eigenvalues of the Dirichlet Laplacian for bounded irregular domains*, Internat. J. Appl. Sci. Comput. **2** (1995), no. 2, 261–287. Some aspects of fractals in mathematics and physics (Toulouse, 1993). MR1369783

[CarlJonYoc] L. Carleson, P. W. Jones, and J.-C. Yoccoz, *Julia and John*, Bol. Soc. Brasil. Mat. (N.S.) **25** (1994), no. 1, 1–30, DOI 10.1007/BF01232933. MR1274760

[Cart] P. Cartier, *A mad day's work: From Grothendieck to Connes and Kontsevich. The evolution of concepts of space and symmetry [in Les relations entre les mathématiques et la physique théorique, 23–42, Inst. Hautes Études Sci., Bures-sur-Yvette, 1998; MR1667896 (2000c:01028)]*, Bull. Amer. Math. Soc. (N.S.) **38** (2001), no. 4, 389–408, DOI 10.1090/S0273-0979-01-00913-2. Translated from the French by Roger Cooke. MR1848254

[ChrIvLap] E. Christensen, C. Ivan, and M. L. Lapidus, *Dirac operators and spectral triples for some fractal sets built on curves*, Adv. Math. **217** (2008), no. 1, 42–78, DOI 10.1016/j.aim.2007.06.009. MR2357322

[ChuOsgPomm] M. Chuaqui, B. Osgood, and Ch. Pommerenke, *John domains, quasidisks, and the Nehari class*, J. Reine Angew. Math. **471** (1996), 77–114. MR1374919

[CipGIS] F. Cipriani, D. Guido, T. Isola, and J.-L. Sauvageot, *Spectral triples for the Sierpinski gasket*, J. Funct. Anal. **266** (2014), no. 8, 4809–4869, DOI 10.1016/j.jfa.2014.02.013. MR3177323

[CobLap1] T. Cobler and M. L. Lapidus, *Towards a fractal cohomology: Spectra of Polya-Hilbert operators, regularized determinants and Riemann zeros*, Exploring the Riemann zeta function: 190 years from Riemann's birth, Springer, Cham, 2017, pp. 35–65. MR3700037

[CobLap2] T. Cobler and M. L. Lapidus, Zeta functions and Weierstrass' factorization theorem via regularized determinants and infinitesimal shifts on weighted Bergman spaces (tentative title), in preparation, 2019.

[Coh] D. L. Cohn, *Measure theory*, Birkhäuser, Boston, Mass., 1980. MR578344

[Con1] A. Connes, *Noncommutative geometry*, Academic Press, Inc., San Diego, CA, 1994. MR1303779

[Con2] A. Connes, *Trace formula in noncommutative geometry and the zeros of the Riemann zeta function*, Selecta Math. (N.S.) **5** (1999), no. 1, 29–106, DOI 10.1007/s000290050042. MR1694895

[Conw] J. B. Conway, *Functions of one complex variable*, 2nd ed., Graduate Texts in Mathematics, vol. 11, Springer-Verlag, New York-Berlin, 1978. MR503901

[CouHil] R. Courant and D. Hilbert, *Methods of mathematical physics. Vol. I*, Interscience Publishers, Inc., New York, N.Y., 1953. MR0065391

[CranMH] A. Crannell, S. May, and L. Hilbert, *Shifts of finite type and Fibonacci Harps*, Appl. Math. Lett. **20** (2007), no. 2, 138–141, DOI 10.1016/j.aml.2006.03.007. MR2283900

[DaMcCS] G. Dafni, R. J. McCann, and A. Stancu (eds.), *Analysis and geometry of metric measure spaces*, CRM Proceedings & Lecture Notes, vol. 56, American Mathematical Society, Providence, RI, 2013. Lecture notes from the 50th Séminaire de Mathématiques Supérieures (SMS) held at the Université de Montréal, Montréal, June 27–July 8, 2011. MR3075414

[Dav] E. B. Davies, *Two-dimensional Riemannian manifolds with fractal boundaries*, J. London Math. Soc. (2) **49** (1994), no. 2, 343–356, DOI 10.1112/jlms/49.2.343. MR1260117

[Del1] P. Deligne, *La conjecture de Weil. I* (French), Inst. Hautes Études Sci. Publ. Math. **43** (1974), 273–307. MR0340258

[Del2] P. Deligne, *La conjecture de Weil. II* (French), Inst. Hautes Études Sci. Publ. Math. **52** (1980), 137–252. MR601520

[Del3] P. Deligne, *et al.* (eds.), *Quantum fields and strings: A course for mathematicians*, vols. 1 & 2, Amer. Math. Soc., Providence, RI, 1999.

[DemDenKoÜ] B. Demír, A. Deníz, Ş. Koçak, and A. E. Üreyen, *Tube formulas for graph-directed fractals*, Fractals **18** (2010), no. 3, 349–361, DOI 10.1142/S0218348X10004919. MR2683935

[DemKoÖÜ] A. Deniz, Ş. Koçak, Y. Özdemir, and A. E. Üreyen, *Tube formula for self-similar fractals with non-Steiner-like generators*, Proceedings of the Gökova Geometry-Topology Conference 2012, Int. Press, Somerville, MA, 2013, pp. 123–145. MR3203360

[Den1] C. Deninger, *Local L-factors of motives and regularized determinants*, Invent. Math. **107** (1992), no. 1, 135–150, DOI 10.1007/BF01231885. MR1135468

[Den2] C. Deninger, *Lefschetz trace formulas and explicit formulas in analytic number theory*, J. Reine Angew. Math. **441** (1993), 1–15, DOI 10.1515/crll.1993.441.1. MR1228608

[Den3] C. Deninger, *Evidence for a cohomological approach to analytic number theory*, First European Congress of Mathematics, vol. I (Paris, 1992), Progr. Math., vol. 119, Birkhäuser, Basel, 1994, pp. 491–510. MR1341834

[Den4] C. Deninger, *L-functions of mixed motives*, Motives (Seattle, WA, 1991), Proc. Sympos. Pure Math., vol. 55, Amer. Math. Soc., Providence, RI, 1994, pp. 517–525. MR1265542

[Den5] C. Deninger, *Motivic L-functions and regularized determinants*, Motives (Seattle, WA, 1991), Proc. Sympos. Pure Math., vol. 55, Part 1, Amer. Math. Soc., Providence, RI, 1994, pp. 707–743. MR1265547

[Den6] C. Deninger, *Some analogies between number theory and dynamical systems on foliated spaces*, Proceedings of the International Congress of Mathematicians, vol. I (Berlin, 1998), Doc. Math. **Extra vol. I** (1998), 163–186. MR1648030

[DenSchr] C. Deninger and M. Schröter, *A distribution-theoretic proof of Guinand's functional equation for Cramér's V-function and generalizations*, J. London Math. Soc. (2) **52** (1995), no. 1, 48–60, DOI 10.1112/jlms/52.1.48. MR1345713

[DeniKoÖÜ] A. Deniz, Ş. Koçak, Y. Özdemir, and A. E. Üreyen, *Tube volumes via functional equations*, J. Geom. **106** (2015), no. 1, 153–162, DOI 10.1007/s00022-014-0241-3. MR3320885

[DeniKoÖRÜ] A. Deniz, M. Ş. Koçak, Y. Özdemir, A. Ratiu, and A. E. Üreyen, *On the Minkowski measurability of self-similar fractals in $\mathbb{R}^d$*, Turkish J. Math. **37** (2013), no. 5, 830–846. MR3105496

[dV1] C.-J. de la Vallée Poussin, Recherches analytiques sur la théorie des nombres; Première partie: La fonction $\zeta(s)$ de Riemann et les nombres premiers en général, *Ann. Soc. Sci. Bruxelles Sér. I* **20** (1896), 183–256.

[dV2] C.-J. de la Vallée Poussin, Sur la fonction $\zeta(s)$ de Riemann et le nombre des nombres premiers inférieurs à une limite donnée, *Mém. Couronnés et Autres Mém. Publ. Acad. Roy. Sci., des Lettres, Beaux-Arts Belg.* **59** (1899–1900).

[DerGrVo] G. Derfel, P. J. Grabner, and F. Vogl, *The zeta function of the Laplacian on certain fractals*, Trans. Amer. Math. Soc. **360** (2008), no. 2, 881–897, DOI 10.1090/S0002-9947-07-04240-7. MR2346475

[deSLapRRo] R. de Santiago, M. L. Lapidus, S. A. Roby, and J. A. Rock, Multifractal analysis via scaling zeta functions and recursive structure of lattice strings, in: *Fractal geometry and dynamical systems in pure and applied mathematics I: Fractals in pure mathematics* (D. Carfi, M. L. Lapidus, E. P. J. Pearse and M. van Frankenhuijsen, eds.), Contemporary Mathematics, vol. 600, Amer. Math. Soc., Providence, RI, 2013, pp. 205–238. (dx.doi.org/10.1090/conm/600/11930.) (Also, e-print, `arXiv:1207.6680v3 [math-ph]`, 2013; IHES preprint, IHES/M/12/19, 2013.)

[DieRuzSchu] L. Diening, M. Růžička, and K. Schumacher, *A decomposition technique for John domains*, Ann. Acad. Sci. Fenn. Math. **35** (2010), no. 1, 87–114, DOI 10.5186/aasfm.2010.3506. MR2643399

[Dieu1] J. Dieudonné, On the history of the Weil conjectures, *Math. Intelligencer* **10** (1975), 7–21.

[Dieu2] J. Dieudonné, *A history of algebraic and differential topology. 1900–1960*, Birkhäuser Boston, Inc., Boston, MA, 1989. MR995842

[DubSep] E. Dubon and J. M. Sepulcre, *On the complex dimensions of nonlattice fractal strings in connection with Dirichlet polynomials*, Exp. Math. **23** (2014), no. 1, 13–24, DOI 10.1080/10586458.2013.853630. MR3177454

[DunSch] N. Dunford and J. T. Schwartz, *Linear operators*, Parts I–III, Wiley Classics Library, John Wiley & Sons, Hoboken, 1971 and 1988. (Part I: *General theory.* Part II: *Spectral theory.* Part III: *Spectral operators.*)

[Dur-LopGar] R. G. Durán and F. López García, *Solutions of the divergence and Korn inequalities on domains with an external cusp*, Ann. Acad. Sci. Fenn. Math. **35** (2010), no. 2, 421–438, DOI 10.5186/aasfm.2010.3527. MR2731700

[DureSchu] P. Duren and A. Schuster, *Bergman spaces*, Mathematical Surveys and Monographs, vol. 100, American Mathematical Society, Providence, RI, 2004. MR2033762

[Ebe] W. Ebeling, *Functions of several complex variables and their singularities*, Graduate Studies in Mathematics, vol. 83, American Mathematical Society, Providence, RI, 2007. Translated from the 2001 German original by Philip G. Spain. MR2319634

[EdmHar] W. D. Evans and D. J. Harris, *Fractals, trees and the Neumann Laplacian*, Math. Ann. **296** (1993), no. 3, 493–527, DOI 10.1007/BF01445117. MR1225988

[Edw] H. M. Edwards, *Riemann's zeta function*, Dover Publications, Inc., Mineola, NY, 2001. Reprint of the 1974 original [Academic Press, New York; MR0466039 (57 #5922)]. MR1854455

[ElLapMcRo] K. E. Ellis, M. L. Lapidus, M. C. Mackenzie and J. A. Rock, Partition zeta functions, multifractal spectra, and tapestries of complex dimensions, in: *Benoit Mandelbrot: A life in many dimensions*, The Mandelbrot memorial volume (M. Frame and N. Cohen, eds.), World Scientific Publisher, Singapore and London, 2015, pp. 267–312. (Also: e-print, `arXiv:1007.1467v2 [math-ph]`, 2011; IHES preprint, IHES/M/12/15, 2012.)

[Es1] D. Essouabri, *Singularité des séries de Dirichlet associées à des polynômes de plusieurs variables et applications en théorie analytique des nombres* (French, with English summary), Ann. Inst. Fourier (Grenoble) **47** (1997), no. 2, 429–483. MR1450422

[Es2] D. Essouabri, *Zeta functions associated to Pascal's triangle mod p*, Japan. J. Math. (N.S.) **31** (2005), no. 1, 157–174, DOI 10.4099/math1924.31.157. MR2153729

[EsLi1] D. Essouabri and B. Lichtin, *Zeta functions of discrete self-similar sets*, Adv. Math. **232** (2013), 142–187, DOI 10.1016/j.aim.2012.09.011. MR2989980

[EsLi2] D. Essouabri and B. Lichtin, k-point configurations of discrete self-similar sets, in: *Fractal geometry and dynamical systems in pure and applied mathematics I:Fractals in pure mathematics* (D. Carfi, M. L. Lapidus and M. van Frankenhuijsen, eds.), Contemporary Mathematics, vol. 600, Amer. Math. Soc, Providence, RI, 2013, pp. 21–50.

[Fa1] K. Falconer, *Fractal geometry: Mathematical foundations and applications*, 3rd ed., John Wiley & Sons, Ltd., Chichester, 2014. MR3236784

[Fa2] K. J. Falconer, *On the Minkowski measurability of fractals*, Proc. Amer. Math. Soc. **123** (1995), no. 4, 1115–1124, DOI 10.2307/2160708. MR1224615

[Fed1] H. Federer, *Curvature measures*, Trans. Amer. Math. Soc. **93** (1959), 418–491, DOI 10.2307/1993504. MR0110078

[Fed2] H. Federer, *Geometric measure theory*, Die Grundlehren der mathematischen Wissenschaften, Band 153, Springer-Verlag New York Inc., New York, 1969. MR0257325

[Fel] W. Feller, *An introduction to probability theory and its applications. Vol. II*, John Wiley & Sons, Inc., New York-London-Sydney, 1966. MR0210154

[FlLap] J. Fleckinger and M. L. Lapidus, *Remainder estimates for the asymptotics of elliptic eigenvalue problems with indefinite weights*, Arch. Rational Mech. Anal. **98** (1987), no. 4, 329–356, DOI 10.1007/BF00276913. MR872751

[FlVa] J. Fleckinger-Pellé and D. G. Vassiliev, *An example of a two-term asymptotics for the "counting function" of a fractal drum*, Trans. Amer. Math. Soc. **337** (1993), no. 1, 99–116, DOI 10.2307/2154311. MR1176086

[Fo] G. B. Folland, *Real analysis: Modern techniques and their applications; A Wiley-Interscience Publication*, 2nd ed., Pure and Applied Mathematics (New York), John Wiley & Sons, Inc., New York, 1999. MR1681462

[Fr] M. Frantz, *Lacunarity, Minkowski content, and self-similar sets in* $\mathbb{R}$, Fractal geometry and applications: A jubilee of Benoît Mandelbrot. Part 1, Proc. Sympos. Pure Math., vol. 72, Amer. Math. Soc., Providence, RI, 2004, pp. 77–91. MR2112101

[FreiKom] U. Freiberg and S. Kombrink, *Minkowski content and local Minkowski content for a class of self-conformal sets*, Geom. Dedicata **159** (2012), 307–325, DOI 10.1007/s10711-011-9661-5. MR2944534

[Fred] I. Fredholm, *Sur une classe d'équations fonctionnelles* (French), Acta Math. **27** (1903), no. 1, 365–390, DOI 10.1007/BF02421317. MR1554993

[FukShi] M. Fukushima and T. Shima, *On a spectral analysis for the Sierpiński gasket*, Potential Anal. **1** (1992), no. 1, 1–35, DOI 10.1007/BF00249784. MR1245223

[GarSteu] R. Garunkštis and J. Steuding, *On the roots of the equation* $\zeta(s) = a$, Abh. Math. Semin. Univ. Hambg. **84** (2014), no. 1, 1–15, DOI 10.1007/s12188-014-0093-7. MR3197009

[Gat] D. Gatzouras, *Lacunarity of self-similar and stochastically self-similar sets*, Trans. Amer. Math. Soc. **352** (2000), no. 5, 1953–1983, DOI 10.1090/S0002-9947-99-02539-8. MR1694290

[GelSh] I. M. Gelfand and G. E. Shilov, *Generalized functions*, vols. I–III, Academic Press, new edition, 1986.

[Ger] J. Gerling, *Untersuchungen zur Theorie von Weyl–Berry–Lapidus*, Graduate thesis (diplomarbeit), Dept. of Physics, Universität Osnabrück, Germany, 1992.

[GerScm1] J. Gerling and H.-J. Schmidt, *Self-similar drums and generalized Weierstrass functions: Fractals and disorder*, Phys. A **191** (1992), no. 1-4, 536–539, DOI 10.1016/0378-4371(92)90578-E. (Hamburg, 1992). MR1199397

[GerScm2] J. Gerling and H.-J. Schmidt, *Three-term asymptotics of the spectrum of self-similar fractal drums*, J. Math. Sci. Univ. Tokyo **6** (1999), no. 1, 101–126. MR1683321

[Gi] P. B. Gilkey, *Invariance theory, the heat equation, and the Atiyah-Singer index theorem*, 2nd ed., Studies in Advanced Mathematics, CRC Press, Boca Raton, FL, 1995. MR1396308

[Go] J. A. Goldstein, *Semigroups of linear operators and applications*, Oxford Mathematical Monographs, The Clarendon Press, Oxford University Press, New York, 1985. MR790497

[GohKre] I. C. Gohberg and M. G. Kreĭn, *Introduction to the theory of linear nonselfadjoint operators*, Translated from the Russian by A. Feinstein. Translations of Mathematical Monographs, Vol. 18, American Mathematical Society, Providence, RI, 1969. MR0246142

[Gra] A. Gray, *Tubes*, 2nd ed., Progress in Mathematics, vol. 221, Birkhäuser Verlag, Basel, 2004. With a preface by Vicente Miquel. MR2024928

[Gro1] A. Grothendieck, *The cohomology theory of abstract algebraic varieties*, Proc. Internat. Congress Math. (Edinburgh, 1958), Cambridge Univ. Press, New York, 1960, pp. 103–118. MR0130879

[Gro2] A. Grothendieck, Formule de Lefschetz et rationalité des fonctions L (French), *Séminaire N. Bourbaki*, vol. 9, Soc. Math. France, Paris, Exp. 279, 1964–1966, pp. 41–55 MR1608788

[Gro3] A. Grothendieck, *Standard conjectures on algebraic cycles*, Algebraic Geometry (Internat. Colloq., Tata Inst. Fund. Res., Bombay, 1968), Oxford Univ. Press, London, 1969, pp. 193–199. MR0268189

[Gro4] A. Grothendieck, *Eléments de Géométrie Algébrique*, Inst. Hautes Études Sci. Publ. Math., vols. 4, 11, 17, 20, 24, 28 and 32. (With the collaboration of J. Dieudonné.) [Reprinted in 1999 by the Institut des Hautes Études Scientifiques (IHES), Bures-sur-Yvette, France.]

[GunRos] R. C. Gunning and H. Rossi, *Analytic functions of several complex variables*, AMS Chelsea Publishing, Providence, RI, 2015. Reprint of the 1965 original. MR2568219

[Had1] J. Hadamard, Étude sur les propriétés des fonctions entières et en particulier d'une fonction considérée par Riemann, *J. Math. Pures Appl.* (4) **9** (1893), 171–215. (Reprinted in [Had3, pp. 103–147].)

[Had2] J. Hadamard, *Sur la distribution des zéros de la fonction $\zeta(s)$ et ses conséquences arithmétiques* (French), Bull. Soc. Math. France **24** (1896), 199–220. MR1504264

[Had3] J. Hadamard, *Œuvres de Jacques Hadamard. Tome I* (French), Comité de publication des oeuvres de Jacques Hadamard: M. Fréchet, P. Levy, S. Mandelbrojt, L. Schwartz, Éditions du Centre National de la Recherche Scientifique, Paris, 1968. MR0230598

[HajKosTu1] P. Hajłasz, P. Koskela, and H. Tuominen, *Sobolev embeddings, extensions and measure density condition*, J. Funct. Anal. **254** (2008), no. 5, 1217–1234, DOI 10.1016/j.jfa.2007.11.020. MR2386936

[HajKosTu2] P. Hajłasz, P. Koskela, and H. Tuominen, *Measure density and extendability of Sobolev functions*, Rev. Mat. Iberoam. **24** (2008), no. 2, 645–669, DOI 10.4171/RMI/551. MR2459208

[Ham1] B. M. Hambly, *Brownian motion on a random recursive Sierpinski gasket*, Ann. Probab. **25** (1997), no. 3, 1059–1102, DOI 10.1214/aop/1024404506. MR1457612

[Ham2] B. M. Hambly, *On the asymptotics of the eigenvalue counting function for random recursive Sierpinski gaskets*, Probab. Theory Related Fields **117** (2000), no. 2, 221–247, DOI 10.1007/s004400050005. MR1771662

[HamLap] B. M. Hambly and M. L. Lapidus, *Random fractal strings: their zeta functions, complex dimensions and spectral asymptotics*, Trans. Amer. Math. Soc. **358** (2006), no. 1, 285–314, DOI 10.1090/S0002-9947-05-03646-9. MR2171234

[Har1] S. Haran, *Riesz potentials and explicit sums in arithmetic*, Invent. Math. **101** (1990), no. 3, 697–703, DOI 10.1007/BF01231521. MR1062801

[Har2] M. J. S. Haran, *The mysteries of the real prime*, London Mathematical Society Monographs. New Series, vol. 25, The Clarendon Press, Oxford University Press, New York, 2001. MR1872029

[Har3] S. Haran, *On Riemann's zeta function*, Dynamical, spectral, and arithmetic zeta functions (M. L. Lapidus and M. van Frankenhuysen, eds.), Contemporary Mathematics, vol. 290, Amer. Math. Soc., Providence, RI, 2001, pp. 93–112.

[Har4] M. J. S. Haran, *New foundations for geometry—two non-additive languages for arithmetical geometry*, Mem. Amer. Math. Soc. **246** (2017), no. 1166, x+200, DOI 10.1090/memo/1166. MR3605614

[Hart] R. Hartshorne, *Algebraic geometry*, Graduate Texts in Mathematics, vol. 52, Springer-Verlag, New York-Heidelberg, 1977. MR0463157

[HarvPol] R. Harvey and J. Polking, *Removable singularities of solutions of linear partial differential equations*, Acta Math. **125** (1970), 39–56, DOI 10.1007/BF02838327. MR0279461

[Has] H. Hasse, *Abstrakte Begründung der Komplexen Multiplikation und Riemannsche vermutung in Funktionenkörpern* (German), Abh. Math. Sem. Univ. Hamburg **10** (1934), no. 1, 325–348, DOI 10.1007/BF02940685. MR3069636

[Hat] M. Hata, *On some properties of set-dynamical systems*, Proc. Japan Acad. Ser. A Math. Sci. **61** (1985), no. 4, 99–102. MR796477

[HeLap] C. Q. He and M. L. Lapidus, *Generalized Minkowski content, spectrum of fractal drums, fractal strings and the Riemann zeta-function*, Mem. Amer. Math. Soc. **127** (1997), no. 608, x+97, DOI 10.1090/memo/0608. MR1376743

[HedKonZhu] H. Hedenmalm, B. Korenblum, and K. Zhu, *Theory of Bergman spaces*, Graduate Texts in Mathematics, vol. 199, Springer-Verlag, New York, 2000. MR1758653

[HerLap1] H. Herichi and M. L. Lapidus, *Quantized Number Theory, Fractal Strings and the Riemann Hypothesis: From Spectral Operators to Phase Transitions and Universality*, research monograph, World Scientific Publisher, Singapore and London, 2019, in press, approx. 450 pages. ISBN: 978-981-3230-79-8.

[HerLap2] H. Herichi and M. L. Lapidus, *Riemann zeros and phase transitions via the spectral operator on fractal strings*, J. Phys. A **45** (2012), no. 37, 374005, 23 pp., DOI 10.1088/1751-8113/45/37/374005. MR2970522

[HerLap3] H. Herichi and M. L. Lapidus, *Fractal complex dimensions, Riemann hypothesis and invertibility of the spectral operator*, Fractal geometry and dynamical systems in pure and applied mathematics. I. Fractals in pure mathematics, Contemp. Math., vol. 600, Amer. Math. Soc., Providence, RI, 2013, pp. 51–89, DOI 10.1090/conm/600/11948. MR3203399

[HerLap4] H. Herichi and M. L. Lapidus, *Truncated infinitesimal shifts, spectral operators and quantized universality of the Riemann zeta function* (English, with English and French summaries), Ann. Fac. Sci. Toulouse Math. (6) **23** (2014), no. 3, 621–664, DOI 10.5802/afst.1419. MR3266708

[HerLap5] H. Herichi and M. L. Lapidus, Quantized Riemann zeta function and other L-functions: Their convergent operator-valued Dirichlet series, Euler product and analytic continuation, in preparation, 2019.

[HilPh] E. Hille and R. S. Phillips, *Functional analysis and semi-groups*, American Mathematical Society Colloquium Publications, vol. 31, American Mathematical Society, Providence, RI, 1957. rev. ed. MR0089373

[Ho1] L. Hörmander, *The spectral function of an elliptic operator*, Acta Math. **121** (1968), 193–218, DOI 10.1007/BF02391913. MR0609014

[Ho2] L. Hörmander, *The analysis of linear partial differential operators. I: Distribution theory and Fourier analysis*, 2nd ed., Springer Study Edition, Springer-Verlag, Berlin, 1990. MR1065136

[Ho3] L. Hörmander, *The analysis of linear partial differential operators*, vols. II-IV, Springer-Verlag, Berlin, 1983 & 1985.

[HugLasWeil] D. Hug, G. Last, and W. Weil, *A local Steiner-type formula for general closed sets and applications*, Math. Z. **246** (2004), no. 1-2, 237–272, DOI 10.1007/s00209-003-0597-9. MR2031455

[Hut] J. E. Hutchinson, *Fractals and self-similarity*, Indiana Univ. Math. J. **30** (1981), no. 5, 713–747, DOI 10.1512/iumj.1981.30.30055. MR625600

[Ing] A. E. Ingham, *The distribution of prime numbers*, Cambridge Mathematical Library, Cambridge University Press, Cambridge, 1990. Reprint of the 1932 original; With a foreword by R. C. Vaughan. MR1074573

[Ivi] A. Ivić, *The Riemann zeta-function: The theory of the Riemann zeta-function with applications*, A Wiley-Interscience Publication, John Wiley & Sons, Inc., New York, 1985. MR792089

[Ivr1] V. Ja. Ivrii, Second term of the spectral asymptotic expansion of the Laplace–Beltrami operator on manifolds with boundary, *Functional Anal. Appl.* **14** (1980), 98–106.

[Ivr2] V. Ja. Ivriĭ, *Precise spectral asymptotics for elliptic operators acting in fiberings over manifolds with boundary*, Lecture Notes in Mathematics, vol. 1100, Springer-Verlag, Berlin, 1984. MR771297

[Ivr3] V. Ja. Ivrii, *Microlocal analysis and precise spectral asymptotics*, Springer Monographs in Mathematics, Springer-Verlag, Berlin, 1998. MR1631419

[JoLap] G. W. Johnson and M. L. Lapidus, *The Feynman Integral and Feynman's Operational Calculus*, Oxford Science Publications, Oxford Mathematical Monographs, Oxford Univ. Press, Oxford and New York, 2000. (Corrected printing and paperback edition, 2002.)

[JoLapNie] G. W. Johnson, M. L. Lapidus, and L. Nielsen, *Feynman's operational calculus and beyond: Noncommutativity and time-ordering*, Oxford Mathematical Monographs, Oxford University Press, Oxford, 2015. MR3381096

[John] F. John, *Rotation and strain*, Comm. Pure Appl. Math. **14** (1961), 391–413, DOI 10.1002/cpa.3160140316. MR0138225

[Jon] P. W. Jones, *Quasiconformal mappings and extendability of functions in Sobolev spaces*, Acta Math. **147** (1981), no. 1-2, 71–88, DOI 10.1007/BF02392869. MR631089

[Kac] M. Kac, *Can one hear the shape of a drum?*, Amer. Math. Monthly **73** (1966), no. 4, 1–23, DOI 10.2307/2313748. MR0201237

[Kah] B. Kahn, *Motifs*, in: Leçons de mathématiques d'aujourd'hui, vol. 3 (É. Charpentier and N. Nikolski, eds.), Cassini, Paris, 2007, pp. 359-390, (in French). [*Motives*, in: Lessons of today's mathematics.]

[KarVo] A. A. Karatsuba and S. M. Voronin, *The Riemann zeta-function*, De Gruyter Expositions in Mathematics, vol. 5, Walter de Gruyter & Co., Berlin, 1992. Translated from the Russian by Neal Koblitz. MR1183467

[Kat] T. Kato, *Perturbation theory for linear operators*, Classics in Mathematics, Springer-Verlag, Berlin, 1995. Reprint of the 1980 edition. MR1335452

[Katz] N. M. Katz, *An overview of Deligne's proof of the Riemann hypothesis for varieties over finite fields*, Mathematical developments arising from Hilbert problems (Proc. Sympos. Pure Math., Vol. XXVIII, Northern Illinois Univ., De Kalb, Ill., 1974), Amer. Math. Soc., Providence, RI, 1976, pp. 275–305. MR0424822

[Katzn] Y. Katznelson, *An introduction to harmonic analysis*, 3rd ed., Cambridge Mathematical Library, Cambridge University Press, Cambridge, 2009. MR2039503

[KeKom] M. Kesseböhmer and S. Kombrink, *Fractal curvature measures and Minkowski content for self-conformal subsets of the real line*, Adv. Math. **230** (2012), no. 4-6, 2474–2512, DOI 10.1016/j.aim.2012.04.023. MR2927378

[Ki] J. Kigami, *Analysis on fractals*, Cambridge Tracts in Mathematics, vol. 143, Cambridge University Press, Cambridge, 2001. MR1840042

[KiLap1] J. Kigami and M. L. Lapidus, *Weyl's problem for the spectral distribution of Laplacians on p.c.f. self-similar fractals*, Comm. Math. Phys. **158** (1993), no. 1, 93–125. MR1243717

[KiLap2] J. Kigami and M. L. Lapidus, *Self-similarity of volume measures for Laplacians on p.c.f. self-similar fractals*, Comm. Math. Phys. **217** (2001), no. 1, 165–180, DOI 10.1007/s002200000326. MR1815029

[KlRot] D. A. Klain and G.-C. Rota, *Introduction to Geometric Probability*, Accademia Nazionale dei Lincei, Cambridge Univ. Press, Cambridge, 1999.

[KoRati] Ş. Koçak and A. V. Ratiu, *Inner tube formulas for polytopes*, Proc. Amer. Math. Soc. **140** (2012), no. 3, 999–1010, DOI 10.1090/S0002-9939-2011-11307-8. MR2869084

[Kom] S. Kombrink, *A survey on Minkowski measurability of self-similar and self-conformal fractals in* $\mathbb{R}^d$, Fractal geometry and dynamical systems in pure and applied mathematics. I. Fractals in pure mathematics, Contemp. Math., vol. 600, Amer. Math. Soc., Providence, RI, 2013, pp. 135–159, DOI 10.1090/conm/600/11931. MR3203401

[KomPeWi] S. Kombrink, E. P. J. Pearse, and S. Winter, *Lattice-type self-similar sets with pluriphase generators fail to be Minkowski measurable*, Math. Z. **283** (2016), no. 3-4, 1049–1070, DOI 10.1007/s00209-016-1633-x. MR3519995

[Kor] J. Korevaar, *Tauberian theory: A century of developments*, Grundlehren der Mathematischen Wissenschaften [Fundamental Principles of Mathematical Sciences], vol. 329, Springer-Verlag, Berlin, 2004. MR2073637

[Lag] J. C. Lagarias, *Number theory zeta functions and dynamical zeta functions*, Spectral problems in geometry and arithmetic (Iowa City, IA, 1997), Contemp. Math., vol. 237, Amer. Math. Soc., Providence, RI, 1999, pp. 45–86, DOI 10.1090/conm/237/1710789. MR1710789

[LalLap1] N. Lal and M. L. Lapidus, *Hyperfunctions and spectral zeta functions of Laplacians on self-similar fractals*, J. Phys. A **45** (2012), no. 36, 365205, 14pp., DOI 10.1088/1751-8113/45/36/365205. MR2967908

[LalLap2] N. Lal and M. L. Lapidus, *The decimation method for Laplacians on fractals: Spectra and complex dynamics*, Fractal geometry and dynamical systems in pure and applied mathematics. II. Fractals in applied mathematics, Contemp. Math., vol. 601, Amer. Math. Soc., Providence, RI, 2013, pp. 227–249, DOI 10.1090/conm/601/11959. MR3203865

[Lall1] S. P. Lalley, *The packing and covering functions of some self-similar fractals*, Indiana Univ. Math. J. **37** (1988), no. 3, 699–710, DOI 10.1512/iumj.1988.37.37034. MR962930

[Lall2] S. P. Lalley, *Renewal theorems in symbolic dynamics, with applications to geodesic flows, non-Euclidean tessellations and their fractal limits*, Acta Math. **163** (1989), no. 1-2, 1–55, DOI 10.1007/BF02392732. MR1007619

[Lall3] S. P. Lalley, *Probabilistic methods in certain counting problems of ergodic theory*, Ergodic theory, symbolic dynamics, and hyperbolic spaces (Trieste, 1989), Oxford Sci. Publ., Oxford Univ. Press, New York, 1991, pp. 223–258. MR1130178

[Lap1] M. L. Lapidus, *Fractal drum, inverse spectral problems for elliptic operators and a partial resolution of the Weyl-Berry conjecture*, Trans. Amer. Math. Soc. **325** (1991), no. 2, 465–529, DOI 10.2307/2001638. MR994168

[Lap2] M. L. Lapidus, *Spectral and fractal geometry: From the Weyl-Berry conjecture for the vibrations of fractal drums to the Riemann zeta-function*, Differential equations and mathematical physics (Birmingham, AL, 1990), Math. Sci. Engrg., vol. 186, Academic Press, Boston, MA, 1992, pp. 151–181, DOI 10.1016/S0076-5392(08)63379-2. MR1126694

[Lap3] M. L. Lapidus, *Vibrations of fractal drums, the Riemann hypothesis, waves in fractal media and the Weyl-Berry conjecture*, Ordinary and partial differential equations, Vol. IV (Dundee, 1992), Pitman Res. Notes Math. Ser., vol. 289, Longman Sci. Tech., Harlow, 1993, pp. 126–209. MR1234502

[Lap4] M. L. Lapidus, *Fractals and vibrations: can you hear the shape of a fractal drum?*, Fractals **3** (1995), no. 4, 725–736, DOI 10.1142/S0218348X95000643. Symposium in Honor of Benoit Mandelbrot (Curaçao, 1995). MR1410291

[Lap5] M. L. Lapidus, *Analysis on fractals, Laplacians on self-similar sets, noncommutative geometry and spectral dimensions*, Topol. Methods Nonlinear Anal. **4** (1994), no. 1, 137–195, DOI 10.12775/TMNA.1994.025. MR1321811

[Lap6] M. L. Lapidus, *Towards a noncommutative fractal geometry? Laplacians and volume measures on fractals*, Harmonic analysis and nonlinear differential equations (Riverside, CA, 1995), Contemp. Math., vol. 208, Amer. Math. Soc., Providence, RI, 1997, pp. 211–252, DOI 10.1090/conm/208/02742. MR1467009

[Lap7] M. L. Lapidus, *In search of the Riemann zeros: Strings, fractal membranes and noncommutative spacetimes*, American Mathematical Society, Providence, RI, 2008. MR2375028

[Lap8] M. L. Lapidus, *Towards quantized number theory: spectral operators and an asymmetric criterion for the Riemann hypothesis*, Philos. Trans. Roy. Soc. A **373** (2015), no. 2047, 20140240, 24, DOI 10.1098/rsta.2014.0240. MR3367510

[Lap9] M. L. Lapidus, *The sound of fractal strings and the Riemann hypothesis*, Analytic number theory, Springer, Cham, 2015, pp. 201–252. MR3467400

[Lap10] M. L. Lapidus, *From Complex Fractal Dimensions and Quantized Number Theory To Fractal Cohomology: A Tale of Oscillations, Unreality and Fractality*, book in preparation, World Scientific Publisher, Singapore and London, 2019.

[LapLevyRo] M. L. Lapidus, J. Lévy-Véhel, and J. A. Rock, *Fractal strings and multifractal zeta functions*, Lett. Math. Phys. **88** (2009), no. 1-3, 101–129, DOI 10.1007/s11005-009-0302-y. MR2512142

[LapLu1] M. L. Lapidus and H. Lũ', *Nonarchimedean Cantor set and string*, J. Fixed Point Theory Appl. **3** (2008), no. 1, 181–190, DOI 10.1007/s11784-008-0062-9. MR2402916

[LapLu2] M. L. Lapidus and H. Lũ', *Self-similar p-adic fractal strings and their complex dimensions*, p-Adic Numbers Ultrametric Anal. Appl. **1** (2009), no. 2, 167–180, DOI 10.1134/S2070046609020083. MR2566062

[LapLu3] M. L. Lapidus and H. Lũ', *The geometry of p-adic fractal strings: A comparative survey*, Advances in non-Archimedean analysis, Contemp. Math., vol. 551, Amer. Math. Soc., Providence, RI, 2011, pp. 163–206, DOI 10.1090/conm/551/10893. MR2882397

[LapLu-vF1] M. L. Lapidus, H. Lu and M. van Frankenhuijsen, *Minkowski measurability and exact fractal tube formulas for p-adic self-similar strings*, Fractal geometry and dynamical systems in pure and applied mathematics I: Fractals in pure mathematics (D. Carfi, M. L. Lapidus, E. P. J. Pearse and M. van Frankenhuijsen, eds.), Contemporary Mathematics, vol. 600, Amer. Math. Soc., Providence,

RI, 2013, pp. 161–184. (dx.doi.org/10.1090/conm/600/11949.) (Also: e-print, `arXiv:1209.6440v1 [math.MG]`, 2012; IHES preprint, IHES/M/12/23, 2012.)

[LapLu-vF2] M. L. Lapidus, H. Lu and M. van Frankenhuijsen, Minkowski dimension and explicit tube formulas for p-adic fractal strings, *Fractal and Fractional* No. 4, **2** (2018), 26th paper. (DOI: 10.3390/fractalfract2040026.) (Also: e-print, `arXiv:1603.09409v2 [math-ph]`, 2018.)

[LapMa1] M. L. Lapidus and H. Maier, *Hypothèse de Riemann, cordes fractales vibrantes et conjecture de Weyl-Berry modifiée* (French, with English summary), C. R. Acad. Sci. Paris Sér. I Math. **313** (1991), no. 1, 19–24. MR1115940

[LapMa2] M. L. Lapidus and H. Maier, *The Riemann hypothesis and inverse spectral problems for fractal strings*, J. London Math. Soc. (2) **52** (1995), no. 1, 15–34, DOI 10.1112/jlms/52.1.15. MR1345711

[LapNRG] M. L. Lapidus, J. W. Neuberger, R. J. Renka, and C. A. Griffith, *Snowflake harmonics and computer graphics: Numerical computation of spectra on fractal drums*, Internat. J. Bifur. Chaos Appl. Sci. Engrg. **6** (1996), no. 7, 1185–1210, DOI 10.1142/S0218127496000680. MR1412217

[LapNes] M. L. Lapidus and R. Nest, *Fractal membranes as the second quantization of fractal strings*, in preparation, 2019.

[LapPa] M. L. Lapidus and M. M. H. Pang, *Eigenfunctions of the Koch snowflake domain*, Comm. Math. Phys. **172** (1995), no. 2, 359–376. MR1350412

[LapPe1] M. L. Lapidus and E. P. J. Pearse, *A tube formula for the Koch snowflake curve, with applications to complex dimensions*, J. London Math. Soc. (2) **74** (2006), no. 2, 397–414, DOI 10.1112/S0024610706022988. MR2269586

[LapPe2] M. L. Lapidus and E. P. J. Pearse, *Tube formulas for self-similar fractals*, Analysis on graphs and its applications, Proc. Sympos. Pure Math., vol. 77, Amer. Math. Soc., Providence, RI, 2008, pp. 211–230, DOI 10.1090/pspum/077/2459871. MR2459871

[LapPe3] M. L. Lapidus and E. P. J. Pearse, *Tube formulas and complex dimensions of self-similar tilings*, Acta Appl. Math. **112** (2010), no. 1, 91–136, DOI 10.1007/s10440-010-9562-x. MR2684976

[LapPeWi1] M. L. Lapidus, E. P. J. Pearse, and S. Winter, *Pointwise tube formulas for fractal sprays and self-similar tilings with arbitrary generators*, Adv. Math. **227** (2011), no. 4, 1349–1398, DOI 10.1016/j.aim.2011.03.004. MR2799798

[LapPeWi2] M. L. Lapidus, E. P. J. Pearse, and S. Winter, *Minkowski measurability results for self-similar tilings and fractals with monophase generators*, Fractal geometry and dynamical systems in pure and applied mathematics. I. Fractals in pure mathematics, Contemp. Math., vol. 600, Amer. Math. Soc., Providence, RI, 2013, pp. 185–203, DOI 10.1090/conm/600/11951. MR3203403

[LapPo1] M. L. Lapidus and C. Pomerance, *Fonction zêta de Riemann et conjecture de Weyl-Berry pour les tambours fractals* (French, with English summary), C. R. Acad. Sci. Paris Sér. I Math. **310** (1990), no. 6, 343–348. MR1046509

[LapPo2] M. L. Lapidus and C. Pomerance, *The Riemann zeta-function and the one-dimensional Weyl-Berry conjecture for fractal drums*, Proc. London Math. Soc. (3) **66** (1993), no. 1, 41–69, DOI 10.1112/plms/s3-66.1.41. MR1189091

[LapPo3] M. L. Lapidus and C. Pomerance, *Counterexamples to the modified Weyl-Berry conjecture on fractal drums*, Math. Proc. Cambridge Philos. Soc. **119** (1996), no. 1, 167–178, DOI 10.1017/S0305004100074053. MR1356166

[LapRaRo] M. L. Lapidus, G. Radunović and J. A. Rock, *An Invitation to Fractal Geometry: Dimension Theory, Zeta Functions and Applications* I & II, books in preparation, 2019. (Two-volume set for publication in the book series *Student Mathematical Library*, Amer. Math. Soc., Providence, RI.)

[LapRaZu1] M. L. Lapidus, G. Radunović, and D. Žubrinić, *Fractal zeta functions and fractal drums: Higher-dimensional theory of complex dimensions*, Springer Monographs in Mathematics, Springer, New York, 2017. MR3587778

[LapRaZu2] M. L. Lapidus, G. Radunović and D. Žubrinić, Distance and tube zeta functions of fractals and arbitrary compact sets, *Adv. in Math.* **307** (2017), 1215–1267. (dx.doi.org/10.1016/j.aim2016.11.034). (Also: e-print, `arXiv:1506.03525v3 [math-ph]`, 2016; IHES preprint, M/15/15, 2015.)

[LapRaZu3] M. L. Lapidus, G. Radunović, and D. Žubrinić, *Complex dimensions of fractals and meromorphic extensions of fractal zeta functions*, J. Math. Anal. Appl. **453** (2017), no. 1, 458–484, DOI 10.1016/j.jmaa.2017.03.059. MR3641785

[LapRaZu4] M. L. Lapidus, G. Radunović, and D. Žubrinić, *Zeta functions and complex dimensions of relative fractal drums: Theory, examples and applications*, Dissertationes Math. **526** (2017), 1-105, DOI 10.4064/dm757-4-2017. MR3764036

[LapRaZu5] M. L. Lapidus, G. Radunović, and D. Žubrinić, *Fractal tube formulas and a Minkowski measurability criterion for compact subsets of Euclidean spaces*, Discrete Contin. Dyn. Syst. Ser. S **12** (2019), no. 1, 105–117, DOI 10.3934/dcdss.2019007. MR3836595

[LapRaZu6] M. L. Lapidus, G. Radunović, and D. Žubrinić, *Fractal tube formulas for compact sets and relative fractal drums: Oscillations, complex dimensions and fractality*, J. Fractal Geom. **5** (2018), no. 1, 1–119, DOI 10.4171/JFG/57. MR3760302

[LapRaZu7] M. L. Lapidus, G. Radunović and D. Žubrinić, *Minkowski measurability criteria for compact sets and relative fractal drums in Euclidean spaces*, Proceedings of the 6th Cornell Conference on Analysis, Probability and Mathematical Physics on Fractals (Ithaca, June 2017), World Scientific Publisher, Singapore and London, 2019, in press. (Also: e-print, `arXiv:1609.04498v2 [math-ph]`, 2018.)

[LapRaZu8] M. L. Lapidus, G. Radunović, and D. Žubrinić, *Fractal zeta functions and complex dimensions of relative fractal drums*, J. Fixed Point Theory Appl. **15** (2014), no. 2, 321–378, DOI 10.1007/s11784-014-0207-y. Festschrift issue in honor of Haim Brezis' 70th birthday. MR3298004

[LapRaZu9] M. L. Lapidus, G. Radunović, and D. Žubrinić, *Fractal zeta functions and complex dimensions: A general higher-dimensional theory*, Fractal geometry and stochastics V, Progr. Probab., vol. 70, Birkhäuser/Springer, Cham, 2015, pp. 229–257, DOI 10.1007/978-3-319-18660-3_13. MR3558159

[LapRaZu10] M. L. Lapidus, G. Radunović and D. Žubrinić, *Fractal zeta functions and logarithmic gauge Minkowski measurability* (tentative title), in preparation, 2019.

[LapRo] M. L. Lapidus and J. A. Rock, *Towards zeta functions and complex dimensions of multifractals*, Complex Var. Elliptic Equ. **54** (2009), no. 6, 545–559, DOI 10.1080/17476930802326758. MR2537255

[LapRoZu] M. L. Lapidus, J. A. Rock and D. Žubrinić, *Box-counting fractal strings, zeta functions, and equivalent forms of Minkowski dimension*, Fractal geometry and dynamical systems in pure and applied mathematics I: Fractals in pure mathematics (D. Carfi, M. L. Lapidus, E. P. J. Pearse and M. van Frankenhuijsen, eds.), Contemporary Mathematics, vol. 600, Amer. Math. Soc., Providence, RI, 2013, pp. 239–271. (Also: e-print, `arXiv:1207.6681v3 [math-ph]`, 2013; IHES preprint, IHES/M/12/22, 2012.)

[LapSar] M. Lapidus and J. Sarhad, *Dirac operators and geodesic metric on the harmonic Sierpinski gasket and other fractal sets*, J. Noncommut. Geom. **8** (2014), no. 4, 947–985, DOI 10.4171/JNCG/174. MR3310935

[Lap-vF1] M. L. Lapidus and M. van Frankenhuysen, *Complex dimensions of fractal strings and oscillatory phenomena in fractal geometry and arithmetic*, Spectral problems in geometry and arithmetic (Iowa City, IA, 1997), Contemp. Math., vol. 237, Amer. Math. Soc., Providence, RI, 1999, pp. 87–105, DOI 10.1090/conm/237/1710790. MR1710790

[Lap-vF2] M. L. Lapidus and M. van Frankenhuysen, *Fractal geometry and number theory: Complex dimensions of fractal strings and zeros of zeta functions*, Birkhäuser Boston, Inc., Boston, MA, 2000. MR1726744

[Lap-vF3] M. L. Lapidus and M. van Frankenhuijsen, *Fractal geometry, complex dimensions and zeta functions: Geometry and spectra of fractal strings*, Springer Monographs in Mathematics, Springer, New York, 2006. MR2245559

[Lap-vF4] M. L. Lapidus and M. van Frankenhuijsen, *Fractal geometry, complex dimensions and zeta functions: Geometry and spectra of fractal strings*, 2nd ed., Springer Monographs in Mathematics, Springer, New York, 2013. MR2977849

[Lap-vF5] M. L. Lapidus and M. van Frankenhuysen, *Complex dimensions of self-similar fractal strings and Diophantine approximation*, Experiment. Math. **12** (2003), no. 1, 41–69. MR2002673

[Lap-vF6] M. L. Lapidus and M. van Frankenhuijsen (eds.), *Fractal geometry and applications: A jubilee of Benoit Mandelbrot*, Proc. Symp. Pure Math., vol. 72, Parts 1 & 2, Amer. Math Soc., Providence, RI, 2004. (Part 1: *Analysis, number theory, and dynamical systems.* Part 2: *Multifractals, probability and statistical mechanics, applications.*)

[Lap-vF7] M. L. Lapidus and M. van Frankenhuijsen, *Fractality, self-similarity and complex dimensions*, Fractal geometry and applications: A jubilee of Benoît Mandelbrot. Part 1, Proc. Sympos. Pure Math., vol. 72, Amer. Math. Soc., Providence, RI, 2004, pp. 349–372. MR2112111

[LapWat] M. L. Lapidus and S. Watson, Ahlfors metric measure spaces, fractal zeta functions and complex dimensions (tentative title), in preparation, 2019.

[Lau] A. Laurinčikas, *Limit theorems for the Riemann zeta-function*, Mathematics and its Applications, vol. 352, Kluwer Academic Publishers Group, Dordrecht, 1996. MR1376140

[LeviVa] M. Levitin and D. Vassiliev, *Spectral asymptotics, renewal theorem, and the Berry conjecture for a class of fractals*, Proc. London Math. Soc. (3) **72** (1996), no. 1, 188–214, DOI 10.1112/plms/s3-72.1.188. MR1357092

[LevyMen] J. Lévy-Véhel and F. Mendivil, *Multifractal and higher-dimensional zeta functions*, Nonlinearity **24** (2011), no. 1, 259–276, DOI 10.1088/0951-7715/24/1/013. MR2746147

[LiRadz] X. Li and M. Radziwiłł, *The Riemann zeta function on vertical arithmetic progressions*, Int. Math. Res. Not. IMRN **2** (2015), 325–354, DOI 10.1093/imrn/rnt197. MR3340323

[Lids] V. B. Lidskiĭ, *Non-selfadjoint operators with a trace* (Russian), Dokl. Akad. Nauk SSSR **125** (1959), 485–487. MR0105023

[LopGar1] F. López García, *A decomposition technique for integrable functions with applications to the divergence problem*, J. Math. Anal. Appl. **418** (2014), no. 1, 79–99, DOI 10.1016/j.jmaa.2014.03.080. MR3198867

[LopGar2] F. López García, *Weighted generalized Korn inequalities on John domains*, Mathematical Methods in the Applied Sciences, **41**, (2018) no. 17, 8003 8018.

[MacL] S. Mac Lane, *Homology*, Classics in Mathematics, Springer, New York, 1995. (Original edition: 1963.)

[Man] B. B. Mandelbrot, *The Fractal Geometry of Nature*, revised and enlarged edition (of the 1977 French edn.), W. H. Freeman, New York, 1983.

[Mani] Y. Manin, *Lectures on zeta functions and motives (according to Deninger and Kurokawa)*, Astérisque **228** (1995), 4, 121–163. Columbia University Number Theory Seminar (New York, 1992). MR1330931

[ManFra] B. B. Mandelbrot and M. Frame, *A primer of negative test dimensions and degrees of emptiness for latent sets*, Fractals **17** (2009), no. 1, 1–14, DOI 10.1142/S0218348X09004211. MR2498744

[MartVuo] O. Martio and M. Vuorinen, *Whitney cubes, p-capacity, and Minkowski content*, Exposition. Math. **5** (1987), no. 1, 17–40. MR880256

[Mat] P. Mattila, *Geometry of sets and measures in Euclidean spaces: Fractals and rectifiability*, Cambridge Studies in Advanced Mathematics, vol. 44, Cambridge University Press, Cambridge, 1995. MR1333890

[Maz] V. G. Maz'ja, *Sobolev spaces*, Springer Series in Soviet Mathematics, Springer-Verlag, Berlin, 1985. Translated from Russian by T. O. Shaposhnikova. MR817985

[McMul] C. T. McMullen, *Kleinian groups and John domains*, Topology **37** (1998), no. 3, 485–496, DOI 10.1016/S0040-9383(97)00039-6. MR1604874

[Mel1] R. B. Melrose, *Weyl's conjecture for manifolds with concave boundary*, Geometry of the Laplace operator (Proc. Sympos. Pure Math., Univ. Hawaii, Honolulu, Hawaii, 1979), Proc. Sympos. Pure Math., XXXVI, Amer. Math. Soc., Providence, RI, 1980, pp. 257–274. MR573438

[Mel2] R. Melrose, *The trace of the wave group*, Microlocal analysis (Boulder, Colo., 1983), Contemp. Math., vol. 27, Amer. Math. Soc., Providence, RI, 1984, pp. 127–167, DOI 10.1090/conm/027/741046. MR741046

[Met1] G. Métivier, *Etude asymtotique des valeurs propres et de la fonction spectrale de problèmes aux limites*, Thèse de Doctorat d'Etat, Mathématiques, Université de Nice, France, 1976.

[Met2] G. Métivier, *Valeurs propres de problèmes aux limites elliptiques irrégulières* (French), Bull. Soc. Math. France Suppl. Mém. **51–52** (1977), 125–219. MR0473578

[Mink] H. Minkowski, Theorie der konvexen Körper, insbesondere Begründung ihres Oberflächenbegriffs, in: *Gesammelte Abhandlungen von Hermann Minkowski* (part II, Chapter XXV), Chelsea, New York, 1967, pp. 131–229. (Originally reprinted in: Gesamm. Abh., vol. II, Leipzig, 1911.)

[MolVai] S. Molchanov and B. Vainberg, *On spectral asymptotics for domains with fractal boundaries*, Comm. Math. Phys. **183** (1997), no. 1, 85–117, DOI 10.1007/BF02509797. MR1461952

[MorSep] G. Mora and J. M. Sepulcre, *Privileged regions in critical strips of non-lattice Dirichlet polynomials*, Complex Anal. Oper. Theory **7** (2013), no. 4, 1417–1426, DOI 10.1007/s11785-012-0248-4. MR3079865

[MorSepVi1] G. Mora, J. M. Sepulcre, and T. Vidal, *On the existence of exponential polynomials with prefixed gaps*, Bull. Lond. Math. Soc. **45** (2013), no. 6, 1148–1162, DOI 10.1112/blms/bdt043. MR3138483

[MorSepVi2] G. Mora, J. M. Sepulcre, and T. Vidal, *On the existence of fractal strings whose set of dimensions of fractality is not perfect*, Rev. R. Acad. Cienc. Exactas Fís. Nat. Ser. A Math. RACSAM **109** (2015), no. 1, 11–14, DOI 10.1007/s13398-014-0164-8. MR3315698

[Mora] P. A. P. Moran, *Additive functions of intervals and Hausdorff measure*, Proc. Cambridge Philos. Soc. **42** (1946), 15–23. MR0014397

[Murt1] M. R. Murty, *A motivated introduction to the Langlands program*, Advances in number theory (Kingston, ON, 1991), Oxford Sci. Publ., Oxford Univ. Press, New York, 1993, pp. 37–66. MR1368410

[Murt2] M. R. Murty, *Selberg's conjectures and Artin L-functions*, Bull. Amer. Math. Soc. (N.S.) **31** (1994), no. 1, 1–14, DOI 10.1090/S0273-0979-1994-00479-3. MR1242382

[Murt3] M. R. Murty, *Selberg's conjectures and Artin L-functions. II*, Current trends in mathematics and physics, Narosa, New Delhi, 1995, pp. 154–168. MR1354178

[Ol1] L. Olsen, *Multifractal tubes: Multifractal zeta-functions, multifractal Steiner formulas and explicit formulas*, Fractal geometry and dynamical systems in pure and applied mathematics. I. Fractals in pure mathematics, Contemp. Math., vol. 600, Amer. Math. Soc., Providence, RI, 2013, pp. 291–326, DOI 10.1090/conm/600/11920. MR3203407

[Ol2] L. Olsen, *Multifractal tubes*, Further developments in fractals and related fields, Trends Math., Birkhäuser/Springer, New York, 2013, pp. 161–191, DOI 10.1007/978-0-8176-8400-6_9. MR3185037

[Oort] F. Oort, *The Weil conjectures*, Nieuw Arch. Wiskd. (5) **15** (2014), no. 3, 211–219. MR3243075

[Ove] M. Overholt, *A course in analytic number theory*, Graduate Studies in Mathematics, vol. 160, American Mathematical Society, Providence, RI, 2014. MR3290245

[Oxl] J. G. Oxley, *Matroid theory*, Oxford Science Publications, The Clarendon Press, Oxford University Press, New York, 2006. MR1207587

[ParrPol1] W. Parry and M. Pollicott, *An analogue of the prime number theorem for closed orbits of Axiom A flows*, Ann. of Math. (2) **118** (1983), no. 3, 573–591, DOI 10.2307/2006982. MR727704

[ParrPol2] W. Parry and M. Pollicott, *Zeta functions and the periodic orbit structure of hyperbolic dynamics* (English, with French summary), Astérisque **187-188** (1990), 268. Soc. Math. France, Paris. MR1085356

[ParsSh1] I. R. Shafarevich (ed.), *Number theory. I*, Encyclopaedia of Mathematical Sciences, vol. 49, Springer-Verlag, Berlin, 1995. Fundamental problems, ideas and theories; A translation of *Number theory. 1* (Russian), Akad. Nauk SSSR, Vsesoyuz. Inst. Nauchn. i Tekhn. Inform., Moscow, 1990 [MR1056487 (91j:11001a)]; Translation edited by A. N. Parshin and I. R. Shafarevich. MR1366204

[ParsSh2] I. R. Shafarevich (ed.), *Number theory. II*, Encyclopaedia of Mathematical Sciences, vol. 62, Springer-Verlag, Berlin, 1992. Algebraic number theory; A translation of *Number theory, 2 (Russian)*, Akad. Nauk SSSR, Vsesoyuz. Inst. Nauchn. i Tekhn. Inform., Moscow, 1990 [MR1100312 (92a:11118a)]; Translation edited by A. N. Parshin and I. R. Shafarevich. MR1218887

[Pat] S. J. Patterson, *An introduction to the theory of the Riemann zeta-function*, Cambridge Studies in Advanced Mathematics, vol. 14, Cambridge University Press, Cambridge, 1988. MR933558

[Pe] E. P. J. Pearse, *Canonical self-affine tilings by iterated function systems*, Indiana Univ. Math. J. **56** (2007), no. 6, 3151–3169, DOI 10.1512/iumj.2007.56.3220. MR2375713

[PeWi] E. P. J. Pearse and S. Winter, *Geometry of canonical self-similar tilings*, Rocky Mountain J. Math. **42** (2012), no. 4, 1327–1357, DOI 10.1216/RMJ-2012-42-4-1327. MR2981047

[Ph] Phạm The Lại, *Meilleures estimations asymptotiques des restes de la fonction spectrale et des valeurs propres relatifs au laplacien* (French), Math. Scand. **48** (1981), no. 1, 5–38. MR621413

[Pit] H. R. Pitt, *Tauberian theorems*, Tata Institute of Fundamental Research, Monographs on Mathematics and Physics, vol. 2, Oxford University Press, London, 1958. MR0106376

[PitWie] H. R. Pitt and N. Wiener, A generalization of Ikehara's theorem, *J. Math. and Phys. M.I.T.* **17** (1939), 247–258.

[Plem] J. Plemelj, *Zur Theorie der Fredholmschen Funktionalgleichung* (German), Monatsh. Math. Phys. **15** (1904), no. 1, 93–128, DOI 10.1007/BF01692293. MR1547272

[Poin1] H. Poincaré, *Analysis situs, Journal de l'École Polytechnique* (2) **1** (1895), 1–123.

[Poin2] H. Poincaré, Complément à L'Analysis Situs, *Rendiconti del Circolo Matermatico di Palermo* **13** (1899), 285–343.

[Poin3] H. Poincaré, *Second Complement a l'Analysis Situs*, Proc. Lond. Math. Soc. **32** (1900), 277–308, DOI 10.1112/plms/s1-32.1.277. MR1576227

[Poin4] H. Poincaré, *Papers on topology: Analysis situs and its five supplements*, History of Mathematics, vol. 37, American Mathematical Society, Providence, RI; London Mathematical Society, London, 2010. Translated and with an introduction by John Stillwell. MR2723194

[Poin5] H. Poincaré, *Remarques diverses sur l'équation de Fredholm* (French), Acta Math. **33** (1910), no. 1, 57–86, DOI 10.1007/BF02393212. MR1555056

[Pomm] Ch. Pommerenke, *Boundary behaviour of conformal maps*, Grundlehren der Mathematischen Wissenschaften [Fundamental Principles of Mathematical Sciences], vol. 299, Springer-Verlag, Berlin, 1992. MR1217706

[Pos] A. G. Postnikov, *Tauberian theory and its applications*, Proc. Steklov Inst. Math. **2** (1980), v+138. A translation of Trudy Mat. Inst. Steklov. **144** (1979). MR603991

[Put1] C. R. Putnam, *On the non-periodicity of the zeros of the Riemann zeta-function*, Amer. J. Math. **76** (1954), 97–99, DOI 10.2307/2372402. MR0058703

[Put2] C. R. Putnam, *Remarks on periodic sequences and the Riemann zeta-function*, Amer. J. Math. **76** (1954), 828–830, DOI 10.2307/2372656. MR0064865

[Ra1] G. Radunović, *Fractal analysis of unbounded sets in Euclidean spaces and Lapidus zeta functions*, Ph. D. Thesis, University of Zagreb, Zagreb, Croatia, 2015.

[Ra2] G. Radunović, *Fractality and Lapidus zeta functions at infinity*, Math. Commun. **21** (2016), no. 2, 141–162. MR3517485

[Ram] R. Rammal, *Spectrum of harmonic excitations on fractals*, J. Physique **45** (1984), no. 2, 191–206, DOI 10.1051/jphys:01984004502019100. MR737523

[RamTo] R. Rammal and G. Toulouse, Random walks on fractal structures and percolation cluster, *J. Physique Lettres* **44** (1983), L13–L22.

[RatWi] J. Rataj and S. Winter, *Characterization of Minkowski measurability in terms of surface area*, J. Math. Anal. Appl. **400** (2013), no. 1, 120–132, DOI 10.1016/j.jmaa.2012.10.059. MR3003969

[ReSi1] M. Reed and B. Simon, *Methods of Modern Mathematical Physics*, vol. I, *Functional Analysis*, revised and enlarged edition (of the 1975 edn.), Academic Press, New York, 1980.

[ReSi2] M. Reed and B. Simon, *Methods of modern mathematical physics. II. Fourier analysis, self-adjointness*, Academic Press [Harcourt Brace Jovanovich, Publishers], New York-London, 1975. MR0493420

[ReSi3] M. Reed and B. Simon, *Methods of modern mathematical physics. IV. Analysis of operators*, Academic Press [Harcourt Brace Jovanovich, Publishers], New York-London, 1978. MR0493421

[Rei1] A. Reich, *Universelle Werteverteilung von Eulerprodukten* (German), Nachr. Akad. Wiss. Göttingen Math.-Phys. Kl. II **1** (1977), 1–17. MR0567687

[Rei2] A. Reich, *Werteverteilung von Zetafunktionen* (German), Arch. Math. (Basel) **34** (1980), no. 5, 440–451, DOI 10.1007/BF01224983. MR593771

[Res] M. Resman, *Invariance of the normalized Minkowski content with respect to the ambient space*, Chaos Solitons Fractals **57** (2013), 123–128, DOI 10.1016/j.chaos.2013.10.001. MR3128607

[Rie] B. Riemann, Ueber die Anzahl der Primzahlen unter einer gegebenen Grösse, *Monatsb. der Berliner Akad.* 1858/60, pp. 671–680. (English translation in [**Edw**, Appendix, pp. 229–305].)

[Ru1] W. Rudin, *Real and complex analysis*, 3rd ed., McGraw-Hill Book Co., New York, 1987. MR924157

[Ru2] W. Rudin, *Functional analysis*, 2nd ed., International Series in Pure and Applied Mathematics, McGraw-Hill, Inc., New York, 1991. MR1157815

[Rue1] D. Ruelle, *Generalized zeta-functions for Axiom A basic sets*, Bull. Amer. Math. Soc. **82** (1976), no. 1, 153–156, DOI 10.1090/S0002-9904-1976-14003-7. MR0400311

[Rue2] D. Ruelle, *Zeta-functions for expanding maps and Anosov flows*, Invent. Math. **34** (1976), no. 3, 231–242, DOI 10.1007/BF01403069. MR0420720

[Rue3] D. Ruelle, *Thermodynamic formalism*, Encyclopedia of Mathematics and its Applications, vol. 5, Addison-Wesley Publishing Co., Reading, Mass., 1978. The mathematical structures of classical equilibrium statistical mechanics. With a foreword by Giovanni Gallavotti and Gian-Carlo Rota. MR511655

[Rue4] D. Ruelle, *Dynamical zeta functions for piecewise monotone maps of the interval*, CRM Monograph Series, vol. 4, American Mathematical Society, Providence, RI, 1994. MR1274046

[Sab1] C. Sabot, *Integrated density of states of self-similar Sturm-Liouville operators and holomorphic dynamics in higher dimension* (English, with English and French summaries), Ann. Inst. H. Poincaré Probab. Statist. **37** (2001), no. 3, 275–311, DOI 10.1016/S0246-0203(00)01068-2. MR1831985

[Sab2] C. Sabot, *Spectral properties of self-similar lattices and iteration of rational maps* (English, with English and French summaries), Mém. Soc. Math. Fr. (N.S.) **92** (2003), vi+104, DOI 10.24033/msmf.405. MR1976877

[Sab3] C. Sabot, *Spectral analysis of a self-similar Sturm-Liouville operator*, Indiana Univ. Math. J. **54** (2005), no. 3, 645–668, DOI 10.1512/iumj.2005.54.2490. MR2151229

[Sarn] P. Sarnak, *L-functions*, Proceedings of the International Congress of Mathematicians, Vol. I (Berlin, 1998), Doc. Math. **Extra Vol. I** (1998), 453–465. MR1648042

[Sc] M. Schechter, *Operator methods in quantum mechanics*, Dover Publications, Inc., Mineola, NY, 2002. Reprint of the 1981 original. MR1969612

[Schl] W. Schlag, *A course in complex analysis and Riemann surfaces*, Graduate Studies in Mathematics, vol. 154, American Mathematical Society, Providence, RI, 2014. MR3186310

[Schm] F. K. Schmidt, *Analytische Zahlentheorie in Körpern der Charakteristik p* (German), Math. Z. **33** (1931), no. 1, 1–32, DOI 10.1007/BF01174341. MR1545199

[Schn] R. Schneider, *Convex bodies: The Brunn-Minkowski theory*, Encyclopedia of Mathematics and its Applications, vol. 44, Cambridge University Press, Cambridge, 1993. MR1216521

[Schw] L. Schwartz, *Théorie des Distributions*, revised and enlarged edition (of the 1951 edn.), Hermann, Paris, 1996.

[See1] R. T. Seeley, *Complex powers of an elliptic operator*, Singular Integrals (Proc. Sympos. Pure Math., Chicago, vol. 10, Ill., 1966), Amer. Math. Soc., Providence, RI, 1967, pp. 288–307. MR0237943

[See2] R. T. Seeley, *A sharp asymptotic remainder estimate for the eigenvalues of the Laplacian in a domain of* $\mathbf{R}^3$, Adv. in Math. **29** (1978), no. 2, 244–269, DOI 10.1016/0001-8708(78)90013-0. MR506893

[See3] R. T. Seeley, *An estimate near the boundary for the spectral function of the Laplace operator*, Amer. J. Math. **102** (1980), no. 5, 869–902, DOI 10.2307/2374196. MR590638

[Sel1] A. Selberg, *Old and new conjectures and results about a class of Dirichlet series*, Proceedings of the Amalfi Conference on Analytic Number Theory (Maiori, 1989), Univ. Salerno, Salerno, 1992, pp. 367–385. (Reprinted in [**Sel2**, pp. 47-63].) MR1220477

[Sel2] A. Selberg, *Collected papers. Vol. II*, Springer-Verlag, Berlin, 1991. With a foreword by K. Chandrasekharan. MR1295844

[Ser] J.-P. Serre, *A course in arithmetic*, Springer-Verlag, New York-Heidelberg, 1973. Translated from the French; Graduate Texts in Mathematics, No. 7. MR0344216

[Shi] T. Shima, *On eigenvalue problems for Laplacians on p.c.f. self-similar sets*, Japan J. Indust. Appl. Math. **13** (1996), no. 1, 1–23, DOI 10.1007/BF03167295. MR1377456

[Shis] M. Shishikura, *The Hausdorff dimension of the boundary of the Mandelbrot set and Julia sets*, Ann. of Math. (2) **147** (1998), no. 2, 225–267, DOI 10.2307/121009. MR1626737

[Sim1] B. Simon, *Notes on infinite determinants of Hilbert space operators*, Advances in Math. **24** (1977), no. 3, 244–273, DOI 10.1016/0001-8708(77)90057-3. MR0482328

[Sim2] B. Simon, *Trace ideals and their applications*, 2nd ed., Mathematical Surveys and Monographs, vol. 120, American Mathematical Society, Providence, RI, 2005. MR2154153

[Sim3] B. Simon, *Operator theory*, A Comprehensive Course in Analysis, Part 4, American Mathematical Society, Providence, RI, 2015. MR3364494

[Smit] F. Smithies, *The Fredholm theory of integral equations*, Duke Math. J. **8** (1941), 107–130. MR0004699

[Sta] L. L. Stachó, *On the volume function of parallel sets*, Acta Sci. Math. (Szeged) **38** (1976), no. 3–4, 365–374. MR0442202

[Stein] J. Steiner, Über parallele Flächen, *Monatsb. preuss. Akad. Wiss.*, Berlin, 1840, pp. 114–118. (Reprinted in: *Gesamm. Werke*, vol. II, pp. 173–176.)

[Steu] J. Steuding, *Value-distribution of L-functions*, Lecture Notes in Mathematics, vol. 1877, Springer, Berlin, 2007. MR2330696

[Str] R. S. Strichartz, *Differential equations on fractals: A tutorial*, Princeton University Press, Princeton, NJ, 2006. MR2246975

[Tep1] A. Teplyaev, *Spectral zeta function of symmetric fractals*, Fractal geometry and stochastics III, Progr. Probab., vol. 57, Birkhäuser, Basel, 2004, pp. 245–262. MR2087144

[Tep2] A. Teplyaev, *Spectral zeta functions of fractals and the complex dynamics of polynomials*, Trans. Amer. Math. Soc. **359** (2007), no. 9, 4339–4358, DOI 10.1090/S0002-9947-07-04150-5. MR2309188

[Tha1] K. Thas (ed.), *Absolute arithmetic and* $\mathbb{F}_1$*-geometry*, European Mathematical Society (EMS), Zürich, 2016. MR3702023

[Tha2] K. Thas, *The combinatorial-motivic nature of* $\mathbb{F}_1$*-schemes*, Absolute arithmetic and $\mathbb{F}_1$-geometry, Eur. Math. Soc., Zürich, 2016, pp. 83–159. MR3701900

[Tit] E. C. Titchmarsh, *The theory of the Riemann zeta-function*, 2nd ed., The Clarendon Press, Oxford University Press, New York, 1986. Edited and with a preface by D. R. Heath-Brown. MR882550

[Tri1] C. Tricot, *Mesures et Dimensions*, Thèse de Doctorat d'Etat Es Sciences Mathématiques, Université Paris-Sud, Orsay, France, 1983.

[Tri2] C. Tricot, *Dimensions aux bords d'un ouvert* (French, with English summary), Ann. Sci. Math. Québec **11** (1987), no. 1, 205–235. MR912170

[Tri3] C. Tricot, *Curves and fractal dimension*, Springer-Verlag, New York, 1995. With a foreword by Michel Mendès France. Translated from the 1993 French original. MR1302173

[Tri4] C. Tricot, *General Hausdorff functions, and the notion of one-sided measure and dimension*, Ark. Mat. **48** (2010), no. 1, 149–176, DOI 10.1007/s11512-008-0087-8. MR2594591

[vB-Gi] M. van den Berg and P. B. Gilkey, *A comparison estimate for the heat equation with an application to the heat content of the s-adic von Koch snowflake*, Bull. London Math. Soc. **30** (1998), no. 4, 404–412, DOI 10.1112/S0024609398004469. MR1620833

[Vel-San] A. Vélez-Santiago, *Global regularity for a class of quasi-linear local and nonlocal elliptic equations on extension domains*, J. Funct. Anal. **269** (2015), no. 1, 1–46, DOI 10.1016/j.jfa.2015.04.016. MR3345603

[vF1] M. van Frankenhuijsen, *Arithmetic progressions of zeros of the Riemann zeta function*, J. Number Theory **115** (2005), no. 2, 360–370, DOI 10.1016/j.jnt.2005.01.002. MR2180508

[vF2] M. van Frankenhuijsen, *Riemann zeros in arithmetic progression*, Fractal geometry and dynamical systems in pure and applied mathematics. I. Fractals in pure mathematics, Contemp. Math., vol. 600, Amer. Math. Soc., Providence, RI, 2013, pp. 365–380, DOI 10.1090/conm/600/11921. MR3203410

[vM1] H. von Mangoldt, *Auszug aus einer Arbeit unter dem Titel: Zu Riemann's Abhandlung 'Über die Anzahl der Primzahlen unter einer gegebenen Grösse'*, Sitzungsberichte preuss. Akad. Wiss., Berlin, 1894, pp. 883–896.

[vM2] H. von Mangoldt, *Zu Riemanns Abhandlung "Ueber die Anzahl der Primzahlen unter einer gegebenen Grösse"* (German), J. Reine Angew. Math. **114** (1895), 255–305, DOI 10.1515/crll.1895.114.255. MR1580379

[Vor] S. M. Voronin, *A theorem on the "universality" of the Riemann zeta-function* (Russian), Izv. Akad. Nauk SSSR Ser. Mat. **39** (1975), no. 3, 475–486, 703. MR0472727

[Wat] S. R. Watson, *Fractal Zeta Functions: To Ahlfors Spaces and Beyond*, ProQuest LLC, Ann Arbor, MI, 2017. Thesis (Ph.D.)–University of California, Riverside. MR3755432

[Wei1] A. Weil, *On the Riemann hypothesis in function-fields*, Proc. Nat. Acad. Sci. U. S. A. **27** (1941), 345–347. (Reprinted in [**Wei6**, vol. I, pp. 277-279].) MR0004242

[Wei2] A. Weil, Sur les courbes algébriques et les variétés qui s'en déduisent, *Pub. Inst. Math. Strasbourg* VII (1948), pp. 1–85. (Reprinted in: *Courbes algébriques et variétés abéliennes*, Hermann, Paris, 1971.)

[Wei3] A. Weil, *Numbers of solutions of equations in finite fields*, Bull. Amer. Math. Soc. **55** (1949), 497–508, DOI 10.1090/S0002-9904-1949-09219-4. (Reprinted in [**Wei6**, vol. I, pp. 399-410].) MR0029393

[Wei4] A. Weil, *Fonction zêta et distributions* (French), Séminaire Bourbaki, vol. 9, Soc. Math. France, Paris, 1995, pp. 523–531. (Reprinted in [**Wei6**, vol. III, pp. 158-163].) MR1610983

[Wei5] A. Weil, *Sur les formules explicites de la théorie des nombres* (French, with Russian summary), Izv. Akad. Nauk SSSR Ser. Mat. **36** (1972), 3–18. English translation in: Math. USSR Izv. **6** (1973), 1-17 (Reprinted in [**Wei6**, vol. III, pp. 249-264].) MR0379440

[Wei6] A. Weil, *André Weil: Oeuvres Scientifiques* (Collected Papers), vols. I, II and III, second edition (with corrected printing), Springer-Verlag, Berlin and New York, 1980.

[Wein] S. Weinberg, *The quantum theory of fields. vol. III*, Cambridge University Press, Cambridge, 2000. Supersymmetry. MR1737296

[Wel] D. J. A. Welsh, *Matroid theory*, Academic Press [Harcourt Brace Jovanovich, Publishers], London-New York, 1976. L. M. S. Monographs, No. 8. MR0427112

[Wey1] H. Weyl, *Über die Abhängigkeit der Eigenschwingungen einer Membran und deren Begrenzung* (German), J. Reine Angew. Math. **141** (1912), 1–11, DOI 10.1515/crll.1912.141.1. (Reprinted in [**Wey4**, vol. I, pp. 431-441].) MR1580843

[Wey2] H. Weyl, *Das asymptotische Verteilungsgesetz der Eigenwerte linearer partieller Differentialgleichungen (mit einer Anwendung auf die Theorie der Hohlraumstrahlung)* (German), Math. Ann. **71** (1912), no. 4, 441–479, DOI 10.1007/BF01456804. (Reprinted in [**Wey4**, vol. I, pp. 393-430].) MR1511670

[Wey3] H. Weyl, *On the Volume of Tubes*, Amer. J. Math. **61** (1939), no. 2, 461–472, DOI 10.2307/2371513. (Reprinted in [**Wey4**, vol. III, pp. 658-669].) MR1507388

[Wey4] H. Weyl, *Hermann Weyl: Gesammelte Abhandlungen. Bände I, II, III, IV* (German), Herausgegeben von K. Chandrasekharan, Springer-Verlag, Berlin-New York, 1968. MR0230597

[Wi] S. Winter, Curvature measures and fractals, *Dissertationes Math. (Rozprawy Mat.)* **453** (2008), 1–66.

[WiZä] S. Winter and M. Zähle, *Fractal curvature measures of self-similar sets*, Adv. Geom. **13** (2013), no. 2, 229–244, DOI 10.1515/advgeom-2012-0026. MR3038703

[Wils] K. G. Wilson, Renormalization group and critical phenomena, I & II, *Phys. Rev.* **B4** (1971), 3174–3183 & 3184–3205.

[Yo] K. Yosida, *Functional analysis*, 5th ed., Springer-Verlag, Berlin-New York, 1978. Grundlehren der Mathematischen Wissenschaften, Band 123. MR0500055

[Zä1] M. Zähle, *Integral and current representation of Federer's curvature measures*, Arch. Math. (Basel) **46** (1986), no. 6, 557–567, DOI 10.1007/BF01195026. MR849863

[Zä2] M. Zähle, *Curvatures and currents for unions of sets with positive reach*, Geom. Dedicata **23** (1987), no. 2, 155–171, DOI 10.1007/BF00181273. MR892398

[Zä3] M. Zähle, *Lipschitz-Killing curvatures of self-similar random fractals*, Trans. Amer. Math. Soc. **363** (2011), no. 5, 2663–2684, DOI 10.1090/S0002-9947-2010-05198-0. MR2763731

[Zä4] M. Zähle, *Curvature measures of fractal sets*, Fractal geometry and dynamical systems in pure and applied mathematics. I. Fractals in pure mathematics, Contemp. Math., vol. 600, Amer. Math. Soc., Providence, RI, 2013, pp. 381–399, DOI 10.1090/conm/600/11953. MR3203411

[Zu1] D. Žubrinić, *Minkowski content and singular integrals*, Chaos Solitons Fractals **17** (2003), no. 1, 169–177, DOI 10.1016/S0960-0779(02)00441-1. MR1960771

[Zu2] D. Žubrinić, *Analysis of Minkowski contents of fractal sets and applications*, Real Anal. Exchange **31** (2005/06), no. 2, 315–354. MR2265777

Department of Mathematics, University of California, Riverside, CA 92521, USA
Email address: lapidus@math.ucr.edu

Contemporary Mathematics
Volume **731**, 2019
https://doi.org/10.1090/conm/731/14678

Eigenvalues of the Laplacian on domains with fractal boundary

Paul Pollack and Carl Pomerance

For Michel Lapidus on his 60th birthday

Abstract. Consider the Laplacian operator on a bounded open domain in Euclidean space with Dirichlet boundary conditions. We show that for each number D with $1 < D < 2$, there are two bounded open domains in $\mathbf{R}^2$ of the same area, with their boundaries having Minkowski dimension D, and having the same content, yet the secondary terms for the eigenvalue counts are not the same. This was shown earlier by Lapidus and the second author, but a possible countable set of exceptional dimensions D were excluded. Here we show that the earlier construction has no exceptions.

1. Introduction

Let Ω be a nonempty, bounded open set in $\mathbf{R}^2$. We consider eigenvalues for the Laplacian operator $\Delta = \partial^2/\partial x^2 + \partial^2/\partial y^2$ for the closure in the Sobolev space of smooth functions with compact support, which are 0 on $\partial\Omega$. It is well-known that the nonzero eigenvalues are negative, forming a discrete multiset, with each multiplicity bounded. By convention we consider the absolute value of these eigenvalues and label them $0 < \lambda_1 \le \lambda_2 \le \dots$. Let

$$N(\lambda;\Omega) = \sum_{\substack{m\ge 1\\ \lambda_m \le \lambda}} 1$$

denote the counting function of the λ_m's.

Weyl's classical asymptotic formula for $N(\lambda;\Omega)$,

$$N(\lambda;\Omega) \sim \frac{|\Omega|_2}{4\pi}\lambda, \quad \lambda \to \infty, \tag{1.1}$$

is now known for arbitrary Ω, see [**1**], [**12**]. Weyl conjectured that if $\partial\Omega$ is sufficiently regular then there is a secondary term in (1.1) that is asymptotically a constant depending on the one-dimensional measure of $\partial\Omega$ times $\lambda^{1/2}$. Ivrii (see [**7**]) essentially proved this conjecture.

There remains the issue of when $\partial\Omega$ is not sufficiently regular, in particular if the boundary has a fractal dimension larger than 1. In 1979, Berry suggested a modified Weyl conjecture with a secondary term for $N(\lambda;\Omega)$ proportional to a constant times $\lambda^{d/2}$, where d is the Hausdorff dimension of $\partial\Omega$.

2010 *Mathematics Subject Classification.* 58J50, 35P20, 11M06.

However, the Hausdorff measure of $\partial\Omega$ depends on the relative placement of the connected components of Ω in the ambient space, yet the eigenvalues do not care about this placement, so the Berry conjecture cannot strictly be true, see Lapidus [**8**]. Brossard and Carmona [**2**] had earlier demonstrated a specific counterexample to the Berry conjecture, and suggested instead that the Minkowski dimension D of $\partial\Omega$ is the more appropriate parameter. In fact, it was shown in [**8**] (also see [**9**]) that if $\partial\Omega$ has Minkowski dimension D with $1 < D < 2$ and finite upper Minkowski content in this dimension, then the error in (1.1) is $O(\lambda^{D/2})$. This led Lapidus [**8**] to conjecture that if in addition it was assumed that $\partial\Omega$ is Minkowski measurable in dimension D, then there would be a secondary term in (1.1) of the form $c\lambda^{D/2}$, with c a positive constant depending on the Minkowski content of $\partial\Omega$.

The analogue of this modified Weyl–Berry conjecture for regions in $\mathbf{R}^1$ was subsequently proved in [**10**], with a simplified proof given in [**5**]. However, for dimension 2 the conjecture is false in general. If a set of Newtonian capacity zero is removed from a given domain, the eigenvalues are not changed, yet the Minkowski dimension and content can be altered by strategically choosing which set of capacity zero to remove. This idea was developed in [**6**] and [**11**]. However, this behavior seems to be simple to bar with a further modification of the Weyl–Berry conjecture by stating it in terms of the "intrinsic" Minkowski dimension of the boundary, where we take the infimum of Minkowski dimensions of the boundaries of domains that agree up to a set of Newtonian capacity 0, and also the "intrinsic" Minkowski content, defined in the same way.

A more compelling counterexample was given in [**11**], involving sprays of the 1×1 unit square and of the 1×2 rectangle. (A "spray" is a disjoint union of similar copies of some given simple region with bounded total area.) However, the argument in [**11**] was not sufficient to give a counterexample for all Minkowski dimensions D with $1 < D < 2$, but rather for all but a possible countable set. In this note we show that the construction in [**11**] actually works for every D with $1 < D < 2$: there are no exceptions.

As in [**11**], the counterexample extends in a natural way to higher-dimensional ambient spaces $\mathbf{R}^n$ for $n \geq 2$.

Our argument involves getting precise formulations of the coefficient of the secondary terms for $N(\lambda; \Omega)$ for our two domains Ω in terms of the Riemann zeta-function and the Dedekind zeta-function for the quadratic field $\mathbf{Q}(i)$. We then show that these two functions are different for all D with $1 < D < 2$. Along the way we prove some results of perhaps independent interest about the zeta-functions involved. For example, we show that $(s-1)\zeta(s)$ is monotone for real $s \geq 1/2$. We show the same for $L(s, \chi)$, where χ is the quadratic Dirichlet character mod 4.

2. Our domains

Fix an arbitrary number D with $1 < D < 2$. Our first domain, denoted Ω_1, is the disjoint union of the $j^{-1/D} \times j^{-1/D}$ open squares for $j = 1, 2, \ldots$ arranged in the plane so that they don't overlap and sit inside a large disc. (It is routine to show that such an arrangement is possible; for a particularly efficient packing, see Moon and Moser [**13**].)

Let a be the positive real number $(2/(D+2))^{1/D}$ and let Ω_2' be the disjoint union of the $aj^{-1/D} \times 2aj^{-1/D}$ rectangles for $j = 1, 2, \ldots$ also arranged in the plane so that they don't overlap and sit inside a large disc.

The area of Ω_1 is $\zeta(2/D)$, where ζ is the Riemann zeta-function. The area of Ω_2' is $2a^2\zeta(2/D)$. Note that since $1 < D < 2$, we have $2a^2 < 1$, so that the area of Ω_2' is smaller than the area of Ω_1. Let Ω_2 be the disjoint union of Ω_2' and a square of area $(1-2a^2)\zeta(2/D)$. Thus, Ω_1 and Ω_2 have the same area.

As in [**11**], we have that the boundaries of Ω_1 and Ω_2 both have Minkowski dimension D with Minkowski content in dimension D of $2^{3-D}(2-D)^{-1}(D-1)^{-1}$.

Let $\zeta_1(s)$ be the spectral zeta-function for the 1×1 square and let $\zeta_2(s)$ be the spectral zeta-function for the $a \times 2a$ rectangle. Also let $N(\lambda;\Omega_i)$ denote the counting function of the eigenvalues for the Dirichlet Laplacian on Ω_i, for $i = 1, 2$. From [**11**, Theorem 3.2] we have

$$N(\lambda;\Omega_i) = \frac{1}{4\pi}\zeta(2/D)\lambda + (\zeta_i(D/2) + o(1))\lambda^{D/2}, \quad \lambda \to \infty, \tag{2.1}$$

for $i = 1, 2$. Note that the eigenvalues for the additional square tacked on to Ω_2' affect the main term for $N(\lambda;\Omega_2)$ (and is taken into account in (2.1)) and create an error of $O(\lambda^{1/2})$, which is negligible. That is, these eigenvalues are invisible to the secondary term.

The argument in [**11**] depended on ζ_1, ζ_2 being non-identical analytic functions, and so the secondary term coefficients in (2.1) could agree for at most countably many D in $(1, 2)$. Our goal in this paper is to show that they actually are unequal for all D in $(1, 2)$. Towards this end, we obtain explicit descriptions of the spectral zeta-functions ζ_1, ζ_2.

3. Our spectral zeta-functions

The eigenvalues for the 1×1 square are the numbers $\pi^2(m^2+n^2)$ where m, n run over positive integers. Thus,

$$\zeta_1(s) = \sum_{m,n>0} \frac{1}{\pi^{2s}(m^2+n^2)^s}.$$

This function resembles the Dedekind zeta-function for the Gaussian field $\mathbf{Q}(i)$, namely

$$\zeta_{\mathbf{Q}(i)}(s) = \sum_{\mathfrak{I}\neq 0} \frac{1}{N(\mathfrak{I})^s} = \frac{1}{4}\sum_{(m,n)\neq(0,0)} \frac{1}{(m^2+n^2)^s},$$

where $\mathfrak{I}$ runs over the nonzero ideals of $\mathbf{Z}[i]$. For a pair $m, n > 0$ that we see in $\zeta_1(s)$, there are 4 corresponding terms $(\pm m, \pm n)$ giving the same value to m^2+n^2, and this 4-fold appearance in the last sum is compensated by the $\frac{1}{4}$ in front of it. In addition, $\zeta_{\mathbf{Q}(i)}(s)$ has terms coming from pairs $(\pm m, 0)$ and $(0, \pm n)$ that have no counterpart in $\zeta_1(s)$. These extra terms contribute $\zeta(2s)$ to $\zeta_{\mathbf{Q}(i)}(s)$. Thus,

$$\pi^{2s}\zeta_1(s) = \zeta_{\mathbf{Q}(i)}(s) - \zeta(2s). \tag{3.1}$$

Let χ be the Dirichlet character mod 4; that is χ is defined on all integers n, with $\chi(n) = 0, 1, -1$ depending, respectively, on whether n is even, $n \equiv 1 \pmod 4$, $n \equiv -1 \pmod 4$. Consider the L-function

$$L(s,\chi) = \sum_{n=1}^{\infty} \frac{\chi(n)}{n^s}.$$

We know that $L(s,\chi)$ is an entire function, and the series for it converges uniformly on compact subsets of $\Re s > 0$. It is of interest to us via the formula

$$\zeta_{\mathbf{Q}(i)}(s) = \zeta(s)L(s,\chi). \tag{3.2}$$

Now we look at $\zeta_2(s)$. We take a as in the last section. We have the eigenvalues $a^{-2}\pi^2(m^2+n^2/4)$, where $m,n>0$ are integers. We find it more convenient to work with $4^{-s}a^{-2s}\pi^{2s}\zeta_2(s)$, giving us

$$4^{-s}a^{-2s}\pi^{2s}\zeta_2(s) = \sum_{m,n>0} \frac{1}{(4m^2+n^2)^s}.$$

Let $r_2(k)$ be the number of representations of k as $4m^2+n^2$ with $m,n>0$. Further, let $r(k)$ denote the number of representations of k as m^2+n^2, where m,n are any integers. We have

$$\sum_{m,n>0} \frac{1}{(4m^2+n^2)^s} = \sum_{k=1}^{\infty} \frac{r_2(k)}{k^s}, \qquad \sum_{(m,n)\neq(0,0)} \frac{1}{(m^2+n^2)^s} = \sum_{k=1}^{\infty} \frac{r(k)}{k^s}.$$

Here are some observations on $r_2(k)$. First note that for $k \equiv 2 \pmod 4$, we have $r_2(k) = 0$, since squares are never $2 \pmod 4$. Next note that for k odd, a representation of k as a sum of two squares must have one of the squares even and one odd. From this we see that

$$r_2(k) = \begin{cases} \frac{1}{8}r(k), & k \text{ not a square,} \\ \frac{1}{8}r(k) - \frac{1}{2}, & k \text{ is a square.} \end{cases}$$

Indeed, if $m,n>0$ and $k = 4m^2+n^2$ with n odd, then there are 8 corresponding representations of k as a sum of two squares, namely $(\pm 2m, \pm n), (\pm n, \pm 2m)$. The expression $\frac{1}{8}r(k)$ also counts an additional $\frac{1}{2}$ if k is a square, so the formula holds for odd k.

Now consider even values of k, so as we have seen, we may assume that $4 \mid k$. We claim in this case that we have

$$r_2(k) = \begin{cases} \frac{1}{4}r(k/4), & k \text{ not a square,} \\ \frac{1}{4}r(k/4) - 1, & k \text{ is a square.} \end{cases}$$

Indeed, a representation of k as $4m^2+n^2$ has n even, so that $k/4 = m^2 + (n/2)^2$. Further, a pair m,n with $m,n>0$ gives rise to 4 signed representations of $k/4$. In addition, there are 4 additional representations of $k/4$ as a sum of two squares when k is a square. So $\frac{1}{4}r(k)$ needs to be decreased by 1 in this case.

Putting these thoughts together, we have

$$\begin{aligned} 4^{-s}a^{-2s}\pi^{2s}\zeta_2(s) &= \sum_{k=1}^{\infty} \frac{r_2(k)}{k^s} \\ &= \sum_{\substack{k>0 \\ k \text{ odd}}} \left(\frac{r(k)/8}{k^s} - \frac{1/2}{k^{2s}} \right) + \sum_{\substack{k>0 \\ 4|k}} \left(\frac{r(k/4)/4}{k^s} - \frac{1}{(k/2)^{2s}} \right) \\ &= \frac{1}{2} \sum_{\substack{k>0 \\ k \text{ odd}}} \frac{r(k)/4}{k^s} + 4^{-s}\zeta_{\mathbf{Q}(i)}(s) - \frac{1}{2}\zeta(2s) - \frac{1}{2}2^{-2s}\zeta(2s). \end{aligned}$$

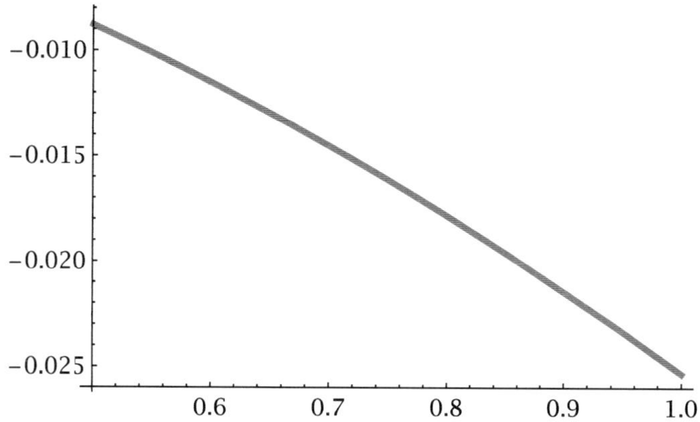

FIGURE 1. Mathematica plot of the right-hand side of (4.1) on $(1/2, 1)$.

Now, $r(k)/4$ is multiplicative, and the local factor corresponding to the prime 2 in the Euler product for $\zeta_{\mathbf{Q}(i)}(s)$ is $(1-2^{-s})^{-1}$, so that

$$\frac{1}{2}\sum_{\substack{k>0\\ k \text{ odd}}}\frac{r(k)/4}{k^s} = \frac{1}{2}(1-2^{-s})\zeta_{\mathbf{Q}(i)}(s).$$

Thus, with the above calculation, we have

$$4^{-s}a^{-2s}\pi^{2s}\zeta_2(s) = \left(\frac{1}{2} - 2^{-1-s} + 4^{-s}\right)\zeta_{\mathbf{Q}(i)}(s) - \frac{1}{2}(1+4^{-s})\zeta(2s),$$

so

$$a^{-2s}\pi^{2s}\zeta_2(s) = \left(2^{2s-1} - 2^{s-1} + 1\right)\zeta_{\mathbf{Q}(i)}(s) - \frac{1}{2}(4^s+1)\zeta(2s).$$

We have proved the following result.

PROPOSITION 3.1. *With the notation defined earlier, we have*

$$\pi^{2s}\zeta_1(s) = \zeta_{\mathbf{Q}(i)}(s) - \zeta(2s),$$
$$a^{-2s}\pi^{2s}\zeta_2(s) = (2^{2s-1} - 2^{s-1} + 1)\zeta_{\mathbf{Q}(i)}(s) - \frac{1}{2}(4^s+1)\zeta(2s).$$

4. Are they equal?

Our task is to show that $\zeta_1(D/2) \neq \zeta_2(D/2)$ for $1 < D < 2$. With $s = D/2$ we have $a^{-2s} = D/2 + 1 = s + 1$. So, from Proposition 3.1, we would like to show that

$$(4.1)\quad (s+1)\pi^{2s}(\zeta_1(s)-\zeta_2(s)) = (s-2^{2s-1}+2^{s-1})\zeta_{\mathbf{Q}(i)}(s) - \left(s + \frac{1}{2} - 2^{2s-1}\right)\zeta(2s)$$

is nonzero for $\frac{1}{2} < s < 1$. In Figure 1 we present a Mathematica plot of the expression in (4.1), and though it is close to 0, one can plainly see that it is not 0. Is this a proof? Not quite, since there may conceivably be some wild gyrations of the functions between the discrete points used by Mathematica to form the plot. In this section we give the details necessary to prove that the expression in (4.1) is negative for s in $(1/2, 1)$.

We begin with the following result.

PROPOSITION 4.1. *The function $(s-1)\zeta(s)$ is increasing on $[1/2, \infty)$.*

PROOF. For $\Re s > 0, s \neq 1$, we have

$$\zeta(s) = \frac{s}{s-1} - s\int_1^\infty x^{-1-s}\{x\}\,dx, \tag{4.2}$$

where $\{x\} = x - \lfloor x \rfloor$ is the fractional part of x. For $\Re s > 1$, this well-known formula follows from the definition of $\zeta(s)$ as a Dirichlet series and partial summation; for $\Re s > 0, s \neq 1$, it follows by analytic continuation. The same argument applied to

$$\zeta'(s) = \sum_{n=1}^{\infty} -n^{-s}\log n,$$

gives us

$$\zeta'(s) = \frac{-1}{(s-1)^2} - \int_1^\infty (-sx^{-1-s}\log x + x^{-1-s})\{x\}\,dx$$

for $\Re s > 0, s \neq 1$. Thus,

$$\begin{aligned}((s-1)\zeta(s))' &= \zeta(s) + (s-1)\zeta'(s)\\ &= 1 - \int_1^\infty (-s(s-1)x^{-1-s}\log x + (2s-1)x^{-1-s})\{x\}\,dx.\end{aligned}$$

(This identity can also be obtained by differentiating $s-1$ times the equation in (4.2).) The integrand is positive for $s \in (1/2, 1)$, so replacing $\{x\}$ with 1 gives a lower bound on this interval. That is,

$$((s-1)\zeta(s))' > 1 - \int_1^\infty -s(s-1)x^{-1-s}\log x + (2s-1)x^{-1-s}\,dx = 0,$$

and the proposition is proved for the interval $[1/2, 1]$.

We now deal with the range $s \geq 1$. We have, as is easy to see,

$$(1-2^{1-s})\zeta(s) = \sum_{n=1}^{\infty}(-1)^{n-1}n^{-s}.$$

Let $h(s)$ be the sum of the first 5 terms of this series, and let $h_k(s) = -k^{-s} + (k+1)^{-s}$, so that

$$(1-2^{1-s})\zeta(s) = h(s) + \sum_{k=3}^{\infty} h_{2k}(s).$$

It is easy to see that for $k \geq 3$, the function $h_k(s)$ is increasing for $s \geq 1$. (The derivative is $k^{-s}\log k - (k+1)^{-s}\log(k+1)$ and this is positive on $s \geq 1$ when $(k+1)/k > \log(k+1)/\log k$, which holds for $k \geq 3$.) We now show that $h(s)$ is also increasing for $s \geq 1$.

We have

$$3^s h'(s) = \left(\frac{3}{2}\right)^s \log 2 - \log 3 + \left(\frac{3}{4}\right)^s \log 4 - \left(\frac{3}{5}\right)^s \log 5.$$

Call this function $g(s)$. Then

$$g'(s) = \left(\frac{3}{2}\right)^s \log\frac{3}{2}\log 2 - \left(\frac{3}{4}\right)^s \log\frac{4}{3}\log 4 + \left(\frac{3}{5}\right)^s \log\frac{5}{3}\log 5.$$

It is easy to check that the sum of the first two terms here is positive for $s \geq 1$, it only involves checking that $(\log\frac{4}{3}\log 4)/(\log\frac{3}{2}\log 2) < 2$. So, $g'(s) > 0$ for $s \geq 1$, which in turn implies that $g(s)$ is increasing for $s \geq 1$. But $g(1) > 0$, so we have

$g(s) > 0$ for $s \geq 1$, which in turn implies that $h(s)$ is increasing for $s \geq 1$. Thus, $(1 - 2^{1-s})\zeta(s)$ is increasing for $s \geq 1$.

It suffices now to show that $(s-1)/(1-2^{1-s})$ is increasing and positive for $s > 1$. It is clearly positive. Letting $x = s - 1$ and taking the derivative, we get

$$\frac{1 - 2^{-x} - x2^{-x}\log 2}{(1-2^{-x})^2} = \frac{2^x - 1 - x\log 2}{2^x(1-2^{-x})^2},$$

which is seen to be positive by using the Taylor expansion for 2^x in the numerator. This completes the proof. □

We also need a monotonicity result for $L(s,\chi)$.

PROPOSITION 4.2. The function $L(s,\chi)$ is increasing on $[1/2,\infty)$.

PROOF. Write $L(s,\chi) = -f_1(s) + f_2(s) + f_3(s)$, where $f_1(s) = -(1 + 1/5^s + 1/9^s + \cdots + 1/49^s)$, $f_2(s) = -(1/3^s + 1/7^s + \cdots + 1/47^s)$, and

$$f_3(s) = \sum_{j=13}^{\infty}\left(\frac{-1}{(4j-1)^s} + \frac{1}{(4j+1)^s}\right).$$

Let $s \geq 1/2$. We have

$$f_3'(s) = \sum_{j=13}^{\infty}\left(\frac{\log(4j-1)}{(4j-1)^s} - \frac{\log(4j+1)}{(4j+1)^s}\right).$$

The function $x \mapsto (\log x)x^{-s}$ is decreasing for $x > e^{1/s}$, and thus for all $x \geq 8$. Hence, each summand in $f_3'(s)$ is positive, and so $f_3'(s) > 0$. Therefore, the proposition will follow if $f_2'(s) - f_1'(s) > 0$ for all $s \in [1/2, 1]$.

We give a computer-assisted proof of this last inequality. Observe that

$$f_2'(s) = \log(3)/3^s + \log(7)/7^s + \cdots + \log(47)/47^s$$

while

$$f_1'(s) = \log(5)/5^s + \log(9)/9^s + \cdots + \log(49)/49^s.$$

In particular, both $f_1'(s)$ and $f_2'(s)$ are decreasing on $[1/2,1]$. To verify that $f_2'(s) - f_1'(s) > 0$ on all of $[1/2,1]$, partition $[1/2,1]$ into $N := 10^4$ equal-length subintervals $[x_i, x_{i+1}]$ for $i = 0,\ldots,N-1$, where each $x_i = 1/2 + i/(2N)$. The minimum of $f_2'(s) - f_1'(s)$ on $[x_i,x_{i+1}]$ is bounded below by $f_2'(x_{i+1}) - f_1'(x_i)$. Using `gp/pari`, one can easily check that $f_2'(x_{i+1}) - f_1'(x_i) > 0.004$ for all $i = 0,\ldots,N-1$.

This shows that $L(1,\chi)$ is increasing on $[1/2,1]$. To complete the proof, fix $s \geq 1$ and note that since $\log(x)/x^s$ is decreasing for $x \geq 3$, it follows that $L'(s,\chi)$ is positive. Hence $L(s,\chi)$ is increasing on $[1/2,\infty)$. □

REMARK. There is an alternative approach to Propositions 4.1 and 4.2 based on the Hadamard product decompositions of $\zeta(s)$ and $L(s,\chi)$. We discuss how this goes for $L(s,\chi)$ first, since the argument is slightly more involved than for $(s-1)\zeta(s)$.

We start from the formula for $\frac{L'}{L}$ found as equation (17) on p. 83 of [**4**]. This gives that for all real s,

$$\frac{L'(s,\chi)}{L(s,\chi)} - \frac{L'(0,\chi)}{L(0,\chi)} = \frac{1}{2}\frac{\Gamma'(1/2)}{\Gamma(1/2)} - \frac{1}{2}\frac{\Gamma'(\frac{s}{2}+\frac{1}{2})}{\Gamma(\frac{s}{2}+\frac{1}{2})} + \sum_{\rho}\Re\left(\frac{1}{s-\rho} + \frac{1}{\rho}\right), \tag{4.3}$$

where ρ runs over the zeros of $L(s,\chi)$ in the critical strip $0 \le \Re(s) \le 1$. (If we did not take real parts in the last summand, then (4.3) would hold for all complex s.) When $s > 0$, we have $\Re(\frac{1}{s-\rho} + \frac{1}{\rho}) \ge 0$ as long as

$$\Re(\rho)(s - \Re(\rho)) + \Im(\rho)^2 > 0. \tag{4.4}$$

Restrict now to real $s > 0$. Then $\Re(\rho)(s - \Re(\rho)) \ge -1$. To get a handle on $\Im(\rho)$, we compare eq. (17) of [**4**, p. 83], taken at $s = 0$, with eq. (18) from the same page; this yields

$$\frac{L'(0,\chi)}{L(0,\chi)} + \frac{1}{2}\log\frac{4}{\pi} + \frac{1}{2}\frac{\Gamma'(1/2)}{\Gamma(1/2)} = -\sum_{\rho}\Re\frac{1}{\rho}.$$

From [**3**, Corollary 10.3.2, p. 188 and Proposition 10.3.5, pp. 189–190], we have $L'(0,\chi) = \log\frac{\Gamma(1/4)}{2\cdot\Gamma(3/4)}$ and $L(0,\chi) = \frac{1}{2}$. It follows that

$$\sum_{\rho}\Re\frac{1}{\rho} = 0.0777839\ldots. \tag{4.5}$$

Note that each term in this sum is nonnegative. We now take the subsum of (4.5) where $\Re\rho \ge \frac{1}{2}$. If ρ is any zero in the critical strip, then $1-\overline{\rho}$ is also a zero with the same imaginary part. So our subsum consists of all zeros on the critical line and for each zero in the critical strip not on the critical line, we take the member of the pair $\rho, 1-\overline{\rho}$ with the larger real part. Now

$$\Re\frac{1}{\overline{\rho}} + \Re\frac{1}{\rho} = \frac{2\cdot\Re\rho}{(\Re\rho)^2 + (\Im\rho)^2} \ge \frac{1}{1+(\Im\rho)^2}.$$

Since ρ and $\overline{\rho}$ are both nontrivial zeros of $L(s,\chi)$, (4.5) implies that $|\Im\rho| > 3.4$. Thus, (4.4) holds for $s > 0$ (for every ρ). Consequently, the sum on ρ in (4.3) is nonnegative for these values of s. Turning to the digamma terms, recall that

$$-\frac{\Gamma'(z)}{\Gamma(z)} = \frac{1}{z} + \gamma + \sum_{k=1}^{\infty}\left(\frac{1}{z+k} - \frac{1}{k}\right);$$

this follows, e.g., by logarithmically differentiating equation (2) on p. 73 of [**4**]. Hence, $-\frac{\Gamma'(z)}{\Gamma(z)}$ is a decreasing function of z for real $z > 0$. Therefore, when $0 < s \le 1$,

$$\frac{1}{2}\frac{\Gamma'(1/2)}{\Gamma(1/2)} - \frac{1}{2}\frac{\Gamma'(\frac{s}{2}+\frac{1}{2})}{\Gamma(\frac{s}{2}+\frac{1}{2})} \ge \frac{1}{2}\left(\frac{\Gamma'(1/2)}{\Gamma(1/2)} - \frac{\Gamma'(1)}{\Gamma(1)}\right) = \log\frac{1}{2}.$$

(Here the final equality can be obtained from the partial fraction expansion of digamma given above.) Plugging back into (4.3), we deduce that for $0 < s \le 1$,

$$\frac{L'(s,\chi)}{L(s,\chi)} \ge 2\log\frac{\Gamma(1/4)}{2\cdot\Gamma(3/4)} + \log\frac{1}{2}.$$

This last expression is larger than 0.09, and in particular is positive. Since $L(s,\chi) > 0$ for $s > 0$, it follows that $L'(s,\chi) > 0$ on $(0,1]$. Thus, with the final step in the proof of Proposition 4.2 we have $L(s,\chi)$ increasing on $[0,\infty)$.

A similar method will show that $(s-1)\zeta(s)$ is increasing on $(0,\infty)$. Notice that for $s>0$, we have $((s-1)\zeta(s))'>0$ exactly when $\frac{\zeta'(s)}{\zeta(s)}+\frac{1}{s-1}>0$. Eq (7) on p. 80 of [**4**], combined with the expression for B on p. 81, shows that for all real s,

$$\frac{\zeta'(s)}{\zeta(s)}+\frac{1}{s-1}=-\frac{1}{2}\gamma-1+\log(2\pi)-\frac{1}{2}\frac{\Gamma'(\frac{s}{2}+1)}{\Gamma(\frac{s}{2}+1)}+\sum_{\rho}\Re\left(\frac{1}{s-\rho}+\frac{1}{\rho}\right),$$

where ρ now runs over the nontrivial Riemann zeta zeros. As remarked on p. 82 of [**4**], $|\Im\rho|>6$ for all ρ. (It is known in fact that $|\Im\rho|>14$.) It follows from our earlier arguments that the the sum on ρ is nonnegative for all $s>0$. Since $-\frac{\Gamma'(z)}{\Gamma(z)}$ is decreasing for real $z>0$, we deduce that for $0<s\leq 4$,

$$\frac{\zeta'(s)}{\zeta(s)}+\frac{1}{s-1}\geq-\frac{1}{2}\gamma-1+\log(2\pi)-\frac{1}{2}\frac{\Gamma'(3)}{\Gamma(3)}>0.08.$$

It remains to show that $(s-1)\zeta(s)$ is increasing for $s>4$. Using the idea at the end of the proof of Proposition 4.1, it suffices to show that $(1-2^{1-s})\zeta(s)$ is increasing in this range of s. Now $((1-2^{1-s})\zeta(s))'=\log(2)/2^s-\log(3)/3^s+\dots$. When $s>4$, the terms $\log(x)/x^s$ are decreasing for $x\geq 2$. Hence, $((1-2^{1-s})\zeta(s))'>0$.

REMARK. Concerning specifically Proposition 4.1, Harold Diamond has shown us a proof by somewhat different methods that $(s-1)\zeta(s)$ is monotone for $s\geq -2.5$, which is nearly best possible.

Referring back to (4.1), we see that

$$(s+1)\pi^{2s}(\zeta_1(s)-\zeta_2(s))=(s-2^{2s-1}+2^{s-1})\zeta_{\mathbf{Q}(i)}(s)-\left(s+\frac{1}{2}-2^{2s-1}\right)\zeta(2s)$$

$$(4.6)\qquad =\frac{2^{2s-1}-s-\frac{1}{2}}{s-\frac{1}{2}}\cdot\frac{1}{2}(2s-1)\zeta(2s)-\frac{(s-2^{2s-1}+2^{s-1})}{1-s}(s-1)\zeta(s)L(s,\chi).$$

Each of $\frac{1}{2}(2s-1)\zeta(2s)$, $(s-1)\zeta(s)$, and $L(s,\chi)$ is positive on $[1/2,1]$, and our work above shows that these functions are increasing there. The next proposition supplies the corresponding results for the remaining factors in (4.6).

PROPOSITION 4.3. Both of the functions

$$\frac{s-2^{2s-1}+2^{s-1}}{1-s}\quad\text{and}\quad\frac{2^{2s-1}-s-\frac{1}{2}}{s-\frac{1}{2}}$$

are positive and increasing on $(0,\infty)$. (We assume here that the discontinuities have been filled in to make the functions continuous.)

PROOF. We first prove that both functions are increasing on the entire real line. We begin by recalling a fact from calculus about convex functions: *Suppose that g is a $\mathcal{C}^2$ function on an open interval I. For each $x,y\in I$, put*

$$S(x,y)=\begin{cases}\frac{g(x)-g(y)}{x-y} & \text{if } x\neq y,\\ g'(x) & \text{if } x=y.\end{cases}$$

If $g''>0$ on I, then $S(x,y)$ is increasing separately in both x,y. Applying this with $g(x)=2^{2x-1}-x-2^{x-1}$ and $x=1,y=s$ shows that the first function is increasing. To handle the second function, take $g(x)=2^{2x-1}-x-1/2$, and look at $x=s,y=1/2$. Since both functions vanish at 0, their positivity on $(0,\infty)$ is now immediate. We remark that this calculus fact could also have been used for the last step in the proof of Proposition 4.1. □

We can now prove our main result.

PROOF THAT $\zeta_1(s) \neq \zeta_2(s)$ FOR $1/2 < s < 1$. Let $F(s)$ denote the first term on the right-hand side of (4.6), and let $G(s)$ denote the second, subtracted term. Then F and G are positive, increasing functions on $[1/2, 1]$. We will prove that $F - G < 0$ on $[1/2, 1]$ by the same method employed in the proof of Proposition 4.2. We partition $[1/2, 1]$ into $N := 50$ equal length intervals $[x_i, x_{i+1}]$ for $i = 0, 1, \ldots, N-1$, with each $x_i = 1/2 + i/(2N)$. The maximum of $F - G$ on $[x_i, x_{i+1}]$ is at most $F(x_{i+1}) - G(x_i)$. Using `Mathematica`, one easily computes that each of these differences is smaller than -0.001. □

5. Acknowledgments

We gratefully acknowledge some helpful discussions with and comments from Harold Diamond, Michel Lapidus, Micah Milinovich, and Craig Sutton. This project was completed during a very enjoyable visit of the first-named author to Dartmouth College. The first-named author is supported by NSF award DMS-1402268.

References

[1] M. Š. Birman and M. Z. Solomjak, *The principal term of the spectral asymptotics for "non-smooth" elliptic problems* (Russian), Funkcional. Anal. i Priložen. **4** (1970), no. 4, 1–13. MR0278126

[2] Jean Brossard and René Carmona, *Can one hear the dimension of a fractal?*, Comm. Math. Phys. **104** (1986), no. 1, 103–122. MR834484

[3] Henri Cohen, *Number theory. Vol. II. Analytic and modern tools*, Graduate Texts in Mathematics, vol. 240, Springer, New York, 2007. MR2312338

[4] Harold Davenport, *Multiplicative number theory*, 3rd ed., Graduate Texts in Mathematics, vol. 74, Springer-Verlag, New York, 2000. Revised and with a preface by Hugh L. Montgomery. MR1790423

[5] K. J. Falconer, *On the Minkowski measurability of fractals*, Proc. Amer. Math. Soc. **123** (1995), no. 4, 1115–1124, DOI 10.2307/2160708. MR1224615

[6] Jacqueline Fleckinger-Pellé and Dmitri G. Vassiliev, *An example of a two-term asymptotics for the "counting function" of a fractal drum*, Trans. Amer. Math. Soc. **337** (1993), no. 1, 99–116, DOI 10.2307/2154311. MR1176086

[7] Victor Ivrii, *Microlocal analysis and precise spectral asymptotics*, Springer Monographs in Mathematics, Springer-Verlag, Berlin, 1998. MR1631419

[8] Michel L. Lapidus, *Fractal drum, inverse spectral problems for elliptic operators and a partial resolution of the Weyl-Berry conjecture*, Trans. Amer. Math. Soc. **325** (1991), no. 2, 465–529, DOI 10.2307/2001638. MR994168

[9] Michel L. Lapidus and Jacqueline Fleckinger-Pellé, *Tambour fractal: vers une résolution de la conjecture de Weyl-Berry pour les valeurs propres du laplacien* (French, with English summary), C. R. Acad. Sci. Paris Sér. I Math. **306** (1988), no. 4, 171–175. MR930556

[10] Michel L. Lapidus and Carl Pomerance, *The Riemann zeta-function and the one-dimensional Weyl-Berry conjecture for fractal drums*, Proc. London Math. Soc. (3) **66** (1993), no. 1, 41–69, DOI 10.1112/plms/s3-66.1.41. MR1189091

[11] Michel L. Lapidus and Carl Pomerance, *Counterexamples to the modified Weyl-Berry conjecture on fractal drums*, Math. Proc. Cambridge Philos. Soc. **119** (1996), no. 1, 167–178, DOI 10.1017/S0305004100074053. MR1356166

[12] Guy Métivier, *Valeurs propres de problèmes aux limites elliptiques irrégulières* (French), Bull. Soc. Math. France Suppl. Mém. **51–52** (1977), 125–219. MR0473578

[13] J. W. Moon and L. Moser, *Some packing and covering theorems*, Colloq. Math. **17** (1967), 103–110, DOI 10.4064/cm-17-1-103-110. MR0215197

MATHEMATICS DEPARTMENT, UNIVERSITY OF GEORGIA, ATHENS, GA 30602, USA
Email address: pollack@uga.edu

MATHEMATICS DEPARTMENT, DARTMOUTH COLLEGE, HANOVER, NH 03755, USA
Email address: carl.pomerance@dartmouth.edu

Contemporary Mathematics
Volume **731**, 2019
https://doi.org/10.1090/conm/731/14679

Forward integrals and SDE with fractal noise

Martina Zähle and Erik Schneider

Abstract. The approach to stochastic integrals introduced in **[25]** and **[26]** was first applied in **[26]** and **[14]** to SDE with fractal noise. Here we complete our former local results as follows. The main tool are uniform local contraction properties of related integral operators which lead to the global statements: We consider SDE in $\mathbb{R}^n$ with time dependent (not necessarily adapted) non-linear random coefficients, where one continuous driving process Z^0 admits a generalized quadratic variation. The other driving processes $Z^1, \dots, Z^m$ possess sample paths in fractional Sobolev spaces of order $> 1/2$. In particular, one Brownian motion and m multifractional Brownian motions can be treated. The corresponding stochastic integrals are determined as generalized forward integrals. For $Z^1, \dots, Z^m$ they agree with the integrals via fractional derivatives and with the Young integrals. A higher-dimensional variant of the Doss-Sussman approach to pathwise global solutions is presented. Moreover, we show that the unique solutions have the same Hölder regularity as that of the worst noise. Finally, random Weierstrass functions, which are Hölder continuous of order $1/2$, provide an example for our Z^0 where rough path analysis is not needed, but former versions of stochastic forward integrals do not work.

0. Introduction

In **[25]** an extension of the Lebesgue-Stieltjes integral for functions of unbounded variation was introduced by means of fractional calculus. In particular, it has led to a pathwise approach to stochastic integrals where the summed order of fractional differentiability of integrand and integrator is at least 1. A two-parameter version is presented in **[12]**. The explicit representation by means of fractional derivatives and integration-by-parts rules provides useful tools for applications to various problems of stochastic analysis. First steps in this direction were done in **[26]** and in **[14]**, where an extension of this approach was applied to ODE driven by fractal functions and to pathwise solutions of associated anticipate SDE.
The relationships to other types of stochastic forward integrals were discussed in **[27]**. Note that all these notions on the joint domains of definition lead to the same values. It turned out that an averaged version of the stochastic integrals via regularization introduced in **[16]**, which was first considered in **[26]** includes all former notions and can be applied to more general processes. In the case of semimartingales it agrees with the Itô integral. (A fractional calculus approach to the integrals of controlled paths against β-Hölder rough paths with $\beta \in (1/3, 1/2]$ can be found in **[9]**, see also **[10]** for a special case.)

2010 *Mathematics Subject Classification.* 60H05, 60H10, 34F05, 26A33.

In **[26]** and **[29]** only local solutions of the corresponding stochastic differential equations were obtained. The aim of the present paper is to complete our approach by proving existence and uniqueness results for global solutions. Moreover, we derive some optimal regularity properties: The solutions possess fractional derivatives of all orders less than that of the noises and the degrees of Hölder continuity agree with each other.

For simplicity we here consider stochastic processes with values in $\mathbb{R}^n$. All results can be extended to the Banach space setting (cf. **[8]** for the integral notions).

In Section 1 we recall the definition of the fractional derivatives D^α_{a+} and D^α_{b-} on a finite interval $[a,b]$ from **[19]** and the integral notion from **[25]** given by

$$\int_a^b f\,dg := (-1)^\alpha \int_a^b D^\alpha_{a+} f_{a+}(x) D^{1-\alpha}_{b-} g_{b-}(x)\,dx + f(a+)(g(b-)-g(a+))$$

for functions f and g whose boundary corrections f_{a+} and g_{b-} are elements of the Liouville spaces $I^\alpha_{a+}(L_2(a,b))$ and $I^{1-\alpha}_{b-}(L_2(a,b))$, resp., where $0<\alpha<1$. Moreover, some essential norm estimates of $\int_a^{(\cdot)} f\,dg$ in the function spaces

$$W^\beta_{2,\infty}(a,b) := W^\beta_2(a,b) \cap L_\infty(a,b)$$

from **[26]** are summarized. Here $W^\beta_2(a,b) = B^\beta_{2,2}(a,b)$ is the fractional Sobolev space of order β on the interval $[a,b]$.

Section 2 contains the first main result: For a sufficiently regular coefficient function a and a parameter function $\varphi \in W^{1/2-}_{2,\infty}(0,T)$ we prove a *uniform* local contraction principle on sufficiently small intervals $[t_0,t] \subset [0,T]$ for the integral operator

$$Af := x_0 + \int_{t_0}^{(\cdot)} a(f,\varphi)\,dg$$

in the space $W^{\beta'}_{2,\infty}(t_0,t)$, provided $g_{T-} \in I^\beta_{T-}(L_2(0,T))$ for some $1/2<\beta'<\beta$. (In **[26]** a non-uniform version was proved for almost all t_0 under the slightly weaker assumption $g \in W^\beta_2(0,T)$.) A higher dimensional extension to the case

$$Af := x_0 + \sum_{j=1}^{l} \int_{t_0}^{(\cdot)} a_j(f,\varphi)\,dg^j$$

with vector functions a_j and coordinatewise definition of the integrals is straightforward (Theorems 2.4 and 2.5).

Moreover, a function f on $[0,T]$ belonging to $W^\beta_{2,\infty}(t_0,t_1)$ for overlapping small intervals of the same size is an element of $W^\beta_{2,\infty}(0,T)$ (Proposition 2.7). This allows a gluing procedure for obtaining global solutions of related stochastic differential equations in Sections 4 and 5.

In Section 3 the above mentioned notion of *generalized stochastic forward integral* introduced in **[26]** and its relationships to the former versions are recalled. For simplicity we here apply it only to continuous stochastic processes Z as integrators, with respect to which the integral of a process Y is defined by:

$$\int_0^t Y\,dZ := \lim_{\epsilon\to 0} \epsilon \int_0^1 u^{\epsilon-1} \int_0^t Y(s) \frac{Z_{t-}(s+u) - Z_{t-}(s)}{u}\,ds\,du$$

whenever the limit exists in the sense of *uniform convergence in probability*, (ucp) for short. For functions f and g as above and $Y := f$ and $Z := g$ the limit exists and agrees with the above integral via fractional derivatives. (The same holds true for the other notions from the literature.) It was shown in **[26]** that for processes with finite *generalized quadratic variation*

$$[Z](t) := \lim_{\epsilon\to 0} \epsilon \int_0^1 u^{\epsilon-1} \int_0^t \frac{(Z_{t-}(s+u) - Z_{t-}(s))^2}{u}\, ds\, du$$

i.e., for which the (ucp)-limit exists, the corresponding Itô formula remains valid, in particular, the related stochastic integral is determined. This goes back to **[6]** in the special case of Riemann-Stieltjes approach. In the ordinary version of the above limits (i.e., for $u \to 0$ not averaged by the outer integral) it was worked out in **[17]**. However, the classical *Weierstrass-type functions* W or self-affine functions and their stochastic versions with scaling exponent $1/2$ are examples, where the ordinary quadratic variation (in the Riemann-Stieltjes or the Russo-Vallois sense) does not exist because of high oscillations, whereas the above average convergence holds true. This can be shown with the methods from **[24]** and **[27]** (cf. the Appendix). It implies existence (resp. non-existence) of the above stochastic integrals of the form $\int_0^t a(W)\, dW$ for regular functions a in the sense of average (resp. ordinary) convergence.

In Section 4 the results of Section 2 are applied to *vector-valued* differential equations on $[0,T]$ with fractional noises of the following form:

$$\begin{aligned} dx(t) &= \sum_{j=1}^{l} a_j(x(t), \varphi(t))\, dz^j(t) \\ x(0) &= x_0\,, \end{aligned}$$

with *real-valued* driving functions z^j such that $z^j_{T-} \in I^\beta_{T-}(L_2(0,T))$ for some $\beta > 1/2$. φ is an $\mathbb{R}^k$-valued parameter function whose coordinate functions are elements of $W^{1/2-}_{2,\infty}(0,T)$. Under some regularity assumptions on the vector-valued coefficient functions a_j we prove existence and uniqueness of a global solution $x \in W^{\beta-}_{2,\infty}(0,T)$ (Theorem 4.1). It can be obtained by Picard's iteration method because of the uniform local contraction principle.

A stochastic version of the above differential equation is investigated in Section 5. Here the functions z^j are replaced by stochastic processes Z^j, a random noise term Z^0 is added which corresponds to a continuous martingale in the classical case, and the vector-valued a.s. regular random coefficient functions can be anticipating:

$$\begin{aligned} dX(t) &= \sum_{j=0}^{m} a_j(X(t), t)\, dZ^j(t) + b(X(t), t)\, dt \\ X(0) &= X_0\,, \end{aligned}$$

where Z^0 is a process with finite generalized quadratic variation $[Z^0]$ in the above sense, and $[Z^0]$ and $Z^1, \ldots, Z^m$ are processes with sample paths in $I^\beta_{T-}(L_2(0,T))$ for some $\beta > 1/2$. The corresponding integrals with respect to $[Z^0], Z^1, \ldots Z^m$ are understood in the sense of Section 1 and that with respect to Z^0 in the sense of Section 3. In order to prove existence and uniqueness of a global pathwise solution in $W^{1/2-}_{2,\infty}(0,T)$ in the context of Itô-type calculus (Theorem 5.2) we use ideas from **[29]** for deriving a higher-dimensional version of the Doss-Sussman approach (cf.

[4], [22]). The global solution has a *pathwise representation* of the form

$$X(t) = h(Y(t), Z^0(t), t)\,.$$

Here a classical parameter dependent differential equation for the random vector function h combined with Taylor expansion, our notion of stochastic integral and the fractional calculus approach from Section 4 determining the process Y are the main tools.

In [18] the special case was treated, where $n = m = 1$, the coefficient functions are deterministic, Z^0 has finite quadratic variation in their sense, and Z^1 is a process of locally bounded variation. Then fractional calculus is not needed.

Applications of such SDE for $m = 1$ to option pricing in mathematical finance can be found e.g. in [28], [12], [2] and the references therein with different approaches, in particular, when Z^0 is the Wiener process and the noise Z^1 can generate long range dependence. Under certain natural restrictions on the hedging strategies the models are shown to be arbitrage free.

In Section 6 first the fractional ODE from Section 4 are examined for the case of integrators with Hölder exponent $\beta > 1/2$. (Note that in general $g_{T-} \in I^{\beta}_{T-}(L_2(0,T))$ implies only Hölder continuity of order $\beta - 1/2$.) Under an additional assumption on such z^j it was proved in [14] existence and uniqueness of a global solution in the space of Hölder continuous functions of order β for slightly different conditions on the coefficient functions a_j than in Section 4. Instead of the fractional Sobolev spaces these authors worked with modified versions using L_1 and L_∞ instead of L_2, which avoids the local approach. Moreover, they applied this pathwise to fractional Brownian motions B^H with Hurst exponents $H > 1/2$. Using the contraction argument from Section 4 we can derive that the solution has generally the same Hölder regularity as the noises (Theorem 6.3).

Finally, we consider the SDE from Section 5 under the condition that almost surely Z^0 is Hölder continuous with degree not exceeding 1/2 and $[Z^0], Z^1, \ldots, Z^m$ are Hölder continuous of order greater than 1/2. In this case we show that the pathwise unique solution obtained from the Doss-Sussman type approach has the same Hölder regularity as Z^0 (Theorem 6.5).

In particular, this can be applied to the situation where Z^0 is Brownian motion and for $j = 1, \ldots m$, $Z^j = B^{H_j}$ with $H_j > 1/2$. The case $m = 1$ for less regular, but adapted coefficient functions was investigated in [7], combining the former methods with Itô integration. The authors of [13] treated this case for certain deterministic coefficient functions with different methods. Moreover, they replaced B^H by an arbitrary Hölder continuous noise. In [21] this was extended to adapted random coefficients under weaker regularity assumptions. For other approaches to such situations see also the survey [15]. If the stochastic forward integral is replaced by the Skorohod integral or determined by means of Wick products (cf. [3]), then one obtains different solutions (see, e.g., the discussion in [12])).

In the Appendix certain *random Weierstrass-type functions* are presented as examples for Z^0, where the other approaches to stochastic forward integrals from the literature do not work. They are Hölder continuous of order 1/2 and possess the generalized quadratic variation processes $[Z^0](t) = ct$ for some constants c.

Throughout the whole paper C denotes a finite positive constant whose exact value is not important and may differ from one occurence to the other.

1. Fractional integrals and Sobolev type spaces

A detailed survey on fractional calculus can be found in the book of Samko, Kilbas and Marichev [**19**]. Here we just want to recall some important notions and results. Let us consider a finite interval $[a,b] \subset \mathbb{R}$. For a real-valued $f \in L_1(a,b)$ and $\alpha > 0$ the *left-sided* and *right-sided fractional Riemann-Liouville integrals of f of order α* are defined at almost all x by

$$I^\alpha_{a+}f(x) := \frac{1}{\Gamma(\alpha)} \int_a^x (x-y)^{\alpha-1} f(y)\, dy$$

and

$$I^\alpha_{b-}f(x) := \frac{(-1)^{-\alpha}}{\Gamma(\alpha)} \int_x^b (y-x)^{\alpha-1} f(y)\, dy.$$

The class of functions f which are representable as I^α_{a+} (resp. I^α_{b-})-integrals of some real-valued L_p-function φ $(1 \le p \le \infty)$ is denoted by $I^\alpha_{a+}(L_p(a,b))$ (resp. $I^\alpha_{b-}(L_p(a,b))$). These sets $I^\alpha_{\substack{a+\\(b-)}}(L_p(a,b))$ become Banach spaces by the norms

$$\|f\|_{I^\alpha_{\substack{a+\\(b-)}}(L_p(a,b))} := \|f\|_{L_p(a,b)} + \|D^\alpha_{\substack{a+\\(b-)}} f\|_{L_p(a,b)} \sim \|D^\alpha_{\substack{a+\\(b-)}} f\|_{L_p(a,b)}.$$

For $0 < \alpha < 1$ the function φ agrees at almost all x with the *Weyl-Marchaud derivative of f of order α*

$$D^\alpha_{a+}f(x) := \frac{1}{\Gamma(1-\alpha)} \left(\frac{f(x)}{(x-a)^\alpha} + \alpha \int_a^x \frac{f(x)-f(y)}{(x-y)^{\alpha+1}}\, dy \right)$$

and

$$D^\alpha_{b-}f(x) := \frac{(-1)^\alpha}{\Gamma(1-\alpha)} \left(\frac{f(x)}{(b-x)^\alpha} + \alpha \int_x^b \frac{f(x)-f(y)}{(y-x)^{\alpha+1}}\, dy \right),$$

where convergence of the integrals at the singularity $y = x$ holds pointwise for almost all x if $p = 1$ and in the L_p-sense if $p > 1$.
Denote $f_{a+}(x) = 1_{(a,b)}(x)(f(x) - f(a+))$ and $f_{b-}(x) = 1_{(a,b)}(x)(f(x) - f(b-))$ assuming that the one-sided limits of f at the interval ends exist. In [**26**] there was introduced an extension of Lebesgue-Stieltjes integrals by

DEFINITION 1.1.

$$\int_a^b f\, dg = (-1)^\alpha \int_a^b D^\alpha_{a+} f_{a+}(x) D^{1-\alpha}_{b-} g_{b-}(x)\, dx + f(a+)(g(b-) - g(a+)) \tag{1.1}$$

provided that $f_{a+} \in I^\alpha_{a+}(L_p(a,b)), g_{b-} \in I^{1-\alpha}_{b-}(L_q(a,b))$ for some $\frac{1}{p} + \frac{1}{q} \le 1$ and $0 \le \alpha \le 1$.

It can be shown that the definition of the integral is independent of the choice of α. If $\alpha p < 1$ this integral equals

$$\int_a^b f\, dg = (-1)^\alpha \int_a^b D^\alpha_{a+} f(x) D^{1-\alpha}_{b-} g_{b-}(x)\, dx \tag{1.2}$$

which is determined for general $f \in I^\alpha_{a+}(L_p(a,b))$ with $\limsup_{x \searrow a} f(x) < \infty$. For the special case where f and g are Hölder continuous of summed order greater than 1 our integral (1.1) agrees with the Riemann-Stieltjes integral.

In this section we consider for vector functions $f : [a,b] \to \mathbb{R}^n$ the following *fractional Sobolev- (or Slobodeckij-) type spaces* given by their (semi) norms

$$\|f\|_{\widetilde{W}_2^\alpha(a,b)} := \left(\int_a^b \int_a^b \frac{|f(y)-f(x)|^2}{|y-x|^{2\alpha+1}}\,dx\,dy\right)^{1/2}$$

$$\|f\|_{W_2^\alpha(a,b)} := \|f\|_{L_2(a,b)} + \|f\|_{\widetilde{W}_2^\alpha(a,b)}$$

$$\|f\|_{W_{2,\infty}^\alpha(a,b)} := \|f\|_{L_\infty(a,b)} + \|f\|_{\widetilde{W}_2^\alpha(a,b)}$$

$$\|f\|_{W_2^\alpha(b-)(a,b)} := \left(\int_a^b \frac{|f(x)|^2}{|b-x|^{2\alpha}}\,dx\right)^{1/2} + \|f\|_{\widetilde{W}_2^\alpha(a,b)},$$

for $0 < \alpha < 1$.
Furthermore, we denote

$$I^{\beta-}_{\substack{a+\\(b-)}}(L_2(a,b)) := \bigcap_{\alpha<\beta} I^{\alpha}_{\substack{a+\\(b-)}}(L_2(a,b))$$

and

$$W^{\beta-}_{2,\infty}(a,b) := \bigcap_{\alpha<\beta} W^{\alpha}_{2,\infty}(a,b).$$

Then we have the embeddings

$$W_2^{\alpha+\delta}(b-)(a,b) \hookrightarrow I^\alpha_{b-}(L_2(a,b)), \quad \delta > 0$$

(see **[26]**, Theorem 1.1 (ii)) and

$$I^{\alpha+\delta}_{\substack{a+\\(b-)}}(L_2(a,b)) \hookrightarrow \widetilde{W}_2^\alpha(a,b), 0 < \delta < 1-\alpha$$

(see Feyel, de La Pradelle **[5]**, Theorem 27).
The following theorem (cf. **[26]**, Thm. 1.3 for the case where f is real-valued) is essential for the *contraction property* in the next section.

THEOREM 1.2. *Suppose* $0 < \alpha < 1/2$, $0 < \beta < 1$. *Then we get for an* $\mathbb{R}^n$*-valued function* f *and a real-valued function* g

(i)

$$\left\|\int_a^{(\cdot)} f\,dg\right\|_{W^\beta_{2,\infty}(a,b)} \le C\Big(\|f\|_{W_{2,\infty}^{\max(\alpha,\alpha+\beta-1/2)}(a,b)}\|g_{b-}\|_{I_{b-}^{1-\alpha}(L_2(a,b))} + \|f\|_{L_\infty(a,b)}\|g\|_{\widetilde{W}_2^\beta(a,b)}\Big)$$

and

(ii)

$$\left\|\int_a^{(\cdot)} f\,dg\right\|_{W^\beta_{2,\infty}(a,b)} \le C\|f\|_{W^\beta_{2,\infty}(a,b)}\|g_{b-}\|_{W_2^\beta(b-)(a,b)}$$

if $\beta > 1/2$. *Here the integral is understood in the coordinatewise sense.*

2. An integral operator and its contraction property

Instead of the interval $[a,b]$ we now use the time interval $[0,T]$ because of the related differential equations. We fix a real-valued integrator function g with $g_{T-} \in I^{\beta}_{T-}(L_2(0,T))$ for some $1/2 < \beta < 1$. Later g will be replaced by a stochastic process with sample paths in $I^{\beta}_{T-}(L_2(0,T))$, for example a fractional Brownian motion. We also consider a vector-valued *parameter function* $\varphi : [0,T] \to \mathbb{R}^k$ with $\varphi \in W^{1/2-}_{2,\infty}(0,T)$ and a transformation mapping $a : \mathbb{R}^n \times \mathbb{R}^k \to \mathbb{R}^n$. For vector functions $f : [0,T] \to \mathbb{R}^n$ with $f \in W^{\beta'}_{2,\infty}(0,T)$, where $1/2 < \beta' < \beta < 1$, we introduce for fixed $x_0 \in \mathbb{R}^n$ the non-linear integral operator A by

$$Af := x_0 + \int_0^{(\cdot)} a(f,\varphi)\, dg. \tag{2.1}$$

with coordinatewise definition of the integral.

If the transformation mapping a fulfills the hypothesis $(\boldsymbol{H1})$ or $(\boldsymbol{H2})$ below we infer a certain contraction property for the integral operator. This is the essential tool for solving the related (anticipative stochastic) differential equations in Section 5.

$(\boldsymbol{H1})$**:** The function $a(x,\tau)$ is differentiable in the first variable. Moreover $\frac{\partial a}{\partial x}$ is Lipschitz in both arguments, i.e. there exist some constants $\mathcal{L}_1$ and $\mathcal{L}_2$ such that:

$$\left|\frac{\partial a}{\partial x}(x,\tau) - \frac{\partial a}{\partial x}(y,\tau)\right| \le \mathcal{L}_1|x-y|,$$
$$\left|\frac{\partial a}{\partial x}(x,\tau) - \frac{\partial a}{\partial x}(x,\sigma)\right| \le \mathcal{L}_2|\tau-\sigma|.$$

for all $x,y \in \mathbb{R}^n$ and $\tau,\sigma \in \mathbb{R}^k$.

$(\boldsymbol{H2})$**:** The function $a(x,\tau)$ is differentiable in both variables. Moreover $\frac{\partial a}{\partial x}$ and $\frac{\partial a}{\partial \tau}$ are Lipschitz in the first argument, i.e. there exist some constants $\mathcal{L}_1$ and $\mathcal{L}_2$ such that:

$$\left|\frac{\partial a}{\partial x}(x,\tau) - \frac{\partial a}{\partial x}(y,\tau)\right| \le \mathcal{L}_1|x-y|,$$
$$\left|\frac{\partial a}{\partial \tau}(x,\tau) - \frac{\partial a}{\partial \tau}(y,\tau)\right| \le \mathcal{L}_2|x-y|.$$

for all $x,y \in \mathbb{R}^n$ and $\tau,\sigma \in \mathbb{R}^k$.

($|\cdot|$ on the left hand sides means the operator norms, e.g. in matrix representation.)

In a first step we consider the behaviour of the mapping

$$f \mapsto a(f,\varphi).$$

Denote $\mathcal{L}_0 := \|\left(\frac{\partial a^i}{\partial x^j}\right)_{i,j=1,\ldots,n}\|_{L_\infty}$. Then for arbitrary $\mathbb{R}^n$-valued vector functions f,h and $\mathbb{R}^k$-valued parameter functions φ with $f,h,\varphi \in W^{\mu}_{2,\infty}(0,T)$, where $0<\mu<1$, and any mapping a satisfying hypothesis $(\boldsymbol{H1})$ or $(\boldsymbol{H2})$ we get the following.

PROPOSITION 2.1.

$$\|a(f,\varphi) - a(h,\varphi)\|_{L_\infty(0,T)} \le \mathcal{L}_0\|f-h\|_{L_\infty(0,T)}$$

and

$$\begin{aligned}\|a(f,\varphi)-a(h,\varphi)\|_{\widetilde{W}_2^{\mu}(0,T)} \leq\ & \mathcal{L}_0\|f-h\|_{\widetilde{W}_2^{\mu}(0,T)}\\ &+\mathcal{L}_1\|f-h\|_{L_\infty(0,T)}\Big(\|f\|_{\widetilde{W}_2^{\mu}(0,T)}+\|h\|_{\widetilde{W}_2^{\mu}(0,T)}\Big)\\ &+\mathcal{L}_2\|f-h\|_{L_\infty(0,T)}\|\varphi\|_{\widetilde{W}_2^{\mu}(0,T)}.\end{aligned}$$

For the case $n=k=1$ this was shown in **[26]**. A detailed proof for higher dimensions can be found in **[20]**.
From this we infer a corresponding integral estimate.

THEOREM 2.2. *Let $1/2<\beta'<\beta<1$, f,h are $\mathbb{R}^n$-valued vector functions with $f,h\in W_{2,\infty}^{\beta'}(0,T)$. The parameter function φ takes values in $\mathbb{R}^k$ with $\varphi\in W_{2,\infty}^{1/2-}(0,T)$ and $g_{T-}\in I_{T-}^{\beta}(L_2(0,T))$. Then we get for any a satisfying **(H1)** or **(H2)** and any $0\le t_0<t<T$,*

$$\begin{aligned}\left\|\int_{t_0}^{(\cdot)}a(f,\varphi)\,dg-\int_{t_0}^{(\cdot)}a(h,\varphi)\,dg\right\|_{W_{2,\infty}^{\beta'}(t_0,t)} &\le C\Big[\mathcal{L}_0\|f-h\|_{W_{2,\infty}^{\beta'}(t_0,t)}\\ &+\mathcal{L}_1\|f-h\|_{L_\infty(t_0,t)}\Big(\|f\|_{\widetilde{W}_2^{\beta'}(t_0,t)}+\|h\|_{\widetilde{W}_2^{\beta'}(t_0,t)}\Big)\\ &+\mathcal{L}_2\|f-h\|_{L_\infty(t_0,t)}\|\varphi\|_{W_{2,\infty}^{1/2-}(t_0,t)}\Big]\|g_{t-}\|_{W_2^{\gamma}(t-)(t_0,t)}\end{aligned}$$

where $\frac{\beta+\beta'}{2}<\gamma<\beta$.

PROOF. By Theorem 1.2 and Proposition 2.1 we have

$$\begin{aligned}&\left\|\int_{t_0}^{(\cdot)}a(f,\varphi)\,dg-\int_{t_0}^{(\cdot)}a(h,\varphi)\,dg\right\|_{W_{2,\infty}^{\beta'}(t_0,t)}\\ &\le C\Big[\mathcal{L}_0\|f-h\|_{W_{2,\infty}^{\max(\alpha,\alpha+\beta'-1/2)}(t_0,t)}\\ &+\mathcal{L}_1\|f-h\|_{L_\infty(t_0,t)}\Big(\|f\|_{\widetilde{W}_2^{\max(\alpha,\alpha+\beta'-1/2)}(t_0,t)}+\|h\|_{\widetilde{W}_2^{\max(\alpha,\alpha+\beta'-1/2)}(t_0,t)}\Big)\\ &+\mathcal{L}_2\|f-h\|_{L_\infty(t_0,t)}\|\varphi\|_{W_{2,\infty}^{\max(\alpha,\alpha+\beta'-1/2)}}\Big]\|g_{t-}\|_{I_{t-}^{1-\alpha}(L_2(t_0,t))}\\ &+\mathcal{L}_0\|f-h\|_{L_\infty(t_0,t)}\|g\|_{\widetilde{W}_2^{\beta'}(t_0,t)}\end{aligned}$$

for all $0<\alpha<1/2$.
Finally, choosing $\alpha:=1-\beta+\frac{\beta-\beta'}{2}$ so that $\alpha<\alpha+\beta'-1/2<1/2$ we can use the estimates

$$\|g_{t-}\|_{I_{t-}^{1-\alpha}(L_2(t_0,t))}\le C\|g_{t-}\|_{W_2^{\gamma}(t-)(t_0,t)}$$

and

$$\|g\|_{\widetilde{W}_2^{\beta'}(t_0,t)}\le C\|g_{t-}\|_{W_2^{\gamma}(t-)(t_0,t)}$$

(cf. **[26]**, proof of Theorem 1.1 (ii) and (iv)). □

LEMMA 2.3. *Let $g_{T-}\in I_{T-}^{\beta}(L_2(0,T))$ and $1/2<\gamma<\beta$. Then we get*

$$\lim_{\Delta\to 0}\sup_{t_0\in[0,T]}\|g_{(t_0+\Delta)-}\|_{W_2^{\gamma}((t_0+\Delta)-)(t_0,t_0+\Delta)}=0.$$

PROOF. By definition,

$$\|g_{(t_0+\Delta)-}\|_{W_2^\gamma((t_0+\Delta)-)(t_0,t_0+\Delta)} = \left(\int_{t_0}^{t_0+\Delta} \frac{|g(t_0+\Delta)-g(s)|^2}{|t_0+\Delta-s|^{2\gamma}}\,ds\right)^{1/2} + \left(\int_{t_0}^{t_0+\Delta}\int_{t_0}^{t_0+\Delta} \frac{|g(r)-g(s)|^2}{|r-s|^{2\gamma+1}}\,ds\,dr\right)^{1/2}$$

First note that if g_{T-} is in $I_{T-}^\beta(L_2(0,T))$ then g is $(\beta-1/2)$ Hölder continuous on $(0,T)$ (cf. [**19**], Theorem 3.6). So the first integral can be estimated as follows

$$\left(\int_{t_0}^{t_0+\Delta} \frac{|g(t_0+\Delta)-g(s)|^2}{|t_0+\Delta-s|^{2\gamma}}\,ds\right)^{1/2} \le C\left(\int_{t_0}^{t_0+\Delta} \frac{|t_0+\Delta-s|^{2\beta-1}}{|t_0+\Delta-s|^{2\gamma}}\,ds\right)^{1/2} \le C\Delta^{\beta-\gamma}.$$

For the second integral we use the embedding $I_{T-}^\beta(L_2(0,T)) \hookrightarrow \widetilde{W}_2^\gamma(0,T)$ (cf. [**5**], Theorem 27). Therefore we can write with the notation $f(r,s) := \frac{|g(r)-g(s)|}{(r-s)^{\gamma+1/2}}$,

$$\begin{aligned}
&\int_{t_0}^{t_0+\Delta}\int_{t_0}^{t_0+\Delta} \frac{|g(r)-g(s)|^2}{|r-s|^{2\gamma+1}}\,ds\,dr \\
&= \int_0^T\int_0^T \mathbf{1}\{(r,s)\in[t_0,t_0+\Delta]^2\} f(r,s)^2\,ds\,dr \\
&\le \int_0^T\int_0^T \mathbf{1}\{f^2\ge N_\varepsilon\} f(r,s)^2\,ds\,dr \\
&\quad + \int_0^T\int_0^T \mathbf{1}\{(r,s)\in[t_0,t_0+\Delta]^2\}\mathbf{1}\{f^2<N_\varepsilon\} f(r,s)^2\,ds\,dr.
\end{aligned}$$

If we choose now N_ε large enough such that the first integral on the right-hand side is less than $\frac{\varepsilon}{2}$ we get

$$\begin{aligned}
&\int_0^T\int_0^T \mathbf{1}\{f^2\ge N_\varepsilon\} f(r,s)^2\,ds\,dr \\
&+ \int_0^T\int_0^T \mathbf{1}\{(r,s)\in[t_0,t_0+\Delta]^2\}\mathbf{1}\{f^2<N_\varepsilon\} f(r,s)^2\,ds\,dr \\
&\le \frac{\varepsilon}{2} + N_\varepsilon \int_{t_0}^{t_0+\Delta}\int_{t_0}^{t_0+\Delta} ds\,dr \\
&= \frac{\varepsilon}{2} + N_\varepsilon\Delta^2 < \varepsilon.
\end{aligned}$$

for $\Delta^2 < \frac{\varepsilon}{2N_\varepsilon}$.

Hence, we have proved that

$$\int_{t_0}^{t_0+\Delta}\int_{t_0}^{t_0+\Delta} \frac{|g(r)-g(s)|^2}{|r-s|^{2\gamma+1}}\,ds\,dr \xrightarrow[\Delta\to 0]{} 0$$

uniformly in t_0. □

Now we can formulate a *uniform local contraction principle* for the integral operator.

THEOREM 2.4. *Let $x_0 \in \mathbb{R}^n, 1/2 < \beta' < \beta < 1, g_{T-} \in I_{T-}^{\beta}(L_2(0,T))$. Suppose that the parameter function $\varphi \in W_{2,\infty}^{1/2-}(0,T)$ takes values in $\mathbb{R}^k$ and the mapping a fulfills hypothesis* **(*H*1)** *or* **(*H*2)**. *Then for every constant $c > 0$ there exists a $\Delta > 0$ such that for every interval $[t_0, t_1] \subset [0,T]$ with $t_1 - t_0 < \Delta$ the integral operator A given by*

$$Af := x_0 + \int_{t_0}^{(\cdot)} a(f, \varphi)\, dg$$

with coordinatewise definition of the integral maps $W_{2,\infty}^{\beta'}(t_0, t_1)$ into itself and we have

$$\|Af - Ah\|_{W_{2,\infty}^{\beta'}(t_0,t_1)} \leq c\|f - h\|_{W_{2,\infty}^{\beta'}(t_0,t_1)}$$

for all vector functions $f, h \in W_{2,\infty}^{\beta'}(t_0, t_1)$.

PROOF. Theorem 2.2 implies for all $0 \leq t_0 < t \leq T$,

$$\left\| \int_{t_0}^{(\cdot)} a(f,\varphi)\, dg - \int_{t_0}^{(\cdot)} a(h,\varphi)\, dg \right\|_{W_{2,\infty}^{\beta'}(t_0,t)} \leq C\|f - h\|_{W_{2,\infty}^{\beta'}(t_0,t)} \|g_{t-}\|_{W_2^{\gamma}(t-)(t_0,t)}$$

where $\frac{\beta+\beta'}{2} < \gamma < \beta$. From the last lemma we know that we can choose some small $\Delta > 0$ (independent of t_0) such that

$$C\|g_{t_1-}\|_{W_2^{\gamma}(t_1-)(t_0,t_1)} \leq c$$

for any $t_0 < t_1 \leq T$ with $t_1 - t_0 < \Delta$. □

The following contraction theorem is a straightforward extension.

THEOREM 2.5. *The statement of Theorem 2.4 remains valid if $x_0 \in \mathbb{R}^n, g_{T-}^j \in I_{T-}^{\beta}(L_2(0,T))$, the functions a_j fullfill hypothesis* **(*H*1)** *or* **(*H*2)** *for $j = 1, ..., l$, φ takes values in $\mathbb{R}^k$, and f and h in $\mathbb{R}^n$ with coordinate functions as before and the operator is replaced by*

$$Af := x_0 + \sum_{j=1}^{l} \int_{t_0}^{(\cdot)} a_j(f, \varphi)\, dg^j$$

with coordinatewise definition of the integrals.

REMARK 2.6. If $g^j(t) = t$ for some j then we can weaken the condition on $a_j(x, \tau)$ to measurability, boundedness and Lipschitz continuity in x uniformly in τ.

The next proposition is needed in Section 4 in order to continue the solution to the whole interval $[0,T]$.

PROPOSITION 2.7. *Let $\Delta > 0$ be fixed, $0 < \beta < 1$ and $f : [0,T] \to \mathbb{R}^n$. If the restriction $f|_{[t_0,t_1]}$ of f belongs to $W_{2,\infty}^{\beta}(t_0, t_1)$ for any interval $[t_0, t_1] \subset [0,T]$ with $t_1 - t_0 < \Delta$, then the function f belongs to the space $W_{2,\infty}^{\beta}(0,T)$.*

The proof easily follows from a splitting procedure and considering the function on overlapping intervals of this type.

3. Stochastic forward integrals, quadratic variation and Itô formula

If the real-valued functions f and g fulfill the conditions in Definition 1.1 then the integral (1.1) can be approximated as follows

$$\int_a^b f\,dg = \lim_{\varepsilon\to 0}\int_a^b I_{a+}^{\varepsilon} f\,dg.$$

Moreover,

$$\int_a^b I_{a+}^{\varepsilon} f\,dg = \frac{1}{\Gamma(\varepsilon)}\int_0^{\infty} u^{\varepsilon-1}\int_a^b f(s)\frac{g_{b-}(s+u)-g_{b-}(s)}{u}\,ds\,du,$$

if $f\in I_{a+}^{\alpha-\varepsilon}(L_p(a,b))$, $g_{b-}\in I_{b-}^{1-\alpha}(L_q(a,b))$ with $\frac{1}{p}+\frac{1}{q}\le 1$, $\varepsilon>0$ and $\alpha p\neq 1$. This suggested the following extension of the integral (1.1) (cf. [**26**], Section 4):

DEFINITION 3.1.

$$\int_a^b f\,dg := \lim_{\varepsilon\to 0}\varepsilon\int_0^1 u^{\varepsilon-1}\int_a^b f(s)\frac{g_{b-}(s+u)-g_{b-}(s)}{u}\,ds\,du \tag{3.1}$$

whenever the right-hand side exists.

In the stochastic calculus below the upper boundary b in the integrals may be replaced by any $t\in(a,b]$ and convergence holds uniformly in t. The stochastic version reads as follows. Let Y be a real-valued stochastic *càglàd process*, i.e. left continuous with right limits, and Z a real-valued stochastic *càdlàg process* (right continuous with left limits) on $[0,T]$. The integral of Y w.r.t. Z is defined by

DEFINITION 3.2.

$$\begin{aligned}\int_0^t Y\,dZ := \lim_{\substack{\varepsilon\to 0\\(ucp)}}\varepsilon\int_0^1 u^{\varepsilon-1}\int_0^t Y(s)\frac{Z_{t-}(s+u)-Z_{t-}(s)}{u}\,ds\,du\\ +Y(t)(Z(t)-Z(t-))\end{aligned} \tag{3.2}$$

whenever the right-hand side is determined, where $\lim_{(ucp)}$ stands for *uniform convergence in probability* and $\int_0^1$ for $\lim_{\delta\searrow 0}\int_\delta^1$ with probability 1. Recall that $Z_{t-}(s)=\mathbf{1}_{(0,t)}(s)(Z(s)-Z(t-))$.

This stochastic integral is an extension of the one introduced by Russo and Vallois (cf. [**16**]). Notice that if Z is a semimartingale and Y is an adapted càglàd process then the integral (3.2) agrees with the usual Itô integral (cf. [**17**]).
Furthermore we say that a real-valued càdlàg process Z admits a (finite) *generalized quadratic variation process*

$$[Z](t) := \lim_{\substack{\varepsilon\to 0\\(ucp)}}\varepsilon\int_0^1 u^{\varepsilon-1}\int_0^t\frac{(Z_{t-}(s+u)-Z_{t-}(s))^2}{u}\,ds\,du+(Z(t)-Z(t-))^2 \tag{3.3}$$

if the limit exists. For the special case of semimartingales this notion also agrees with the classical one. Fractional Brownian motion with Hurst exponent $1/2<H<1$ is an example for a process with quadratic variation $[B^H]\equiv 0$.
A consequence of the definition of $[Z]$ is that any continuous process Z admitting a generalized quadratic variation belongs to the space $W_{2,\infty}^{1/2-}(0,T)$ with probability 1.
As in the classical semimartingale theory and the extensions in [**6**] and [**17**] one obtains the following *Itô formula*.

THEOREM 3.3. *(cf. [26], Thm. 5.8) Let Z be a real-valued continuous process with generalized quadratic variation $[Z]$. Then we get for any random C^1-function $F(z,t) : \mathbb{R} \times [0,T] \to \mathbb{R}$ with continuous $\frac{\partial^2 F}{\partial z^2}$ and $0 \le s < t \le T$,*

$$\begin{aligned} F(Z(t),t) - F(Z(s),s) &= \int_s^t \frac{\partial F}{\partial z}(Z(u),u)\, dZ(u) + \int_s^t \frac{\partial F}{\partial t}(Z(u),u)\, du \\ &\quad + \frac{1}{2}\int_s^t \frac{\partial^2 F}{\partial z^2}(Z(u),u)\, d[Z](u) \end{aligned} \tag{3.4}$$

and the stochastic integral is determined in the sense of (3.2).

Further, by definition, we say that a process X with

$$X(t) = \int_0^t H\, dZ$$

for some càglàd process H *satisfies the general Itô formula* if for any F as above

$$\begin{aligned} F(X(t),t) - F(X(s),s) &= \int_s^t \frac{\partial F}{\partial x}(X(u),u)H(u)\, dZ(u) + \int_s^t \frac{\partial F}{\partial t}(X(u),u)\, du \\ &\quad + \frac{1}{2}\int_s^t \frac{\partial^2 F}{\partial x^2}(Z(u),u)H(u)^2\, d[Z](u). \end{aligned} \tag{3.5}$$

Higher-dimensional extensions are straightforward. Later we will need the following special version. Suppose that $F(y,z) : \mathbb{R}^m \times \mathbb{R} \to \mathbb{R}^n$ is a random C^1-function with continuous $\frac{\partial^2 F}{\partial z^2}$ and $\frac{\partial F}{\partial y}$ is Lipschitz in y. The $\mathbb{R}^m$-valued random process Y has sample paths in $W_{2,\infty}^{\beta'}(0,T)$ for some $\beta' > 1/2$. Then we obtain for Z as above the *higher-dimensional Itô formula*

$$\begin{aligned} F(Y(t),Z(t)) - F(Y(s),Z(s)) &= \sum_{i=1}^m \int_s^t \frac{\partial F}{\partial y^i}(Y(u),Z(u))\, dY^i(u) \\ + \int_s^t \frac{\partial F}{\partial z}(Y(u),Z(u))\, dZ(u) &+ \frac{1}{2}\int_s^t \frac{\partial^2 F}{\partial z^2}(Y(u),Z(u))\, d[Z](u). \end{aligned} \tag{3.6}$$

Here the first m integrals are determined in the sense of (1.1) and the integral w.r.t. Z is given by the stochastic forward integral (3.2). (For a detailed proof see [20].) Note that we may choose, in particular, $Y^m(t) = t$ in order to apply the formula to SDE with time dependent coefficients.
The corresponding higher-dimensional version of the general Itô formula (3.5) concerning different integral representations with respect to the processes Y^i and Z is analogous.

4. Differential equations driven by functions with fractional smoothness of order greater than 1/2

We now consider the differential equation

$$\begin{aligned} dx(t) &= \sum_{j=1}^m a_j(x(t),\varphi(t))dz^j(t) \\ x(0) &= x_0 \end{aligned} \tag{4.1}$$

for $t \in [0,T], x_0 \in \mathbb{R}^n$ and driving functions $z^1, \dots, z^m$ with $z^j_{T-} \in I^{\beta}_{T-}(L_2(0,T))$, $1/2 < \beta < 1$. The parameter function φ takes values in $\mathbb{R}^k$ with coordinate functions in $W^{1/2-}_{2,\infty}(0,T)$ and the a_j are $\mathbb{R}^n$-valued vector fields on $\mathbb{R}^n \times \mathbb{R}^k$ and satisfy the hypothesis **(*H*1)** or **(*H*2)**, see Section 2. We are looking for a global $\mathbb{R}^n$-valued solution of (4.1) in the space $W^{\beta-}_{2,\infty}(0,T)$. The equation becomes precise via integration

$$x(t) = x_0 + \sum_{j=1}^{m} \int_0^t a_j(x(s), \varphi(s))\, dz^j(s)$$

where the integrals may be interpreted in the sense of (1.1).
To this aim we apply the Contraction theorem of Section 3 which leads to the following.

THEOREM 4.1. *Under the above conditions there exists a global solution in* $W^{\beta-}_{2,\infty}(0,T)$ *to equation (4.1). It is unique in* $W^{\beta'}_{2,\infty}(0,T)$ *for any* $\beta' < \beta$.

PROOF. According to Theorem 2.5 there exists a Δ such that for any interval $[r,s] \subset [0,T]$ with $s - r < \Delta$ the integral operator A given by (2.1) is a contraction in the space $W^{\beta'}_{2,\infty}(r,s)$ for any $1/2 < \beta' < \beta$. Let $0 = t_0 < t_1 < \dots < t_N = T$ be a partition of the interval $[0,T]$ with $t_{i+1} - t_i < \Delta$. Setting $x(0) = x_0$ a global solution x to equation (4.1) can be determined successively by the contraction principle providing a unique solution $x(t), t \in [t_i, t_{i+1}]$, in the space $W^{\beta'}_{2,\infty}(t_i, t_{i+1})$ with initial condition $x(t_i)$. Let now $[r,s] \subset [0,T]$ with $s - r < \Delta$ and $t_{i-1} < r < t_i < s < t_{i+1}$ and y be the solution to (4.1) on $[r,s]$ with initial condition $y(r) = x_i(r)$. Then y and x are two solutions for (4.1) on $[r, t_i]$ with same initial condition and hence by the uniqueness from the contraction principle we have $y(t) = x_i(t)$ for all $t \in [r, t_i]$. By the same arguments we can show that $y(t) = x_{i+1}(t)$ for all $t \in [t_i, s]$ and therefore $y = 1_{[r,s]}x$. So the solution of (4.1) is in $W^{\beta'}_{2,\infty}(r,s)$ for any interval $[r,s] \subset [0,T]$ with $s - r < \Delta$ and hence $x \in W^{\beta'}_{2,\infty}(0,T)$ by Proposition 2.7. The uniqueness of the global solution follows from the uniqueness for the local version. □

5. Stochastic differential equations with mixed fractal noise

We now apply the previous results to SDE's.
Througout this section let $Z^0, Z^1, \dots, Z^m$ be processes on $[0,T]$ such that Z^0 is a continuous process with generalized quadratic variation process $[Z^0]$ and the processes $[Z^0]_T, Z^1_T, \dots, Z^m_T$ have sample paths in $I^{\beta}_{T-}(L_2(0,T))$ for some $1/2 < \beta < 1$. (Here we denote $Z_T := \mathbf{1}_{[0,T]}(t)(Z(t) - Z(T))$. We consider the stochastic differential equation in $\mathbb{R}^n$,

$$\begin{aligned} dX(t) &= \sum_{j=0}^{m} a_j(X(t),t)dZ^j(t) + b(X(t),t)dt \\ X(0) &= X_0 \end{aligned} \tag{5.1}$$

for certain random vector fields $a_0, a_1, \dots a_m, b : \mathbb{R}^n \times [0,T] \to \mathbb{R}^n$ and an arbitrary random initial vector $X_0 \in \mathbb{R}^n$.

DEFINITION 5.1. A solution of (5.1) is a continuous process $X = (X^1, \ldots, X^n)$ admitting a generalized quadratic variation processes $[X]$ which satisfies the multidimensional generalized version of the Itô formula (3.5) with respect to its coordinatewise integral representation

$$X(t) = X_0 + \sum_{j=0}^{m} \int_0^t a_j(X(s), s) dZ^j(s) + \int_0^t b(X(s), s) ds$$

where the first $m+1$ integrals are defined in the sense of (3.2).

For the vector fields a_j and b we assume (with probability 1)

(C1): $a_j : \mathbb{R}^n \times [0,T] \to \mathbb{R}^n$ and satisfy hypothesis $(\boldsymbol{H1})$ or $(\boldsymbol{H2})$ from Section 2, $j = 1, ..., m$,

(C2): $a_0 \in C^1(\mathbb{R}^n \times [0,T], \mathbb{R}^n)$, the partial derivatives $\frac{\partial a_0}{\partial x}(x,t)$ and $\frac{\partial a_0}{\partial t}(x,t)$ are Lipschitz in x, and $\frac{\partial a_0}{\partial x} a_0$ also satisfy hypothesis $(\boldsymbol{H1})$ or $(\boldsymbol{H2})$ [1],

(C3): $b \in C(\mathbb{R}^n \times [0,T], \mathbb{R}^n)$, $b(x,t)$ is bounded and Lipschitz in $x \in \mathbb{R}^n$.

To determine a pathwise global solution we consider the vector-valued auxiliary partial differential equation on $\mathbb{R}^n \times K \times [0,T]$

$$\begin{aligned} \frac{\partial h}{\partial z}(y,z,t) &= a_0(h(y,z,t),t) \\ h(y, Z_0, t) &= y \end{aligned} \tag{5.2}$$

where K is a compact subset of $\mathbb{R}$ containing $Z_0 = Z^0(0)$. If a_0 is a random function then the above equation holds true almost surely. The classical theory of parameter dependent ODE (cf. e.g. **[23]**) provides the existence of a C^1-solution which satisfies

$$h(y,z,t) = y + \int_{Z_0}^{z} a_0(h(y,u,t),t)\, du$$

and in matrix representation

$$\frac{\partial h}{\partial y}(y,z,t) = \exp\left\{ \int_{Z_0}^{z} \frac{\partial a_0}{\partial x}(h(y,u,t),t)\, du \right\},$$

Then, the mapping $h(\cdot, z, t) : \mathbb{R}^n \to \mathbb{R}^n$ is invertible with inverse function $u(x,z,t)$, i.e.,

$$u(h(y,z,t),z,t) = y.$$

Moreover,

$$\frac{\partial^2 h}{\partial z^2}(y,z,t) = \frac{\partial a_0}{\partial x}(h(y,z,t),t) a_0(h(y,z,t),t).$$

We will seek the solution X of (5.1) in the form

$$X(t) = h(Y(t), Z^0(t), t)$$

[1] This condition is missing in **[29]**

for some random $W^{\beta-}_{2,\infty}$-process Y with $Y(0) = X_0$.
Applying the Itô formula (3.6) to the function h we obtain

$$\begin{aligned} dX(t) = & \frac{\partial h}{\partial z}(Y(t), Z^0(t), t)dZ^0(t) + \sum_{k=1}^{n} \frac{\partial h}{\partial y^k}(Y(t), Z^0(t), t)dY^k(t) \\ & + \frac{\partial h}{\partial t}(Y(t), Z^0(t), t)dt + \frac{1}{2}\frac{\partial^2 h}{\partial z^2}(Y(t), Z^0(t), t)d[Z^0](t) \\ = & \; a_0(X(t), t)dZ^0(t) + \sum_{k=1}^{n} \frac{\partial h}{\partial y^k}(Y(t), Z^0(t), t)dY^k(t) + \frac{\partial h}{\partial t}(Y(t), Z^0(t), t)dt \\ & + \frac{1}{2}\sum_{k=1}^{n} \frac{\partial a_0}{\partial x^k}(h(Y(t), Z^0(t), t), t)a_0^k(h(Y(t), Z^0(t), t), t)d[Z^0](t). \end{aligned}$$

Comparing this with (5.1) we are led to a second auxiliary SDE, now for the process Y (in matrix representation):

$$\begin{aligned} dY(t) = & \left(\frac{\partial h}{\partial y}(Y(t), Z^0(t), t)\right)^{-1} \left[\sum_{j=1}^{m} a_j(h(Y(t), Z^0(t), t), t)dZ^j(t)\right. \\ & + \left(b(h(Y(t), Z^0(t), t), t) - \frac{\partial h}{\partial t}(Y(t), Z^0(t), t)\right)dt \\ & \left. - \frac{1}{2}\frac{\partial a_0}{\partial x}(h(Y(t), Z^0(t), t), t)a_0(h(Y(t), Z^0(t), t), t)d[Z^0](t)\right]. \end{aligned} \tag{5.3}$$

Using Theorem 4.1 we know that this differential equation has a pathwise unique solution $Y \in W^{\beta-}_{2,\infty}(0, T)$ which may be determined via Picard's iteration method.

THEOREM 5.2. *Under the conditions* **(C1)** - **(C3)** *for the random vector fields the process*

$$X(t) = h(Y(t), Z^0(t), t)$$

with h and Y satisfying (5.2) *and* (5.3) *provides a pathwise global solution of the SDE* (5.1).
If X is an arbitrary solution of the SDE (5.1) *in the sense of Definition* 5.1 *then it agrees with the above representations on the whole interval* $[0, T]$.

PROOF. By construction the process $h(Y(t), Z^0(t), t)$ is indeed a solution to the SDE (5.1). For uniqueness we take another solution $X(t)$ and $h(Y(t), Z^0(t), t)$ from above. As mentioned before the mapping

$$(y, z, t) \mapsto (h(y, z, t), z, t)$$

is invertible (a.s.). If we denote the inverse mapping by $(u(x, z, t), z, t)$, i.e.,

$$u(h(y, z, t), z, t) = y,$$

then the matrix equality

$$\left(\frac{\partial u}{\partial x}(x, z, t)\right) = \left(\frac{\partial h}{\partial y}(u(x, z, t), z, t)\right)^{-1}$$

holds true and furthermore,

$$\begin{aligned}
\frac{\partial u}{\partial z}(x,z,t) &= -\sum_{k=1}^{n} \frac{\partial u}{\partial x^k}(x,z,t) a_0^k(x,t) \\
\frac{\partial u}{\partial t}(x,z,t) &= -\sum_{k=1}^{n} \frac{\partial u}{\partial x^k}(x,z,t) \frac{\partial h^k}{\partial t}(u(x,z,t),z,t) \\
\frac{\partial^2 u}{\partial z^2}(x,z,t) &= -\sum_{k,l=1}^{n} \frac{\partial^2 u}{\partial x^k \partial x^l}(x,z,t) a_0^k(x,t) a_0^l(x,t) \\
&\quad - \sum_{k=1}^{n} \frac{\partial^2 u}{\partial z \partial x^k}(x,z,t) a_0^k(x,t) - \sum_{k,l=1}^{n} \frac{\partial u}{\partial x^k}(x,z,t) \frac{\partial a^k}{\partial x^l}(x,t) a_0^l(x,t).
\end{aligned}$$

Applying the Itô formula to the function $u(x,z,t)$ and the process $(X(t), Z^0(t))$ given by the integral representation

$$X(t) = X_0 + \sum_{j=0}^{m} \int_0^t a_j(X(s),s)\, dZ^j(s) + \int_0^t b(X(s),s)\, ds,$$

leads to

$$\begin{aligned}
u(X(t),Z^0(t),t) - X_0 &= \int_0^t \frac{\partial u}{\partial z}(X(s),Z^0(s),s)\, dZ^0(s) + \int_0^t \frac{\partial u}{\partial t}(X(s),Z^0(s),s)\, ds \\
&\quad + \sum_{k=1}^{n} \int_0^t \frac{\partial u}{\partial x^k}(X(s),Z^0(s),s) b^k(X(s),s)\, ds \\
&\quad + \sum_{j=0}^{m} \sum_{k=1}^{n} \int_0^t \frac{\partial u}{\partial x^k}(X(s),Z^0(s),s) a_j^k(X(s),s)\, dZ^j(s) \\
&\quad + \frac{1}{2} \sum_{l,k=1}^{n} \int_0^t \frac{\partial^2 u}{\partial x^l \partial x^k}(X(s),Z^0(s),s) a_0^l(X(s),s) a_0^k(X(s),s)\, d[Z^0](s) \\
&\quad + \frac{1}{2} \int_0^t \frac{\partial^2 u}{\partial z^2}(X(s),Z^0(s),s)\, d[Z^0](s) \\
&\quad + \frac{1}{2} \sum_{k=1}^{n} \int_0^t \frac{\partial^2 u}{\partial x^k \partial z}(X(s),Z^0(s),s) a_0^k(X(s),s)\, d[Z^0](s).
\end{aligned}$$

Substituting the expressions from above we infer

$$\begin{aligned}
u(X(t),Z^0(t),t) - X_0 &= \sum_{j=1}^{m} \sum_{k=1}^{n} \int_0^t \frac{\partial u}{\partial x^k}(X(s),Z^0(s),s) a_j^k(X(s),s)\, dZ^{j,k}(s) \\
&\quad + \sum_{k=1}^{n} \int_0^t \frac{\partial u}{\partial x^k}(X(s),Z^0(s),s) b^k(X(s),s)\, ds \\
&\quad - \sum_{k=1}^{n} \int_0^t \frac{\partial u}{\partial x^k}(X(s),Z^0(s),s) \frac{\partial h^k}{\partial t}(u(X(s),Z^0(s),s),Z^0(s),s)\, ds \\
&\quad - \frac{1}{2} \sum_{l,k=1}^{n} \int_0^t \frac{\partial u}{\partial x^k}(X(s),Z^0(s),s) \frac{\partial a_0^k}{\partial x^l}(X(s),s) a_0^l(X(s),s)\, d[Z^0](s).
\end{aligned}$$

Together with the relation $\frac{\partial h}{\partial z}(y,z,t) = a_0(h(y,z,t),z,t)$ it follows that the process

$$\widetilde{Y}(t) := u(X(t), Z^0(t), t)$$

has sample paths in $W_{2,\infty}^{1/2-}(0,T)$ and satisfies the SDE (5.3):

$$\begin{aligned} d\widetilde{Y}(t) =& \left(\frac{\partial h}{\partial y}(\widetilde{Y}(t), Z^0(t), t)\right)^{-1} \Bigg[\sum_{j=1}^{m} a_j(h(\widetilde{Y}(t), Z^0(t), t), t) dZ^j(t) \\ &+ \left(b(h(\widetilde{Y}(t), Z^0(t), t), t) - \frac{\partial h}{\partial t}(\widetilde{Y}(t), Z^0(t), t)\right) dt \\ &- \frac{1}{2}\frac{\partial a_0}{\partial x}(h(\widetilde{Y}(t), Z^0(t), t), t) a_0(h(\widetilde{Y}(t), Z^0(t), t), t) d[Z^0](t)\Bigg] \\ \widetilde{Y}(0) =& X_0. \end{aligned}$$

Since the solution of (5.3) is unique in $W_{2,\infty}^{1/2-}(0,T)$ we get $\widetilde{Y}(t) = u(X(t), Z^0(t), t) = Y(t)$ and hence $X(t) = h(Y(t), Z^0(t), t)$. □

6. Hölder continuity of the solution

In this section we are first interested in the degree of Hölder continuity of the solution of Section 4 when the integrator functions are Hölder continuous of order $\beta > 1/2$ and compare it with some other results. In **[14]** the following deterministic differential equation on $\mathbb{R}^n$ was considered:

$$x(t) = x_0 + \sum_{j=1}^{m} \int_0^t \sigma_j(x(s), s)\, dg^j(s) + \int_0^t b(x(s), s)\, ds, \tag{6.1}$$

where the functions σ_j satisfy

$$\begin{aligned} |\sigma_j(x,t) - \sigma_j(y,t)| &\le C|x-y|, \\ \left|\frac{\partial \sigma_j}{\partial x^i}(x,t) - \frac{\partial \sigma_j}{\partial y^i}(y,t)\right| &\le C|x-y|^{\delta_1}, \\ |\sigma_j(x,t) - \sigma_j(x,s)| &\le C|t-s|^{\delta_2}, \\ \left|\frac{\partial \sigma_j}{\partial x^i}(x,t) - \frac{\partial \sigma_j}{\partial x^i}(x,s)\right| &\le C|t-s|^{\delta_3}, \end{aligned} \tag{6.2}$$

and the function b is bounded and fulfills

$$|b(x,t) - b(y,t)| \le C|x-y| \tag{6.3}$$

for all $x, y \in \mathbb{R}^n$, $i = 1, ..., n$, $t, s \in [0,T]$, $j = 1, ..., m$, and some constants $0 \le \delta_1, \delta_2, \delta_3 \le 1$, C. Here the integrals are understood in the same sense as in (1.1). For $0 < \alpha < 1/2$ those authors considered the space $W_T^{1-\alpha,\infty}(0,T)$ of vector functions $f : [0,T] \to \mathbb{R}^n$ such that

$$\|f\|_{1-\alpha,\infty,T} := \sup_{0<s<t<T} \left(\frac{|f(t) - f(s)|}{(t-s)^{(1-\alpha)}} + \int_s^t \frac{|f(y) - f(s)|}{(y-s)^{2-\alpha}}\, dy\right) < \infty$$

and the Banach space $W_0^{\alpha,\infty}$ with respect to the norm

$$\|f\|_{\alpha,\infty} := \sup_{t\in[0,T]} \left(|f(t)| + \int_0^t \frac{|f(t) - f(s)|}{(t-s)^{\alpha+1}}\, ds\right).$$

For the space $W_T^{1-\alpha,\infty}(0,T)$ we have the embeddings

$$H^{1-\alpha+\varepsilon}(0,T) \hookrightarrow W_T^{1-\alpha,\infty}(0,T) \hookrightarrow H^{1-\alpha}(0,T),$$

for every $\varepsilon > 0$ and H^λ denotes the space of Hölder continuous functions of order λ. Nualart and Răşcanu proved the following.

THEOREM 6.1. *([**14**], Thm. 2.1) Let $g^j \in W_T^{1-\alpha,\infty}(0,T)$, the functions b and σ_j satisfy the conditions (6.2) for $j = 1, ..., m$, and $0 < \alpha < \alpha_0 = \min\left(\frac{1}{2}, \delta_2, \delta_3, \frac{\delta_1}{1+\delta_1}\right)$. Then Equation (6.1) has a unique solution $x \in W_0^{\alpha,\infty}$. Moreover, the solution is $(1-\alpha)$-Hölder continuous.*

So for Hölder continuous functions g^j of order $\beta > 1/2$ the above theorem implies (if $1-\beta < \min(\frac{1}{2}, \delta_2, \delta_3, \frac{\delta_1}{\delta_1+1})$) that (6.1) has a β-Hölder continuous solution x if the functions g^j are in the smaller space $W_T^{\beta,\infty}(0,T)$, and we get a $(\beta-\varepsilon)$-Hölder continuous solution in the other case. The aim of this section is a slight extension of this result by showing that the solution is β-Hölder continuous for all integrators $g^j \in C^\beta(0,T)$. In Section 4 we considered a similar equation but under slightly different conditions on the coefficients. We will show that the solution of (6.1) is Hölder continuous of the same order as the integrator functions. The basic tool to this aim are the following relationships.

PROPOSITION 6.2. *Let g be a real-valued β-Hölder continuous function with $1/2 < \beta \le 1$ and $f \in W_2^\gamma(0,T)$ for some $\gamma > 1-\beta$ (and hence $f_{0+} \in I_{0+}^{\gamma-}(L_2(0,T))$). The function h is defined by*

$$h(t) := \int_0^t f(s)\, dg(s) \tag{6.4}$$

and the integral is determined in the sense of (1.1). Then we have the following:

(i) *The function h is η-Hölder continuous for all $\eta < \beta - 1/2$.*
(ii) *If $\gamma > 1/2$ and f is δ-Hölder continuous for some $0 < \delta \le 1$, then h is β-Hölder continuous.*
(iii) *If $\gamma < 1/2$ and f is δ-Hölder continuous for some $0 < \delta \le 1$, then h is Hölder continuous of all orders $\mu < \beta + \gamma - 1/2$.*

PROOF. (i) Theorem 2.5 in [**25**] implies

$$\begin{aligned}
|h(t) - h(s)| &= \left| \int_s^t D_{s+}^\alpha f(x) D_{t-}^{1-\alpha} g_{t-}(x)\, dx \right| \\
&\le \sup_{0 \le s < x < t \le T} |D_{t-}^{1-\alpha} g_{t-}(x)| \int_s^t |D_{s+}^\alpha f(x)|\, dx \\
&\le \|g_{t-}\|_{I_{T-}^{1-\alpha}(L_\infty(0,T))} \int_s^t |D_{s+}^\alpha f(x)|\, dx
\end{aligned}$$

for all $1 - \beta < \alpha < \gamma$.
The integral does not exceed

$$\begin{aligned}
\int_s^t |D_{s+}^\alpha f(x)|\, dx &= C \int_s^t \frac{|f(x)|}{(x-s)^\alpha}\, dx + \int_s^t \int_s^x \frac{|f(x) - f(u)|}{(x-u)^{\alpha+1}}\, du\, dx \\
&=: S_1 + S_2.
\end{aligned}$$

Using the Hölder inequality and assuming that $\alpha < 1/2$ it follows

$$S_1 \leq C\|f\|_{L_2(0,T)}\Big(\int_s^t (x-s)^{-2\alpha}\,dx\Big)^{1/2} = C\|f\|_{L_2(0,T)}|t-s|^{1/2-\alpha}.$$

For S_2 one infers

$$\begin{aligned} S_2 &= C\int_s^t\int_s^x \frac{|f(x)-f(u)|}{(x-u)^{\alpha+1}}\,du\,dx \\ &= C\int_s^t\int_s^x \frac{|f(x)-f(u)|}{(x-u)^{\gamma+1/2}}\frac{1}{(x-u)^{\alpha-\gamma+1/2}}\,du\,dx \\ &\leq C\Big(\int_s^t\int_s^x \frac{(f(x)-f(u))^2}{(x-u)^{2\gamma+1}}\,du\,dx\Big)^{1/2}\Big(\int_s^t\int_s^x \frac{1}{(x-u)^{2\alpha-2\gamma+1}}\,du\,dx\Big)^{1/2} \\ &= C\|f\|_{\widetilde{W}_2^\gamma(s,t)}(t-s)^{1/2+\gamma-\alpha} \\ &\leq C\|f\|_{W_2^\gamma(0,T)}(t-s)^{1/2+\gamma-\alpha} \end{aligned}$$

and thus,

$$\begin{aligned} |h(t)-h(s)| &\leq C\|f\|_{W_2^\gamma(0,T)}\|g_{t-}\|_{I_{T-}^{1-\alpha}(L_\infty(0,T))}|t-s|^{\min(1/2-\alpha,1/2+\gamma-\alpha)} \\ &= C\|f\|_{W_2^\gamma(0,T)}\|g_{t-}\|_{I_{T-}^{1-\alpha}(L_\infty(0,T))}|t-s|^{1/2-\alpha} \end{aligned}$$

for every $1-\beta < \alpha < \min(1/2,\gamma)$. Hence, the function h is Hölder continuous of all orders less than $\beta - 1/2$.

(ii)

The function h can be rewritten as

$$h(t) = \int_0^t D_{0+}^\alpha f_{0+}(x)D_{t-}^{1-\alpha}g_{t-}(x)\,dx + f(0+)(g(t)-g(0))$$

for all $1-\beta < \alpha < \gamma$ and so we get

$$|h(t)-h(s)| \leq \Big|\int_s^t D_{s+}^\alpha f_{s+}(x)D_{t-}^{1-\alpha}g_{t-}(x)\,dx\Big| + \Big|f(s+)(g(t)-g(s))\Big| =: S_3 + S_4.$$

The second summand can be estimated by

$$S_4 \leq C\|f\|_{L_\infty(0,T)}|t-s|^\beta$$

because of the Hölder continuity of g. For the first summand we have like in (i)

$$S_3 \leq \sup_{0\leq s<x<t\leq T}|D_{t-}^{1-\alpha}g_{t-}(x)|\int_s^t |D_{s+}^\alpha f_{s+}(x)|\,dx.$$

The last integral does not exceed

$$\begin{aligned} \int_s^t |D_{s+}^\alpha f_{s+}(x)|\,dx &\leq C\int_s^t \frac{|f(x)-f(s)|}{(x-s)^\alpha}\,dx + C\int_s^t\int_s^x \frac{|f(x)-f(u)|}{(x-u)^{\alpha+1}}\,du\,dx \\ &=: S_5 + S_6. \end{aligned}$$

Using the fact that f is Hölder continuous of order δ we obtain

$$\begin{aligned} S_5 &= C\int_s^t \frac{|f(x)-f(s)|}{(x-s)^\alpha}\,dx \\ &\leq C\int_s^t (x-s)^{\delta-\alpha}\,dx \\ &\leq C(t-s)^{1+\delta-\alpha} \end{aligned}$$

and recall that

$$S_6 \leq C\|f\|_{W_2^\gamma(0,T)}(t-s)^{1/2+\gamma-\alpha}.$$

Therefore we get

$$|h(t)-h(s)| \leq C\|f\|_{W_{2,\infty}^\gamma(0,T)}\|g_{t-}\|_{I_{T-}^{1-\alpha}(L_\infty(0,T))}|t-s|^{\min(\beta,1+\delta-\alpha,1/2+\gamma-\alpha)}.$$

for every $1-\beta < \alpha < \gamma$. Choosing α such that the minimum of $1+\delta-\alpha$ and $1/2+\gamma-\alpha$ is greater than β, this leads to the assertion.
(iii)
With the same estimates for the function h as in the proof of (ii) we conclude that

$$\begin{aligned}|h(t)-h(s)| &\leq C\|f\|_{W_{2,\infty}^\gamma(0,T)}\|g_{t-}\|_{I_{T-}^{1-\alpha}(L_\infty(0,T))}|t-s|^{\min(\beta,1+\delta-\alpha,1/2+\gamma-\alpha)} \\ &= C\|f\|_{W_{2,\infty}^\gamma(0,T)}\|g_{t-}\|_{I_{T-}^{1-\alpha}(L_\infty(0,T))}|t-s|^{1/2+\gamma-\alpha}\end{aligned}$$

for all $1-\beta < \alpha < \gamma < 1/2$. Hence h is Hölder continuous of order μ for all $\mu < \beta+\gamma-1/2 < \beta$. □

By means of this lemma we can prove the statement about the *optimal degree of Hölder continuity.*

THEOREM 6.3. *Let $g^1,...,g^m$ be real-valued Hölder continuous functions of order $\beta > 1/2$ and consider Equation (6.1) under hypothesis **(H1)** or **(H2)** from Section 2 for the vector-valued coefficient functions σ_j, $j=1,...,m$. Then there exists a unique solution x to (6.1) in the sense of Theorem 4.1. Moreover, this is also the unique solution in the space of Hölder continuous functions of order β.*

PROOF. The Hölder continuity of $g^1,...,g^m$ implies $g_{T-}^j \in I_{T-}^{\beta'}(L_2(0,T))$, $j=1,...,m$, for any $\beta' < \beta$. Then Theorem 4.1 provides a unique solution $x \in W_{2,\infty}^{\beta-}(0,T)$. Therefore the functions $\sigma_j(x(\cdot),\cdot)$ are also elements of the space $W_{2,\infty}^{\beta-}(0,T)$. Using the representation for the solution

$$x(t) = x_0 + \sum_{j=1}^m \int_0^t \sigma_j(x(s),s)\,dg^j(s)$$

and applying Proposition 6.2 (i) to these integrals we conclude that the function x is Hölder continuous of order $\widetilde{\beta} < \beta - 1/2$. Thus the functions $\sigma_j(x(\cdot),\cdot)$ are also Hölder continuous of order $\widetilde{\beta}$. We now apply Proposition 6.2 (ii) to the representation for the solution and obtain that x is Hölder continuous of order β.
Finally, a β-Hölder continuous function belongs to the space $W_{2,\infty}^{\beta-}(0,T)$. Therefore uniqueness follows from that in Theorem 4.1. □

REMARK 6.4. The statement of Theorem 6.3 remains valid if the condition **(H1)** or **(H2)** for the vector fields σ_j is replaced by (6.2) with $\delta_2 \geq \beta$ and $1-\beta < \min(\frac{1}{2},\delta_3,\frac{\delta_1}{\delta_1+1})$. The proof is similar.

In a second step we consider again the SDE (5.1) from Section 5, but now under the condition that w.p.1 the process Z^0 admits a generalized quadratic variation process $[Z^0]$ and has Hölder exponent $\theta \leq 1/2$ and is not Hölder continuous of some order larger than 1/2. The real-valued stochastic processes $Z^1,...,Z^m$ and $[Z^0]$ are Hölder continuous of order $\beta > 1/2$. For the vector fields $a_0,...,a_m,b$ we suppose the conditions **(C1)** - **(C3)** with hypothesis **(H1)** or **(H2)** in **(C1)** and **(C2)** from Section 5.

THEOREM 6.5. *Under the above conditions Equation* (5.1) *provides a pathwise unique solution* X *in the sense of Definition* 5.1. *Moreover, the solution is Hölder continuous of the same order as the process* Z^0.

PROOF. Again the Hölder continuity implies that $Z^j_{T-} \in I^{\beta'}_{T-}(L_2(0,T))$, $j = 1,...,m$, for any $\beta' < \beta$. Therefore the proof for the existence and uniqueness of a solution is the same as in Theorem 5.2. This solution can be represented by $X(t) = h(Y(t), Z^0(t), t)$ where h is a C^1-function that solves the ODE (5.2) and Y has sample paths in $W^{\beta-}_{2,\infty}(0,T)$ and solves the auxiliary SDE (5.3). Similar as in the proof of Lemma 6.3 we can apply Lemma 6.2 (i) and (iii) to show that Y is Hölder continuous of some order $\widetilde{\beta}$ with $1/2 < \widetilde{\beta} < \beta$. This yields Hölder continuity of X with the same order as that of Z^0. □

For the particular case when $m = 1$, Z^0 is a Brownian motion and the process Z^1 is a fractional Brownian motion with Hurst parameter $H > 1/2$ the result of Theorem 6.5 was obtained by Guerra and Nualart **[7]** under weaker regularity conditions on the coefficient functions a_j.
An example for a process Z^0 being Hölder continuous of order 1/2 and $[Z^0](t) = Ct$ will be considered below.

7. Appendix: Stochastic integrals with respect to random Weierstrass-type functions

Here we give examples fitting into the approach of Section 3, for which quadratic variations and stochastic forward integrals via regularization in the sense of Russo and Vallois (**[16]** and **[17]**) and hence, in the Riemann-Stieltjes sense do not exist. Let $\{\xi_k\}_{k=1}^\infty$ be a stationary sequence of random variables with second moments such that $\sup_k |\xi_k| < \infty$ w.p.1. Then define the *random Weierstrass-type function*

$$W(s) := \sum_{k=1}^{\infty} \xi_k \, 2^{-\frac{k}{2}} \sin(2^k s) \tag{7.1}$$

on $\mathbb{R}$ which has period 2π.
As in the proof of **[24]**, Proposition 1 one can show the following (taking there the square of the integrand).

PROPOSITION 7.1. *There exists a constant* $c > 0$ *such that w.p.1,*

$$\lim_{\delta\to 0} \frac{1}{|\ln\delta|} \int_0^\delta \frac{(W(s+u) - W(s))^2}{u} \frac{1}{u}\, du = c$$

for Lebesgue almost all $s \in [0, 2\pi)$.

REMARK 7.2. Since the averaging kernel $\varepsilon u^{\varepsilon-1}$ arises from the last kernel $\frac{1}{|\ln\delta|}\mathbf{1}_{(\delta,1)}(u)\frac{1}{u}$ by means of a second averaging kernel $\varepsilon^2 v^{\varepsilon-1}|\ln v|$ (cf. **[27]**, Proposition 3.1) this implies

$$\lim_{\varepsilon\to 0} \varepsilon \int_0^1 u^{\varepsilon-1} \frac{(W(s+u) - W(s))^2}{u}\, du = c \tag{7.2}$$

for almost all s w.p.1.

COROLLARY 7.3.

$$\lim_{\varepsilon\to 0}\left(\varepsilon\int_0^1 u^{\varepsilon-1}\int_0^t \frac{(W_{t-}(s+u)-W_{t-}(s))^2}{u}\,ds\,du - ct\right)=0$$

uniformly in t w.p.1, i.e., $[W](t)=ct$. The same holds true for W instead of W_{t-}.

PROOF. Since $\sup_k |\xi_k|<\infty$ w.p.1 one can show as in the classical case with constant ξ_k that W is Hölder contionuous of order 1/2 w.p.1, i.e., the integrand is uniformly bounded. Therefore the statement follows from (7.2) and the fact that $\varepsilon\int_0^1 u^{\varepsilon-1}\,du=1$. $\square$

Below we outline that the limit of

$$\mathbb{E}\int_0^{2\pi}\frac{(W(s+\varepsilon)-W(s))^2}{\varepsilon}\,ds$$

as $\varepsilon\to 0$ does not exist. Therefore

$$I(\varepsilon):=\int_0^{2\pi}\frac{(W(s+\varepsilon)-W(s))^2}{\varepsilon}\,ds$$

cannot converge in probability, provided the random variables $I(\varepsilon)$ are integrable uniformly in ε. The latter is guaranteed, since $\{\xi_k\}_{k=1}^\infty$ is a stationary sequence of square integrable random variables. Hence, the process $W(t)$ does not fit into the approach of **[17]** and **[18]**.
In order to show the high oscillation behavior of $\mathbb{E}I(\varepsilon)$ note that because of the orthogonality of the involved trigonometric functions we get

$$I(\varepsilon)=\frac{1}{\varepsilon}\sum_{k=1}^\infty \xi_k^2\,2^{-k}\left(\int_0^{2\pi}(\sin(2^k(s+\varepsilon))-\sin(2^k s))\,ds\right)^2$$

so that for $m:=\mathbb{E}\xi_k^2$,

$$\varepsilon\mathbb{E}I(\varepsilon)=m\sum_{k=1}^\infty 2^{-k}\left(\int_0^{2\pi}(\sin(2^k(s+\varepsilon))-\sin(2^k s))\,ds\right)^2.$$

Using trigonometric identities straightforward calculations lead to the equivalent expression

$$C\sum_{k=1}^\infty 2^{-k}\sin^2(2^k\varepsilon).$$

This is again a Weierstrass-type function which is known to be nowhere differentiable. Therefore the limit of

$$\mathbb{E}I(\varepsilon)=C\frac{1}{\varepsilon}\sum_{k=1}^\infty 2^{-k}\sin^2(2^k\varepsilon)$$

as $\varepsilon\to 0$ cannot exist.

REMARK 7.4. For deterministic Hölder continuous functions of order 1/2, such that the convergence of Proposition 7.1 holds true, a corresponding calculus has been developed in Bedford and Kamae **[1]**. By Remark 7.2 concerning the averaging kernels it may be considered as a special case of the approach from Section 3. (See also the discussion in **[27]**.)

References

[1] Tim Bedford and Teturo Kamae, *Stieltjes integration and stochastic calculus with respect to self-affine functions*, Japan J. Indust. Appl. Math. **8** (1991), no. 3, 445–459, DOI 10.1007/BF03167146. MR1137651

[2] Christian Bender, Tommi Sottinen, and Esko Valkeila, *Fractional processes as models in stochastic finance*, Advanced mathematical methods for finance, Springer, Heidelberg, 2011, pp. 75–103, DOI 10.1007/978-3-642-18412-3_3. MR2792076

[3] Francesca Biagini, Yaozhong Hu, Bernt Øksendal, and Tusheng Zhang, *Stochastic calculus for fractional Brownian motion and applications*, Probability and its Applications (New York), Springer-Verlag London, Ltd., London, 2008. MR2387368

[4] Halim Doss, *Liens entre équations différentielles stochastiques et ordinaires* (French, with English summary), Ann. Inst. H. Poincaré Sect. B (N.S.) **13** (1977), no. 2, 99–125. MR0451404

[5] Denis Feyel and Arnaud de La Pradelle, *On fractional Brownian processes*, Potential Anal. **10** (1999), no. 3, 273–288, DOI 10.1023/A:1008630211913. MR1696137

[6] H. Föllmer, *Calcul d'Itô sans probabilités* (French), Seminar on Probability, XV (Univ. Strasbourg, Strasbourg, 1979/1980), Lecture Notes in Math., vol. 850, Springer, Berlin, 1981, pp. 143–150. MR622559

[7] João Guerra and David Nualart, *Stochastic differential equations driven by fractional Brownian motion and standard Brownian motion*, Stoch. Anal. Appl. **26** (2008), no. 5, 1053–1075, DOI 10.1080/07362990802286483. MR2440915

[8] Michael Hinz and Martina Zähle, *Gradient-type noises. I. Partial and hybrid integrals*, Complex Var. Elliptic Equ. **54** (2009), no. 6, 561–583, DOI 10.1080/17476930802669652. MR2537256

[9] Ito Y.: A fractional calculus approach to rough integration. Kyoto J. Math. (to appear) (2016)

[10] Yaozhong Hu and David Nualart, *Rough path analysis via fractional calculus*, Trans. Amer. Math. Soc. **361** (2009), no. 5, 2689–2718, DOI 10.1090/S0002-9947-08-04631-X. MR2471936

[11] Terry Lyons, *Differential equations driven by rough signals. I. An extension of an inequality of L. C. Young*, Math. Res. Lett. **1** (1994), no. 4, 451–464, DOI 10.4310/MRL.1994.v1.n4.a5. MR1302388

[12] Yuliya S. Mishura, *Stochastic calculus for fractional Brownian motion and related processes*, Lecture Notes in Mathematics, vol. 1929, Springer-Verlag, Berlin, 2008. MR2378138

[13] Yuliya Mishura and Georgiy Shevchenko, *Mixed stochastic differential equations with long-range dependence: Existence, uniqueness and convergence of solutions*, Comput. Math. Appl. **64** (2012), no. 10, 3217–3227, DOI 10.1016/j.camwa.2012.03.061. MR2989350

[14] David Nualart and Aurel Răşcanu, *Differential equations driven by fractional Brownian motion*, Collect. Math. **53** (2002), no. 1, 55–81. MR1893308

[15] David Nualart, *Stochastic calculus with respect to fractional Brownian motion* (English, with English and French summaries), Ann. Fac. Sci. Toulouse Math. (6) **15** (2006), no. 1, 63–78. MR2225747

[16] Francesco Russo and Pierre Vallois, *Forward, backward and symmetric stochastic integration*, Probab. Theory Related Fields **97** (1993), no. 3, 403–421, DOI 10.1007/BF01195073. MR1245252

[17] Francesco Russo and Pierre Vallois, *The generalized covariation process and Itô formula*, Stochastic Process. Appl. **59** (1995), no. 1, 81–104, DOI 10.1016/0304-4149(95)93237-A. MR1350257

[18] Francesco Russo and Pierre Vallois, *Stochastic calculus with respect to continuous finite quadratic variation processes*, Stochastics Stochastics Rep. **70** (2000), no. 1-2, 1–40, DOI 10.1080/17442500008834244. MR1785063

[19] Stefan G. Samko, Anatoly A. Kilbas, and Oleg I. Marichev, *Fractional integrals and derivatives*, Gordon and Breach Science Publishers, Yverdon, 1993. Theory and applications; Edited and with a foreword by S. M. Nikol′skiĭ; Translated from the 1987 Russian original; Revised by the authors. MR1347689

[20] Schneider E.: Stochastic differential equations driven by fractal processes. Thesis, University of Jena (2015)

[21] G. Shevchenko, *Mixed stochastic delay differential equations* (English, with English, Russian and Ukrainian summaries), Teor. Ĭmovīr. Mat. Stat. **89** (2013), 167–180, DOI 10.1090/S0094-9000-2015-00944-3; English transl., Theory Probab. Math. Statist. **89** (2014), 181–195. MR3235184

[22] Héctor J. Sussmann, *On the gap between deterministic and stochastic ordinary differential equations*, Ann. Probability **6** (1978), no. 1, 19–41. MR0461664

[23] Wolfgang Walter, *Gewöhnliche Differentialgleichungen* (German, with German summary), 5th ed., Springer-Lehrbuch. [Springer Textbook], Springer-Verlag, Berlin, 1993. Eine Einführung. [An introduction]. MR1231977

[24] M. Zähle and H. Ziezold, *Fractional derivatives of Weierstrass-type functions*, J. Comput. Appl. Math. **76** (1996), no. 1-2, 265–275, DOI 10.1016/S0377-0427(96)00110-0. MR1423522

[25] M. Zähle, *Integration with respect to fractal functions and stochastic calculus. I*, Probab. Theory Related Fields **111** (1998), no. 3, 333–374, DOI 10.1007/s004400050171. MR1640795

[26] M. Zähle, *Integration with respect to fractal functions and stochastic calculus. II*, Math. Nachr. **225** (2001), 145–183, DOI 10.1002/1522-2616(200105)225:1⟨145::AID-MANA145⟩3.3.CO;2-S. MR1827093

[27] M. Zähle, *Forward integrals and stochastic differential equations*, Seminar on Stochastic Analysis, Random Fields and Applications, III (Ascona, 1999), Progr. Probab., vol. 52, Birkhäuser, Basel, 2002, pp. 293–302. MR1958824

[28] M. Zähle, *Long range dependence, no arbitrage and the Black-Scholes formula*, Stoch. Dyn. **2** (2002), no. 2, 265–280, DOI 10.1142/S0219493702000406. MR1912144

[29] M. Zähle, *Stochastic differential equations with fractal noise*, Math. Nachr. **278** (2005), no. 9, 1097–1106, DOI 10.1002/mana.200310295. MR2150381

Institute of Mathematics, Friedrich Schiller University Jena, 07737 Jena, Germany

Email address: `martina.zaehle@uni-jena.de`

Email address: `erik-schneider@freenet.de`

Selected Published Titles in This Series

731 **Robert G. Niemeyer, Erin P. J. Pearse, John A. Rock, and Tony Samuel, Editors,** Horizons of Fractal Geometry and Complex Dimensions, 2019

728 **Nicolás Andruskiewitsch and Dmitri Nikshych, Editors,** Tensor Categories and Hopf Algebras, 2019

727 **André Leroy, Christian Lomp, Sergio López-Permouth, and Frédérique Oggier, Editors,** Rings, Modules and Codes, 2019

726 **Eugene Plotkin, Editor,** Groups, Algebras and Identities, 2019

725 **Shijun Zheng, Marius Beceanu, Jerry Bona, Geng Chen, Tuoc Van Phan, and Avy Soffer, Editors,** Nonlinear Dispersive Waves and Fluids, 2019

724 **Lubjana Beshaj and Tony Shaska, Editors,** Algebraic Curves and Their Applications, 2019

723 **Donatella Danielli, Arshak Petrosyan, and Camelia A. Pop, Editors,** New Developments in the Analysis of Nonlocal Operators, 2019

722 **Yves Aubry, Everett W. Howe, and Christophe Ritzenthaler, Editors,** Arithmetic Geometry: Computation and Applications, 2019

721 **Petr Vojtěchovský, Murray R. Bremner, J. Scott Carter, Anthony B. Evans, John Huerta, Michael K. Kinyon, G. Eric Moorhouse, and Jonathan D. H. Smith, Editors,** Nonassociative Mathematics and its Applications, 2019

720 **Alexandre Girouard, Editor,** Spectral Theory and Applications, 2018

719 **Florian Sobieczky, Editor,** Unimodularity in Randomly Generated Graphs, 2018

718 **David Ayala, Daniel S. Freed, and Ryan E. Grady, Editors,** Topology and Quantum Theory in Interaction, 2018

717 **Federico Bonetto, David Borthwick, Evans Harrell, and Michael Loss, Editors,** Mathematical Problems in Quantum Physics, 2018

716 **Alex Martsinkovsky, Kiyoshi Igusa, and Gordana Todorov, Editors,** Surveys in Representation Theory of Algebras, 2018

715 **Sergio R. López-Permouth, Jae Keol Park, S. Tariq Rizvi, and Cosmin S. Roman, Editors,** Advances in Rings and Modules, 2018

714 **Jens Gerlach Christensen, Susanna Dann, and Matthew Dawson, Editors,** Representation Theory and Harmonic Analysis on Symmetric Spaces, 2018

713 **Naihuan Jing and Kailash C. Misra, Editors,** Representations of Lie Algebras, Quantum Groups and Related Topics, 2018

712 **Nero Budur, Tommaso de Fernex, Roi Docampo, and Kevin Tucker, Editors,** Local and Global Methods in Algebraic Geometry, 2018

711 **Thomas Creutzig and Andrew R. Linshaw, Editors,** Vertex Algebras and Geometry, 2018

710 **Raphaël Danchin, Reinhard Farwig, Jiří Neustupa, and Patrick Penel, Editors,** Mathematical Analysis in Fluid Mechanics, 2018

709 **Fernando Galaz-García, Juan Carlos Pardo Millán, and Pedro Solórzano, Editors,** Contributions of Mexican Mathematicians Abroad in Pure and Applied Mathematics, 2018

708 **Christian Ausoni, Kathryn Hess, Brenda Johnson, Ieke Moerdijk, and Jérôme Scherer, Editors,** An Alpine Bouquet of Algebraic Topology, 2018

707 **Nitya Kitchloo, Mona Merling, Jack Morava, Emily Riehl, and W. Stephen Wilson, Editors,** New Directions in Homotopy Theory, 2018

706 **Yeonhyang Kim, Sivaram K. Narayan, Gabriel Picioroaga, and Eric S. Weber, Editors,** Frames and Harmonic Analysis, 2018

For a complete list of titles in this series, visit the AMS Bookstore at **www.ams.org/bookstore/conmseries/**.